Reagents for Organic Synthesis

THE WILEY BICENTENNIAL–KNOWLEDGE FOR GENERATIONS

Each generation has its unique needs and aspirations. When Charles Wiley first opened his small printing shop in lower Manhattan in 1807, it was a generation of boundless potential searching for an identity. And we were there, helping to define a new American literary tradition. Over half a century later, in the midst of the Second Industrial Revolution, it was a generation focused on building the future. Once again, we were there, supplying the critical scientific, technical, and engineering knowledge that helped frame the world. Throughout the 20th Century, and into the new millennium, nations began to reach out beyond their own borders and a new international community was born. Wiley was there, expanding its operations around the world to enable a global exchange of ideas, opinions, and know-how.

For 200 years, Wiley has been an integral part of each generation's journey, enabling the flow of information and understanding necessary to meet their needs and fulfill their aspirations. Today, bold new technologies are changing the way we live and learn. Wiley will be there, providing you the must-have knowledge you need to imagine new worlds, new possibilities, and new opportunities.

Generations come and go, but you can always count on Wiley to provide you the knowledge you need, when and where you need it!

William J. Pesce
PRESIDENT AND CHIEF EXECUTIVE OFFICER

Peter Booth Wiley
CHAIRMAN OF THE BOARD

Fiesers'

Reagents for Organic Synthesis

VOLUME TWENTY THREE

Tse-Lok Ho

National Chiao Tung University
Republic of China

and

Shanghai Institute of Organic Chemistry
China

WILEY-INTERSCIENCE
A John Wiley & Sons, Inc., Publication

Published by John Wiley & Sons, Inc., Hoboken, New Jersey
Published simultaneously in Canada

Library of Congress Cataloging-in-Publication Data:

ISBN 13: 978-0-471-68243-1
ISBN 10: 0-471-68243-8

ISSN: 0271-616X

Printed in the United States of America

10 9 8 7 6 5 4 3 2 1

INDEXER
Honor Ho

PREFACE

子曰：志於道。。。游於藝

Confucius says: Purpose toward the Way

Practice toward artistry

The current volume covers synthetic literature of mostly the 2003-2004 period. Rapid advances in the area still concentrate in the method development for asymmetric synthesis and organometallic catalysis. There have been remarkable discoveries in the use of gold and ruthenium compounds. More emphasis on environmental issues has led to the exploration of solventless reaction systems and/or aqueous media. Fascinating applications of synthetic reactions continue to appear.

The author completed this work while on campus of the Shanghai Institute of Organic Chemistry, Shanghai, China. The Institute was built up by Dr. Huang Minlon who discovered a valuable synthetic method that bears his name in the laboratory of Professor L. F. Fieser at Harvard University. I am grateful to Professor Li-xin Dai, Professor Xue-Long Hou, and Professor Henry N.C. Wong for their hospitality during the stay in the Institute and the Class of 2005 of SIOC for help with the ChemDraw graphics.

Tse-Lok Ho

CONTENTS

GENERAL ABBREVIATIONS

Ac	acetyl
acac	acetylacetonate
ADDP	1,1′-(azodicarbonyl)dipiperidine
AIBN	2,2′-azobisisobutyronitrile
An	*p*-anisyl
aq	aqueous
Ar	aryl
ATPH	aluminum tris(2,6-diphenylphenoxide)
9-BBN	9-borabicyclo[3.3.1]nonane
BINOL	1,1′-binaphthalene-2,2′-diol
Bn	benzyl
Boc	*t*-butoxycarbonyl
bpy	2,2′-bipyridyl
BSA	*N*,*O*-bis(trimethylsilyl)acetamide
Bt	benzotriazol-1-yl
Bu	*n*-butyl
Bz	benzoyl
18-c-6	18-crown-6
c-	cyclo
CAN	cerium(IV)ammonium nitrate
cat	catalytic
Cbz	benzyloxycarbonyl
Chx	cyclohexyl
cod	1,5-cyclooctadiene
cot	1,3,5-cyclooctatriene
Cp	cyclopentadienyl
Cp*	1,2,3,4,5-pentamethylcyclopentadienyl
CSA	10-camphorsulfonic acid
Cy	cyclohexyl
cyclam	1,4,8,11-tetraazacyclotetradecane
DABCO	1,4-diazobicyclo[2.2.2]octane
DAST	(diethylamino)sulfur trifluoride
dba	dibenzylideneacetone
DBN	1,5-diazobicyclo[4.3.0]non-5-ene
DBU	1,8-diazobicyclo[5.4.0]undec-7-ene
DCC	*N*,*N*′-dicyclohexylcarbodiimide

DDQ	2,3-dichloro-5,6-dicyano-1,4-benzoquinone
de	diastereomer excess
DEAD	diethyl azodicarboxylate
DIAD	diisopropyl azodicarboxylate
Dibal-H	diisobutylaluminum hydride
DMA	*N*,*N*-dimethylacetamide
DMAD	dimethyl acetylenedicarboxylate
DMAP	4-dimethylaminopyridine
DMD	dimethyldioxirane
DME	1,2-dimethoxyethane
DMF	*N*,*N*-dimethylformamide
DMPU	*N*,*N*′-dimethylpropyleneurea
DMSO	dimethyl sulfoxide
dpm	dipivaloylmethane
dppb	1,4-bis(diphenylphosphino)butane
dppe	1,2-bis(diphenylphosphino)ethane
dppf	1,2-bis(diphenylphosphino)ferrocene
dppp	1,3-bis(diphenylphosphino)propane
dr	diastereomer ratio
DTTB	4,4′-di-*t*-butylbiphenyl
E	COOMe
ee	enantiomer excess
en	ethylenediamine
er	enantiomer ratio
Et	ethyl
EVE	ethyl vinyl ether
Fc	ferrocenyl
Fmoc	9-fluorenylmethoxycarbonyl
Fu	furanyl
HMDS	hexamethyldisilazane
HMPA	hexamethylphosphoric amide
hv	light
Hx	*n*-hexyl
i	iso
Ipc	isopinocampheyl
kbar	kilobar
L	ligand
LAH	lithium aluminum hydride
LDA	lithium diisopropylamide
LHMDS	lithium hexamethyldisilazide
LTMP	lithium 2,2,6,6-tetramethylpiperidide
LN	lithium naphthalenide

lut	2,6-lutidine
M	metal
MAD	methylaluminum bis(2,6-di-*t*-butyl-4-methylphenoxide)
MCPBA	*m*-chloroperoxybenzoic acid
Me	methyl
MEM	methoxyethoxymethyl
Men	menthyl
Mes	mesityl
MOM	methoxymethyl
Ms	methanesulfonyl (mesyl)
MS	molecular sieves
MTO	methyltrioxorhodium
MVK	methyl vinyl ketone
NBS	*N*-bromosuccinimide
NCS	*N*-chlorosuccinimide
NIS	*N*-iodosuccinimide
NMO	*N*-methylmorpholine *N*-oxide
NMP	*N*-methylpyrrolidone
Np	naphthyl
Ns	*p*-nitrobenzenesulfonyl
Nu	nucleophile
Oc	octyl
PCC	pyridinium chlorochromate
PDC	pyridinium dichromate
PEG	poly(ethylene glycol)
Ph	phenyl
phen	1,10-phenenthroline
Pht	phthaloyl
Piv	pivaloyl
PMB	*p*-methoxybenzyloxymethyl
PMHS	poly(methylhydrosiloxane)
PMP	*p*-methoxyphenyl
Pr	*n*-propyl
py	pyridine
Q^+	quaternary onium ion
RAMP	(*R*)-1-amino-2-methoxymethylpyrrolidine
RaNi	Raney nickel
RCM	ring closure metathesis
R^f	perfluoroalkyl
ROMP	ring opening metathesis polymerization
s-	secondary
(s)	solid

salen	*N,N′*-ethylenebis(salicylideneiminato)
SAMP	(*S*)-1-amino-2-methoxymethylpyrrolidine
sc	supercritical
SDS	sodium dodecyl sulfate
sens.	sensitizer
SEM	2-(trimethylsilyl)ethoxymethyl
SES	2-[(trimethylsilyl)ethyl]sulfonyl
TASF	tris(dimethylamino)sulfur(trimethylsilyl)difluoride
TBAF	tetrabutylammonium fluoride
TBDPS	*t*-butyldiphenylsilyl
TBDMS	*t*-butyldimethylsilyl
TBS	*t*-butyldimethylsilyl
TEMPO	2,2,6,6-tetramethylpiperidinooxy
Tf	trifluoromethanesulfonyl
THF	tetrahydrofuran
THP	tetrahydropyranyl
Thx	*t*-hexyl
TIPS	triisopropylsilyl
TMEDA	*N,N,N′,N′*-tetramethylethylenediamine
TMS	trimethylsilyl
Tol	*p*-tolyl
Tp	tris(1-pyrazolyl)borato
tpp	tetraphenylporphyrin
Ts	tosyl (*p*-toluenesulfonyl)
TSE	2-(trimethylsilyl)ethyl
TTN	thallium trinitrate
Z	benzyloxycarbonyl
Δ	heat
))))	microwave

REFERENCE ABBREVIATIONS

ACIEE	Angew. Chem. Int. Ed. Engl.
ACR	Acc. Chem. Res.
ACS	Acta Chem. Scand.
AJC	Aust. J. Chem.
AOMC	Appl. Organomet. Chem.
ASC	Adv. Syn. Catal.
BC	Bioorg. Chem.
BCSJ	Bull. Chem. Soc. Jpn.
BMCL	Biorg. Med. Chem. Lett.
BRAS	Bull. Russ. Acad. Sci.
CB	Chem. Ber.
CC	Chem. Commun.
CEJ	Chem. Eur. J.
CI	Chem. Ind. (London)
CJC	Can. J. Chem.
CL	Chem. Lett.
CPB	Chem. Pharm. Bull.
CR	Carbohydr. Res.
EJIC	Eur. J. Inorg. Chem.
EJOC	Eur. J. Org. Chem.
H	Heterocycles
HCA	Helv. Chim. Acta.
JACS	J. Am. Chem. Soc.
JCC	J. Carbohydr. Chem.
JCCS(T)	J. Chin. Chem. Soc. (Taipei)
JCR(S)	J. Chem. Res. (Synopsis)
JCS(P1)	J. Chem. Soc. Perkin Trans. 1
JFC	J. Fluorine Chem.
JHC	J. Heterocycl. Chem.
JOC	J. Org. Chem.
JOCU	J. Org. Chem. USSR (Engl. Trans.)
JOMC	J. Organomet. Chem.
MC	Mendeleev Commun.
NJC	New J. Chem.
OL	Organic Letters
OM	Organometallics

PAC	Pure Appl. Chem.
PSS	Phosphours Sulfur Silicon
RCB	Russian Chem. Bull.
RJGC	Russ. J. Gen. Chem.
RJOC	Russian J. Org. Chem.
S	Synthesis
SC	Synth. Commun.
SCI	Science.
SL	Synlett
SOC	Synth. Org. Chem. (Jpn.)
T	Tetrahedron
TA	Tetrahedron: Asymmetry
TL	Tetrahedron Lett.
ZN	Zeitschr. Naturforsch.

4-Acetamido-2,2,6,6-tetramethylpiperidinooxy.

Oxidation.[1] 3,4-Dihydroxy-1-alkenes which can be prepared from 2-alken-4-ols via reaction with singlet oxygen undergo oxidation, initially to 1-alken-3-on-4-ols (with 1 equiv. each of acetamido-TEMPO and TsOH) and then the unsaturated diketones (2.5 equiv. reagents).

[1]Habel, L.W., De Keersmaecker, S., Wahlen, J., Jacobs, P.A., De Vos, D.E. *TL* **45,** 4057 (2004).

Acetic anhydride. 20, 1, **21,** 1; **22,** 1

Acetylation.[1] Hydroxy groups of sugars are protected by reaction with Ac_2O in the presence of 4A-molecular sieves. Conditions permit survival of acid-labile groups. Selectivity for primary alcohols is shown.

Benzannulation.[2] Substituted dibenzofurans are readily synthesized by the following method.

OMe COOEt COOH → (Ac_2O, NaOAc / hydroquinone / ¡÷ 1h) → COOEt, O, OAc

92%

[1]Adinolfi, M., Barone, G., Iadonisi, A., Schiattarella, M. *TL* **44,** 4661 (2003).
[2]Serra, S., Fuganti, C. *SL* 2005 (2003).

Acetonitrile(cyclopentadienyl)triphenylphosphineruthenium hexafluorophosphate.

3-Hydroxyalkanitriles.[1] The cationic Ru complex, $NaPF_6$, and an amine base form a cooperative catalyst that can activate MeCN in nucleophilic addition to aldehydes (11 examples, 77–93%).

[1]Kumagai, N., Matsunaga, S., Shibasaki, M. *JACS* **126**, 13632 (2004).

Acetonyltriphenylphosphonium bromide. 21, 1; **22,** 1

Acetalization.[1] With the reagent as catalyst carbonyl compounds are converted into acetals (and thioacetals). A polymer form can also be used.

[1]Hon, Y.-S., Lee, C.-F., Chen, R.-J., Huang, Y.-F. *SC* **33,** 28291 (2003).

Fiesers' Reagents for Organic Synthesis, Volume 23. Edited by Tse-Lok Ho

Acetylacetonato(dicarbonyl)rhodium. 21, 1; **22**, 2–4

Hydroformylation. High *anti*-selectivivity is achieved in the formation of α-alkyl aldols from allylic *o*-diphenylphosphinylbenzoates.[1] The reaction of unsaturated esters can give rise to 1,3- or 1,4-dicarbonyl products, depending on reaction conditions.[2] The tris(2,4-di-*t*-butylphenyl)phosphate ligand seems important because very low conversion is observed with $(PhO)_3P$.

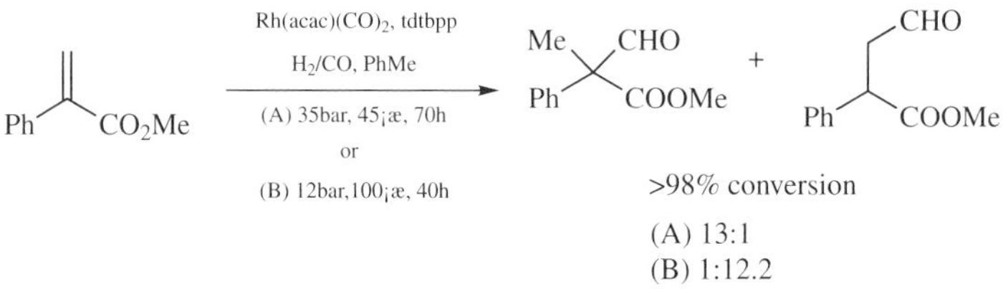

Ph CO$_2$Me

Rh(acac)(CO)$_2$, tdtbpp
H$_2$/CO, PhMe
(A) 35bar, 45¡æ, 70h
or
(B) 12bar,100¡æ, 40h

Me CHO Ph COOMe + CHO Ph COOMe

>98% conversion
(A) 13:1
(B) 1:12.2

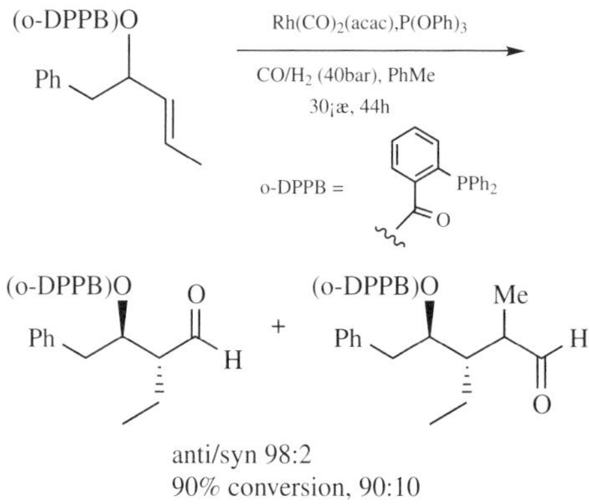

(o-DPPB)O Ph

Rh(CO)$_2$(acac),P(OPh)$_3$
CO/H$_2$ (40bar), PhMe
30¡æ, 44h

o-DPPB = PPh$_2$ O

(o-DPPB)O Ph O H + (o-DPPB)O Ph Me H O

anti/syn 98:2
90% conversion, 90:10

Hydroformylation followed by aldol cyclization with an internal enolborate is stereoselective.[3] It is also possible to form arylhydrazones or indoles from arylhydrazines and alkenes (via homologation of the alkenes).[4]

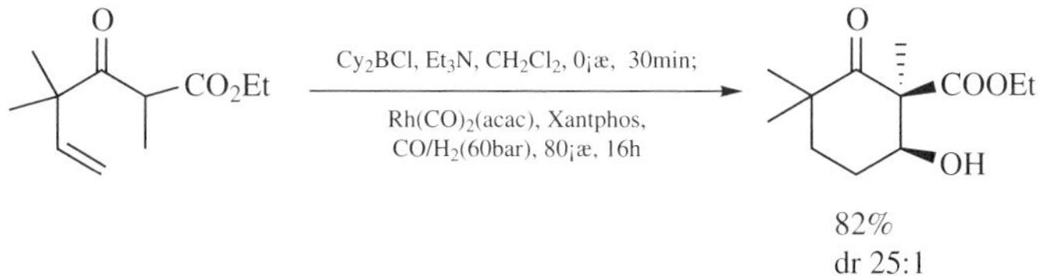

O CO$_2$Et

Cy$_2$BCl, Et$_3$N, CH$_2$Cl$_2$, 0¡æ, 30min;
Rh(CO)$_2$(acac), Xantphos,
CO/H$_2$(60bar), 80¡æ, 16h

O COOEt OH

82%
dr 25:1

Silylative cyclization. Allenynes with suitably separated unsaturated linkages undergo cyclization on a catalyzed hydrosilylation.[5] 5- and 6-membered carbocycles and heterocycles are readily synthesized from allenyl aldehydes.[6]

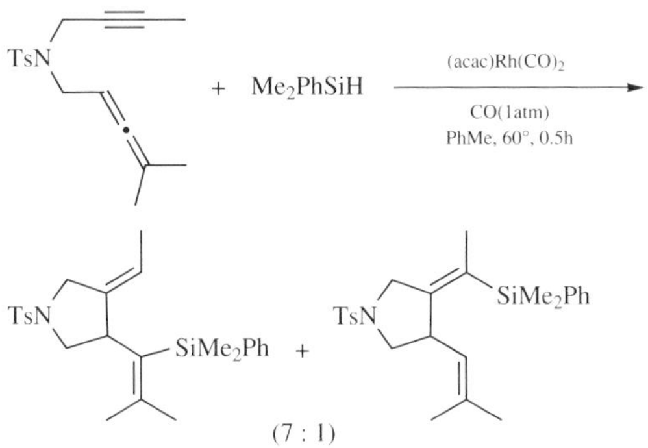

TsN + Me_2PhSiH —(acac)$Rh(CO)_2$, CO(1atm), PhMe, 60°, 0.5h→

TsN ... $SiMe_2Ph$ + TsN ... $SiMe_2Ph$

(7 : 1)
90%

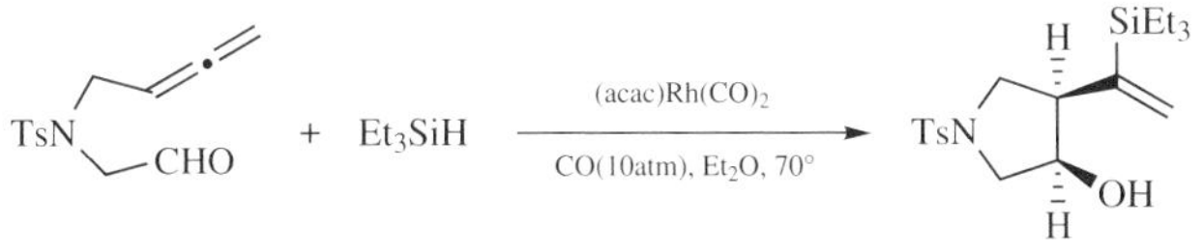

TsN ... CHO + Et_3SiH —(acac)$Rh(CO)_2$, CO(10atm), Et_2O, 70°→ TsN ... H $SiEt_3$ OH H

74%

4-Alkynones.[7] 1-Alkynes add to enones (MVK) to produce 4-alkynones. Alkynyl(hydrido)rhodium species are involved. Enones are weaker ligands than alkynes for Rh, therefore it requires excess of enones to drive the reaction.

[1]Breit, B., Demel, P., Gebert, A. *CC* 114 (2004).
[2]Clarke, M.L. *TL* **45**, 4043 (2004).
[3]Keranen, M.D., Eilbracht, P. *OBC* **2**, 1688 (2004).
[4]Ahmed, M., Jackstell, F., Seayad, A.M., Klein, H., Beller, M. *TL* **45**, 869 (2004).
[5]Shibata, T., Kadowaki, S., Takagi, K. *OM* **23**, 4116 (2004).
[6]Kang, S.-K., Hong, Y.-T., Lee, J.-H., Kim, W.-Y., Lee, I., Yu, C.-M. *OL* **5**, 2813 (2003).
[7]Lerum, R.V., Chisholm, J.D. *TL* **45**, 6591 (2004).

Acetylacetonato(diolefin)rhodium. 21, 2; 22, 4

Reductive acylation.[1] Arylboronic acids react with cyclobutanones in the presence of (acac)Rh(ethene)$_2$ to give aryl ketones.

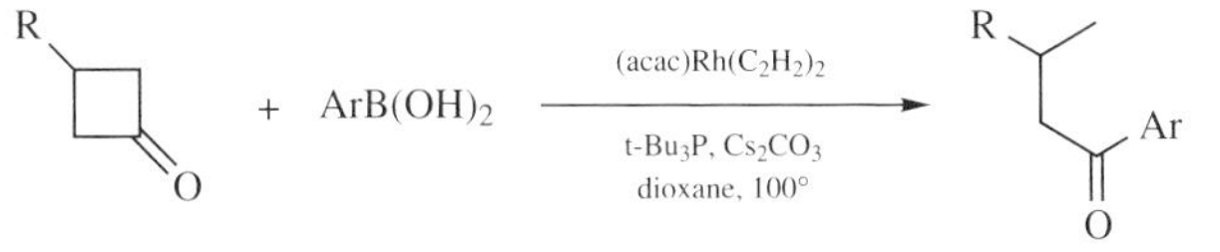

R ... O + $ArB(OH)_2$ —(acac)$Rh(C_2H_2)_2$, t-Bu_3P, Cs_2CO_3, dioxane, 100°→ R ... Ar O

R = Ph 95%

Conjugate additions. Reductive arylation of enones[2] and alkenyl sulfones[3] by $ArB(OH)_2$ is catalyzed by (acac)Rh(ethene)$_2$. Both transformations are subjected to asymmetric induction in the presence of chiral ligands.

[1]Matsuda, T., Makino, M., Murakami, M. *OL* **6**, 1257 (2004).
[2]Ma, Y., Song, C., Ma, C., Sun, Z., Chai, Q., Andrus, M.B. *ACIEE* **42**, 5871 (2003).
[3]Mauleon, P., Carretero, J.C. *OL* **6**, 3195 (2004).

Acetyl chloride. 22, 5

Desilylation.[1] TBS ethers are rapidly cleaved with catalytic amount of AcCl in dry MeOH, typically at room temperature.

Acetylation.[2] The normal trend for acetylation of alcohols (prim.>sec.) is reversed when SiO_2 is present. Thus 1,5-hexanediol gives 5-acetoxyhexanol in 60% yield.

[1]Khan, A.T., Mondal, E. *SL* 694 (2003).
[2]Ogawa, H., Ide, Y., Honda, R., Chihara, T. *JPOC* **16**, 355 (2003).

Acetyl hypofluorite.

Fluorination.[1] Silyl ketene acetals react with AcOF at −45° to give α-fluoroalkanoic esters.

[1]Rozen, S., Hagooly, A., Harduf, R. *JOC* **66**, 7464 (2001).

***N*-Acylbenzotriazoles. 22**,

Preparations.[1] In addition to the conventional use of $SOCl_2$ and BtH, direct reaction of RCOOH with $BtSO_3Me$ also gives these reagents.

Acylations. Acyl transfer to primary and secondary alkyl nitriles to give β-keto nitriles[2] is carried out with these reagents, usually at room temperature (*t*-BuOK/DMSO). Similarly, β-keto sulfones[4] are prepared (BuLi is used to deprotonate the sulfones).

Highly regioselective *C*-acylation of at a β-position of pyrroles and indoles using the title reagents is promoted by $TiCl_4$.[5]

Carboxylic acid derivatives. RCOBt are convenient precursors of hydroxamic acids,[1] thiol esters,[3] and oxazolines/thiazolines.[6]

α-Diketones. Symmetrical α-diketones are produced when RCOBt are exposed to SmI_2.[7]

[1]Katritzky, A.R., Kirichenko, N., Rogovoy, B.V. *S* 2777 (2003).
[2]Katritzky, A.R., Abdel-Fattah, A.A.A., Wang, M. *JOC* **68**, 4932 (2003).
[3]Katritzky, A.R., Shestopalov, A.A., Suzuki, K. *S* 1806 (2004).
[4]Katritzky, A.R., Abdel-Fattah, A.A.A., Wang, M. *JOC* **68**, 1443 (2003).
[5]Katritzky, A.R., Suzuki, K., Singh, S.K., He, H.-Y. *JOC* **68**, 5720 (2003).
[6]Katritzky, A.R., Cai, C., Suzuki, K., Singh, S.K. *JOC* **69**, 811 (2004).
[7]Wang, X., Zhang, Y. *T* **59**, 4201 (2003).

Acyl fluorides. 22, 5

Reductive acylation.[1] Reduction of lactones with Dibal-H followed by kinetic trapping of the hemiacetals with RCOF avoids ring opening. The reaction on esters works well when X = F but not with X = Cl.

[1]Zhang, Y., Rovis, T. *OL* **6**, 1877 (2004).

N-Acyl(methanesulfonamides).

***N*-Acylation.**[1] Amines form amides on heating the mixture with the reagents (via decomposition of the salts). A primary amino group reacts more readily and *N*-acylation takes preference to *O*-acylation in amino alcohols.

NHMs H_2N + NH 120°

H N NH O

90%

[1]Coniglio, S., Aramini, A., Cesta, M.C., Colagioia, S., Curti, R., D'Alessandro, F., D'Anniballe, G., D'Elia, V., Nano, G., Orlando, V., Allegretti, M. *TL* **45**, 5375 (2004).

Alkoxydiphenylphosphines.

Esters and phenyl ethers.[1] Together with a quinone (2,6-dimethylbenzoquinone the best) Ph_2POR forms an adduct which is capable of esterifying carboxylic acids and phenols. Acidic compounds (not ordinary alcohols) react through proton transfer to generate adequate nucleophiles.

[1]Shintou, T., Kikuchi, W., Mukaiyama, T. *BCSJ* **76**, 1645 (2003).

2-Alkoxymethylsulfonylbenzothiazoles.

Vinyl ethers.[1] These reagents are useful for Julia olefination.

O MeO + N S BnO S O O LiHMDS THF, 0°, 1.5h CHOBn MeO

87% E/Z 50:50

[1]Suprenant, S., Chan, W. Y., Berthelette, C. *OL* **5**, 48515 (2003).

Alkylaluminum chlorides. 22, 7–8

*[2 + 2]**Cycloaddition.***[1] Cyclobutanes are obtained when silyl enol ethers and substituted alkenes are brought together in the presence of $EtAlCl_2$.

Hydroxybenzylation.[2] Baylis-Hillman-type reaction of α,β-unsaturated lactones is promoted by Et_2AlI.

Alkylation reactions. $Et_3Al_2Cl_3$ promotes alkyl chloroformates to transfer their alkyl groups to alkenes.[3]

$Et_3Al_2Cl_3$

CH_2Cl_2, –15°

H_2O

67%

N-Alkylation and *C*-allylation of imines are performed in tandem with organoaluminum halides and allyltributylstannane.[4]

Et_2AlCl, $EtAlCl_2$, Bz_2O_2

$H_2C{=}CHCH_2SnBu_3$, EtCN

–20° to rt, 7h

75%

Rearrangement. Ketene-*N,O*-acetals isomerize to the more stable amides[5] by an O- > C alkyl migration process when they are treated with $MeAlCl_2$.

Dehydration. Alkenylsilane synthesis initiated by addition of Me_3SiCH_2Li to carbonyl compounds (best in Et_3N) is completed by heating the adducts directly with Et_2AlCl at 150° as to avoid desilylation.[6]

[1]Takasu, K., Ueno, M., Inanaga, K., Ihara, M. *JOC* **69**, 517 (2004).
[2]Karur, S., Hardin, J., Headley, A., Li, G. *TL* **44**, 2991 (2003).
[3]Biermann, U., Metzger, J.O. *JACS* **126**, 10319 (2004).
[4]Niwa, Y., Shimizu, M. *JACS* **125**, 3720 (2003).
[5]Suzuki, T., Inui, M., Hosokawa, S., Kobayashi, S. *TL* **44**, 3713 (2003).
[6]Kwan, M.L., Battiste, M.E., Macala, M.K., Aybar, S.C., James, N.C., Haoui, J.J. *SC* **34**, 1943 (2004).

Alkyliminotris(dimethylamino)phosphoranes.

Michael reaction.[1] These $RN{=}P(NMe_2)_3$ and a few other nonionic bases are useful catalysts for the Michael reaction of β-keto esters in water.

[1]Bensa, D., Rodriguez, J. *SC* **34**, 1515 (2004).

1-Alkyl-3-methylimidazolium salts. 20, 70; **21,** 85; **22**, 88–91

Common transformations. There are too numerous routine reactions being repeated in ionic liquids, just mentioning them is impractical. Therefore a more critical selection is presented.

Interconversion of alkyl halides and alcohols are more facile in ionic liquids, due to apparent increase in the nucleophilicity of water.[1] Halide ions from the salts enter the TsOH-catalyzed substitution.[2]

Formation of *t*-butyl alkyl ethers from *t*-BuOH and ROH without an acid catalyst is described. Heating the mixture with 1-decyl-3-methylimidazolium tetrafluoroborate accomplishes the task.[3] Glycosylation that employs trichloroacetimidates as glycosyl donors proceeds well at room temperature in an ionic liquid.[4,5] Glycosyl phosphates react with alcohols in another ionic liquid containing Tf_2NH.[6]

Ionic liquid enables milder conditions to be used for the Cu-mediated diaryl synthesis from ArOK and Ar'I.[7]

Wohl-Ziegler bromination that is a free radical reaction can be carried out in ionic liquids.[8]

Formation of three-membered ring compounds in ionic liquids: epoxidation [$MnSO_4$/H_2O_2],[9] aziridination [$Cu(acac)_2$/PhI=NTs],[10] and Corey-Chaykovsky cyclopropanation of enones [KOH/$Me_3S(O)I$][11] have all been successful.

Incorporation of CO_2 into epoxides to form cyclic carbonates[12] can be done in ionic liquids and catalyzed by $ZnCl_2$. Carbodiimide synthesis using amines, CO_2 and ionic liquids[13] is definitely superior to that uses CO or phosgene.

Aromatic substitutions benefit from medium effects of ionic liquids. Metal triflate-catalyzed Friedel-Crafts alkenylation with alkynes is enhanced dramatically,[14] rapid sulfamoylation at room temperature also gives excellent yields.[15]

Couplings. The recoverable Pd complex coordinated by an ionic liquid is highly efficient for catalyzing Heck reaction.[16] Reactions involving electron-rich alkenes proceed better.[17] [Bmim]NTf_2 is a low viscosity medium very suitable for conducting Pd-catalyzed cross-coupling reactions.[18]

Sonagashira coupling[19] and Rh-catalyzed deboronative allylation of arene derivatives with allylic alcohols[20] have also employed the modified media.

Some other reactions. Baylis-Hillman reaction,[21] $InCl_3$-mediated allylation of aldehydes,[22] and IBX oxidation[23] may be mentioned in terms of demonstrating the viability of ionic liquids.

[1]Kim, D.W., Hong, D.J., Seo, J.W., Kim, H.S., Kim, H.K., Song, C.E., Chi, D.Y. *JOC* **69**, 3186 (2004).

[2]Leadbeater, N.E., Torenius, H.M., Tye, H. *T* **59**, 2253 (2003).

[3]Shi, F., Xiong, H., Gu, Y., Guo, S., Deng, Y. *CC* 1054 (2003).

[4]Pakulski, Z. *S* 2074 (2003).

[5]Poletti, L., Rencurosi, A., Lay, L., Russo, G. *SL* 2297 (2003).

[6]Sasaki, K., Nagai, H., Matsumura, S., Toshima, K. *TL* **44**, 5605 (2003).

[7]Chauhan, S.M.S., Jain, N., Kumar, A., Srinivas, K.A. *SC* **33**, 3607 (2003).

[8]Togo, H., Hirai, T. *SL* 702 (2003).

[9]Tong, K.-H., Wong, K.-Y., Chan, T.H. *OL* **5**, 3423 (2003).

[10]Kantam, M.L., Kavita, N.B., Haritha, Y. *SL* 525 (2004).

[11]Chandrasekhar, S., Narasihmulu, C., Jagadeshwar, V., Reddy, K.V. *TL* **44**, 3629 (2003).

[12]Li, F., Xiao, L., Xia, C., Hu, B. *TL* **45**, 8307 (2004).

[13]Shi, F., Deng, Y., SiMa, T., Peng, J., Gu, Y., Qiao, B. *ACIEE* **42**, 3257 (2003).
[14]Song, C., Jung, D., Choung, S., Roh, E., Lee, S. *ACIEE* **43**, 6183 (2004).
[15]Naik, P.U., Harjani, J.R., Nara, S.J., Salunkhe, M.M. *TL* **45**, 1933 (2004).
[16]Xiao, J., Twamley, B., Shreeve, J. *OL* **6**, 3845 (2004).
[17]Mo, J., Xu, L., Xiao, J. *JACS* **127**, 751 (2005).
[18]Liu, S., Fukuyama, T., Sato, M., Ryu, I. *SL* 1814 (2004).
[19]Park, S.B., Alper, H. *CC* 1306 (2004).
[20]Kabalka, G.W., Dong, G., Venkataiah, B. *OL* **5**, 893 (2003).
[21]Hsu, J.-C., Yen, Y.-H., Chu, Y.-H. *TL* **45**, 4673 (2004).
[22]Lu, J., Ji, S.-J., Qian, R., Chen, J.-P., Liu, Y., Loh, T.-P. *SL* 534 (2004).
[23]Chhikara, B.S., Chandra, R., Tandon, V. *TL* **45**, 7585 (2004).

Allylboronates. 22, 8

α-Methylene γ-butyrolactones.[1] α-Boronatomethyl α,β-unsaturated esters are prepared from conjugated addition of organocuprate reagents to alkynoic esters followed by trapping with *B*-iodomethylboronates. Reaction of these allylboronates with aldehydes leads to the methylene lactones.

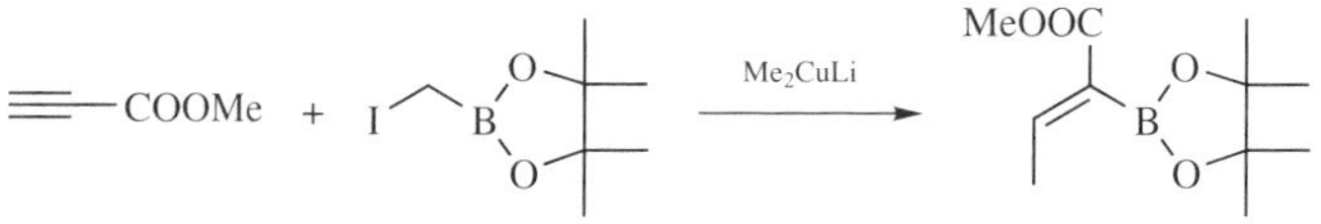

COOMe + I B O O — Me_2CuLi → MeOOC B O O

Homoallylic amines.[2] A 3-component reaction involving an allylboronate, an aldehyde and ammonia gives the products stereoselectively. The process has been applied to the synthesis of amino sugars and an uncommon amino acid.

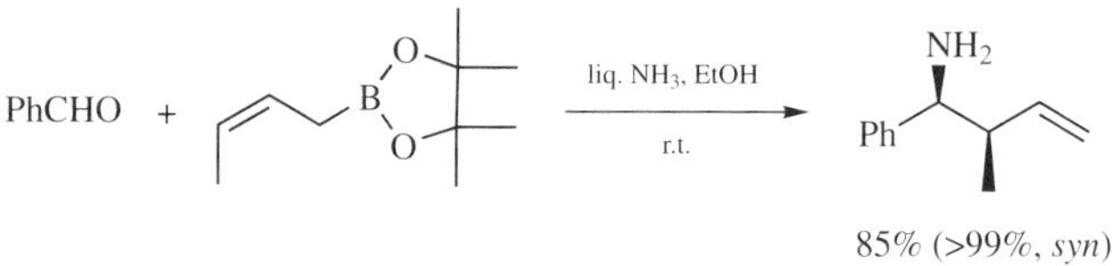

PhCHO + B O O — liq. NH_3, EtOH, r.t. → NH_2 Ph

85% (>99%, *syn*)

Homoallylic Alcohols.[3] Allylboronic acid react with α-epoxy alcohols via rearrangement, due to the Lewis acidity of the reagent.

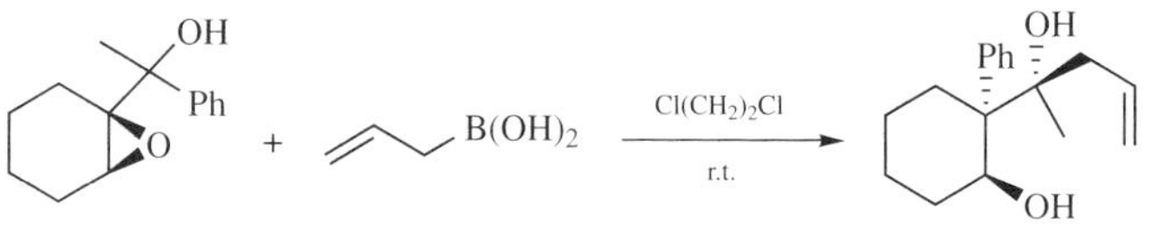

OH Ph O + $B(OH)_2$ — $Cl(CH_2)_2Cl$, r.t. → OH Ph OH

75%

[1]Kennedy, J.W.J., Hall, D.G. *JOC* **69**, 4412 (2004).
[2]Sugiura, M., Hirano, K., Kobayashi, S. *JACS* **126**, 7182 (2004).
[3]Hu, X.-D., Fan, C.-A., Zhang, F.-M., Tu, Y.Q. *ACIEE* **43**, 1702 (2004).

Allyl(chloro)dibutylstannane.

2-Oxazolidinones.[1] Treatment of 4–oxo-2-alkenals with the reagent followed by trapping of the initial adducts with an isocyanate results in the formation of 4-acylmethyl-5-allyl-2-oxazolidinones. Alternatively, a combination of an allylstannane and dichlorodibutylstannane also can be used.

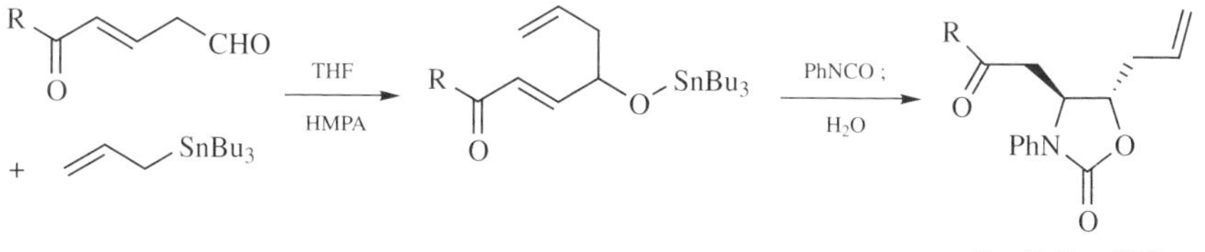

R ... CHO + $SnBu_3$ —THF, HMPA→ R ... O–$SnBu_3$ —PhNCO ; H_2O→ PhN, O, O

$R = C_8H_{17}$ 81%

[1]Shibata, I., Kato, H., Kanazawa, N., Yasuda, M., Baba, A. *JACS* **126**, 466 (2004).

η^3-Allyl(chloro)palladium, carbene complex.

Suzuki coupling.[1] Complex **1** is useful as a coupling catalyst as well as for dehalogenation of ArCl (11 examples, 91-100%) and amination of ArOTf (10 examples, 77-95%).

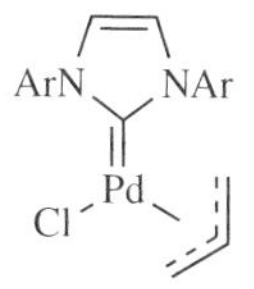

ArN NAr Pd Cl

1

[1]Navarro, O., Kaur, H., Mahjoor, P., Nolan, S.P. *JOC* **69**, 3173 (2004).

η^3-Allyl(chloro)palladium, phosphine complex.

Cyclization.[1] Intramolecular allylic substitution is rendered regioselective by placing a heteroatom in the chain.

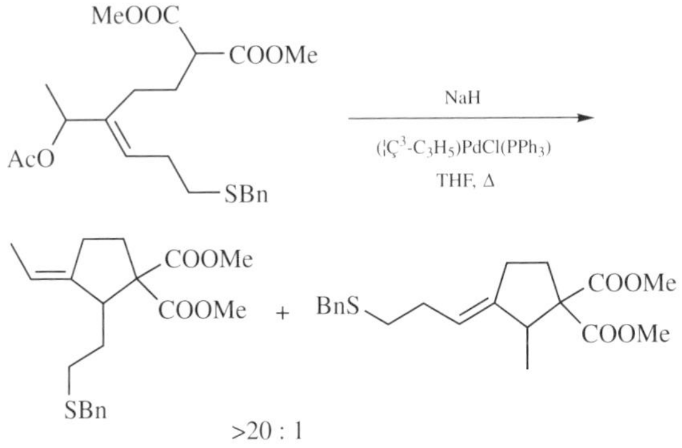

[1]Krafft, M.E., Lucas, M.C. *CC* 1232 (2003).

η^3-Allyl(cyclooctadiene)palladium tetrafluoroborate.

Benzylation.[1] Malonic esters, amines and alcohols are readily benzylated with $ArCH_2OCOOMe$.

[1]Kuwano, R., Kondo, Y., Matsuyama, Y. *JACS* **125,** 12104 (2004).

η^3-Allyl(cyclopentadienyl)palladium.

Benzylation.[1] Active methylene compounds undergo benzylation without additional base in the presence of the Pd catalyst and a phosphine ligand. COD improves the yields remarkably (e.g., 10% -> 98%), presumably by preventing aggregation of the catalyst to form inactive Pd species.

Allylation of ketones.[2] Certain ketones and β-diketones react with conjugated dienes to afford adducts.

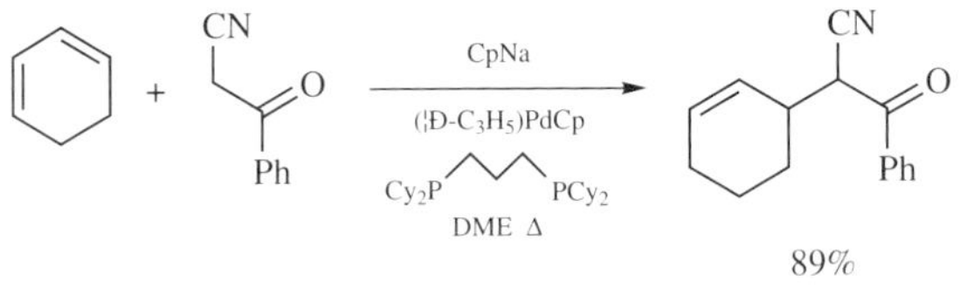

[1]Kuwano, R., Kondo, Y. *OL* **6,** 3545 (2004).
[2]Leitner, A., Larsen, J., Steffens, C., Hartwig, J.F. *JOC* **69,** 7552 (2004).

Allylsilanes. 21, 11; **22**, 9–10

Homoallyl amines. Allylation of *N*-acylhydrazones[1] is found to be very efficient in the presence of phosphine oxides such as that of dppp. Reaction of imines derived from 2-hydroxyaniline with allyltrichlorosilanes is demonstrated.[2] The internal coordination (H-bonding) of the substrates is important.

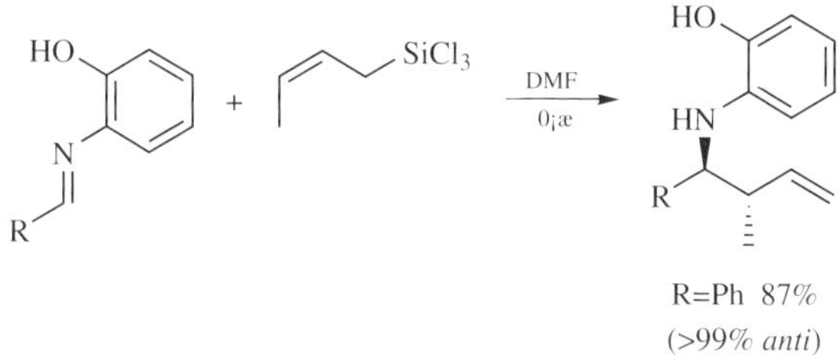

DMF, 0°C

R=Ph 87%
(>99% *anti*)

The silacycle **1** is stable to storage and transfers an allyl group readily to acylhydrazones.[3] Most tested cases afford good yields and ee.

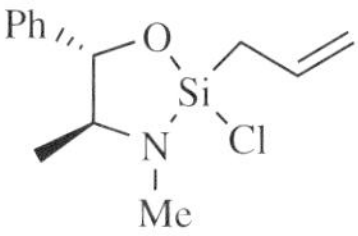

1

Hydrodehalogenation.[4] Silylated cyclohexadienes such as **2** are useful reducing agents in the radical chain process.

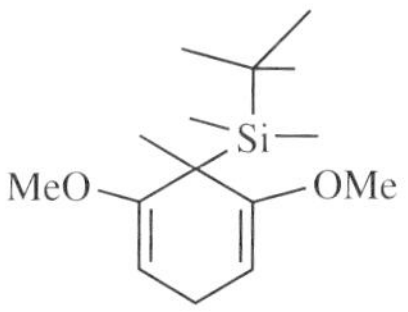

2

[1]Ogawa, C., Konishi, H., Sugiura, M., Kobayashi, S. *OBC* **2,** 4467 (2004).
[2]Sugiura, M., Robvieux, F., Kobayashi, S. *SL* 1749 (2003).
[3]Berger, R., Rabbat, P.M.A., Leighton, J.L. *JACS* **125**, 9596 (2003).
[4]Studer, A., Amrein, S., Schleth, F., Schulte, T., Walton, J.C. *JACS* **125**, 5726 (2003).

Allylstannanes. 21, 13; **22**, 10

Homoallylic alcohols. Tetraallylstannane is fully utilized in its reaction with conjugated carbonyl compounds in refluxing MeOH.[1] Highly selective 1,2-addition is observed.

Use of (*S*)-aspartic acid to induce asymmetric addition to aldehydes gives low ee.[2]

Homoallylic amines. Allylation of imines activated by organoaluminums in the presence of benzoyl peroxide takes place simultaneously with *N*-silylation and *C*-allylation. Thus, in the former case the use of $(Me_3Si)_2AlCl$ is indicated.[3]

[1]Leitch, S.K., McClusky, A. *SL* 699 (2003).
[2]Yanagisawa, A., Nakamura, Y., Arai, T. *TA* **15**, 19096 (2004).
[3]Niwa, Y., Shimizu, M. *JACS* **125**, 3720 (2003).

S-Allylsulfolanium bromide.

2-Vinylaziridines. The ylide generated from the reagent (e.g., with BuLi) reacts with *N*-tosylimines to afford aziridines.

[1]Arini, L. G., Sinclair, A., Szeto, P., Stockman, R. A. *TL* **45**, 1589 (2004).

Allyltriphenylphosphonium peroxodisulfate.

Oxidation.[1] Alcohols and their silyl ethers and THP ethers are converted into carbonyl compounds with this reagent $[(CH_2{=}CHCH_2PPh_3)_2S_2O_8]$.

[1]Tajbakhsh, M., Lakouraj, M.M., Fadavi, A. *SC* **34**, 1173 (2004).

Alumina. 21, 14–15; **22,** 10–11

2-Amino-2-chromenes.[1] A 3-component condensation from phenols, aldehydes and malononitrile is achieved on heating with γ-alumina (9 examples, 83–98%).

C-Alkynylation of pyrroles.[2] Pyrroles react with 1-acyl-2-bromoethynes at room temperature in the presence of alumina. Roles of the reactants are different from the Sonagashira coupling.

[1]Maggi, R., Ballini, R., Sartori, G., Sartorio, R. *TL* **45**, 2297 (2004).
[2]Trofimov, B. A., Stepanova, Z. V., Sobenina, L. N., Mikhaleva, A. I., Ushakov, I. A. *TL* **45**, 6513 (2004).

Aluminum. 22, 11

Diaryl ketone reduction.[1] A system consisting of Al (5 eq.), NaOH (2.5 eq.), and $MeOH\text{-}H_2O$ (2:1) is useful for reduction of diaryl ketones while leaving other types of ketones unchanged.

O O OH O
Ph Al Ph
NaOH
$MeOH\text{-}H_2O$
76%

Quinone reduction.[2] 2,3,9,10-tetrakis(trimethylsilyl)pentacene has been obtained by reduction of the quinone with amalgamated Al in refluxing cyclohexanol in the presence of CBr_4.

[1]Bhar, S., Guha, S. *TL* **45**, 3775 (2004).
[2]Chan, S.H., Lee, H.K., Wang, Y.M., Fu, N.Y., Chen, X.M., Cai, Z.W., Wong, H.N.C. *CC* 66 (2005).

Aluminum alkoxide.

Propargylic alcohols.[1] Alkynylation of aldehydes is accomplished by a group-exchange process using the alkyne-acetone adducts in the presence of Al alkoxide in CH_2Cl_2 at room temperature

[1]Ooi, T., Miura, T., Ohmatsu, K., Saito, A., Maruoka, K. *OBC* **2**, 3312 (2004).

Aluminum biphenyl-2′-perfluorooctanesulfonamide-2-oxide isopropoxide. 22, 13–14

Tishchenko reaction.[1] Various aldehydes are converted into esters by this reagent (9 examples, 68–99%).

[1]Ooi, T., Ohmatsu, K., Sasaki, K., Miura, T., Maruoka, K. *TL* **44**, 3191 (2003).

Aluminum bromide.

Carboxylation.[1] Arylsilanes undergo desilylative carboxylation with CO_2 at room temperature using $AlBr_3$ as mediator. For the conversion of allylsilanes $MeAlCl_2$ is employed.

[1]Hattori, T., Suzuki, Y., Miyano, S. *CL* **32**, 454 (2003).

Aluminum chloride. 20, 12–13; **21,** 15–17; **22,** 11–13

β-Amino ketones. By a Michael addition to conjugated ketones promoted by $AlCl_3$ such adducts are readily formed at room temperature.[1]

Thiiranes.[2] Epoxides are converted to thiiranes on reaction with thiourea in MeCN at room temperature in the presence of polystyrene-supported $AlCl_3$.

3-(2-Oxoalkyl)indoles.[3] Skatole is acylated at the methyl group using $AlCl_3$ as promoter. However, 2-acyl-3-methylindoles are formed by changing $AlCl_3$ to $ZnCl_2$.

Ring expansion.[4] Bicyclic Diels-Alder adducts from enals undergo ring expansion on exposure to $AlCl_3$. Actually the products are formed directly from dienes and dienophiles (to achieve a formal [4 + 3]cycloaddition). However, asymmetric version is best carried out by catalytic Diels-Alder addition and then treatment with $AlCl_3$.

CHO
$AlCl_3$
CH_2Cl_2
−78¡æ~0¡æ
CHO
O
90%

Double addition.[5] Conjugated imines add HN_3 and an allyl group from an allylstannane in the presence of $AlCl_3$. The products contain a homoallyl amine unit.

Thia-Fries rearrangement.[6] Aryl triflinates are induced to rearrange, affording trimethanesulfinylphenols.

Dihydrocinnamides.[7] Acrylamides (*N*-free or *N,N*-disubstituted) react with benzene at room temperature to provide 3-phenylpropanamides.

Aluminacyclopentadienes.[8] Reaction of (*Z,Z*)-1,4-dilithio-1,3-dienes with $AlCl_3$ leads to the metallacycles. Substituted cyclopentadienes are formed when such reactive species are mixed with aldehydes.

3-Cyanoflavones.[9] A route to the cyanoflavones involves a cyclization step.

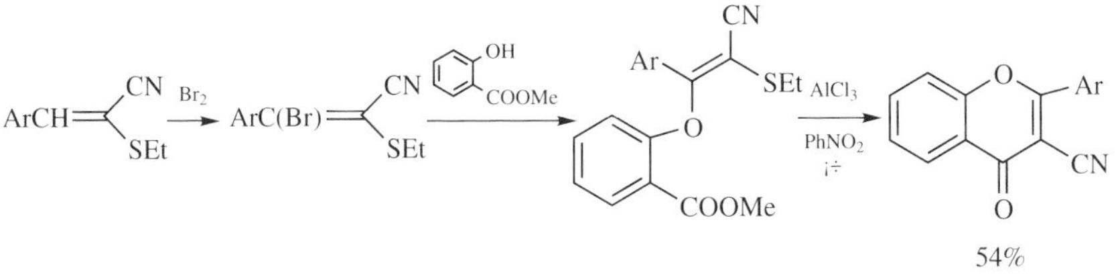

[1]Lee, A. S.-Y., Wang, S.-H., Chu, S.-F. *SL* 2359 (2003).
[2]Tamammi, B., Borujeny, K.P. *SC* **34**, 65 (2004).
[3]Pal, M., Dakarapu, R., Padakanti, S. *JOC* **69**, 2913 (2004).
[4]Davies, H.M.L., Dai, X. *JACS* **126**, 2692 (2004).
[5]Shimizu, M., Yamauchi, C., Ogawa, T. *CL* **33**, 606 (2004).
[6]Chen, X., Tordeux, M., Desmurs, J.-R., Wakselman, C. *JFC* **123**, 51 (2003).
[7]Koltunov, K.Yu., Walspurger, S., Sommer, J. *TL* **45**, 3547 (2004).
[8]Fang, H., Zhao, C., Li, G., Xi, Z. *T* **59**, 3779 (2003).
[9]Lassagne, F., Pochat, F. *TL* **44**, 9283 (2003).

Aluminum dodecatungstophosphate.

Acylation. A solvent-free acetylation of amines, alcohols, and thiols with Ac_2O is catalyzed by $AlPW_{12}O_{40}$. Selective *N*-acetylation of amino alcohols with limited amount of Ac_2O has been demonstrated.[1] Aldehydes are converted into *gem*-diacetates.[2] Acylation of electron-rich aromatic rings[3] with RCOOH-TFAA has been carried out.

[1]Firouzabadi, H., Iranpoor, N., Nowrouzi, F., Amani, K. *CC* 764 (2003).
[2]Firouzabadi, H., Iranpoor, N., Nowrouzi, F., Amani, K. *TL* **44**, 3951 (2003).
[3]Firouzabadi, H., Iranpoor, N., Nowrouzi, F. *TL* **44**, 5343 (2003).

Aluminum tris[2,6-bis(4-alkylphenyl)phenoxide].

Vinylogous aldol reaction.[1] γ-Hydroxyalkylation is accomplished in good yield (7 examples, 70–99%) in the presence of the very bulky aluminum phenoxide

(LTMP as base). Some degree of 1,7-stereocontrol is manifested with (-)-menthyl crotonate.

[1]Takikawa, H., Ishihara, K., Saito, S., Yamamoto, H. *SL* 732 (2004).

Aluminum tris(2,6-diphenylphenoxide), ATPH. 20, 14–15; **21,** 17–18; **22,** 13–14

Asymmetric Michael reaction.[1] Esters containing a chirality center at the β-position are obtained from α,β-unsaturated esters by a reaction exemplified below. Interestingly, a temperature effect of added LiX to the reaction is noted: positive effect at −78° but negative at room temperature.

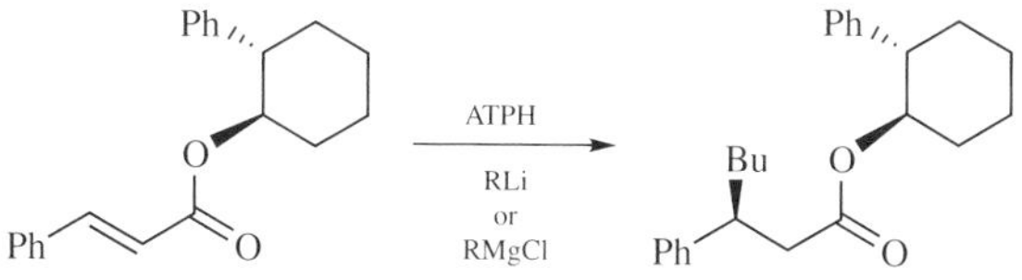

α-Substituted amines.[2] ATPH complexes with 2-substituted oxazolidines to promote ring opening with RLi in a diastereoselective manner. A 4-substituent provides the stereocontrolling element.

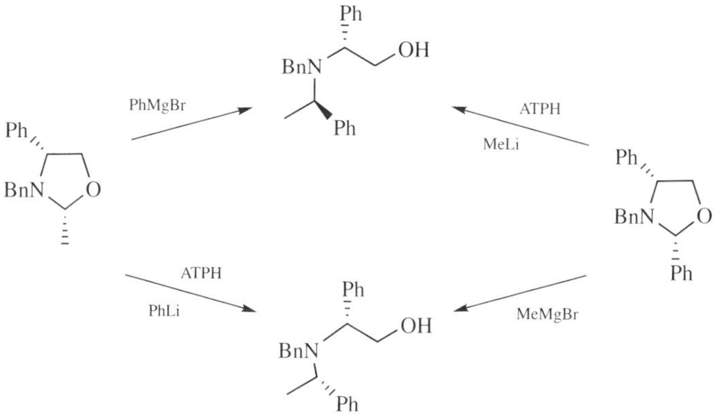

[1]Ito, H., Nagahara, T., Ishihara, K., Saito, S., Yamamoto, H. *ACIEE* **43**, 994 (2004).
[2]Yamauchi, T., Sazanami, H., Sasaki, Y., Higashiyama, K. *T* **61**, 1731 (2005).

Aminomethyltributylstannane.

Azomethine ylide generation.[1] The reagent $H_2NCH_2SnBu_3$ condenses with certain ω-haloalkanals to form precursors of azomethine ylides. In the presence of dipolarophiles indolizidines are obtained.

PhMe, 110°C; 84%

[1]Pearson, W.H., Stoy, P., Mi, Y. *JOC* **69**,19194 (2004).

Ammonia. 22, 15

Debenzyloxycarbonylation.[1] Hydroxyl groups in carbohydrates protected as Cbz-derivatives are liberated by ammonia in MeOH at ice temperature. Some other protecting groups such as TBS survive the conditions.

Homopropargylic amines.[2] Liquid ammonia or NH_4OH can be used in the reaction with aldehydes and allylboronic esters.

[1]Mouffouk, F., Morere, A., Vidal, S., Leydet, A., Montero, J.-L. *SC* **34**, 303 (2004).
[2]Sugiura, M., Hirano, K., Kobayashi, S. *JACS* **126**, 7182 (2004).

Ammonium bicarbonate.

α-Amino phosphonates.[1] The silica-supported reagent combines with aldehydes and dialkyl phosphonates under microwave irradiation.

[1]Kaboudin, B., Rahmani, A. *S* 2705 (2003).

Ammonium formate.

N-Formylamino acid esters.[1] Refluxing the HCl salt of an amino acid ester with $HCOONH_4$ in MeCN leads to the *N*-formyl derivative (7 examples, 66–91%).

[1]Kotha, S., Behera, M., Khedkar, P. *TL* **45**, 7589 (2004).

9-Anthraldehyde dimethyl acetal.

Diol protection.[1] By exchange reaction diols are derivatized. The cyclic acetals are cleaved by $NaBH_3CN$ while leaving benzylidene acetals intact.

[1]Ellervik, U. *TL* **44**, 2279 (2003).

Antimony(V) chloride. 20, 16; **22**, 15

Deoximation.[1] Regeneration of carbonyl compounds is reported by treatment with $SbCl_5$ in CH_2Cl_2.

Desilylation.[2] A system with substoichiometric $SbCl_5$ in MeCN containing 0.1% H_2O serves to cleave silyl ethers at room temperature. Acid-sensitive groups such as acetonide survive the treatment.

[1]Narsaiah, A.V., Nagaiah, K. *S* 1881 (2003).
[2]Gloria, P.M.C., Prabhakar, S., Lobo, A.M., Gomes, M.J.S. *TL* **44**, 8819 (2003).

Arenediazonium *o*-benzenedisulfonamides. 21, 19; **22**, 16

S-Aryl thiol esters.[1] The diazonium salts react with RC(=O)SNa rapidly at room temperature to give RC(=O)SAr (28 examples, 81–100%).

[1]Barbero, M., Degani, I., Dughera, S., Focchi, R. *S* 1225 (2003).

Arylboronic acids. 21, 20; **22,** 16

Monoaryl hydrazines.[1] The Boc-diprotected hydrazines are obtained from reaction of Boc-hydrazine with arylboronic acids in the presence of CuCl.

HN(NH₂)–C(=O)–O–tBu + $ArB(OH)_2$ —CuCl, 3A-MS, Py,DCE→ ArN(Boc)–NH(Boc)

R=Ph, 81%

Petasis reaction. The reaction applying to $ArB(OH)_2$ and 1,3,5-$(RO)_3C_6H_3$ leads to moderate yields of diarylacetic acids.[2] α-Arylamino acid analogues can be synthesized using hydroxylamine derivatives.[3]

R—NHOR′ + CHO–COOH + $ArB(OH)_2$ —CH_2Cl_2, r.t.→ R′O–N(R)–CH(Ar)–COOH

[1]Kabalka, G.W., Guchhait, S.K. *OL* **5**, 4129 (2003).
[2]Naskar, D., Roy, A., Seibel, W.L. *TL* **44**, 8861 (2003).
[3]Naskar, D., Roy, A., Seibel, W.L., Portlock, D.E. *TL* **44**, 8865 (2003).

3-Arylcarboxamidooxaziridines.

Sulfinylimines.[1] The reagents transfer the [$NCONR_2$] group to sulfides at −40°. But only 3 examples (72–99%) are reported.

[1]Armstrong, A., Edmonds, I.D., Swarbrick, M.E. *TL* **44**, 5335 (2003).

2-Arylethenyl(diphenyl)sulfonium triflates.

2-Arylaziridines.[1] Reaction of the salts with primary amines affords the aziridines. *t*-Butylamines is generally added (though not always needed), and for the synthesis of *N*-benzenesulfonylaziridines from $PhSO_2NH_2$, NaH is employed as the base.

A one-pot procedure involves admixture of the styrene, Ph_2SO, and Tf_2O in CH_2Cl_2 at $-78°$, and subsequent addition of an amine.

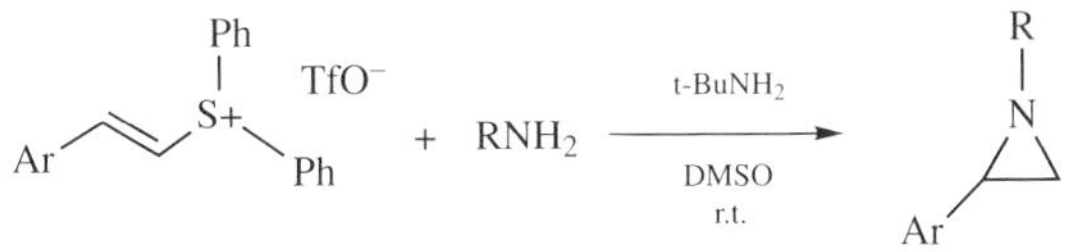

Ph TfO⁻ S+ Ar Ph + RNH_2 t-$BuNH_2$ DMSO r.t. R N Ar

[1]Matsuo, J., Yamanaka, H., Kawana, A., Mukaiyama, T. *CL* **32**, 392 (2003).

Azidoalkanes.

Allylic amines.[1] When phosphites derived from allyl alcohols and 2-chloro-5,5-dimethyl-1,3,2-dioxaphosphane react with azidoalkanes (Staudinger reaction) the products undergo [3,3]sigmatropic rearrangement. Accordingly, the process is valuable for synthesis of transposed allylic amines from allyl alcohols. Allyl azide and benzyl azide are the most desirable reactants because primary amines are much more easily accessible from the initial products.

[1]Chen, B., Mapp, A.K. *JACS* **126,** 5364 (2004).

Azobisisobutyronitrile.

Acyl radicals. AIBN initiates generation of PhS· That adds to the triple bond of 4-pentynylthiol esters. Cyclization releases acyl radicals that give aldehydes.[1] Such radicals can be trapped to form cyclic ketones.[2]

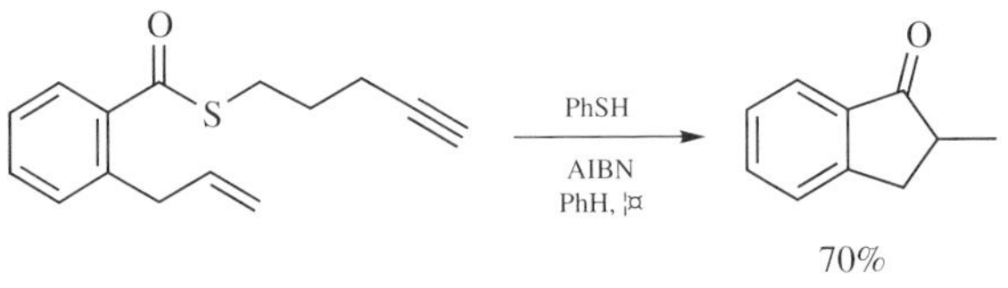

O S PhSH AIBN PhH, Δ O 70%

N-Alkynylimines undergo silylative carbonylation to afford α-tris(trimethylsilyl)-silylmethylene lactams.[3] For 5- and 6-membered lactams only the (*E*)-products are observed. However, a substituent at the propargylic position influences the stereochemical outcome.

1,5-Dicarbonyl compounds.[4] A multicomponent reaction leading to the products involves carbonylation of radicals (from RI) and addition of the resulting acyl radicals to an acceptor and further reaction with tin enolates.

$C_8H_{17}I$ + CO + CN + OSnBu$_3$ Ph → (AIBN; PhH, Δ; 80atm) C_8H_{17} O CN O Ph

78%

A simpler radical addition-fragmentation method for γ-keto ester synthesis[5] uses *t*-butyl vinyl ethers and bromoacetic esters.

[1]Benati, L., Leardini, R., Minozzi, M., Nanni, D., Scialpi, R., Spagnolo, P., Zanardi, G. *SL* **43**, 987 (2004).
[2]Benati, L., Aalestani, G., Leardini, R., Minozzi, M., Nanni, D., Spagnolo, P., Strazzari, S. *OL* **5**,1313 (2003).
[3]Tojino, M., Otsuka, N., Fukuyama, T., Matsubara, H., Schiesser, C.H., Kuriyama, H., Miyazato, H., Minakata, S., Komatsu, M., Ryu, I. *OBC* **1**, 4262 (2003).
[4]Miura, K., Tojino, M., Fujisawa, N., Homosi, A., Ryu, I. *ACIEE* **43**, 2423 (2004).
[5]Cai, Y., Roberts, B.P., Tocher, D.A., Barnett, S.A. *OBC* **2**, 2517 (2004).

B

Barium. 17, 10; **19**, 18

Aldols.[1] Activated Ba promotes enolate formation from α-chloroketones and the subsequent reaction with aldehydes furnishes aldols.

[1]Yanagisawa, A., Takahashi, H., Arai, T. *CC* 580 (2004).

Barium hydroxide. 20, 18; **22**, 18

Hydrolysis.[1] A useful protocol for ester hydrolysis consists of treatment with $Ba(OH)_2$ in MeOH, then acidified with anhydrous HCl. As no aqueous workup is required this method is suitable for parallel synthesis of compound libraries.

[1]Anderson, M.O., Moser, J., Sherill, J. Guy, R.K. *SL* 2391 (2004).

Benzenesulfinic acid.

α-Sulfonylalkylation.[1] Amides are converted into $RCONHCH(SO_2Ph)R'$ on reaction with R′CHO and $PhSO_2H$. The products are reduced by $NaBH_3OAc$ to afford secondary amines (RCH_2NHCH_2R').

[1]Mataloni, M., Petrini, M., Profeta, R. *SL* 1129 (2003).

Benzenesulfonylacetaldehyde diethyl acetal.

Diol protection.[1] Acid-catalyzed alkoxy exchange of the reagent with 1,2-diols leads to 2-substituted dioxolanes which are stable to NaOH, NaOMe, LDA, and DBU. Cleavage is achieved by BuLi or $LiNH_2$.

[1]Chandrasekhar, S., Srinivas, C., Srihari, P. *SC* **33**, 895 (2003).

Benzenesulfonyl azide. 22, 18–19

Azidation. The combination of $PhSO_2N_3$ and Et_3B can be used to convert alkyl iodide into azides. An alkene present will participate in the radical reaction.[1]

R + EtOOC, I, COOEt —($PhSO_2N_3$; Et_3B, H_2O, r.t.)→ R, COOEt, COOEt, N_3

[1]Panchaud, P., Renaud, P. *JOC* **69**, 3205 (2004).

Fiesers' Reagents for Organic Synthesis, Volume 23. Edited by Tse-Lok Ho

Benzotriazole. 21, 24; **22**, 19

N-Acylbenzotriazoles.[1] Carboxylic acids are converted into RCOBt by the $SOCl_2$ and benzotriazole combination.

Nitriles.[2] Dehydration of aldoximes is ra'pidly accomplished with BtH and $SOCl_2$. A secondary alcohol in carbohydrate oximes is said not to be affected.

Baylis-Hillman reaction.[3] Triazoles, benzotriazole as well as diazoles catalyze the Baylis-Hillman reaction in 1M $NaHCO_3$. Presence of the weak inorganic base is important.

[1]Katritzky, A. R., Zhang, Y., Singh, S.K. *S* 2795 (2003)
[2]Telvekar, V.N., Akamanchi, K.G. *SC* **34**, 2331 (2004).
[3]Luo, S., Mi, X., Wang, P.G., Cheng, J.-P. *TL* **45**, 5171 (2004).

Benzoyl cyanide.

Cyanobenzoylation.[1] Aldehydes are converted into either RCH(OBz)CN or RCH (OH)CN by the reagent in DMSO (without catalyst). Anhydrous conditions deliver the benzoates.

[1]Watahiki, T., Ohba, S., Oriyama, T. *OL* **5**, 2679 (2003)

Benzyltriethylammonium tetrathiomolybdate.

Reduction. β-Glycosylamines are obtained from the corresponding glycosyl azides.[1] An azido group at C-2 or C-6 is not affected.

[1]Sridhar, P.R., Prabhu, K.R., Chandrasekaran, S. *JOC* **68**, 5261 (2003).

Benzyltrimethylammonium hydroxide.

Propargylic alcohols.[1] Alkynylation of carbonyl compounds by 1-alkynes is accomplished using the quaternary ammonium hydroxide as base in DMSO.

[1]Ishikawa, T., Mizuta, T., Hagiwara, K., Aikawa, T., Kudo, T., Saito, S. *JOC* **68**, 3702 (2003).

Benzyltrimethylammonium tribromide.

2-Aminobenzothiazoles.[1] Substituted aryl thioureas cyclize on treatment with the title reagent in HOAc at room temperature. A more direct method involves ArN=C=S and amines.

[1]Jordan, A.D., Luo, C., Reitz, A.B. *JOC* **68**, 8693 (2003).

1,1′-Binaphthalenes, 2,2′-heteroaryl.

Coupling reactions. Pinacol coupling of ArCHO in excellent ee is conducted in the presence of a Cr complex of **1**, even an aliphatic aldehyde produces moderate yield of the *syn*-diol with ee approximately 80%.[1] Pd-catalyzed cyclization accompanied by chain extension to provide a building block of vitamin E uses the oxazoline ligand **2**.[2]

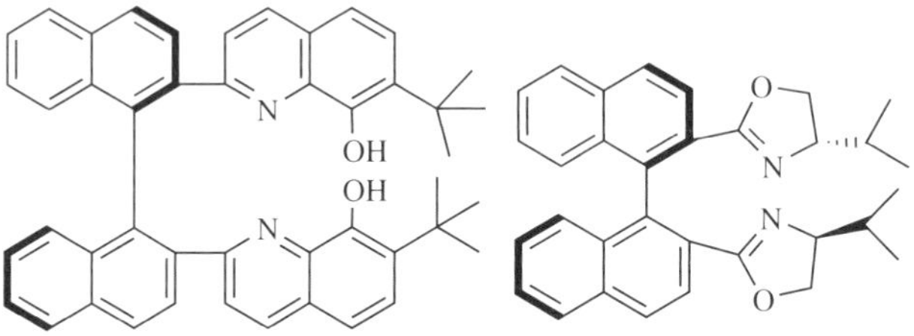

[1]Takenaka, N., Xia, G., Yamamoto, H. *JACS* **126**, 13198 (2004).
[2]Tietze, L., Sommer, K., Zinngrebe, J., Stecker, F. *ACIEE* **44**, 257 (2004).

1,1′-Binaphthalene-2-amine-2′-ol derivatives.

Conjugate addition. The BINAMOL hybrid can be further functionalized, e.g., as both pyridine-2-carboxamide and BINOL phosphite. A derived Cu complex has been used to induce enantioselective addition of organozincs to enones.[1]

Allylic substitution. The dimeric Ag complex in which the amino nitrogen of BINAMOL is also part of imidazolidine carbene segment (also ligated to Ag) is readily converted into the air-stable Cu form. The latter species catalyzes only S_N2' displacement of allylic phosphates by organozinc reagents. High selectivity is observed in forming quaternary carbon centers.[2]

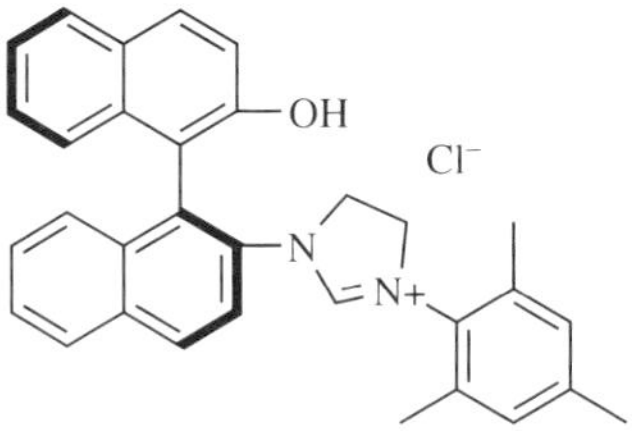

[1]Hu, Y., Liang, X., Wang, J., Zheng, Z., Hu, X. *JOC* **68**, 4542 (2003).
[2]Larsen, A.O., Leu, W., Oberhuber, C.N., Campbell, J.E., Hoveyda, A.H. *JACS* **126**, 11130 (2004).

1,1′-Binaphthalene-2,2′-diamine derivatives. 22, 22–23

Rearrangement. The *N,N′*-bistriflamide complexes with Al to form an effective Lewis acid catalyst for rearrangement of α,α-disubstituted α-amino aldehydes.[1]

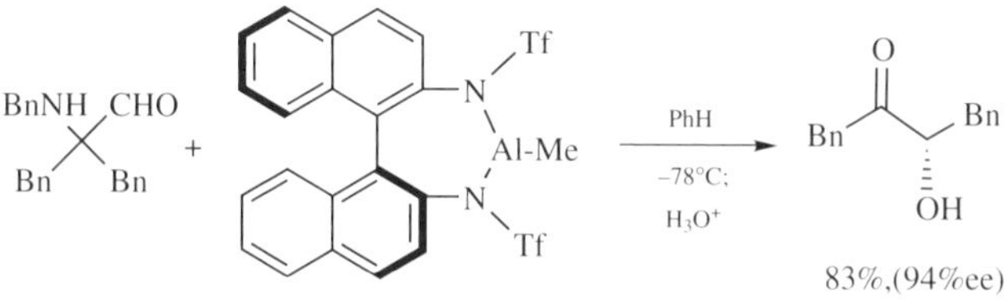

CC Bond formations. Addition of 1-alkynes to imines is assisted with a Cu complex of BINAM-derived imine ligand.[2] Phosphinamides of BINAMINE also show good potential as ligands in the Pd-catalyzed allylic substitutions.[3]

Racemic but chirally flexible BIPHEP ligands (e.g., **1**), when binding to Rh-BINAM, are very useful for inducing enantioselective ene-type cyclization of 1,6-enynes.[4]

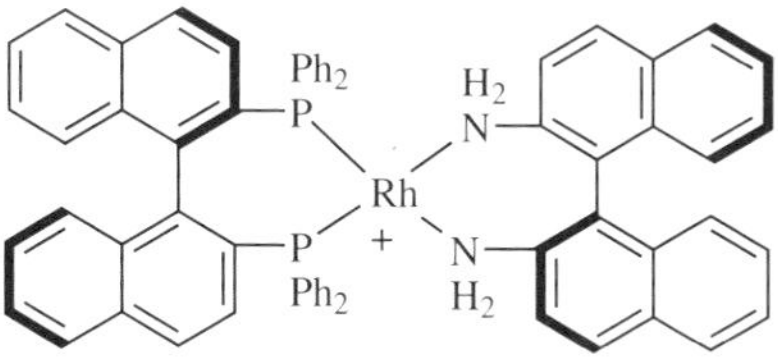

1

Complexation of Cu ion to mono(thiophosphoramido)-BINAM affords chiral catalysts for the addition of organozinc reagents to *N*-tosylimines[5] and enones.[6]

Cu-complexes of several monoalkylated octahydro-BINAM's have been investigated for their ability to effect enantioselective production of BINOL-3,3′-dicarboxylic esters under O_2. The *N*-(3-pentyl) derivative appears to be the best ligand[7] to give reasonable yields and ee when the reaction is carried out at 0°.

Addition to aldehydes. The phosphotriamide incorporating two molecules of *N,N′*-dimethyl-BINAM are chiral Lewis bases that can be used to activate Lewis acids. Applications of such systems include aldol-type condensations[8,9] and carbamoylation[10] of aldehydes.

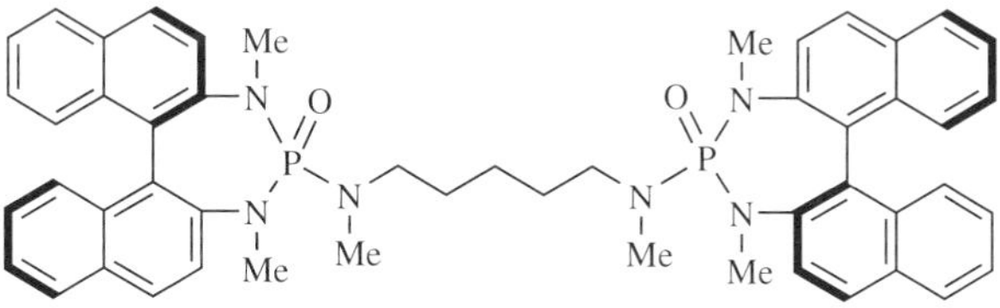

1

Reduction. BINAM-linked carbene complexes of metal ions such as **2** are relatively new. One use of them is in the hydrosilylation of methyl ketones.[11]

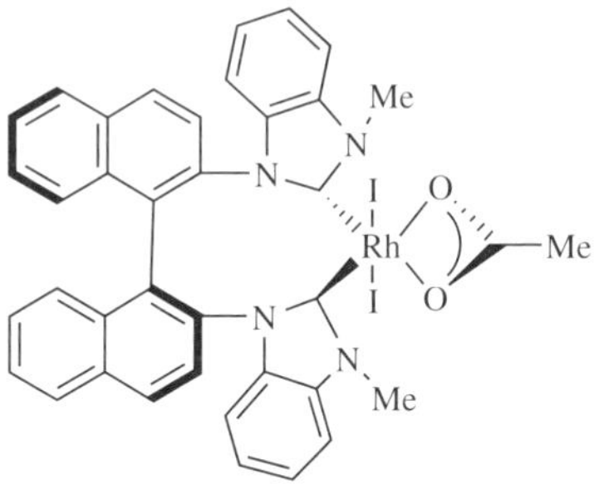

2

Heterocycles. Insertion of CO_2 into epoxides to form 1,3-dioxolan-2-ones is efficiently accomplished using metal complexes of the SALEN-type in which the imino nitrogen atoms are part of BINAM.[12] The bisimine derived from 2,6-dichlorobenzaldehyde promotes asymmetric cyclopropanation and aziridination.[13]

[1]Ooi, T., Saito, A., Maruoka, K. *JACS* **125**, 3220 (2003).
[2]Benaglia, M., Negri, D., Dell'Anna, G. *TL* **45**, 8705 (2004).
[3]Chen, X., Guo, R., Li, Y., Chen, G., Yeung, C.-H., Chan, A.S.C. *TA* **15**, 213 (2004).
[4]Mikami, K., Kataoka, S., Yusa, Y., Aikawa, K. *OL* **6**, 3699 (2004).
[5]Wang, C.-J., Shi, M. *JOC* **68**, 6229 (2003).
[6]Shi, M., Wang, C.-J., Zhang, W. *CEJ* **10**, 5507 (2003).
[7]Kim, K.H., Lee, D.-W., Lee, Y.-S., Ko, D.-H., Ha, D.-C. *T* **60**, 9037 (2004).
[8]Denmark, S.E., Heemstra, Jr., J.R. *OL* **5**, 2303 (2003).
[9]Denmark, S.E., Beutner, G.L. *JACS* **125**, 7800 (2003).
[10]Denmark, S.E., Fan, Y. *JACS* **125**, 7825 (2003).
[11]Duan, W.-L., Shi, M., Rong, G.-B. *CC* 2916 (2003).
[12]Shen, Y.-M., Duan, W.-L., Shi, M. *JOC* **68**, 1559 (2003).
[13]Suga, H., Kakehi, A., Ito, S., Ibata, T., Fudo, T., Watanabe, Y., Kinoshita, Y. *BCSJ* **76**, 189 (2003).

1,1′-Binaphthalene-2,2′-dicarboxylic acid derivatives.

Epoxidation. The dilactone **1** forms a chiral dioxirane for oxygen atom transfer to cinnamic acid derivatives in the asymmetric fashion.

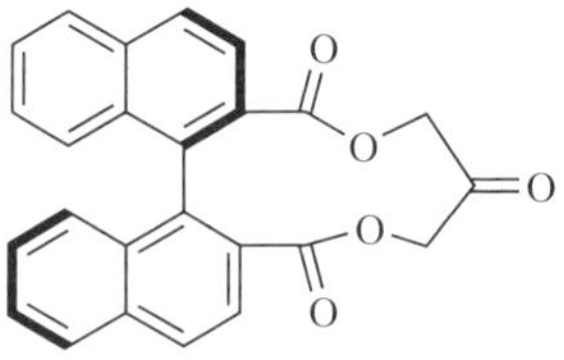

1

[1]Imashiro, R., Seki, M. *JOC* **69**, 4216 (2004).

1,1′-Binaphthalene-2,2′-diol, BINOL. 21, 26; **22**, 23

Review. Synthetic uses are summarized.[1]

Baylis-Hillman reaction. 3,3′-Bisaryl-octahydroBINOL acts as a chiral Bronsted acid to effect the condensation. Aliphatic aldehydes react well when the octahydro-BINOL carries 3,5-bistrifluoromethylphenyl groups, hindered aldehydes are more compatible with the nonfluorinated *m*-xylyl derivative.[2]

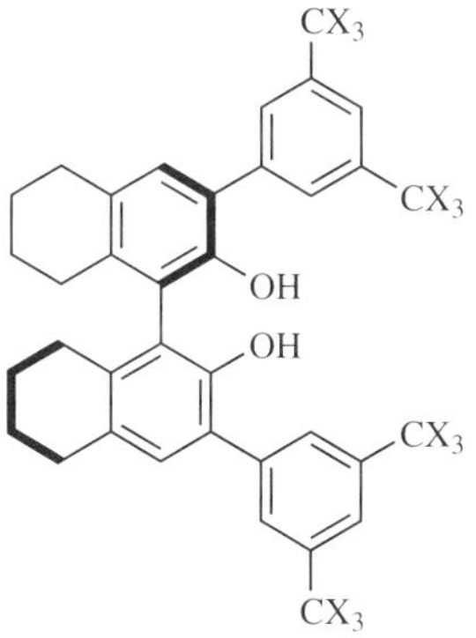

Michael addition. Chiral phase-transfer catalysts are made from BINOL dibenzyl ethers that are linked at the 3,3′-positions with [CH_2NEt_3] units.[3]

Hydrosilylation. *N*-Nosylimines are reduced with $(MeO)_3SiH$ in the presence of BINOL and LiHMDS. Asymmetric induction is moderate.[4]

[1]Brunel, J.M. *CR* **105**, 857 (2005).
[2]McDougal, N.T., Schaus, S.E. *JACS* **125**, 12094 (2003).
[3]Arai, S., Tokumaru, K., Aoyama, T. *CPB* **52**, 646 (2004).
[4]Nishikori, H., Yoshihara, R., Hosomi, A. *SL* 561 (2003).

1,1′-Binaphthalene-2,2′-diol (modified) – aluminum complexes. 21, 27; **22**, 23–24

Addition to C=X bonds. Aluminum complex of a modified BINOL containing a diethylaminomethyl group each at C-3 and C-3′ positions catalyzes asymmetric reaction of aldehydes with NC-COOMe and NC-PO$(OEt)_2$ to give cyanohydrin methyl carbonates[1] and *O*-phosphates,[2] respectively. The complex derived from (*R*)-BINOL and Et_2AlCN is useful for converting imines to α-amino nitriles,[3] but the ee values of the products are only moderate. On the other hand, excellent asymmetric induction is observed in the Reissert reaction[4] on pyridine derivatives using a complex derived from 3,3′-bis(benzenesulfinylmethyl)BINOL.

1,4-Additions. An enantioselective synthesis of (-)-strychnine[5] starts from addition of dimethyl malonate to 2-cyclohexenone, the adduct is obtainable in >99% ee.

[1]Casas, J., Baeza, A., Sansano, J.M., Najera, C., Saa, J.M. *TA* **14**, 197 (2003).
[2]Baeza, A., Casas, J., Najera, C., Sansano, J.M., Saa, J.M. *ACIEE* **42**, 3143 (2003).

[3]Nakamura, S., Sato, N., Sugimoto, M., Toru, T. *TA* **15**, 1513 (2004).
[4]Ichikawa, E., Suzuki, M., Yabu, K., Albert, M., Kanai, M., Shibasaki, M. *JACS* **126**, 11808 (2004).
[5]Ohshima, T., Xu, Y., Takita, R., Shibasaki, M. *T* **60**, 9569 (2004).

1,1′-Binaphthalene-2,2′-diol – boron complex.

Reduction.[1] The borate ester created from BINOL and (*S*)-proline is a new catalyst for borane reduction of ketones.

[1]Liu, D., Shan, Z., Zhou, Y., Wu, Z. *HCA* **87**, 2310 (2004).

1,1′-Binaphthalene-2,2′-diol – gallium complex.

Epoxide opening.[1] The diorganogallium complex derived from monobenzyl ether of BINOL catalyzes the reaction of epoxides with Me_3SiCN to give β-isocyano alcohols, usually in high ee. The corresponding indium complex performs poorer.

[1]Zhu, C., Yuan, F., Gu, W., Pan, Y. *CC* 692 (2003).

1,1′-Binaphthalene-2,2′-diol – lanthanum complex. 22, 24–25

Epoxidation. Further developments in this topic include polymer-support chiral complex derived from BINOL, lanthanum isopropoxide and Ph_3PO to promote epoxidation of conjugated ketones. [1]

Conjugate additions.[2] Two BINOL molecules linked at C-3 by an azabismethylene chain forms ligands for La, and the resultant catalysts promote Michael addition effectively.

Aldol-Tishchenko reactions.[3] Condensation of ketones with aldehydes produce 1,3-diol monoesters diastereoselectively and enantioselectively in one step.

[1]Jayaprakash, D., Kobayashi, Y., Watanabe, S., Arai, T., Sasai, H. *TA* **14**, 1587 (2003).
[2]Majima, K., Takita, R., Okada, A., Ohshima, T., Shibasaki, M. *JACS* **125**, 15837 (2003).
[3]Gnanadesikan, V., Horiuchi, Y., Ohshima, T., Shibasaki, M. *JACS* **126**, 7782 (2004).

1,1′-Binaphthalene-2,2′-diol – lithium complex.

Aldol reaction.[1] Lithium 3,3′-dibromo-BINOL catalyzes aldol reaction between trimethoxysilyl enol ethers with aldehydes in aq. THF with excellent ee.

[1]Nakajima, M., Orito, Y., Ishizuka, T., Hashimoto, S. *OL* **6**, 3763 (2004).

1,1′-Binaphthalene-2,2′-diol – niobium complexes.

Mannich reaction.[1] Smooth Mannich-type reaction proceeds in the presence of a Nb complex of a tridentate BINOL in which C-3 carries a 2-hydroxy-3-isopropylbenzyl group. Enantioselectivities and yields of the products are high.

[1]Kobayashi, S., Arai, K., Shimizu, H., Ihori, Y., Ishitani, H., Yamashita, Y. *ACIEE* **45**, 761 (2005).

1,1′-Binaphthalene-2,2′-diol – samarium complex. 22, 26

Epoxidation. The samarium-BINOL-triphenylarsine oxide complex has been used for epoxidation of conjugated *N*-alkenoylmorpholines. A useful application is synthesis of 1,3-polyol arrays.[1] *N*-Alkenoylpyrroles are similarly epoxidized.[2]

Triphenylphosphine is an effective additive to improve yields and enantioselectivity of epoxidation, conducting sequential Wittig reaction and catalytic asymmetric epoxiation[3] is therefore advantageous.

[1]Tosaki, S., Horiuchi, Y., Nemoto, T., Ohshima, T., Shibasaki, M. *CEJ* **10**, 1527 (2004).
[2]Matsunaga, S., Kinoshita, T., Okada, S., Harada, S., Shibasaki, M. *JACS* **126**, 7559 (2004).
[3]Kinoshita, T., Okada, S., Park, S.-R., Matsunaga, S., Shibasaki, M. *ACIEE* **42**, 4680 (2003).

1,1′-Binaphthalene-2,2′-diol – titanium complexes. 15, 26–27; **16**, 24–25; **17**, 28–30; **18**, 43–44; **19**, 25; **20**, 25–27; **21**, 28–29; **22**, 26–28

Alkylations. Addition to carbonyl compounds with organometallic reagents is rendered enantioselective by Ti complexes of BINOL[1,2] and its various fluorous versions such as the 6,6′-bis(tridecylfluorooctyl)-[3] and 6,6′-bis[(tridecylfluoro-octylsilyl)ethyl] derivatives.[4] The ligands can be conveniently recovered.

Allylation and propargylation with tin reagents using the Ti complex and synergistic agent (hard-soft contrast) show better results.[5] In alkynylation of aldehydes the catalyst is activated by phenols.[6] Diastereoselective epoxidation following allylation of cycloalkenones[7] is a valuable synthetic operation.

BINOL
$(i\text{-PrO})_4Ti/i\text{-PrOH}$,DCM
Sn ,TBHP
4
HO O H
84% (96%ee)

Aldol reactions. A procedure for enantioselective condensation with conjugated silyl ketene acetals is reported.[8] The homo-aldol version is less effective in terms of asymmetric induction.[9]

OEt OSiMe$_3$ + PhCHO
(*R*)-BINOL
TiF_4
SiMe$_3$
MeCN
O O Ph
80% (65%ee)

Cyclization. 4-Alkenols undergo intramolecular iodoetherification with NIS. In the presence of a chiral BINOL complex of titanium moderate asymmetric induction is observed.[10]

[1]Cozzi, P.G., Alesi, S. *CC* 2448 (2004).
[2]Zhou, Y., Wang, R., Xu, Z., Yan, W., Liu, L., Kang, Y., Han, Z. *OL* **6**, 4147 (2004).
[3]Qian, Z., Zhao, G., Yin, Y., Yin, W. *JFC* **120**, 117 (2003).
[4]Takeuchi, S., Nakamura, Y., Ohgo, Y. *JFC* **120**, 121 (2003).
[5]Yu, C.-M., Kim, J.-M., Shin, M.-S., Cho, D. *TL* **44**, 5487 (2003).
[6]Lu, G., Li, X., Chen, G., Chan, W.L., Chan, A.S.C. *TA* **14**, 449 (2003).
[7]Kim, J.-G., Waltz, K.M., Garcia, I.F., Kwiatkowski, D., Walsh, P.J. *JACS* **126**, 12580 (2004).
[8]De Rossa, M., Acocella, M.R., Villano, R., Soriente, A., Scettri, A. *TA* **14**, 2499 (2003).
[9]Burke, E.D., Kim, N.K., Gleason, J.L. *SL* 390 (2003).
[10]Kang, S.H., Park, C.M., Lee, S.B., Kim, M. *SL* 1279 (2004).

1,1′-Binaphthalene-2,2′-diol (modified) - zinc complexes. 21, 29; **22**, 28–29

Aldol and Michael reactions. A Zn-Zn linked –BINOL complex efficiently catalyzes the condensation of α-hydroxy ketones with *N*-phosphonylimines, affording the *anti*-3-amino-2-hydroxy ketones.[1] On the other hand, with *N*-Boc imines the major products are the *syn*-analogues.[2]

Interestingly, α-hydroxy ketones undergo Michael addition to enones to create chiral α-ketol unit in which the hydroxyl group is tertiary.[3]

Propargylic alcohols.[4] Asymmetric addition of 1-alkynes to aldehydes is promoted by a complex of Me_2Zn and **1**.

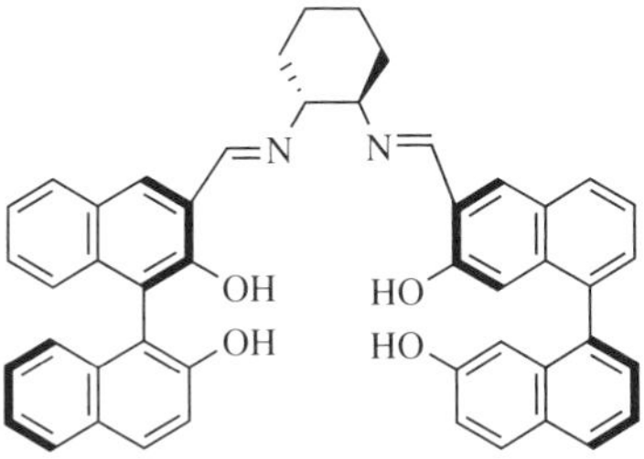

1

Hetero-Diels-Alder reaction. The BINOL-Zn complex catalyzed reaction between an electron-rich diene and electron-poor imine is subjected to solvent and temperature effects. Thus, at room temperature, with 10 mol% catalyst toluene is the best solvent, while CH_2Cl_2 performs best if 100 mol% catalyst is present.[5]

A single catalyst is used in achieving a hetero-Diels-Alder reaction and addition of diorganozinc reagents to a dialdehyde.[6]

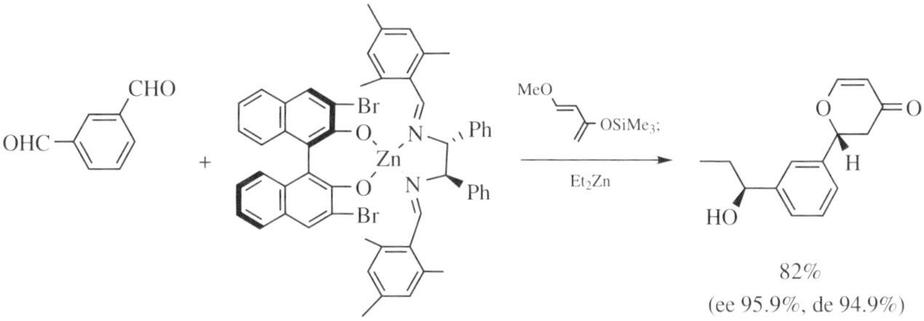

82%
(ee 95.9%, de 94.9%)

[1]Matsunaga, S., Kumagai, N., Harada, S., Shibasaki, M. *JACS* **125**, 4712 (2003).
[2]Matsunaga, S., Yoshida, T., Morimoto, H., Kumagai, N., Shibasaki, M. *JACS* **126**, 8777 (2004).
[3]Harada, S., Kumagai, N., Kinoshita, T., Matsunaga, S., Shibasaki, M. *JACS* **125**, 2582 (2003).
[4]Li, Z.-B., Pu, L. *OL* **6**, 1065 (2004).
[5]Guillarme, S., Whiting, A. *SL* 711 (2004).
[6]Du, H., Ding, K. *OL* **5**, 1091 (2003).

1,1′-Binaphthalene-2,2′-diol (modified) - zirconium complexes. 19, 25–26; **22**, 29

Aldol reaction. Ethyl diazoacetate unites with aldehydes to provide β-hydroxy-α-diazocarboxylates in moderate to good ee, while using the Zr-complex of 6,6′-dibromo-BINOL as catalyst.[1]

An air-stable, storable chiral Zr catalyst for *anti*-selective asymmetric aldol reaction is obtained from 3,3′-diiodo-BINOL and $Zr(OPr)_4 \cdot PrOH$.[2]

[3 + 2]Cycloaddition.[3] Acylhydrazones add to electron-rich alkenes, giving chiral pyrazolidines, apparently in a concerted manner. It is likely the Zr complex with 3,3′-diiodo-BINOL binds to the oxygen and one of the nitrogen atoms of the 1,3-dipoles in the transition state.

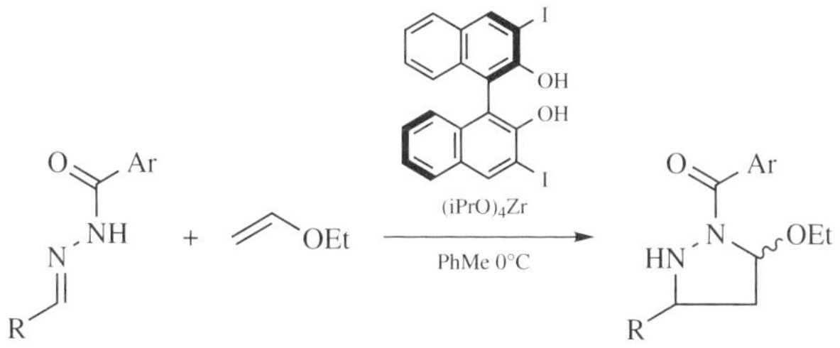

[1]Yao, W., Wang, J. *OL* **5**, 1527 (2003).
[2]Kobayashi, S., Saito, S., Ueno, M., Yamashita, Y. *CC* 2016 (2003).
[3]Yamashita, Y., Kobayashi, S. *JACS* **126**, 11279 (2004).

1,1′-Binaphthalene-2,2′-diol analogues.

Homoallylic alcohols.[1] Asymmetric reaction between ArCHO and allyltrichlorosilane in the presence of Quinox (**1**) reveals the role of arene-arene interactions between the ligand and ArCHO. Accordingly, only electron-poor ArCHO afford products in good ee, electron-rich ArCHO give mostly racemic alcohols.

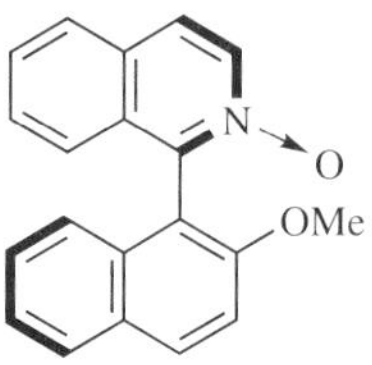

1

[1]Malkov, A.V., Dufkova, L., Farrugia, L., Kocovsky, P. *ACIEE* **42**, 3674 (2003).

1,1′-Binaphthalene-2,2′-diyl heteroesters. 21, 30–31; **21**, 30–31; **22**, 30–31

Hydrogenation. β-Keto esters afford β-hydroxy esters when subjected to hydrogenation in the presence of $RuBr_2$ and a BINOL-phosphite.[1] Hydrogenation of dehydroamino acid derivatives and α-acetamidostyrene has become the testing ground for enantioselectivity, and Rh catalysts based on BINOL-phosphoramidite ligands continue to evolve.[2–5] Ligands such as those containing BINOL and ferrocene moieties[6] have strictly defined spaces.

1,4-Additions. Copper-catalyzed conjugate addition of organozincs to enones is rendered highly enantioselective by a tris-BINOL diphosphite.[7] An aryl group is delivered from $ArB(OH)_2$ to conjugated systems in the presence of Rh-bound tris-BINOL diphosphite[8] or BINOL-phosphoramidite.[9]

A particularly popular ligand is **1** (and its enantiomer) for several metal-catalyzed asymmetric reactions, for example, conjugate addition to unsaturated lactams,[10] nitroalkenes,[11,12] and alkylidenemalonic esters.[13,14]

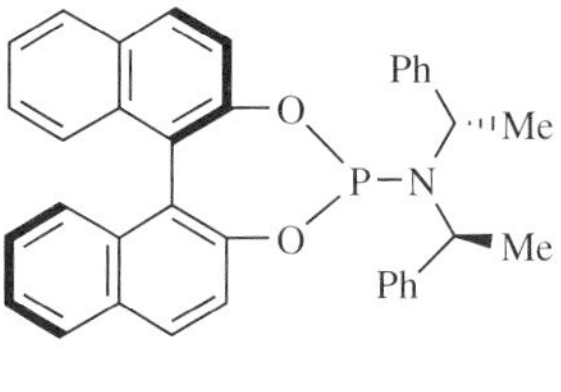

1

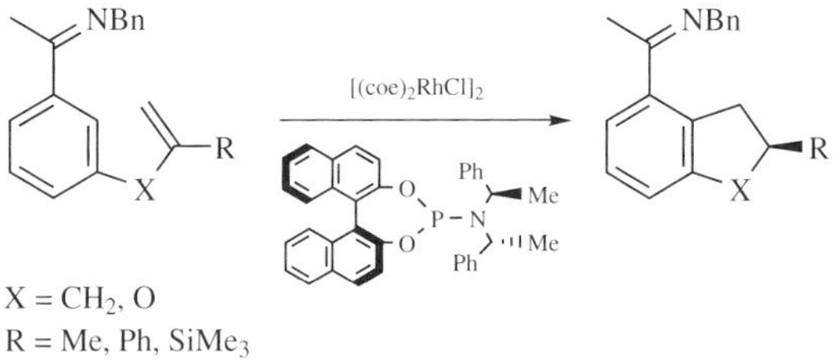

X = CH_2, O
R = Me, Ph, $SiMe_3$

$[(coe)_2RhCl]_2$

Allylic substitutions. Allylic substitutions by *O*- (phenoxides),[15] *C*-,[16–18] and *N*-nucleophiles[17,19,20] employ $[(cod)IrCl]_2$ and **1** (or its slightly modified forms) to great advantages.

β-Branched α-amino acids are synthesized via allylation of the imino derivatives of glycine, using a BINOL phosphite that also contains a divalent sulfur in the other chain as the (bidentate) ligand.[21]

Mannich reaction. Phosphoric esters of 3,3′-diaryl-BINOLs are chiral Bronsted acids that effectively promote the Mannich reaction.[22,23] The analogous aminoalkylation of furans also proceeds well.[24]

Cyclization.[25] Rh-catalyzed cyclization onto an aromatic ring to form a five-membered carbocycle or heterocycle is regioselective with an *meta*-imine to activate the C—H bond at the common *o*-position.

[1]Junge, K., Hagemann, B., Enthaler, S., Oehme, G., Michalik, M., Monsees, X., Riermeier, T., Dingerdissen, U., Beller, M. *ACIEE* **43**, 5066 (2004).
[2]Wang, X., Ding, K. *JACS* **126**, 10524 (2004).
[3]Reetz, M., Ma, J., Goddard, R. *ACIEE* **44**, 412 (2005).
[4]Jia, X., Li, X., Xu, L., Shi, Q., Yao, X., Chan, A.S.C. *JOC* **68**, 4539 (2003).
[5]Li, X., Jia, X., Lu, G., Au-Yeung, T.T.-L., Lam, K.-H., Lo, T.W.H., Chan, A.S.C. *TA* **14**, 2687 (2003).
[6]Hu, X.-P., Zheng, Z. *OL* **6**, 3585 (2004).
[7]Liang, L., Yan, M., Li, Y.-M., Chan, A.S.C. *TA* **15**, 2575 (2004).
[8]Chapman, C.J., Wadsworth, K.J., Frost, C.G. *JOMC* **680**, 206 (2003).
[9]Iguchi, Y., Itooka, R., Miyaura, N. *SL* 1040 (2003).
[10]Pineschi, M., Del Moro, F., Gini, F., Minnaard, A.J., Feringa, B.L. *CC* 1244 (2004).
[11]Duursma, A., Minnaard, A.J., Feringa, B.L. *JACS* **125**, 3700 (2003).
[12]Eilitz, U., Lessmann, F., Seidelmann, O., Wendisch, V. *TA* **14**, 3095 (2003).
[13]Watanabe, T., Knopfel, T.F., Carreira, E.M. *OL* **5**, 4557 (2003).
[14]Schuppan, J., Minnaard, A.J., Feringa, B.L. *CC* 792 (2004).
[15]Lopez, F., Ohmura, T., Hartwig, J.F. *JACS* **125**, 3426 (2003).
[16]Lipowsky, G., Miller, N., Helmchen, G. *ACIEE* **43**, 4595 (2004).
[17]Lipowsky, G., Helmchen, G. *CC* 116 (2004).
[18]Alexakis, A., Polet, D. *OL* **6**, 3529 (2004).
[19]Tissot-Croset, K., Polet, D., Alexakis, A. *ACIEE* **43**, 2426 (2004).
[20]Shu, C., Leitner, A., Hartwig, J.F. *ACIEE* **43**, 47976 (2004).
[21]Kanayama, T., Yoshida, K., Miyabe, H., Kimachi, T., Takemoto, Y. *JOC* **68**, 6197 (2003).
[22]Akiyama, T., Itoh, J., Yokota, K., Fuchibe, K. *ACIEE* **43**, 1566 (2004).
[23]Uraguchi, D., Terada, M. *JACS* **126**, 5356 (2004).
[24]Uraguchi, D., Sorimachi, K., Terada, M. *JACS* **126**, 11804 (2004).
[25]Thalji, R.K., Ellman, J.A., Bergman, R.G. *JACS* **126**, 7192 (2004).

Bis(acetonitrile)(1,5-cyclooctadiene)rhodium(I) tetrafluoroborate. 21, 33; **22**, 32

Aryl transfer. Aryl group transfer from $ArSi(Me)F_2$ to aldehydes works for *N*-tosylimines, as expected.[1] Also, arylstannanes serve in the reaction with aldehydes, diketones, and imines.[2]

[1]Oi, S., Moro, M., Kawanishi, T., Inoue, Y. *TL* **45**, 4855 (2004).
[2]Oi, S., Moro, M., Fukuhara, H., Kawanishi, T., Inoue, Y. *T* **59**, 4351 (2003).

Bis(acetonitrile)dichloropalladium(II). 13, 33, 211, 236; **14**, 35–36; **15**, 28–29; **16**, 25–26; **17**, 30–31; **18**, 44–45; **19**, 26; **21**, 33–35; **22**, 33–34

Hydroxyl protection. Formation of THP ethers from alcohols takes place at room temperature in THF with $(MeCN)_2PdCl_2$ as catalyst. When THP ethers are heated in MeCN with the same catalyst the reaction goes in the reverse direction.[1] Alkenyl ethers and esters suffer O–C bond cleavage in aqueous isopropanol in the air. The method seems applicable to kinetic resolution.[2]

Allylic substitutions. Allenyl sulfoxides are allylated at C-2 with double bond migration.[3]

Dehydroprolines. 2-Tosylamino-4-alkynoic esters cyclize on exposure to $(MeCN)_2PdCl_2$. Reduction of the products leads to 5-substituted proline derivatives.[4]

Rearrangement. A different approach to allylic amines involves formation of (allyloxy)iminodiazaphospholidines and a Pd-catalyzed [3.3]sigmatropic rearrangement.[5]

[1]Wang, Y.-G., Wu, X.-X., Jiang, Z.-Y. *TL* **45**, 2973 (2004).
[2]Aoyama, H., Tokunaga, M., Hiraiwa, S., Shirogane, Y., Obora, Y., Tsuji, Y. *OL* **6**, 509 (2004).
[3]Ma, S., Wei, Q., Ren, H. *TL* **45**, 3517 (2004).
[4]van Esseveldt, B.C.J., van Delft, F.L., Smits, J.M.M., de Gelder, R., Rutjes, F.P.J.T. *SL* **44**, 2354 (2003).
[5]Lee, E.E., Batey, R.A. *ACIEE* **43**, 1865 (2004).

Bis(acetonitrile)dichloropalladium(II) – copper(II) chloride.

Alkylidenation. β-Dicarbonyl compounds react with alkenes (ethene, propene) to give alkylidene derivatives. However, very hindered diketones furnish products containing a vinyl group.[1]

Oxidative cyclization. α-Alkenyl β-diketones undergo Wacker oxidation and cyclization to give substituted furans in one step.[2] The oxidation site can be changed (C-1 vs. C-2) in certain alkenylindoles[3] and 1-alken-5-ones.[4]

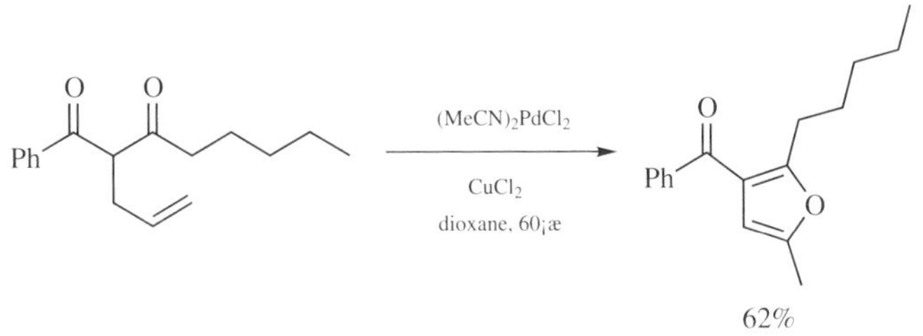

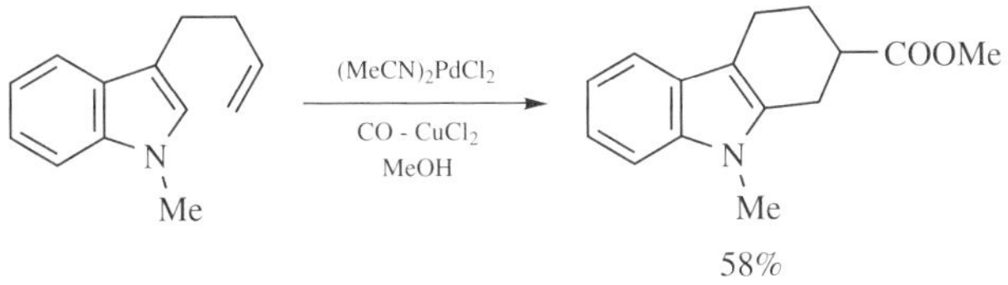

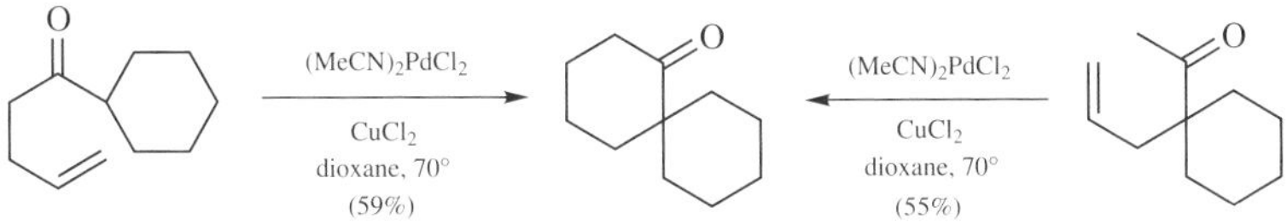

O O
Ph
$(MeCN)_2PdCl_2$
$CuCl_2$
dioxane, 60¡æ
O
Ph
O
62%

$(MeCN)_2PdCl_2$
CO - $CuCl_2$
MeOH
N
Me
COOMe
N
Me
58%

O
$(MeCN)_2PdCl_2$
$CuCl_2$
dioxane, 70°
(59%)
O
$(MeCN)_2PdCl_2$
$CuCl_2$
dioxane, 70°
(55%)
O

Carbonylation.[5] γ-Lactam is formed from *N*-tosyl-butenamine via monocarbonylation. Notably, using CuCl as cocatalyst, (catalyst: $PdCl_2$) dicarbonylation of the double bond occurs to produce the 2-oxopyrrolidine-3-acetic esters.

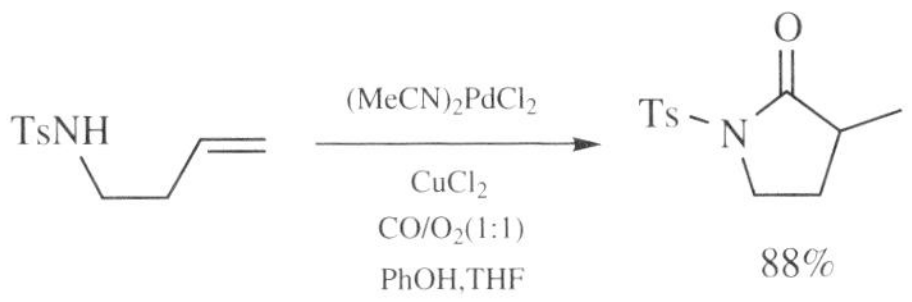

TsNH
$(MeCN)_2PdCl_2$
$CuCl_2$
CO/O_2(1:1)
PhOH,THF
O
Ts ~ N
88%

[1]Wang, X., Widenhoefer, R.A. *CC* 660 (2004).
[2]Han, X., Widenhoefer, R.A. *JOC* **69**, 1738 (2004).
[3]Liu, C., Widenhoefer, R.A.. *JACS* **126**, 10250 (2004).
[4]Wang, X., Pei, T., Han, X., Widenhoefer, R.A. *OL* **5**, 2699 (2003).
[5]Mizutani, T., Ukaji, Y., Inomata, K. *BCSJ* **76**, 1251 (2003).

Bis(acetonitrile)norbornadienerhodium(I) hexafluorophosphate.

Allylation.[1] An allyl group is introduced into electron-rich aromatic compounds by a catalyzed reaction involving allyl tosylate. The reaction is electrophilic in nature yet *o-/p*-ratios differ somewhat from traditional aromatic alkylation.

[1]Tsukada, N., Yagura, Y., Sato, T., Inoue, Y. *SL* 1431 (2003).

Bis(acetonitrile)tris(1-pyrazolyl)borato(triphenylphosphine)ruthenium(I) hexafluorophosphate.

Alkyne degradation. The Ru catalyst effects triple bond cleavage, converting propargyl alcohols into nor-alkenes,[1] and aryl and propargyl ethers into aryl and alkynyl ketones (e.g., 1-propargyloxytetralin into α-tetralone, 90%).[2] Various functional groups are tolerated.

[TpRu(PPh$_3$)(MeCN)$_2$] PF$_6$
LiOTf, PhMe
100¡æ
75%

Rearrangements. Diketones are generated from α,β-epoxy ketones in high yields when heating with the Ru catalyst.[3]

Cyclization. Several types of transformations are observed: aromatic ring formation with group migration,[4] indenes or naphthalenes from *o*-ethynylarylalkenes,[5] phenols/2,4-cyclohexadienones from epoxy-enynes.[6]

CHI
[TpRu(PPh$_3$)(MeCN)$_2$] PF$_6$
PhMe,100¡æ
MeO I
82%

[TpRu(PPh$_3$)(MeCN)$_2$] PF$_6$
X=F, Cl, MeO

[TpRu(PPh$_3$)(MeCN)$_2$] PF$_6$
89%

[TpRu(PPh$_3$)(MeCN)$_2$] PF$_6$
78%

Imines of 2-penten-4-ynes undergo 5-exo-dig cyclization to afford pyrroles.[7]

[1]Datta, S., Chang, C.-L., Yeh, K.-L., Liu, R.-S. *JACS* **125**, 9294 (2003).
[2]Shen, H.-C., Su, H.-L., Hsueh, Y.-C., Liu, R.-S. *OM* **23**, 4332 (2004).
[3]Chang, C.-L., Kumar, M.P., Liu, R.-S. *JOC* **69**, 2793 (2004).

[4]Shen, H.-C., Pal, S., Lian, J.-J., Liu, R.-S. *JACS* **125**, 15762 (2003).
[5]Madhushaw, R., Lo, C.-Y., Hwang, C.-W., Su, M.-D., Shen, H.-C., Pal, S., Shaikh, I.R., Liu, R.-S. *JACS* **126**, 15560 (2004).
[6]Lin, M.-Y., Madhushaw, R.J., Liu, R.-S. *JOC* **69**, 7700 (2004).
[7]Shen, H.-C., Li, C.-W., Liu, R.-S. *TL* **45**, 9245 (2004).

Bis(allyl)dichlorodipalladium. 20, 29; **21**, 35–37; **22**, 34–36

Allylic substitutions. With the Pd complex to catalyze allylic displacement α-isocyanoacetic esters are used successfully as nucleophiles and Cs_2CO_3 as base (also dppe as ligand).[1] Novel ligands for the Pd complex have been prepared from dendrimers,[2] with which allylic amines prepared are mainly linear. A *SAMP* hydrazone attaching to a ferrocene skeleton shows reasonable asymmetric induction.[3] Unusual regioselectivity for *N*-allylation of aziridines is revealed.[4]

The regioselectivity for the allylic substitution is biased when the substrates contain a trifluoromethyl group, contrary to the normal trend, with bond formation even at a neopentyl center (instead of geminal to the CF_3 group).[5] Zn-chelates of protected amino acids also show excellent regioselectivity and diastereoselectivity in the displacement because of their high reactivities that the geometrical isomerization of the π-allylpalladium complexes is suppressed.

NH + OAc —[$[(\eta^3\text{-}C_3H_5)PdCl]_2$, Ph_3P / THF]→ N 89%

CF_3, O, N, O, Zn—O + O, O, OEt —[$[(\eta^3\text{-}C_3H_5)PdCl]_2$, Ph_3P / THF]→ H, N, O, O, F_3C, O

Allylation. Carbonyl allylation in the umpoled sense is achieved by in situ generation of Pd complexes from allylic chlorides and acetates while another organometallic reagent ($Me_3SnSnMe_3$,[7] Et_2Zn [8]) provides activation.

Related is the carbalkoxyallylation of imines employing a related chiral catalyst.[9]

NBn, Ph + Bu_3Sn COOEt —[PdCl dimer, THF, MeOH]→ BnNH, Ph, COOEt 81% (82%ee)

Coupling reactions. The air-stable Pd-complex with *cis,cis,cis*-1,2,3,4-tetrakis-(diphenylphosphinomethyl)cyclopentane catalyzes the Heck reaction of disubstituted alkenes[10] and polyhaloarenes.[11] Conjugated dienes are prepared by the Heck reaction of 1-bromoalkenes.[12] The system also benefits Suzuki coupling[11] including those giving polyarylethenes[13] and diarylmethanes.[14] Alkylarenes are also available from ArBr and $RB(OH)_2$.[15]

A tridentate ferrocenylphosphine (**1**) permits catalyst loading to as low as 10^{-4} mol% in Sonagashira coupling.[16] Alkynylarenes including $ArCCCH_2OH$ are readily obtained by the method (CuI is an additive).[17] However, with other phosphine ligands the Sonagashira coupling is reported to be worse in the presence of a Cu cocatalyst,[18] and the efficiency of the copper-free conditions is demonstrated by accomplishing the reaction of ArBr at room temperature (t-Bu_3P as ligand).[19] With a heterocycle-carbene ligand to assist the Pd complex and in the presence of CuI, alkyl bromides and iodides cross-couple with alkynes.[20]

Regioselectivity switch in the coupling by ligand effect[21] is of enormous synthetic value.

PPh_2 PPh_2 Fe P

1

TBS OMe Bu + IZn COOEt

$[(\pi\text{-}C_3H_5)PdCl]_2$ / [L], THF, rt

TBS COOEt Bu + TBS Bu EtOOC

L=PPh_3, 68%, >97:3
L=DPPHB,49%, 3:97

DPPHB = 2-(diphenylphosphino)-2'-hydroxy-1,1'-binaphthalene

The scope of the Stille coupling is expanded on discovery of ligands such as CyP(pyrrolidinyl)$_2$ which promote reaction of $ArSnBu_3$ with alkyl halides.[22] For cross-coupling of alkyl halides containing a β-hydrogen at room temperature, t-Bu_2PMe is a very effective ligand. Actually, the commercially available and moisture-stable salt $[HP(t\text{-}Bu)_2Me]$ BF_4 can be used.[23]

Cross-couplings involving silicon compounds have been expanded to include arylsilanols.[24] Stereoselective intramolecular arenoformylation of homopropargyl silyl ethers followed by coupling delivers functionalized enals.[25]

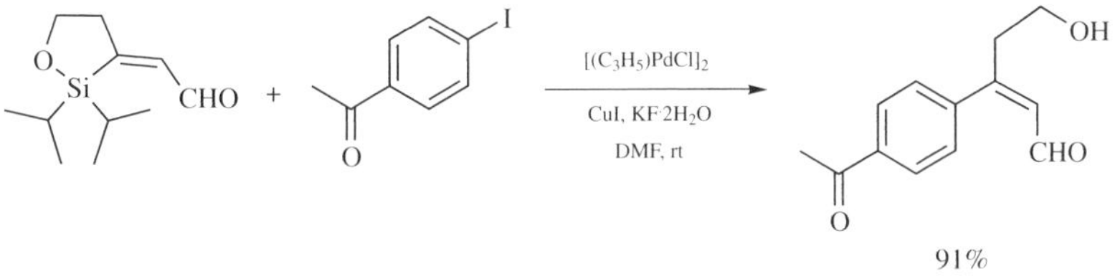

Homocoupling of organostannanes proceeds with allyl acetate or air as oxidant.[26]

By a double coupling with cyclization 1,2-dihydroisoquinolines are synthesized from *o*-alkynylaryl aldimines.[27]

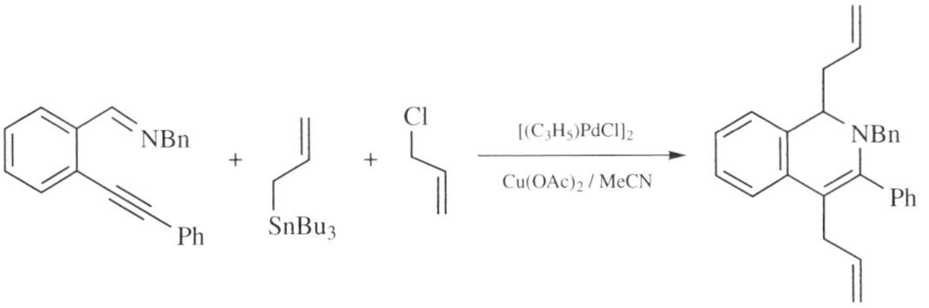

Functionalization-cyclization. Formation of bicyclic lactones via carbonylation[28] and cyclization of bisallenes with simultaneous attachment to each double bond a germyl or stannyl group[29] are new synthetic possibilities.

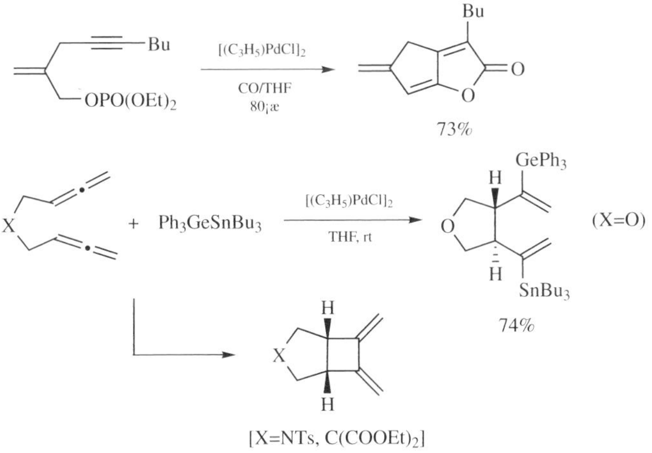

[1]Kazmaier, U., Ackermann, S. *SL* 2576 (2004).

[2]Ooe, M., Murata, M., Takahama, A., Mizugaki, T., Ebitani, K., Kaneda, K. *CL* **32**, 692 (2003).

[3]Mino, T., Segawa, H., Yamashita, M. *JOMC* **689**, 2833 (2004).

[4]Watson, I.D.G., Styler, S.A., Yudin, A.K. *JACS* **126**, 5086 (2004).

[5]Okano, T., Matsubara, H., Kusukawa, T., Fujita, M. *JOMC* **676**, 43 (2003).
[6]Kazmaier, U., Pohlmann, M. *SL* 623 (2004).
[7]Wallner, O.A., Szabo, K.J. *JOC* **68**, 2934 (2003).
[8]Zanoni, G., Gladiali, S., Marchetti, A., Piccinini, P., Tredici, I., Vidari, G. *ACIEE* **43**, 846 (2004).
[9]Fernandes, R.A., Yamamoto, Y. *JOC* **69**, 3562 (2004).
[10]Kondolff, I., Doucet, H., Santelli, M. *TL* **44**, 8487 (2003).
[11]Berthiol, F., Kondolff, I., Doucet, H., Santelli, M. *JOMC* **689**, 2786 (2004).
[12]Berthiol, F., Doucet, H., Santelli, M. *SL* 841 (2003).
[13]Berthiol, F., Doucet, H., Santelli, M. *EJOC* 1091 (2003).
[14]Chahen, L., Doucet, H., Santelli, M. *SL* 1668 (2003).
[15]Kondolff, I., Doucet, H., Santelli, M. *T* **60**, 3813 (2004).
[16]Hierso, J.-C., Fihri, A., Amardeil, R., Meunier, P., Doucet, H., Santelli, M., Ivanov, V.V. *OL* **6**, 3473 (2004).
[17]Feuerstein, M., Doucet, H., Santelli, M. *TL* **45**, 1603 (2004).
[18]Gelman, D., Buchwald, S.L. *ACIEE* **42**, 5993 (2003).
[19]Soheill, A., Albaneze-Walker, J., Murry, J.A., Dormer, P.G., Hughes, D.L. *OL* **5**, 4191 (2003).
[20]Eckhardt, M., Fu, G.C. *JACS* **125**, 13642 (2003).
[21]Ma, S., Wang, G. *ACIEE* **42**, 4215 (2003).
[22]Tang, H., Menzel, K., Fu, G.C. *ACIEE* **42**, 5079 (2003).
[23]Menzel, K., Fu, G.C. *JACS* **125**, 3718 (2003).
[24]Denmark, S.E., Ober, M.H. *OL* **5**, 1357 (2003).
[25]Denmark, S.E., Kobayashi, T. *JOC* **68**, 5153 (2003).
[26]Shirakawa, E., Nakao, Y., Murota, Y., Hiyama, T. *JOMC* **670**, 132 (2003).
[27]Ohtaka, M., Nakamura, H., Yamamoto, Y. *TL* **45**, 7339 (2004).
[28]Kamitani, A., Chatani, N., Murai, S. *ACIEE* **42**, 1397 (2003).
[29]Hong, Y-T., Yoon, S.-K., Kang, S.-K., Yu, C.-M. *EJOC* 4628 (2004).

Bis(benzonitrile)-1,2-bis(diphenylphosphinyl)ethane palladium(II) salts.

Hydroarylation. The cationic Pd complex that can be prepared in situ from $(dba)_2Pd$, DPPE, and $Cu(BF_4)_2$, is a catalyst for conjugate transfer of aryl groups from $ArB(OH)_2$ to conjugated carbonyl compounds (but not acrylates).[1] The analogous reaction of $ArSi(OMe)_3$ with enones such as 2-cyclohexenone is carried out in aqueous acidic media.[2]

[1]Nishikata, T., Yamamoto, Y., Miyaura, N. *ACIEE* **42**, 2768 (2003).
[2]Nishikata, T., Yamamoto, Y., Miyaura, N. *CL* 752 (2003).

Bis(benzonitrile)dichloropalladium(II). **13**, 34; **15**, 29; **18**, 46–47; **19**, 27; **20**, 30–31; **21**, 38–39; **22**, 37

Protecting group maneuvers. Pd-catalyzed removal of an allylic [OTHP] group can induce stereoselective ring formation from an aldehyde.[1]

CHO Ph OTHP —$(PhCN)_2PdCl_2$, ROH/THF→ OR O Ph (major)

Hydrogenolysis of an *N*-benzyloxycarbonyl group from protected amino acids is efficiently performed with a catalyst prepared from $(PhCN)_2PdCl_2$ and calcium hydroxyapatite in acetone.[2]

Coupling reactions. Addition of two aryl groups from ArI and $Ar'B(OH)_2$ to alkynes constitutes a convenient method for the access to tetrasubstituted alkenes.[3]

Bithiophenes are available by a homocoupling at room temperature. In this unusual reaction (2-bromothiophene gives bromine-containing bithiophene) AgF is an additive.[4]

A nonpolar biphasic catalytic system for Sonagashira and Suzuki couplings of ArX (X=Cl, Br) includes $(PhCN)_2PdCl_2$ and CuI with a polymer-bound phosphine.[5]

Carbonylation of unprotected haloindoles to form carboxylic acids and derivatives (esters, amides) is feasible using $(PhCN)_2PdCl_2$, DPPF, Et_3N.[6]

1,4-Additions. Amides add to enones under solvent-free conditions with the Pd catalyst.[7] Addends include acrylamide.

[1]Miyazawa, M., Hirose, Y., Narantsetseg, M., Yokoyama, H., Yamaguchi, S., Hirai, Y. *TL* **45**, 2883 (2004).
[2]Murata, M., Hara, T., Mori, K., Ooe, M., Mizugaki, T., Ebitani, K., Kaneda, K. *TL* **44**, 4981 (2003).
[3]Zhou, C., Emrich, D.E., Larock, R.C. *OL* **5**, 1579 (2003).
[4]Masui, K., Ikegami, H., Mori, A *JACS* **126**, 5074 (2004).
[5]Datta, A., Plenio, H. *CC* 1504 (2003).
[6]Kumar, K., Zapf, A., Michalik, D., Tillack, A., Heinrich, T., Böttcher, H., Arlt, M., Beller, M. *OL* **6**, 7 (2004).
[7]Takasu, K., Nishida, N., Ihara, M. *SL* 1844 (2004).

Bis(benzotriazol-1-yl) methanethione.

Thiocarbamoylbenzotriazoles.[1] The reagent, $Bt_2C{=}S$, is a thiophosgene equivalent. It is superior to $Im_2C{=}S$ in reaction with amines.

[1]Katritzky, A.R., Ledoux, S., Witek, R.M., Nair, S.K. *JOC* **69**, 2976 (2004).

Bis(*t*-butylisonitrile)dichloropalladium(II).

Bis(stannylation).[1] Derivatization of 1-alkynes by the Pd-catalyzed reaction with R_3SnSnR_3 to give *cis*-1,2-bis(trialkylstannyl)-1-alkenes is facile. Internal alkynes do not usually react unless highly activated.

[1]Mancuso, J., Lautens, M. *OL* **5**, 1653 (2003).

Bis[chloro(1,5-cyclooctadiene)iridium(I)]. 21, 39–40; 22, 39–41

Allylic substitutions. The Ir complex tends to mediate substitution at the more highly substituted carbon site of an allylic system, for example, with hydroxylamine derivatives[1] and oximes.[2] It is also possible to create a quaternary carbon center α to an allene unit.[3]

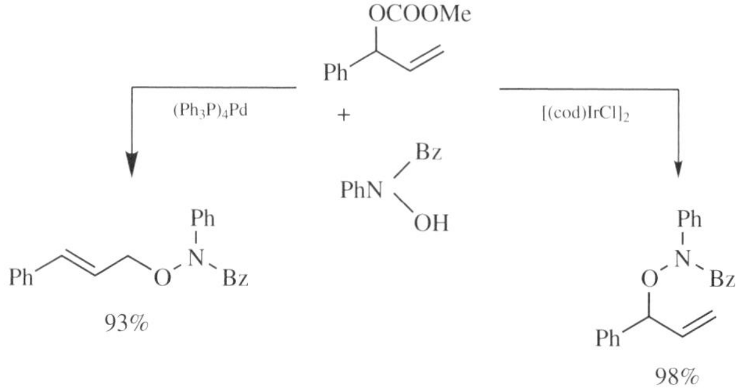

Azacycles are readily obtained by consecutive allylic amination of molecules containing two allylic esters.[4]

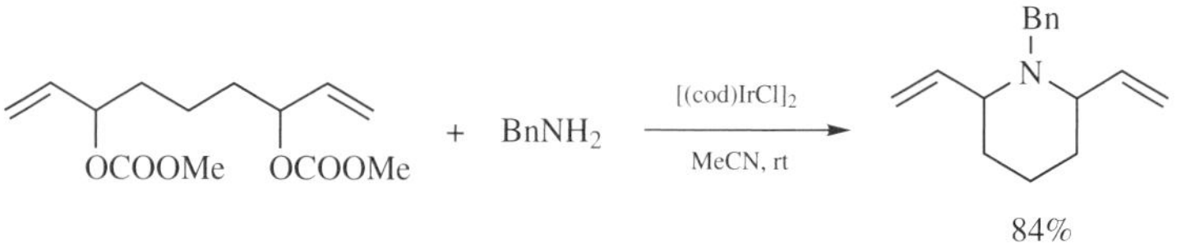

Addition to imines. Trimethylsilylethyne adds to aldimines to furnish propargylic amines under the catalysis of $[(cod)IrCl]_2$ and with MgI_2 as an additive.[5]

Quinolines and isoquinolines also undergo addition when activated with an acyl chloride.[6] On the other hand, the Ir-complex effects condensation of arylamines and aldehydes (1:2 ratio) to afford substituted quinolines.[7]

Alkylations. A solvent-free system for direct α-alkylation of ketones with alcohols at low temperatures contains $[(cod)IrCl]_2$-Ph_3P-KOH.[8]

Formation of $ArNHCH_2R$ from $ArN{=}PPh_3$ and alcohols involves oxidation to aldehydes and an aza-Wittig reaction, followed by in situ reduction (by $[IrH_2]$ species).[9] The more extensive alkylation cyclization events lead to formation of quinolines.[10]

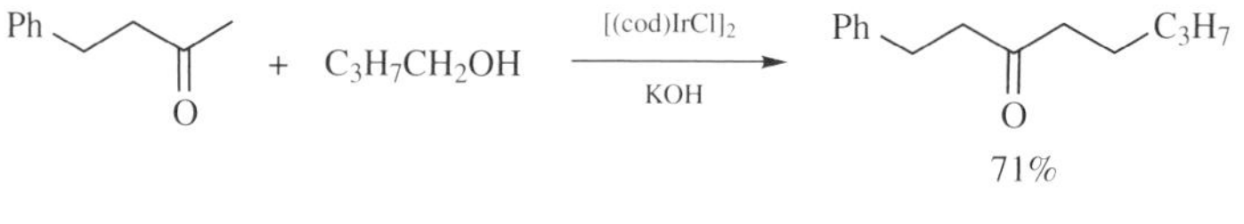

Oxidations. Hydroxamic acids are oxidized to acyl nitroso compounds that can be trapped as ene reaction products. A useful reagent system consists of $[(cod)IrCl]_2$ and aq. H_2O_2.[11]

Organosilanes are efficiently converted into silanols in the air by $[(cod)IrCl]_2$.[12]

Reduction. Both catalytic hydrogenation of imines[13] and transfer hydrogenation of ketones[14] are catalyzed by $[(cod)IrCl]_2$ and both are subject to asymmetric induction in the presence of chiral ligands (diamines and phosphine oxides, respectively).

Isomerization. Chemoselective isomerization of allylamines containing an electron-withdrawing group (Ts, Bz, Boc, etc.) at the *N*-atom to enamine derivatives[15] is effected by the Ir complex.

Heck reaction. Organosilicon reagents participate in Ir-catalyzed Heck reaction with alkenes.[16]

[1]Miyabe, H., Yoshida, K., Matsumura, A., Yamauchi, M., Takemoto, Y. *SL* 567 (2003).
[2]Miyabe, H., Matsumura, A., Yoshida, K., Yamauchi, M., Takemoto, Y. *SL* 2123 (2004).
[3]Kezuka, S., Kanemoto, K., Takeuchi, R. *TL* **45**, 6403 (2004).
[4]Miyabe, H., Yoshida, K., Kobayashi, Y., Matsumura, A., Takemoto, Y. *SL* 1031 (2003).
[5]Fischer, C., Carreira, E.M. *S* 1497 (2004).
[6]Yamazaki, Y., Fujita, K., Yamaguchi, R. *CL* **33**, 1316 (2004).
[7]Igarashi, T., Inada, T., Sekioka, T., Nakajima, T., Shimizu, I. *CL* **34**, 106 (2005).
[8]Taguchi, K., Nakagawa, H., Hirabayashi, T., Sakaguchi, S., Ishii, Y. *JACS* **126**, 72 (2004).
[9]Cami-Kobeci, G., Williams, J.M.J. *CC* 1072 (2004).
[10]Igarashi, T., Inada, T., Sekioka, T., Nakajima, T., Shimizu, I. *CL* **34**, 106 (2005).
[11]Fakhruddin, A., Iwasa, S., Nishiyama, H., Tsutsumi, K. *TL* **45**, 9323 (2004).
[12]Lee, Y., Seomoon, D., Kim, S., Han, H., Chang, S., Lee, P.H. *JOC* **69**, 1741 (2004).
[13]Jiang, X., Minnaard, A.J., Hessen, B., Feringa, B.L., Duchateau, A.L.L., Andrien, J.G.O., Boogers, J.A.F., de Vries, J.G. *OL* **5**, 1503 (2003).
[14]Hartikka, A., Modin, S.A., Andersson, P.G., Arvidsson, P.I. *OBC* **1**, 2522 (2003).
[15]Neugnot, B., Cintrat, J.-C., Rousseau, B. *T* **60**, 3575 (2004).
[16]Koike, T., Du, X., Sanada, T., Danda, Y., Mori, A. *ACIEE* **42**, 89 (2003).

Bis[chloro(1,5-cyclooctadiene)rhodium(I)]. 22, 41–43

Carbonylation. Insertion of a CO unit into the N–S bond of *N*-alkyliso-thiazolidines is accomplished under carbon monoxide by the Rh complex.[1] A synthesis of *N,N′*-diphenylurea from PhN_3 and CO by the Rh-catalyst is suitable for access to compounds with C-isotopes.[2] Benzannulated lactams are formed when 1-aminoalkyl-2-haloarenes are exposed to the Rh catalyst together an aldehyde (e.g., C_6F_5CHO) which is a surrogate for CO.[3]

A Pauson-Khand reaction of enynes[4] uses HCHO as CO source in micellar conditions is a very convenient method. A new reaction pattern is the [2 + 2 + 2 + 1]cycloaddition of enediynes.[5]

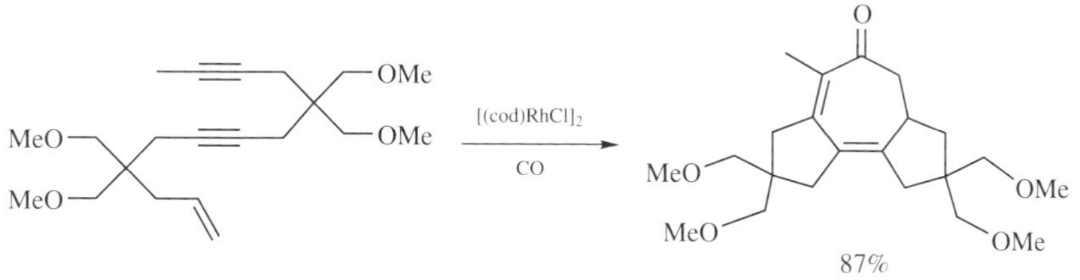

Both the hydroformylation-aldol reactions to give functionalized carbocycles[6] from unsaturated carbonyl compounds and 3-substituted indoles[7] by a hydroformylation-

Fischer indolization tandem are synthetically efficient. In both processes TsOH is present to mediate the second reaction.

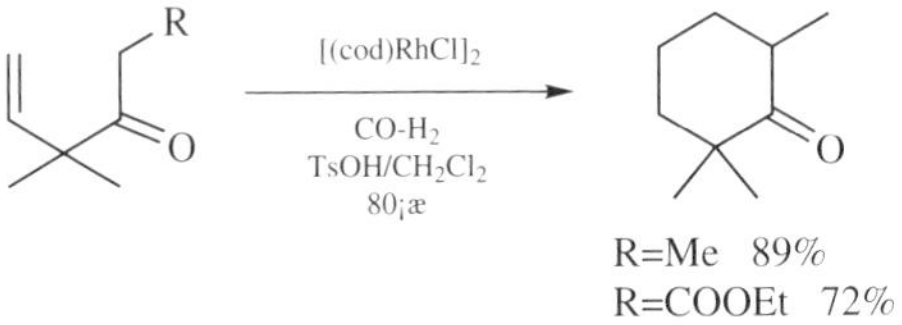

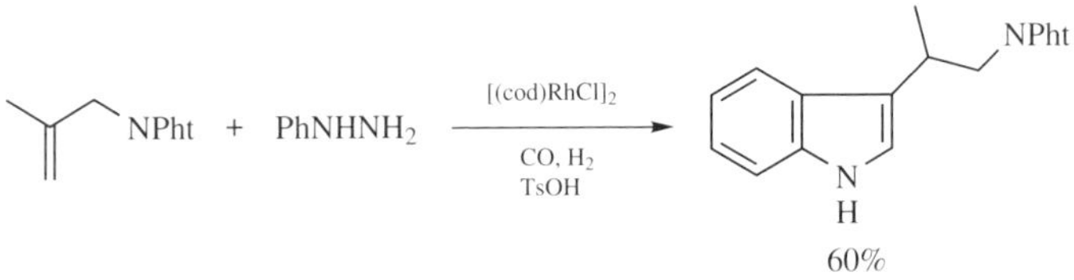

R=Me 89%
R=COOEt 72%

60%

Coupling reactions. Direct *o*-arylation of phenols[8] proceeds in the presence of $[(cod)RhCl]_2$ and HMPA, whereas a Heck-type coupling of $ArB(OH)_2$ with activated alkenes in aq. emulsion involves t-$Bu_2PCH_2CH_2NMe_3Cl$ as the ligand.[9]

The hydroxyl group of Baylis-Hillman adducts is displaced (S_N2'-fashion) by an aryl residue from $ArB(OH)_2$.[10]

Addition reactions. Aryl group transfer from $ArMMe_3$ [M=Sn, Pb] to *N*-tosylimines occurs when they are subjected to ultrasound irradiation together with the Rh catalyst in water.[11] Alkenylzirconocene chlorides add to aldimines, allylic amines are obtained.[12]

Hydroarylation of alkynyl azaaromatic compounds is carried out in water containing a surfactant and a hydrophilic phosphine.[13]

Imination. Thiazolidines form 2-imino derivatives by a Rh-catalyzed reaction with carbodiimides.[14]

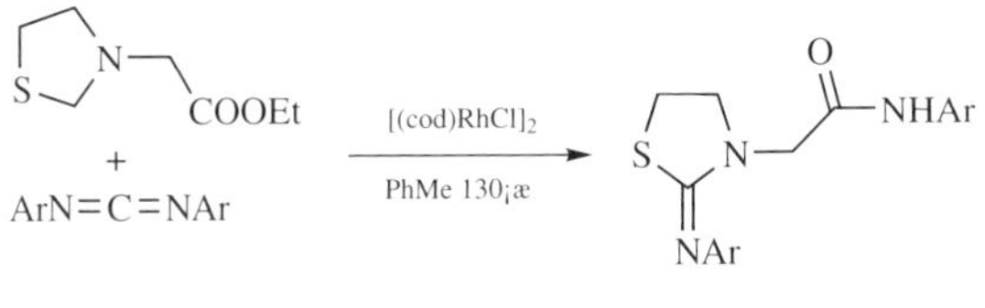

Hydrosilylation. Conversion of $[(cod)RhCl]_2$ to the imidazolidene-Rh complex is by treating with a common ionic liquid [bmim]X and *t*-BuOK. The new complex catalyzes hydrosilylation of 1,6-enynes to 2-methyl-1-silylmethylidenecyclopentane derivatives and heterocyclic analogues.[15]

Cycloisomerization. 1,2,7-Alkatrienes show ligand dependence in the Rh-catalyzed cycloisomerization, giving either 5- or 7-membered ring products.[16]

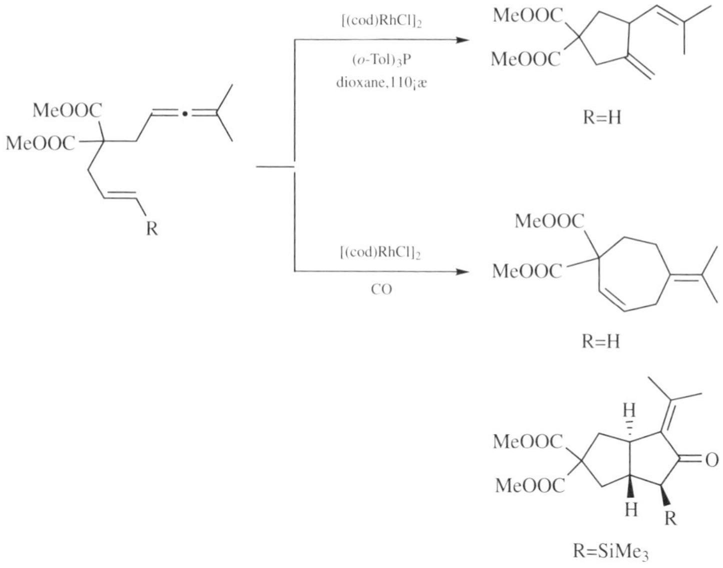

4-Alkynols and 5-alkynols cyclize to 5- and 6-membered cyclic ethers, respectively, by a Rh-catalyzed reaction. The preferred ligand is tris(3,5-difluorophenyl)phosphine.[17]

[1]Dong, C., Alper, H. *OL* **6**, 3489 (2004).
[2]Doi, H., Barletta, M., Suzuki, M., Noyori, R., Watanabe, Y., Langstrom, B. *OBC* **2**, 3063 (2004).
[3]Morimoto, T., Fujioka, M., Fuji, K., Tsutsumi, K., Kakiuchi, K. *CL* 154 (2003).
[4]Fuji, K., Morimoto, T., Tsutsumi, K., Kakiuchi, K. *ACIEE* **42**, 2409 (2003).
[5]Bennacer, B., Fujiwara, M., Ojima, I. *OL* **6**, 3589 (2004).
[6]Keranen, M.D., Kot, K., Hollmann, C., Eilbracht, P. *OBC* **2**, 3379 (2004).
[7]Kohling, P., Schmidt, A.M., Eilbracht, P. *OL* **5**, 3213 (2003).
[8]Oi, S., Watanabe, S., Fukita, S., Inoue, Y. *TL* **44**, 8665 (2003).
[9]Lautens, M., Mancuso, J., Grover, H. *S* 2006 (2004).
[10]Navarre, L., Darses, S., Genet, J.-P. *CC* 1108 (2004).
[11]Ding, R., Zhao, C.H., Chen, Y.J., Liu, L., Wang, D., Li, C.J. *TL* **45**, 2995 (2003).
[12]Kakuuchi, A., Taguchi, T., Hanzawa, Y. *TL* **44**, 923 (2003).
[13]Lautens, M., Yoshida, M. *JOC* **68**, 762 (2003).
[14]Zhou, H.-B., Dong, C., Alper, H. *CEJ* **10**, 6058 (2004).
[15]Park, K.H., Kim, S.Y., Son, S.U., Chung, Y.K. *EJOC* 4341 (2003).
[16]Makino, T., Itoh, K. *JOC* **69**, 395 (2004).
[17]Trost, B.M., Rhee, Y.H. *JACS* **125**, 7482 (2003).

Bis[chloro(dicyclooctene)rhodium(I)].

Alkylation. Aromatic azines (ArCH=NN=CHAr) are alkylated with 1-alkenes at unsubstituted *o*-position(s) by heating with $[Rh(coe)_2Cl]_2$ and DPPF in toluene at 140°. Thus, normal alkyl groups including $CH_2CH_2SiMe_3$ can be introduced.

Arylation. Heterocycles such as benzimidazole are arylated by ArI.[2]

[1]Lim, Y.-G., Koo, B.T. *TL* **46**, 385 (2005).
[2]Lewis, J.C., Wiedemann, S.H., Bergman, R.G., Ellman, J.A. *OL* **6**, 35 (2004).

Bis[chloro(diethene)rhodium(I)].

Heck reaction. With the Rh catalyst ArCOCl can be used instead of ArX in the arylation of alkenes.[1] It is efficient and simple in work up. In the reaction of ArCHO and $Ar'BF_3K$ to deliver ArCOAr′ in reasonably good yields[2] a Heck mechanism is followed by hydride transfer to acetone which is used as part of the solvent system.

Conjugate addition. Asymmetric aryl group transfer from $ArB(OH)_2$ to maleic and fumaric acid derivatives is accomplished with the Rh catalyst and a chiral 2,5-dibenzylnorbornadiene.[3]

[1]Sugihara, T., Satoh, T., Miura, M., Nomura, M. *ACIEE* **42**, 4672 (2003).
[2]Pucheault, M., Darses, S., Genet, J.-P. *JACS* **126**, 15356 (2004).
[3]Shintani, R., Ueyama, K., Yamada, I., Hayashi, T. *OL* **6**, 3425 (2004).

Bis[chloro(norbornadiene)rhodium(I)]. 22, 43

Cycloaddition reactions. Molecules containing a triple bond and a conjugate diene unit that are separated by 3 atoms undergo [4 + 2]cycloaddition in the presence of the cationic Rh catalyst in a micellar system.[1] Remarkably, one such substrate can also give rise to intermolecular [2 + 2 + 2]- and [4 + 2 + 2]cycloadducts. Reaction conditions determine the outcome.

$[(nbd)RhCl]_2$, H_2O,SDS, rt

(nbd = norbornadiene)

[1]Motoda, D., Kinoshita, H., Shinokubo, H., Oshima, K. *ACIEE* **43**, 1860 (2004).

Bis[chloro(pentamethylcyclopentadienyl)hydridoiridium].

Arylation.[1] Direct arylation of arenes with ArI provides biaryls when treated with $[Cp^*IrHCl]_2$ and *t*-BuOK. The base causes reduction of the Ir complex which effects electron transfer to ArI while regenerating the Ir(III) catalyst.

[1]Fujita, K., Nonogawa, M., Yamaguchi, R. *CC* 1926 (2004).

Bis[chloro(pentamethylcyclopentadienyl)methylthioruthenium]. 21, 42

Alkylation reactions. An allenylidene-ene reaction is involved in the condensation of propargylic alcohols with alkenes in the presence of the title reagent **1**.[1] Cinnamylation of arenes with the alcohol, carbonate, or chloride is feasible at 60° using a salt related to **1** as catalyst.[2] Propargylation of electron-rich arenes (including phenol, thiophene, furan, pyrrole, azulene)[3] promoted by the same catalyst (**1** is completely ineffective) represents a more viable method than the multistep process employing the Nicholas reaction that employs stoichiometric $Co_2(CO)_8$.

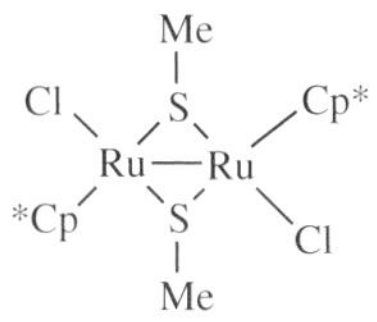

1

Heterocycles. In an annulation of 1,3-cycloalkanediones propargylic alcohols provide a three-carbon unit.[4]

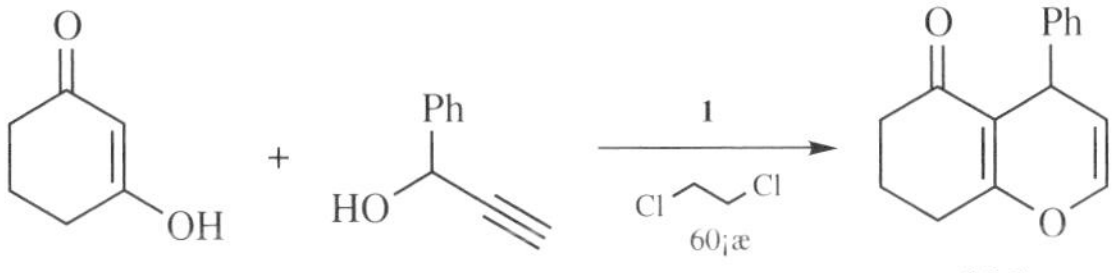

93%

Synthesis of polysubstituted furans from ketones and propargylic alcohols requires **1** and $PtCl_2$. Pyrroles are formed in the presence of anilines.[5] The Ru catalyst performs propargylation of the ketones and $PtCl_2$, hydration of the triple bond and suibsequent cyclodehydration.

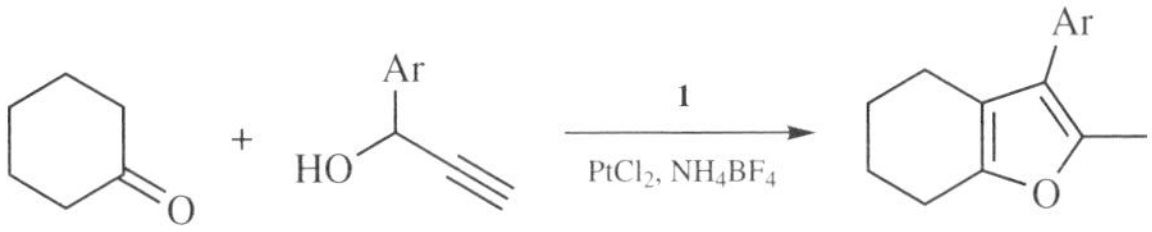

69–75%

Phosponylation.[6] By catalysis of **1**, two phosphonyl groups, one allylic and one alkenyl, are introduced into the skeleton of a propargylic alcohol on reaction with a phosphonic acid.

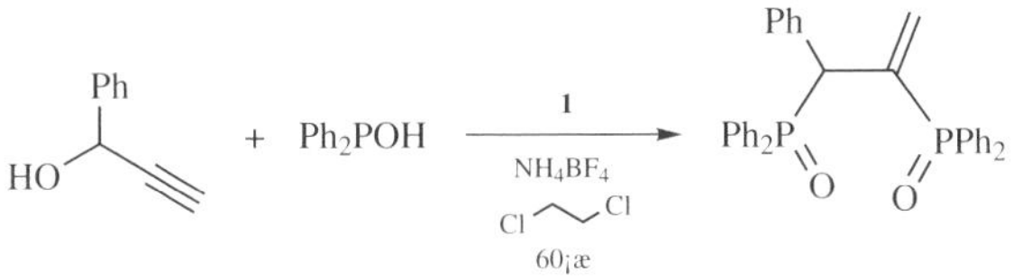

[1]Nishibayashi, Y., Inada, Y., Hidai, M., Uemura, S. *JACS* **125**, 6060 (2003).
[2]Onodera, G., Imajima, H., Yamanashi, M., Nishibayashi, Y., Hidai, M., Uemura, S. *OM* **23**, 5841 (2004).
[3]Nishibayashi, Y., Inada, Y., Yoshikawa, M., Hidai, M., Uemura, S. *ACIEE* **42**, 1495 (2003).
[4]Nishibayashi, Y., Yoshikawa, M., Inada, Y., Hidai, M., Uemura, S. *JOC* **69**, 3408 (2004).
[5]Nishibayashi, Y., Yoshikawa, M., Inada, Y., Milton, M.D., Hidai, M., Uemura, S. *ACIEE* **42**, 2681 (2003).
[6]Milton, M.D., Onodera, G., Nishibayashi, Y., Uemura, S. *OL* **6**, 3993 (2004).

Bis[chlorotris(triphenylphosphine)ruthenium(II)].

Alcohol homologation. Serial oxidoreduction and condensation reactions catalyzed by $(Ph_3P)_3RuCl_2$ under alkaline conditions in the presence of 1-dodecene (hydrogen acceptor) serve to unite 2-alkanols and 1-alkanols to afford secondary alcohols [$RCH(OH)Me + R'CH_2OH$ to $RCH(OH)CH_2CH_2R'$].

[1]Cho, C.S., Kim, B.T., Kim, H.-S., Kim, T.-J., Shim, S.C. *OM* **22**, 3608 (2004).

Bis[(cumene)dichlororuthenium(II)].

Enol esters.[1] The Ru complex catalyzes addition of RCOOH to alkynes. However, phosphine additives have profound effects on the regioselectivity (Markovnikov vs. anti-Markovnikov).

PhCOOH + Bu—≡ → [(*p*-cymene)$RuCl_2$]$_2$, 2-Fu_3P / PhMe → Ph-C(=O)-O-C(=CH$_2$)Bu 93%

PhCOOH + Bu—≡ → [(*p*-cymene)$RuCl_2$]$_2$, (o-ClC_6H_4)$_3$P, DMAP → Ph-C(=O)-O-CH=CH-Bu 89%

[1]Goossen, L.J., Paetzold, J., Koley, D. *CC* 706 (2003).

Bis[(1,5-cyclooctadiene)iridium(I)] tetrafluoroborate.

Allylic substitution. Allylic acetates are converted into other esters and ethers with carboxylic acids and alcohols when their mixtures are heated with $[(cod)_2Ir]BF_4$.

[1]Nakagawa, H., Hirabayashi, T., Sakaguchi, S., Ishii, Y. *JOC* **69**, 3474 (2004).

Bis[(1,5-cyclooctadiene)methoxyiridium(I)]. **22**, 45

Silylation. The iridium complex together with 4,4′-di-*t*-butyl-2,2′-bipyridine activate aromatic C–H bonds toward silylation with halodisilanes.

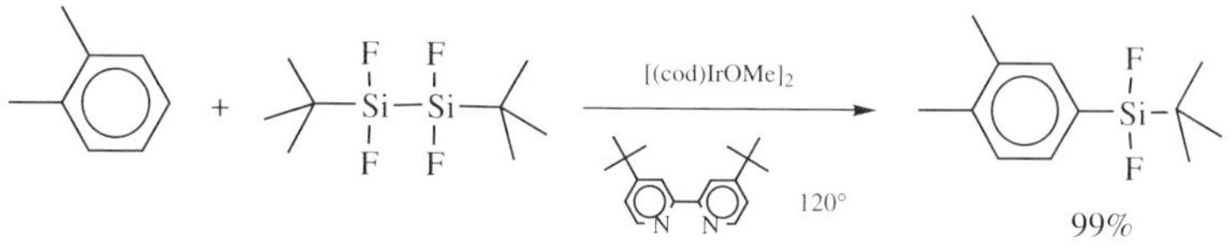

[1]Ishiyama, T., Sato, K., Nishio, Y., Miyaura, N. *ACIEE* **42**, 5346 (2003).

Bis(1,5-cyclooctadiene)nickel(0). **21**, 43–45; **22**, 45–47

Addition reactions. Several interesting additions to alkynes are mediated by $Ni(cod)_2$.[1] In hydrophosphonylation of 1-alkynes, a reaction carried out in the presence of $Ni(cod)_2$ shows different regioselectivity from that of $(Ph_2Me)_4Ni$.[1]

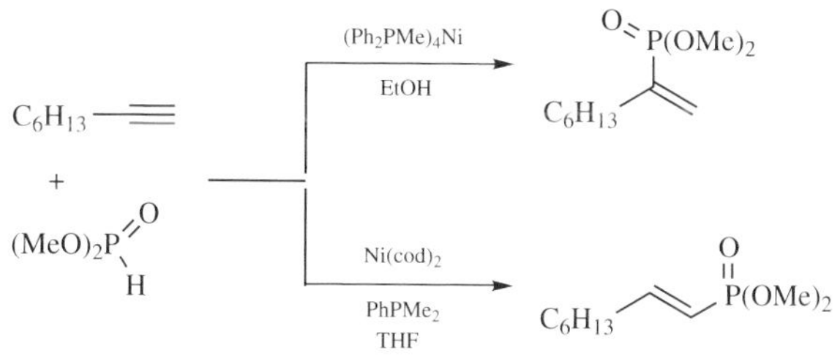

Arylcyanation of an alkyne with ArCN gives products in which the Ar group becomes geminal to the less bulky substituent of the alkyne.[2]

Reductive coupling of alkynes to epoxides gives homoallylic alcohols (intramolecular version gives 3-alkylidenecycloalkanols),[3] while conjugated enynes react with aldehydes and epoxides to furnish dienols (OH group in the allylic and homoallylic positions, respectively) in a highly regioselective manner.[4] This formal reductive hydroxyalkylation introduces a carbon chain at the far end of the double bond.

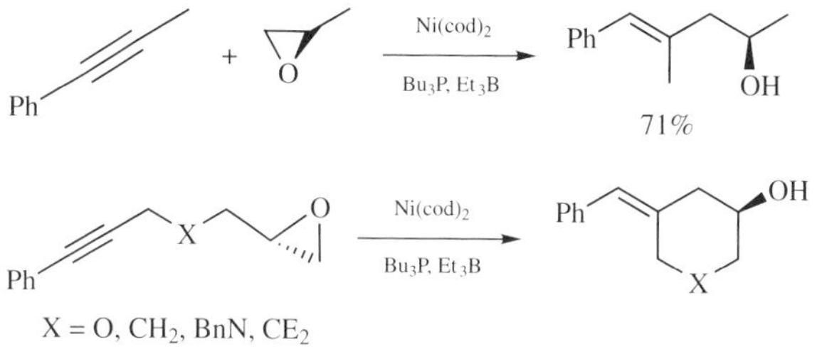

X = O, CH_2, BnN, CE_2

Additives can influence the steric course of reductive hydroxyalkylation of 1- silyl-1,3-dienes.[5] 1-Silylallenes undergo carboxylation under CO_2.[6]

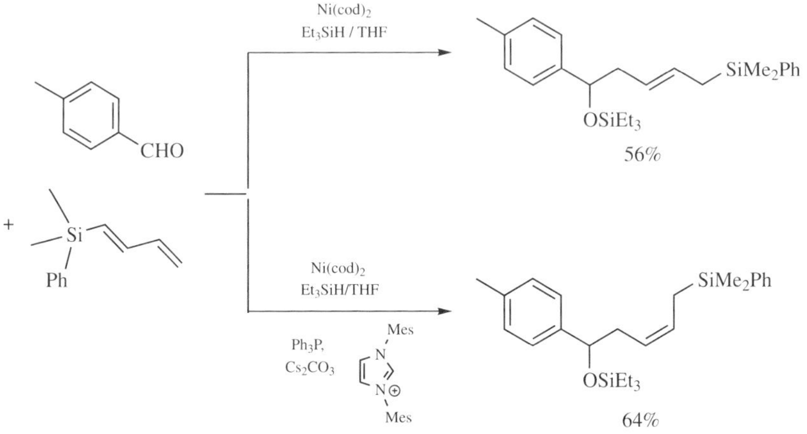

Allylic amines are formed when alkynes, imines, and boranes (or boronic acids) are heated with $Ni(cod)_2$ and a phosphine ligand in MeOH.[7]

cis-β,γ-Disubstituted α-methylene-γ-lactones are synthesized from terminal allenes, CO_2, and aryl aldehydes.[8]

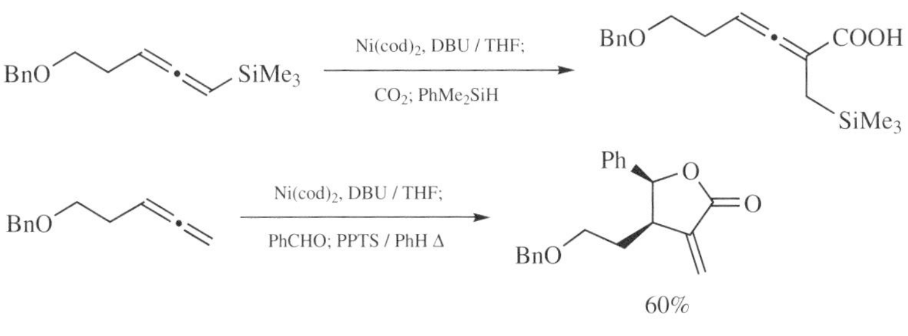

Cycloaddition. Under the influence of $Ni(cod)_2$ ethyl cyclopropylideneacetate combines with 2 equivalents of 1-alkynes to give ethyl 1,3-heptadien-5-ylideneacetates.[9]

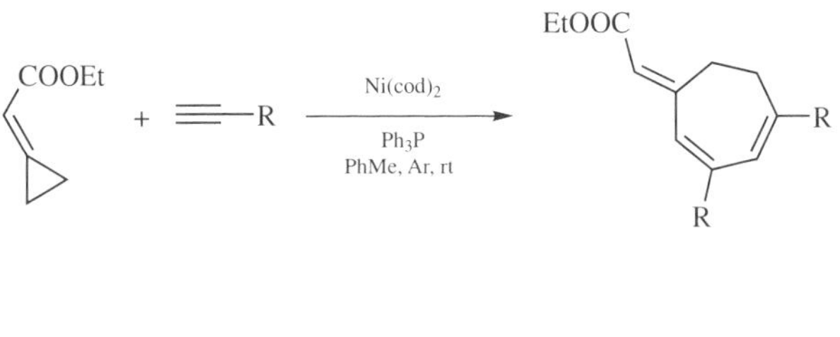

Coupling reactions. The Suzuki coupling is successful with both electron-rich and electron-poor ArOTs at room temperature using the $Ni(cod)_2$ -PCy_3 system.[10] Aryltrimethylammonium salts also can be used when the ligand is 1,3-dimesitylimidazol-2-ylidene.[11]

Negishi coupling of unactivated primary and secondary alkyl bromides and iodides with RZnBr proceeds at room temperature when mediated by $Ni(cod)_2$ [with a pybox ligand added].[12] A similar Suzuki coupling requires a different ligand (4,7-diphenyl-1,10-phenanthroline) and a strong base (t-BuOK) for good results.[13]

Br 74% Ph $Ni(cod)_2$ t-BuOK Ph Ph N N 71% Br

Consecutive catalyzed additions are involved in the new syntheses of 2-benzyl-3-hydroxyalkanoic esters[14] and allylic silyl ethers.[15]

PhCHO + Ph $Ni(cod)_2$ Et_3SiH MesN NMes THF $OSiEt_3$ Ph Ph

PhCHO + PhI + O O $Ni(cod)_2$ Me_2Zn THF OH O Ph O Ph 88% (dr 86:14)

Alkynes through titanium complexes couple with ArI by catalysis of $Ni(cod)_2$ to produce arylalkenes.[16]

Conjugated (E)-enynes.[17] 1-Alkynes are dimerized head-to-head in moderate to excellent yields at room temperature by a mixture of $Ni(cod)_2$ and t-Bu_3P.

Isomerization. Cyclopropylalkenes are isomerized to cyclopentenes. The ligand for the Ni is 1,3-bis(2,6-diisopropyl)imidazol-2-ylidene.[18] However, no reaction occurs when the substrate possesses a 1,2-disubstituted double bond (with the exception of the cyclopropylstyrene.)

On heating with $Ni(cod)_2$ and Ph_3P secondary allylic cyanides undergo isomerization to afford linear isomers, apparently via π-allylnickel complexes.[19]

Homoallylic alcohols.[20] Reductive coupling of alkynes with epoxides is effected by a mixture of $Ni(cod)_2$, Et_3B and Bu_3P. CC bond formation with epoxide ring opening occurs at the least substituted carbon atom.

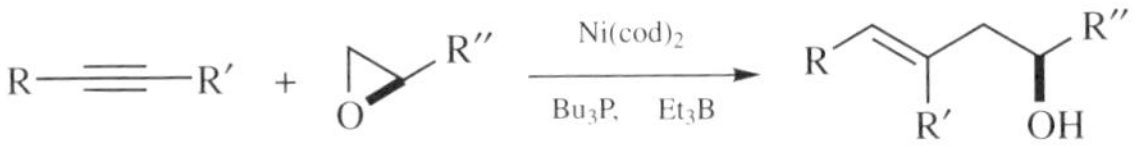

[1]Han, L.-B., Zhang, C., Yazawa, H., Shimada, S. *JACS* **126**, 5080 (2004).
[2]Nakao, Y., Oda, S., Hiyama, T. *JACS* **126**, 13904 (2004).
[3]Molinaro, C., Jamison, T.F. *JACS* **125**, 8076 (2003).
[4]Miller, K.M., Luanphaisarnnont, T., Molinaro, C., Jamison, T.F. *JACS* **126**, 4130 (2004).
[5]Sawaki, R., Sato, Y., Mori, M. *OL* **6**, 1131 (2004).
[6]Takimoto, M., Kawamura, M., Mori, M. *S* 791 (2004).
[7]Patel, S.J., Jamison, T.F. *ACIEE* **42**, 1364 (2003).
[8]Takimoto, M., Kawamura, M., Mori, M. *OL* **5**, 2599 (2003).
[9]Saito, S., Masuda, M., Komagawa, S. *JACS* **126**, 10540 (2004).
[10]Tang, Z.-Y., Hu, Q.-S. *JACS* **126**, 3058 (2004).
[11]Blakey, S.B., MacMillan, D.W.C. *JACS* **125**, 6046 (2003).
[12]Zhou, J., Fu, G.C. *JACS* **125**, 14726 (2003).
[13]Zhou, J., Fu, G.C. *JACS* **126**, 1340 (2004).
[14]Subburaj, K., Montgomery, J. *JACS* **125**, 11210 (2003).
[15]Mahandru, G.M., Liu, G., Montgomery, J. *JACS* **126**, 3698 (2004).
[16]Obora, Y., Moriya, H., Tokunaga, M., Tsuji, Y. *CC* 2820 (2003).
[17]Ogoshi, S., Ueta, M., Oka, M., Kurosawa, H. *CC* 2732 (2004).
[18]Zuo, G., Louie, J. *ACIEE* **43**, 2277 (2004).
[19]Chaumonnot, A., Lamy, F., Sabo-Etienne, S., Donnadieu, B., Chaudret, B., Barthelat, J.-C., Galland, J.-C. *OM* **23**, 3363 (2004).
[20]Molinaro, C., Jamison, T.F. *JACS* **125**, 8076 (2003).

Bis(1,5-cyclooctadiene)rhodium(I) salts. 21, 45–46; 22, 47–48

Reductive coupling. Under catalytic hydrogenation conditions, conjugated diynes,[1] enynes,[2] and dienes[3] couple reductively with 1,2-ketoaldehydes in the presence of $(cod)_2RhPF_6$, affording unsaturated ketols. A variant leads to aldol products.[4]

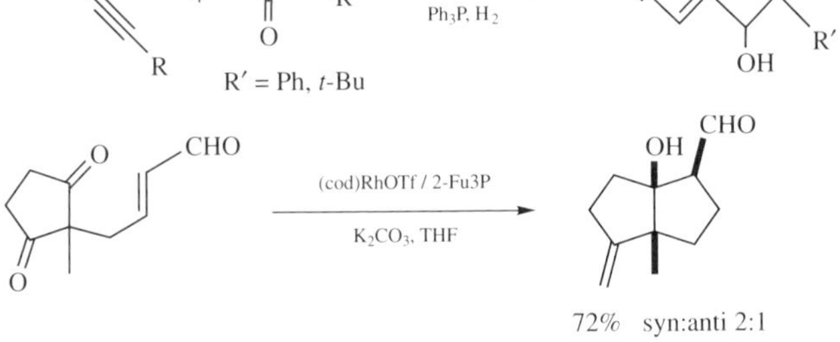

72% syn:anti 2:1

Amino acid derivatives. New routes consist of conjugate addition to dehydroamino acid derivatives[5] using $[ArBF_3]K$, and reductive amination of α-keto acids.[6] In the latter reaction the presence of a chiral ligand [e.g., *(3R,4R)-N*-benzylbis(diphenylphosphino)-pyrrolidine] turns it into an excellent asymmetric synthesis.[6]

Isomerization. Allylic alcohols are converted into aldehydes in moderate yields (3 cases) in water by the cationic Rh catalyst (phosphine ligand present).[7]

Disulfides. Using $(cod)_2RhBF_4$ and Ph_3P to convert RSH into RSSR under inert atmosphere[8] is perhaps of little synthetic value.

Hydroaminomethylation. Extension of 1-alkenes by a $[CH_2NR_2]$ group from reaction with R_2NH, CO and H_2 is readily achieved using $(cod)_2RhBF_4$ and Xantphos.[9] The reaction usually shows >98% regioselectivity and the only byproduct is water.

[1]Huddleston, R.R., Jang, H.-Y., Krische, M.J. *JACS* **125**, 11488 (2003).
[2]Jang, H.-Y., Huddleston, R.R., Krische, M.J. *JACS* **126**, 4664 (2004).
[3]Jang, H.-Y., Huddleston, R.R., Krische, M.J. *ACIEE* **42**, 4074 (2003).
[4]Koech, P.K., Krische, M.J. *OL* **6**, 691 (2004).
[5]Navarre, L., Darses, S., Genet, J.-P. *EJOC* 69 (2004).
[6]Kadyrov, R., Riermeier, T.H., Dingerdissen, U., Tararov, V., Borner, A. *JOC* **68**, 4067 (2003).
[7]Knight, D.A., Schull, T.L. *SC* **33**, 827 (2003).
[8]Tanaka, K., Ajiki, K. *TL* **45**, 25 (2004).
[9]Ahmed, M., Seayad, A.M., Jackstell, R., Beller, M. *JACS* **125**, 10311 (2003).

Bis[(*p*-cymene)chlororuthenium] – *N*-tosyl-1,2-diphenyl-1,2-diamine.

Reduction. Reductive amination of carbonyl compounds under transfer hydrogenation conditions is rendered asymmetric using a chiral diamine ligand.[1]

The chlorine-free complex is also useful for reduction of α-keto ester to the corresponding hydroxyl esters without affecting a β,γ-unsaturation.[2]

[1]Williams, G.D., Pike, R.A., Wade, C.E., Wills, M. *OL* **5**, 4227 (2003).
[2]Guo, M., Li, D., Sun, Y., Zhang, Z. *SL* 741 (2004).

Bis(dibenzylideneacetone)palladium(0). 21, 46–48; **22**, 48–51

Coupling reactions. Homocoupling of ArX with $(dba)_2Pd$ and Bu_4NF has been reported.[1] The Heck reaction is generally improved with sterically bulky thiourea ligands[2] because they are stable to air and moisture. Another report describes *N*-heterocyclic carbene ligands that contain a triarylphosphine group.[3] Formation of 2-arylmethyltetrahydrofurans[4] from ArBr and 4-alkenols is said to involve intramolecular insertion of the alkene linkage to the ArPdOR intermediates. In one operation dihydrofurocoumarins are formed from *o*-iodoacetoxycoumarins which starts with a Heck reaction.[5]

Protocols for the Suzuki coupling using $(dba)_2Pd$ and imidazole carbene are available.[6] Coupling of arylsiloxanes with ArOTf usually fails due to rapid hydrolysis of ArOTf, but a solution to the problem is to form the air-stable triethylammonium aryl(biscatechol)silicates from the siloxanes, catechol, and Et_3N prior to the reaction.[7] Another possibility is to employ siloxanes such as **1**.[8]

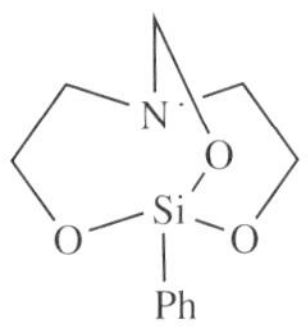

Ketene silyl acetals undergo catalyzed arylation to furnish α-aryl carboxylic acid derivatives when ZnF_2 is added.[9] While zinc enolates are involved in the reaction, many functional groups are tolerated.

Carbogermylation of terminal allenes takes place to afford 2-arylallylgermanes.[10]

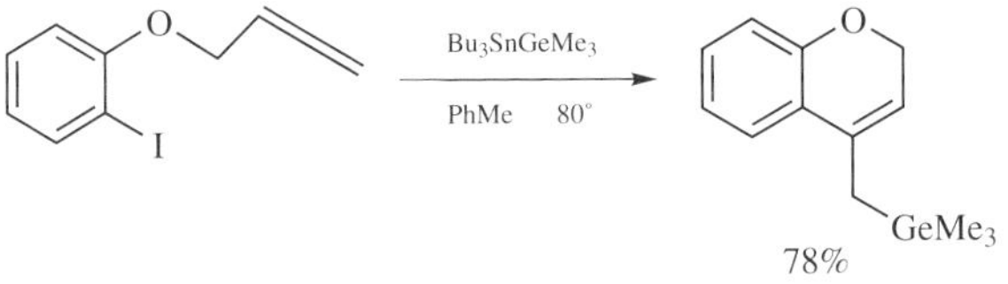

Arylation reactions. An efficient preparative method for arylmalonitriles[11] depends on the $(dba)_2Pd$-imidazole carbene system. For Pd-catalyzed α-arylation of *N*-protected 2-piperidinones[12] a biphenyl bound *N,P*-ligand finds good use.

N-Heteroarylation of amines employs the *t*-Bu_3P- $(dba)_2Pd$ system to advantages.[13]

Oxidation. Primary and secondary benzylic alcohols are oxidized to the carbonyl compounds by PhCl in the presence of $(dba)_2Pd$, 2-dicyclohexylphosphinobiphenyl, and a base (K_2CO_3 or K_3PO_4). For a few cases of bridged ring compounds (e.g., borneol) a stronger base (*t*-BuONa) is used.[14]

[1]Seganish, W.M., Mowery, M.E., Riggleman, S., DeShong, P. *T* **61**, 2117 (2005).
[2]Yang, D., Chen, Y.-C., Zhu, N.-Y. *OL* **6**, 1577 (2004).
[3]Wang, A.-E., Xie, J.-H., Wang, L.-X., Zhou, Q.-L. *T* **61**, 259 (2005).
[4]Wolfe. J.P., Rossi, M.A. *JACS* **126**, 1620 (2004).
[5]Rozhkov, R.V., Larock, R.C. *JOC* **68**, 6314 (2003).
[6]Arentsen, K., Caddick, S., Cloke, F.G.N., Herring, A.P., Hitchcock, P.B. *TL* **45**, 3511 (2004).
[7]Seganish, W.M., DeShong, P.. *JOC* **69**, 1137 (2004).
[8]Riggleman, S., DeShong, P. *JOC* **68**, 8106 (2003).
[9]Liu, X., Hartwig, J.F. *JACS* **126**, 5182 (2004).
[10]Jeganmohan, M., Shanmugasundaram, M., Cheng, C.-H. *CC* 1746 (2003).

[11]Gao, C., Tao, X., Qian, Y., Huang, J. *CC* 1444 (2003).
[12]de Filippis, A., Pardo, D.G., Cossy, J. *T* **60**, 9757 (2004).
[13]Hooper, M.W., Utsunomiya, M., Hartwig, J.F. *JOC* **68**, 2861 (2003).
[14]Guram, A.S., Nei, X., Turner, H.W. *OL* **5**, 2485 (2003).

Bis(di-*t*-butyl-4-methylphenoxy)titanium bis(dimethylamide).

Hydroamination. The hindered Ti catalyst promotes regioselective addition of amines to alkynes in the Markovnikov fashion.[1] The extension to a synthesis of 2-methyltryptamines and homologues[2] is interesting and valuable.

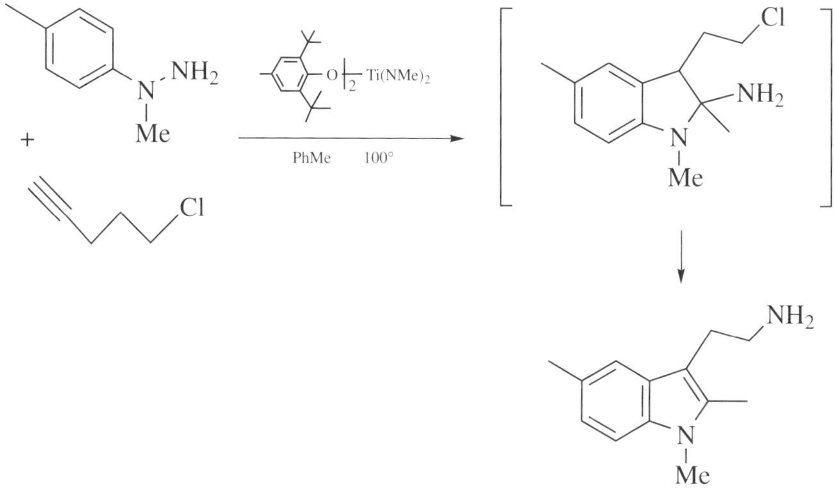

[1]Khedkar, V., Tillack, A., Beller, M. *OL* **5**, 4767 (2003).
[2]Khedkar, V., Tillack, A., Michalik, M., Beller, M. *TL* **45**, 3123 (2004).

Bis[dicarbonylchlororhodium(I)]. 21, 50–51; **22**, 52–53

Cycloadditions. Bicyclic structures are formed by an intramolecular [4 + 2]-cycloaddition in which the stereochemistry is controlled by the substrate.[1] The Rh-catalyzed process requires low reaction temperatures.

Pauson-Khand reaction by Rh catalysis for assembly of tricyclic products has been demonstrated,[2] that the construction of the core structure of guanacasterpene-A[3] is of particular interest. Also related is the [2 + 2 + 1]cycloaddition process that involves an isolated alkene and a double bond of a conjugated diene, in both intermolecular and intramolecular versions.[4]

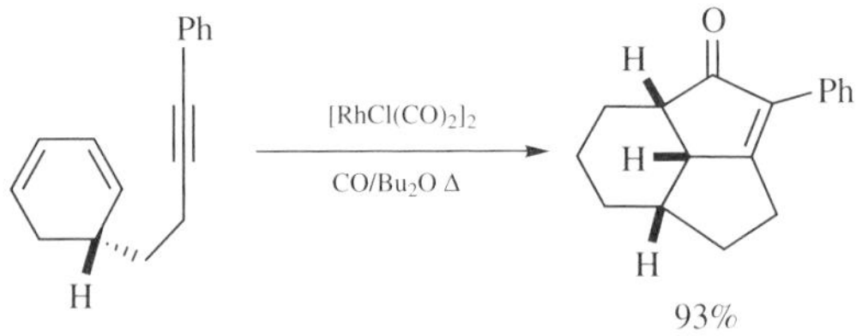

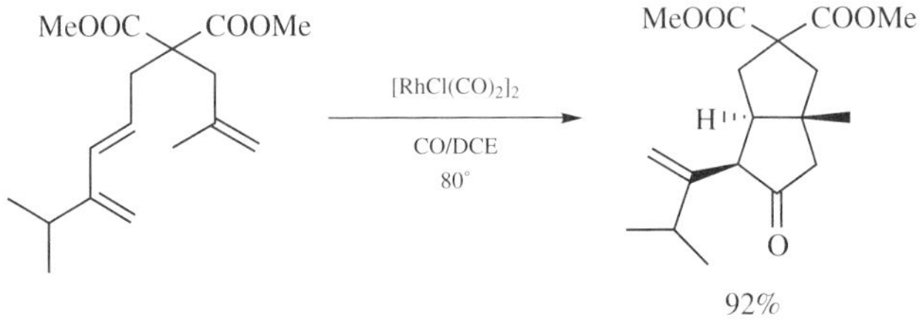

Allenylalkynes may be converted into either bicyclic cyclopentenones or monocyclic trienes, depending on whether the reaction is conducted under CO pressure or in a nitrogen atmosphere.[5]

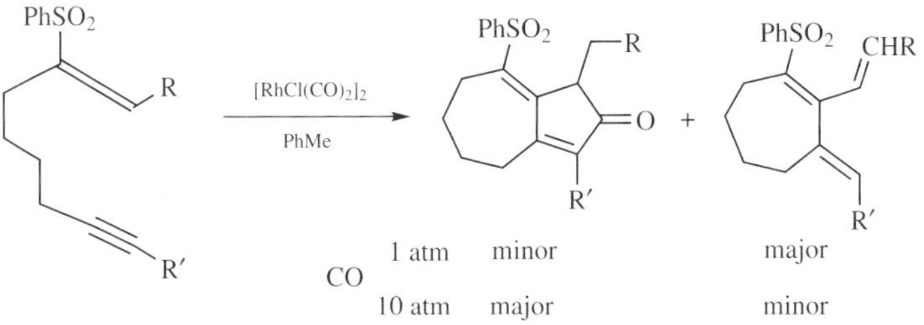

Cyclization. Heterocycles with a 3-hydroxy substituent are obtained from a Rh-induced cyclization of heteroalkyl epoxides.[6] Structural limitations of the substrates include the *trans*-configuration of the epoxide ring and the formation of only 5- and 6-membered cycles.

Substitution reactions. The Rh-catalyzed version is characterized by high regioselectivity (*ipso*-substitution) for unsymmetrical substrates and retention of configuration at the reaction center.[7]

Desilylative acylation on the enol esters of acylsilanes gives monoprotected unsymmetrical 1,2-diketones.[8]

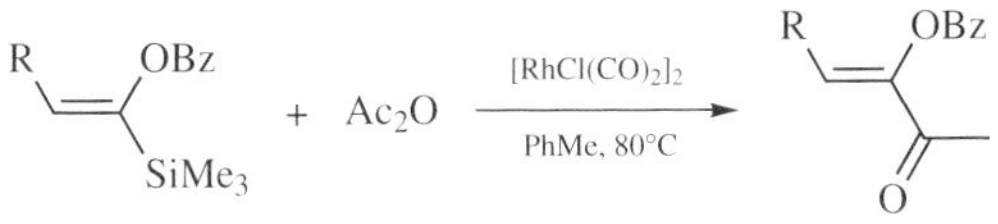

Annulation. Cyclobutenones undergo ring opening and annulate onto alkenes such as norbornene. Formation of cyclopentene or cyclohexenone units occurs, when the reaction is conducted under Ar (with decarbonylation) or CO.[9] When the cyclobutenones are exposed to the catalyst alone, a ring opening -dimerization pathway is followed, resulting in α-pyrones. Interestingly, two isomers different in the configuration of the sidechain double bond are formed on switching the catalyst to $[RuCl_2(CO)_3]_2$.

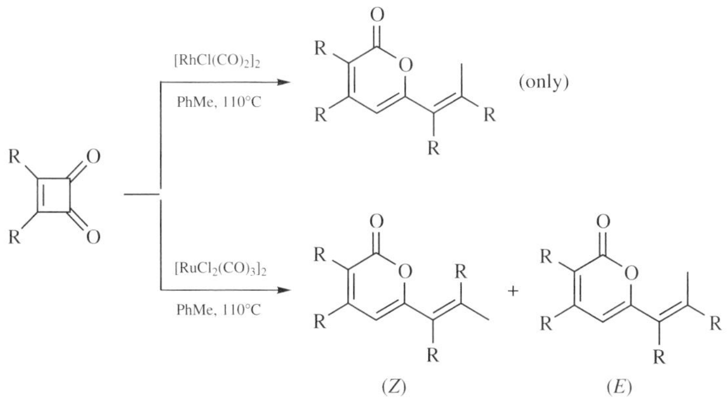

$[RhCl(CO)_2]_2$, PhMe, 110°C (only)

$[RuCl_2(CO)_3]_2$, PhMe, 110°C

(*Z*) + (*E*)

Conjugated dienes use one of the double bonds to participate in Pauson-Khand reaction with alkynes and CO.[10]

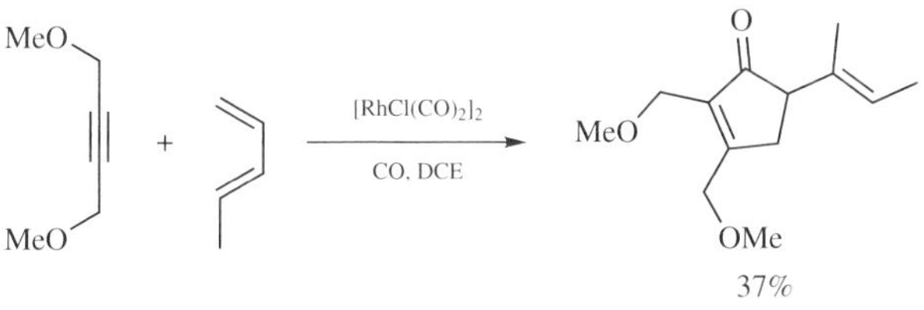

$[RhCl(CO)_2]_2$, CO, DCE

37%

Additions. The H-2 of 4,4-diemthyl-2-oxazoline is replaced by an alkyl group when it is subjected to a Rh-catalyzed reaction with various 1-alkenes.[11]

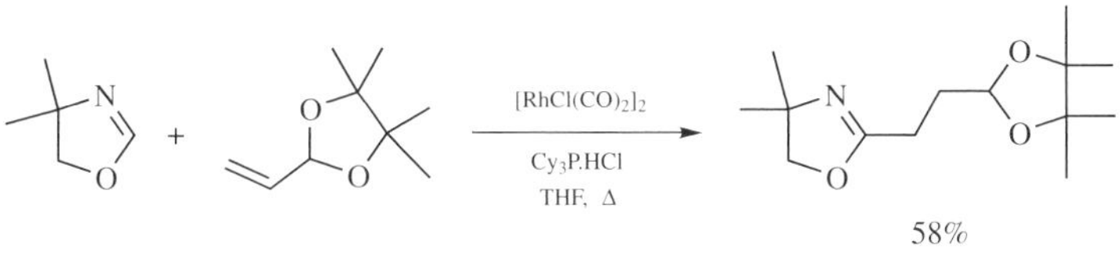

$[RhCl(CO)_2]_2$, $Cy_3P.HCl$, THF, Δ

58%

[1]O'Mahony, D.J.R., Belanger, D.B., Livinghouse, T. *OBC* **1**, 2038 (2003).
[2]Yeh, M.-C.P., Tsao, W.-C., Ho, J.-S., Tai, C.-C., Chiou, D.-Y., Tu, L.-H. *OM* **23**, 792 (2004).

[3]Brummond, K.M., Gao, D. *OL* **5**, 3491 (2003).
[4]Wender, P.A., Croatt, M.P., Deschamps, N.M. *JACS* **126**, 5948 (2004).
[5]Mukai, C., Inagaki, F., Yoshida, T., Kitagaki, S. *TL* **45**, 4117 (2004).
[6]Ha, J.D., Shin, E.Y., Kang, S.K., Ahn, J.H., Choi, J.-K. *TL* **45**, 4193 (2004).
[7]Ashfeld, B.L., Miller, K.A., Martin, S.F. *OL* **6**, 1321 (2004).
[8]Yamane, M., Uera, K., Narasaka, K. *CL* **33**, 424 (2004).
[9]Kondo, T., Taguchi, Y., Kaneko, Y., Niimi, M., Mitsudo, T. *ACIEE* **43**, 5369 (2004).
[10]Wender, P.A., Deschamps, N.M., Williams, T.J. *ACIEE* **43**, 3076 (2003).
[11]Wiedemann, S.H., Bergman, R.G., Ellman, J.A. *OL* **6**, 1685 (2004).

Bis[dicarbonyl(pentamethylcyclopentadienyl)iron]. 22, 53

Reductive cyclization.[1] Heating the acetylated Baylis-Hillman adducts from *o*-nitroarenecarbaldehydes with the Fe complex in dioxane under CO leads to 3-substituted quinolines.

OAc CN NO$_2$ [Cp*Fe(CO)2]2 CO dioxane, 150 °C → CN N 65%

[1]O'Dell, D.K., Nicholas, K.M. *JOC* **68**, 6427 (2003).

Bis[dichloro(ethene)platinum].

Cyclic ethers. Influenced by the Pt complex unsaturated alcohols cyclize to give 5- and 6-membered cyclic ethers.[1]

Amide alkylation. Amides ($RCONH_2$) add to unactivated alkenes to yield *N*-alkyl amides, e.g., with propene, $RCONHCHMe_2$.[2]

[1]Qian, H., Han, X., Widenhoefer, R.A. *JACS* **126**, 9536 (2004).
[2]Wang, X., Widenhoefer, R.A. *OM* **23**, 1649 (2004).

Bis[dichloro(pentamethylcyclopentadienyl)iridium]. 22, 53

N-Alkylation. Amines are alkylated with alcohols in hot toluene.[2] Cyclic amines are formed from primary amines and diols.[2]

NH$_2$ + HO OH [Cp*IrCl]$_2$ NaHCO$_3$ PhMe, 110 °C → N 91%

[1]Fujita, K., Li, Z., Ozeki, N., Yamaguchi, R. *TL* **44**, 2687 (2003).
[2]Fujita, K., Fujii, T., Yamaguchi, R. *OL* **6**, 3525 (2004).

Bis[dichloro(pentamethylcyclopentadienyl)rhodium].

Benzolactams. ω-(*o*-Aminoaryl)alkanols are converted into benzannulated lactams (5-, 6-, and 7-membered) by $[Cp^*RhCl_2]_2$ and K_2CO_3 in acetone.[1] The same substrates form indoles or tetrahydroquinolines when catalyzed by $[Cp^*IrCl_2]_2$.

[1]Fujita, K., Takahashi, Y., Owaki, M., Yamamoto, K., Yamaguchi, R. *OL* **6**, 2785 (2004).

Bis(*N,N*-diisopropylamino)cyanoborane.

Amino(cyano)borates. One of the amino groups of the title reagent can be replaced and the products from reaction with homopropargylic alcohols are valuable precursors of various conjugated nitriles by virtue of Pd-catalyzed intramolecular cyanoboration and subsequent coupling reactions.[1]

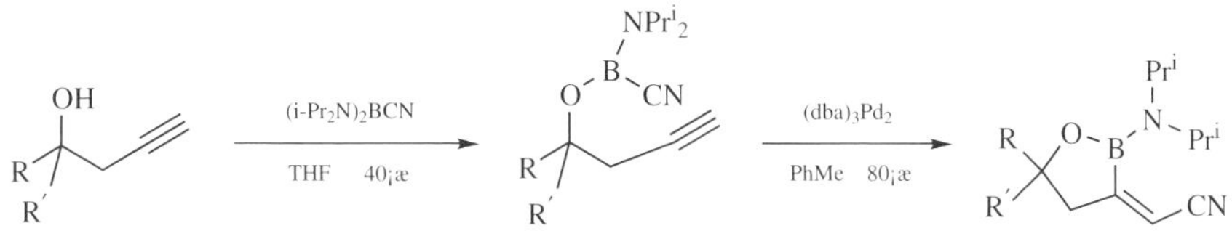

[1]Suginome, M., Yamamoto, A., Murakame, M. *JACS* **125**, 6358 (2003).

2,2′-Bis(diphenylphosphino)-1,1′-binaphthyl, BINAP. 13, 36–37; **14**, 38–44; **15**, 34; **16**, 32–36; **17**, 34–38; **18**, 39–41; **19**, 33–35; **20**, 41–44; **21**, 53–55; **22**, 54–58

Preparation.[1] A method for eduction of phosphine oxides is suitable for preparation of BINAP. It calls for a combination of Ph_3P or $(EtO)_3P$ as sacrificial reagent and $HSiCl_3$ as reducing agent.

Note that BINAP dioxides catalyze asymmetric allylation of α-hydrazono esters with allyltrichlorosilanes.[2]

[1]Wu, H.-C., Yu, J.-Q., Spencer, J.B. *OL* **6**, 4675 (2004).
[2]Ogawa, C., Sugiura, M., Kobayashi, S. *ACIEE* **43**, 6491 (2004).

Copper complexes

Reductions. Hydrosilylation of ketones can employ air and moisture stable $Cu(OAc)_2{\cdot}H_2O$ and (*S*)-BINAP to prepare the catalyst.[1] Conjugate reduction of unsaturated lactones and lactams[2] and nitroalkenes[3] is also performed, although the catalytic systems are somewhat different and may contain a strong base.

[1]Lee, D., Yun, J. *TL* **45**, 5415 (2004).
[2]Hughes, G., Kimura, M., Buchwald, S.L. *JACS* **125**, 11253 (2003).
[3]Czekelius, C., Carreira, E.M. *ACIEE* **42**, 4793 (2003).

Palladium(II) complexes

Allylation. The well-known allylation method has been applied to unsymmetrical 1,3-diketones[1] in the presence of (*R*)-BINAP-Pd catalyst. The allylation using allyl alcohol and *rac*-BINAP ligand is actively promoted by Et_3B.[2]

Allylic substitution. Regioselective substitution is hydrogen bond-directed. 5-Vinyloxazolidin-2-ones are attacked at the terminal carbon atom by malonate, amine, sulfonamide nucleophiles, but succinimide chooses a S_N2 pathway due to hydrogen bonding in the transition state.[3]

$[(C_3H_5)PdCl]_2$ / (*R*)-BINAP / THF; 81%

The scope of alkylative ring opening of oxabridged systems with R_2Zn has been investigated.[4] Ring expansion of vinylaziridines to give 4-vinylimidazolidin-2-ones on reaction with phenyl isocyanate, as catalyzed by Pd complex and promoted by $CeCl_3$, involves π-allylpalladium species.[5]

The acetoxy group of *N*-diphenylmethylene-α-acetoxyglycine esters is allylic and therefore subject to Pd-mediated ionization. Dimers of *N*-diphenylmethyleneglycine esters are formed when the substituted and unsubstitued esters are mixed together with NaH, $Pd(OAc)_2$ and BINAP. The *dl*-diastereoisomers are favored.[6]

Stilbine analogues of BINAPs are found to be useful surrogate ligands in the Pd-catalyzed allylic substitution.[7]

Addition reactions. Chiral β-amino acid derivatives are obtained by conjugate addition of $ArNH_2$ to *N*-alkenyoylcarbamates.[8]

Allenes undergo silaboration via double asymmetric induction. Surprisingly, the monodentate 1-[(naphthyl)naphthyl]diphenylphosphine is the most effective ligand.[9]

B–$SiMe_2Ph$ + allene; $(C_3H_5)PdCp$, PPh_2 ligand; B(OR*), $SiMe_2Ph$; 92% 86% de

Coupling reactions. Creation of quaternary carbon centers bearing two aryl substitutents is accomplishable by using an intramolecular Heck reaction.[10] A synthesis of tetrahydroquinolines spiroannulated at C-3 via an ene-type reaction can be made enantioselective.[11]

N-Arylation has been extended, e.g., to the preparation of tricyclic azacycles,[12] *N*-arylimines in which the imine source can be a fluorous benzophenone derivative,[13] *N*-trialkylsilylimines,[14] or sulfoximines[15] (also the analogous *N*-vinylation[16]). *N*-Arylation of aziridines has been developed.[17]

[1]Kuwano, R., Uchida, K., Ito, Y. *OL* **5**, 2177 (2003).
[2]Kimura, M., Mukai, R., Tanigawa, N., Tanaka, S., Tamaru, Y. *T* **59**, 7767 (2003).
[3]Cook, G.R., Yu, H., Sankaranarayanan, S., Shanker, P.S. *JACS* **125**, 5115 (2003).
[4]Lautens, M., Hiebert, S. *JACS* **126**, 1437 (2004).
[5]Dong, C., Alper, H. *TA* **15**, 1537 (2004).
[6]Chen, Y., Yudin, A.K. *TL* **44**, 4865 (2003).
[7]Yasuike, S., Okajima, S., Yamaguchi, K., Kurita, J. *TL* **44**, 6217 (2003).
[8]Li, K., Cheng, X., Hii, K.K. *EJOC* 959 (2004).
[9]Suginome, M., Ohmura, T., Miyake, Y., Mitani, S., Ito, Y., Murakami, M. *JACS* **125**, 11174 (2003).
[10]Dounay, A.B., Hatanaka, K., Kodanko, J.J., Oestreich, M., Overman, L.E., Pfeifer, L.A., Weiss, M.W. *JACS* **125**, 6261 (2003).
[11]Hatano, M., Mikami, K. *JACS* **125**, 4704 (2003).
[12]Loones, K.T.J., Maes, B.U.W., Dommisse, R.A., Lemiere, G.L.F. *CC* 2466 (2004).
[13]Cioffi, C.L., Berlin, M.L., Herr, R.J. *SL* 841 (2004).
[14]Barluenga, J., Aznar, F., Valdes, C. *ACIEE* **43**, 343 (2004).
[15]Harmata, M., Hong, X., Ghosh, S.K. *TL* **45**, 5233 (2004).
[16]Dehli, J.R., Bolm, C. *JOC* **69**, 8518 (2004).
[17]Sasaki, M., Dalili, S., Yudin, A.K. *JOC* **68**, 2045 (2003).

Rhodium complexes

Hydrogenation.[1] A cationic Rh-BINAP complex that contains also a cod ligand has been used in the enantioselective hydrogenation of butenolides.

Hydroboration. Hydroboration of cyclopropenes shows directing effect of an oxygenated functionality (e.g., ester, $ROCH_2$) at C-3. This effect is critical to achieving high degrees of enantioselectivity.[2]

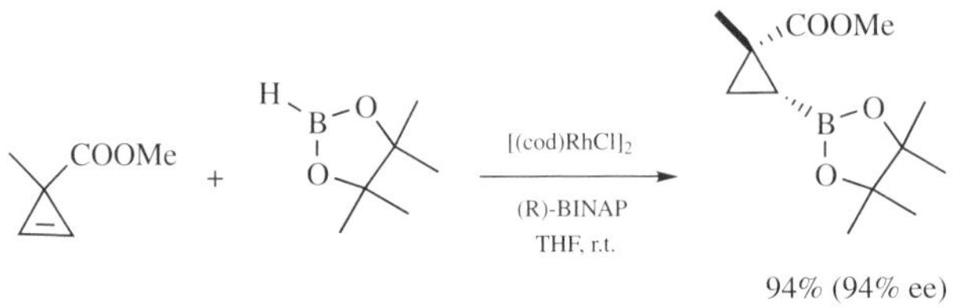

COOMe
+
[(cod)RhCl]2
(R)-BINAP
THF, r.t.
COOMe
94% (94% ee)

Perfluoroalkylethenes give secondary alcohols via hydroboration due to polar control. With a cationic Rh-BINAP catalyst, products with moderate ee have been obtained.[3]

Conjugate additions. Aryl transfer from $[ArBF_3]K$ to α,β-unsaturated esters and amides,[4,5] from $ArSi(OMe)_3$ to enones,[6] and reaction of alkenylzirconium reagents with enones[7] are subjected to chiral induction.

Cyclizations. An asymmetric Pauson-Khand reaction[8] shows stereoselectivity dependent of whether the catalyst is made up of $[Rh(CO)_2Cl]_2$ or $[Ir(cod)Cl]_2$. Such a reaction can be carried out in aqueous media in the presence of a surfactant.[9]

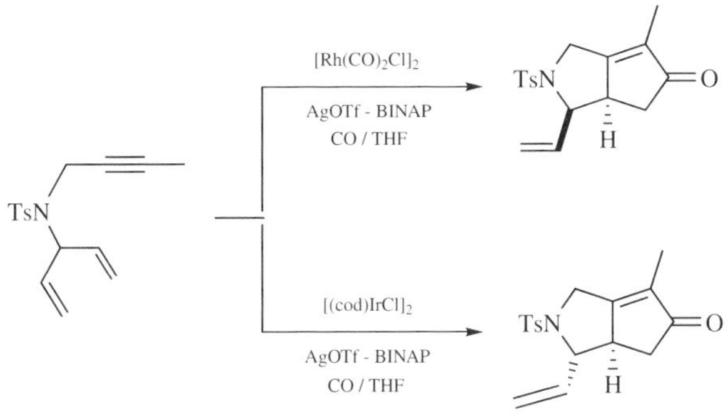

TsN
[Rh(CO)2Cl]2
AgOTf - BINAP
CO / THF
TsN
O
H
[(cod)IrCl]2
AgOTf - BINAP
CO / THF
TsN
O
H

By means of an asymmetric intramolecular ene reaction enynes are kinetically resolved.[10] The cationic $[Rh(binap)]BF_4$ salt promotes 5- and 6-alkynals to cyclize (intramolecular hydroacylation). At room temperature the products are 2-alkylidenecycloalkanones, but at 80°, 2-cycloalkenones are produced.[11]

[2 + 2 + 2]Cycloaddition. A Rh-octahydro-BINAP salt catalyzes a cross- trimerization of alkynes that can be used to synthesize axially chiral phthalides.[12]

O
O
Ar
+
OR
[Rh(S)-BINAP(H8)]BF4
CH2Cl2
O
O
OR
Ar

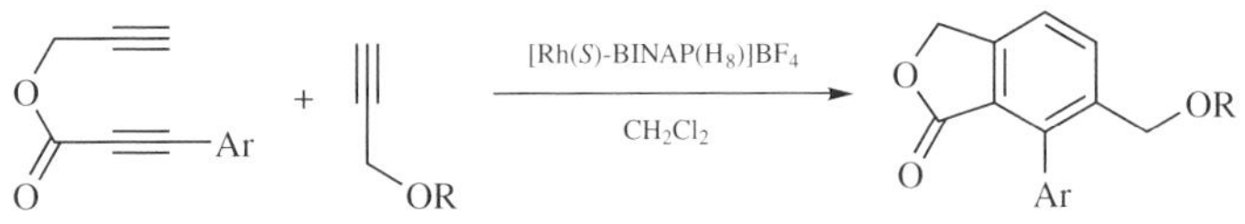

[1]Donate, P.M., Frederico, D., da Silva, R., Constantino, M.G., Del Ponte, G., Bonatto, P.S. *TA* **14**, 3253 (2003).
[2]Rubina, M., Rubin, M., Gevorgyan, V. *JACS* **125**, 7198 (2003).
[3]Segarra, A.M., Claver, C., Fernandez, E. *CC* 4649 (2004).
[4]Navarre, L., Darses, S., Genet, J.-P. *ACIEE* **43**, 719 (2004).
[5]Pucheault, M., Michaut, V., Darses, S., Genet, J.-P. *TL* **45**, 4729 (2004).
[6]Oi, S., Taira, A., Honma, Y., Inoue, Y. *OL* **5**, 97 (2003).
[7]Oi, S., Sato, T., Inoue, Y. *TL* **45**, 5051 (2004).
[8]Jeong, N., Kim, D.H., Choi, J.H. *CC* 1134 (2004).
[9]Suh, W.H., Choi, M., Lee, S.I., Chung, Y.K. *S* 2169 (2003).
[10]Lei, A., He, M., Zhang, X. *JACS* **125**, 11472 (2003).
[11]Takeishi, K., Sugishima, K., Sasaki, K., Tanaka, K. *CEJ* **10**, 5681 (2004).
[12]Tanaka, K., Nishida, G., Wada, A., Noguchi, K. *ACIEE* **43**, 6510 (2004).

Ruthenium(II) complexes

Asymmetric hydrogenation. Good results are obtained in ionic liquids.[1] Increased enantioselectivity for hydrogenation of β-aryl ketoesters[2] with a catalyst in which the BINAP ligand contains bulky substituents at C-4 and C-4′.is due to inhibition of π-π interactions. On the other hand, a ligand bearing diphosphonic acid residues at the 4,4′-positions possesses anchoring sites to Zr, providing another version of hydrogenation catalyst for aromatic ketones.[3]

anti-β-Hydroxy-α-amino acid derivatives are synthesized from the corresponding β-keto compounds via dynamic kinetic resolution during asymmetric hydrogenation.[4]

Reductive amination. Chiral amines result from reaction of ketones with NH_3-$HCOONH_4$ in the presence of the Ru complex of Tol-BINAP.[5]

[1]Nigo, H.L., Hu, A., Lin, W. *CC* 1912 (2003).
[2]Hu, A., Ngo, H.L., Lin, W. *ACIEE* **43**, 2501 (2004).
[3]Hu, A., Ngo, H.L., Lin, W. *JACS* **125**, 11490 (2003).
[4]Makino, K., Goto, T., Hiroki, Y., Hamada, Y. *ACIEE* **43**, 882 (2004).
[5]Kadyrov, R., Riermeier, T.H. *ACIEE* **42**, 5472 (2003).

Silver(I) complexes

Protonation. Silyl enolates give chiral ketones on contact with a Ag-BINAP complex in CH_2Cl_2-MeOH.[1]

Allylation. A catalytic system that promotes asymmetric allyl transfer from allyltrimethoxysilane to aldehydes consists of AgOTf-BINAP [Lewis acid] and KF/18-crown-6 [Lewis base].[2]

Nitroso-aldol reactions. The uncatalyzed reaction of tin enolates with PhN=O forms products with a C-N bond. With Ag-BINAP catalyst, α-aminooxylation occurs.[3] However, there are three Ag-BINAP complexes and the one derived from 2AgOTf and one BINAP molecule turns the chemoselectivity in favoring C-N bond formation.[4]

OSnBu3 + PhN=O —BINAP - AgOTf, EtO⁀OEt, −78°→ (2-oxocyclohexyl)N(OH)Ph >99% ee

OSnMe3 + PhN=O —BINAP - AgOTf, THF −78°→ 2-(ONHPh)cyclohexanone 88% (99% ee)

[1]Yanagisawa, A., Touge, T., Arai, T. *ACIEE* **44**, 1546 (2005).
[2]Wadamoto, M., Ozasa, N., Yanagisawa, A., Yamamoto, H. *JOC* **68**, 5593 (2003).
[3]Momiyama, N., Yamamoto, H. *JACS* **125**, 6038 (2003).
[4]Momiyama, N., Yamamoto, H. *JACS* **126**, 5360 (2003).

Bis(iodomethyl)zinc. 21, 57–58; **22**, 60–61

Methylenation. Homologation with defunctionalization results when α-sulfonylalkylcopper species [RCH(Cu)SOTol] are treated with $Zn(CH_2I)_2$. The products are $RCH{=}CH_2$.[1]

Cyclopropanation. Homoallylic amines derived from several chiral amino alcohols are cyclopropanated in a highly diastereoselective fashion.[2]

[1]Abramovitch, A., Varghese, J.P., Marek, I. *OL* **6**, 621 (2004).
[2]Aggarwal, V.K., Fang, G.Y., Meek, G. *OL* **5**, 4417 (2003).

Bis(2-methoxyethyl)aminosulfur trifluoride. 21, 58; **22**, 62

Amides. Activation of carboxylic acids with the reagent followed by addition of amines leads to amides.[1]

[1]White, J.M., Tunoori, A.R., Turunen, B.J., Georg, G.I. *JOC* **69**, 2573 (2004).

2,4-Bis(4-methoxyphenyl)-1,3,2,4-dithiadiphosphetane 2,4-disulfide, Lawesson's reagent.

Polythia-[n]-peristylanes. The *cis*-cyclobutane-1,2,3,4-tetracarbaldehyde acetals obtained from a cyclooctatetraene dimer is converted to the bowl-shaped tetrasulfide by Lawesson's reagent under ultrasound irradiation.[1] Similarly, trithia-[3]-peristylane is obtained from the ozonolysis product of bullvalene.

MeO MeO OMe O O OMe → S S S S

[1]Mehta, G., Gagliardini, V., Schaefer, C., Gleiter, R. *TL* **44**, 9313 (2003).
[2]Mehta, G., Gagliardini, V., Schaefer, C., Gleiter, R. *OL* **6**, 1617 (2004).

_O_-Bis(4-methoxyphenyl)phosphonylhydroxylamine.

Amination. Stabilized carbanions including those from phenylacetonitrile and alkyl phenylacetates pick up the NH_2 group from the reagent [$Ar_2P(O)ONH_2$].

[1]Smulik, J.A., Vedejs, E. *OL* **5**, 4187 (2003).

Bis(_N_-methylbenzthiazol-2-ylidene)palladium(II) iodide.

Heck reaction. Allylic alcohols and ArX are coupled by means of the Pd-carbene complex to afford carbonyl compounds.

[1]Calo, V., Nacci, A., Monopoli, A., Spinelli, M. *EJOC* 1382 (2003).

Bismuth. 20, 44; **21**, 58; **22**, 63

Homoallylic alcohols. By the ball milling technique (solvent-free conditions) Bi shot causes aldehydes and allyl bromides to condense.[1] Bi nanoparticles (esp. Bi nanotubes) show better reactivity than regular Bi powder in water.[2] The aqueous reaction is carried out in the presence of NH_4Cl[3] or KF.[4]

Triarylbismuthanes.[5] Difficultly accessible *ortho*-functionalized triarylbismuthanes are synthesized in a ball mill from ArI, Bi, Cu, CuI, and $CaCO_3$.

[1]Wada, S., Hayashi, N., Suzuki, H. *OBC* **1**, 2160 (2003).
[2]Xu, X., Zha, Z., Miao, Q., Wang, Z. *SL* 1171 (2004).
[3]Miyamoto, H., Daikawa, N., Tanaka, K. *TL* **44**, 6963 (2003).
[4]Smith, K., Lock, S., El-Hiti, G.A., Wata, M., Miyoshi, N. *OBC* **2**, 935 (2004).
[5]Urano, M., Wada, S., Suzuki, H. *CC* 1202 (2003).

Bismuth(III) bromide. 20, 44–45; **21**, 58; **22**, 63

Tetrahydropyrans.[1] Substituted tetrahydropyrans are formed from δ-siloxy carbonyl compounds by a reaction with nucleophiles in the presence of $BiBr_3$. Although the catalyst is a source of HBr the protonic acid itself is a poor catalyst for the observed transformation.

O H O OSiEt3 + SiMe3 —BiBr3 / MeCN→ H O O

[1]Evans, P.A., Cui, J., Gharpure, S.J., Hinkle, R.J. *JACS* **125**, 11456 (2003).

Bismuth(III) chloride. 15, 37; **18**, 52; **19**, 37; **20**, 45–46; **21**, 59; **22**, 63

Acylation and chlorosulfinylation. Friedel-Crafts acylation[1] and chlorosulfinylation (with $SOCl_2$)[2] on electron-rich aromatics is promoted by $BiCl_3$. In the former process catalyst can be quantitatively recovered on aqueous workup as BiOCl which is reactivated by RCOCl (or treatment with $SOCl_2$ before the next usage).

In situ generation of $BiCl_3$ from BiOCl is involved using the latter compound in acetylation[3] of phenols, anilines and benzenethiol, as well as aldol-Ritter reaction tandem that gives β-acetamido ketones from ketones, aldehydes, AcCl and MeCN.[4]

α-Aminonitriles.[5] Amines, aldehydes, and Me_3SiCN condense in MeCN at room temperature by $BiCl_3$ (11 examples, 81–91%).

1,2-Diamines.[6] Using $BiCl_3$ as catalyst aziridines react with amines to give 1,2-diamines.

[1]Repichet, S., Le Roux, C., Roques, N., Dubac, J. *TL* **44**, 2037 (2003).
[2]Peyronneau, M., Roques, N., Mazieres, S., Le Roux, C. *SL* 631 (2003).
[3]Ghosh, R., Maiti, S., Chakraborty, A. *TL* **45**, 6775 (2004).
[4]Ghosh, R., Maiti, S., Chakraborty, A. *SL* 115 (2005).
[5]De, S.K., Gibbs, R.A. *TL* **45**, 7407 (2004).
[6]Swamy, N.R., Venkateswarlu, Y. *SC* **33**, 547 (2003).

Bismuth(III) iodide.

Allylation.[1] Allylstannanes in which an oxygen functionality is present at the far end to enable formation of oxametallacycles as intermediates show stereoselectivity in the condensation with aldehydes. Notable is the reversal in that aspect when using halides of In and Bi.

PhCHO + Bu_3Sn ... OBn → (BiI_3; MeCN,CH_2Cl_2) Ph ... OBn (OH)

80%

[1]Donnelly, S., Thomas, E.J., Arnott, E.A. *CC* 1460 (2003).

Bismuth(III) nitrate. 20, 46–47; **22**, 64

3,4-Dihydropyrimidin-2(1H)-ones. The heterocyclic compounds are formed in one step from aldehydes (preferably ArCHO), β-keto esters, and urea (Biginelli reaction) using either $Bi(NO_3)_3$ [1] or $BiONO_3$ [2] as promoter.

Iodoarenes.[3] Room temperature, solvent-free conditions for rapid iodination of arenes [I_2- $Bi(NO_3)_3$] have been found.

β-Amino ketones.[4] Amines (including indole) and carbamate esters add to enones at room temperature when catalyzed by $Bi(NO_3)_3$.

[1]Khodaei, M.M., Khosropour, A.R., Beygzadeh, M. *SC* **34**, 1917 (2004).
[2]Reddy, Y.T., Rajitha, B., Reddy, P.N., Kumar, B.S., Rao, V.P. *SC* **34**, 3821 (2004).
[3]Alexander, V.M., Khandekar, A.C., Samant, S.D. *SL* 1895 (2003).
[4]Srivasatava, N., Banik, B.K. *JOC* **68**, 2109 (2003).

Bismuth(III) triflamide.

Friedel-Crafts acylation.[1] 1-Tetralones and heterocyclic analogues such as 4-chromanones and 4-thiochromanones are formed on heating 4-arylbutyric acids (and 3-arylheterapropanoic acids) with $Bi(NTf_2)_3$. But high temperatures are required.

[1]Cui, D.-M., Kawamura, M., Shimada, S., Hayashi, T., Tanaka, M. *TL* **44**, 4007 (2003).

Bismuth(III) triflate. 20, 47; **21**, 59; **22**, 64

Many trivial functional group transformations have been surveyed using $Bi(OTf)_3$ as catalyst: Formation and hydrolysis of THP ethers,[1] deprotection of dithioacetals[2] and MOM ethers,[3] Ferrier rearrangement,[4] epoxide opening[5,6] and Michael addition[7] with amines. Thiols and indoles are effective Michael donors under such conditions.[8] The simple Michael addition-elimination method for the preparation of enaminocarbonyl compounds[9] from β-dicarbonyl substances works as expected. Perhaps achieving the aromatic Claisen rearrangement in refluxing MeCN and Fries rearrangement in hot toluene is of some significance.[10] It represents some improvement in lowering the reaction temperatures. .

Homoallylic amines.[11] The attack of imines (formed in situ) by allylsilanes at room temperature to afford homoallylic amines, as catalyzed by $Bi(OTf)_3$ is a variant of such a well-developed method.

Friedel-Crafts acylation.[12] Carboxylic acids activated by heptafluorobutyric anhydride are induced to acylate arenes in the presence of $Bi(OTf)_3$.

[1]Stephens, J.R., Butler, P.L., Clow, C.H., Oswald, M.C., Smith, R.C., Mohan, R.S. *EJOC* 3827 (2003).
[2]Kamal, A., Reddy, P.S.M.M., Reddy, D.R. *TL* **44**, 2857 (2003).
[3]Reddy, S.V., Rao, R.J., Kumar, U.S., Rao, J.M. *CL* **32**, 1038 (2003).
[4]Banik, B.K., Adler, D., Nguyen, P., Srivastava, N. *H* **61**, 101 (2003).
[5]Ollevier, T., Lavie-Compin, G. *TL* **45**, 49 (2004).
[6]Khosropour, A.R., Khodaei, M.M., Ghozati, K. *CL* **33**, 304 (2004); *TL* **45**, 3525 (2004).
[7]Varala, R., Alam, M.M., Adapa, S.R. *SL* 720 (2003).
[8]Alam, M.M., Varala, R., Adapa, S.R. *TL* **44**, 5115 (2003).
[9]Khosropour, A.R., Khodaei, M.M., Kookhazadeh, M. *TL* **45**, 1725 (2004).
[10]Sridhar, B., Swapna, V., Sridhar, C. *SC* **34**, 1433 (2004).
[11]Ollevier, T., Ba, T. *TL* **44**, 9003 (2003).
[12]Matsushita, Y., Sugamoto, K., Matsui, T. *TL* **45**, 4723 (2004).

1,2-Bis(2-nitrophenyl)ethanediol.

Carbonyl protection. Aldehydes and ketones form acetals with the title reagent in the usual manner (PyHOTs-catalyzed). Photolysis (350 nm) regenerates the carbonyl compounds.

[1]Blanc, A., Bochet, C.G. *JOC* **68**, 1138 (2003).

Bis(pentamethylcyclopentadienylruthenium) *N,N'*-diisopropylacetamidinate.

Radical cyclization. The complex **1** with a wealkyl coordinating PF_6 anion catalyzes atom-transfer reaction such as cyclization of *N*-allyltrichloroacetamide to β-chloromethyl-α,α-dichloropyrrolidinone. Its catalytic activity is comparable to CuCl-bipy.

Cp, Ru, Cp, Ru, N, N, +, PF_6^-

[1]Motoyama, Y., Gondo, M., Masuda, S., Iwashita, Y., Nagashima, H. *CL* **33**, 442 (2004).

Bis(pyridine)iodonium tetrafluoroborate.

Oxidation. The title reagent can be used to oxidize alcohols via an ionic pathway (together with iodine). Under photolysis conditions secondary and tertiary cycloalkanols undergo ring cleavage to afford ω-iodoalkyl carbonyl compounds.

[1]Barluenga, J., Gonzalez-Bobes, F., Murguia, M.C., Ananthoju, S.R., Gonzalez, J.M. *CEJ* **10**, 4206 (2004).

Bis(trialkylphosphine)palladium. 22, 66

Coupling reactions. A simple route to conjugated amides involves coupling of alkenyl halides with $Me_3SiCONR_2$ by catalysis of $(t\text{-}Bu_3P)_2Pd$.[1] The same catalytic system also promotes a one-step indole synthesis from 2-chloroanilines and ketones.[2]

Cl, NH_2, +, O, OMe, $(t\text{-}Bu_3P)_2Pd$, K_3PO_4 - $MgSO_4$, HOAc, DMA, 140°C, N H, OMe, 60%

The rapid Stille coupling of alkynylstannanes with MeI by $(t\text{-}Bu_3P)_2Pd$-KF may find use in the synthesis of short-lived ^{11}C-labelled PET tracers with a 1-propynyl group.[3]

A stereoselective approach to enynes from 1,1-dibromoalkenes relies on two consecutive Negishi couplings: The first in replacing the (*E*)-bromine atom on reaction with an alkynylsilane and then with another organozinc reagent. The latter reaction employs $(t\text{-}Bu_3P)_2Pd$ which is crucial for stereoselectivity and high yields.[4]

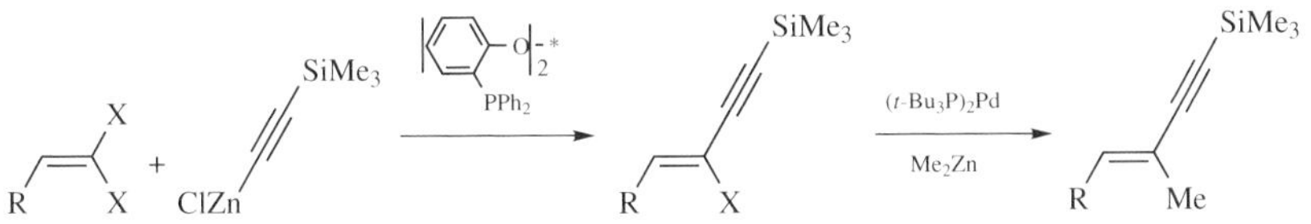

Allylic substitutions. Diesters of 2-substituted 2-butene-1,4-diols undergo Pd-catalyzed substitution at C-1.[5]

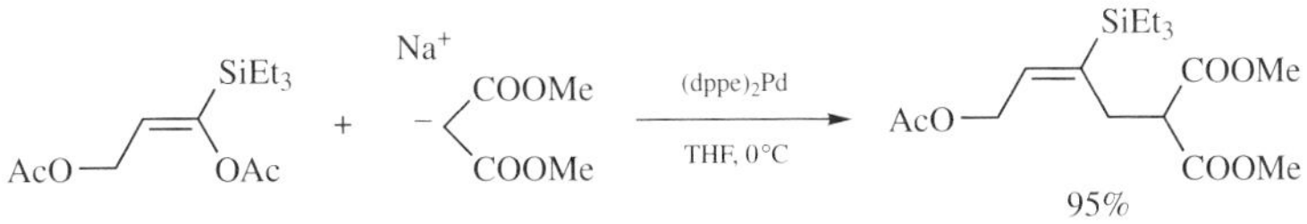

95%

[1]Cunico, R.F., Maity, B.C. *OL* **5**, 4947 (2003).
[2]Nazare, M., Schneider, C., Lindenschmidt, A., Will, D.W. *ACIEE* **43**, 4526 (2004).
[3]Hosoya, T., Wakao, M., Kondo, Y., Doi, H., Suzuki, M. *OBC* **2**, 24 (2004).
[4]Shi, J., Zeng, X., Negishi, E. *OL* **5**, 1825 (2003).
[5]Commandeur, C., Thorimbert, S., Malacria, M. *JOC* **68**, 5588 (2003).

Bis(tricarbonyldichlororuthenium). 22, 67

Vicinal additions. Arylethynes and alkyl propynoates react with sulfenamides in DMF to give alkene derivatives in which the amino group is located at the far terminus.[1]

Propargylic carboxylates undergo rearrangement and cycloaddition to alkenes.[2]

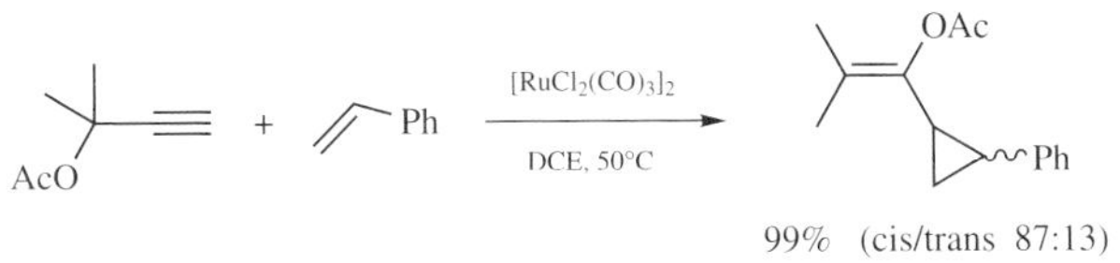

99% (cis/trans 87:13)

[1]Kondo, T., Baba, A., Nishi, Y., Mitsudo, T. *TL* **45**, 1469 (2004).
[2]Miki, K., Ohe, K., Uemura, S. *JOC* **68**, 8505 (2003).

Bis(trichloromethyl) carbonate.

Chlorocarbonylation. The reagent (commonly known as triphosgene) is a solid and much less hazardous than phosgene. It has found use in the preparation of oxime chloroformates[1] and carbamoyl chlorides.[2]

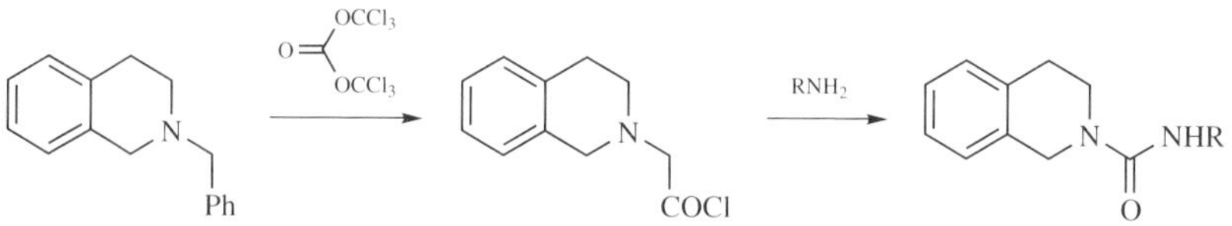

Rearrangement. A Polonovski-type rearrangement[3] of pyridine *N*-oxide occurs to give 2-chloropyridine, and picoline-*N*-oxide to 2-chloromethylpyridine.

[1]Paryzek, Z., Koenig, H. *SC* **33**, 3405 (2003).
[2]Lemoucheux, L., Rouden, J., Ibazizene, M., Sobrio, F., Lasne, M.-C. *JOC* **68**, 7289 (2003).
[3]Narendar, P., Gangadasu, B., Ramesh, C., Raju, B.C., Rao, V.J. *SC* **34**, 1097 (2004).

2-[3,5-Bis(trifluoromethyl)benzenesulfonyl]ethanol.

Carboxyl protection.[1] Esters of this alcohol are easily hydrolyzed, e.g., with $NaHCO_3$ in aq. acetone at room temperature.

[1]Alonso, D.A., Najera, C., Varea, M. *S* 277 (2003).

3,5-Bis(trifluoromethyl)phenylsulfonyl alkyl sulfones.

Julia olefination.[1] These sulfones are quite acidic, they react with aldehydes at room temperature in the presence of bases such as KOH.

[1]Alonso, D.A., Najera, C., Varea, M. *TL* **45**, 573 (2004).

Bis(2-trimethylsilylethanesulfonyl)imide.

Protected amines.[1] Using the Mitsunobu reaction ROH are converted into $RN(SO_2CH_2CH_2SiMe_3)_2$ with this reagent. One of the SES protecting groups can be removed with CsF in MeCN.

[1]Dastrup, D.M., Van Brunt, M.P., Weinreb, S.M. *JOC* **68**, 4112 (2003).

Bis(triphenylphosphine)palladium(II) succinimide.

Coupling reactions. The complex can be prepared from $(Ph_3P)_4Pd$ or $(dba)_3Pd_2 \cdot CHCl_3$ and Ph_3P with succinimide. It is stable in the air, and to light and moisture. With relatively low loading it acts as a catalyst for Suzuki coupling.[1] The halobis(triphenylphosphonine)palladium(II) succinimide complexes are useful for Stille coupling,[2] while showing a preferential selectivity for benzylic and allylic bromides[3] (different from that catalyzed by $(Ph_3P)_2PdBr_2$).

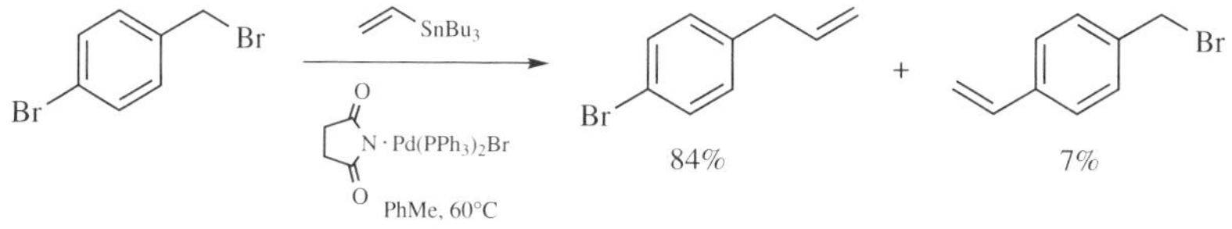

84%

7%

[1]Fairlamb, I.J.S., Kapdi, A.R., Lynam, J.M., Taylor, R.J.K., Whitwood, A.C. *T* **60**, 5711 (2004).
[2]Crawforth, C.M., Burling, S., Fairlamb, I.J.S., Taylor, R.J.K., Whitwood, A.C. *CC* 2194 (2003).
[3]Crawforth, C.M., Fairlamb, I.J.S., Taylor, R.J.K. *TL* **45**, 461 (2004).

9-Borabicyclo[3.3.1]nonane, 9-BBN. 14, 52–53; **15**, 43–44; **17**, 49–50; **20**, 52; **21**, 64; **22**, 68–69

Amino acid protection.[1] Formation of oxaborazolidinone derivatives of α-amino acids permits manipulation of *O*- and *N*-functionalities in the sidechain (e.g., in hydroxylysine).

3-Arylpyrroles.[2] The reduction of unsaturated γ-lactams leads to the substituted pyrroles.

Carboxylic acids.[3] *B*-Alkyl-9-oxa-10-borabicyclo[3.3.2]decanes that are derived from hydroboration of 1-alkenes with 9-BBN, followed by oxidation with Me_3NO react with Cl_2CHOMe by C-B bond insertion in the presence of *t*-BuOLi. Oxidative workup produces carboxylic acids.

N-Alkylation.[4] Upon heterocyclization on treatment with 9-BBN, amino alcohols are readily alkylated with *t*-BuOK and alkylating agents.

[1]Syed, B.M., Gustafsson, T., Kihlberg, J. *T* **60**, 5571 (2004).
[2]Verniest, G., De Kimpe, N. *SL* 2013 (2003).
[3]Soderquist, J.A., Martinez, J., Oyola, Y., Kock, I. *TL* **45**, 5541 (2004).
[4]Bar-Haim, G., Kol, M. *OL* **6**, 3549 (2004).

Borane-amines. 13, 42; **18**, 58; **19**, 43–44; **20**, 53; **21**, 65; **22**, 69

Reductive amination.[1] α-Picoline-borane is another reagent for reductive amination of carbonyl compounds. The reaction can be performed in neat, in H_2O-HOAc, or in MeOH.

Hydrazides.[2] Reduction of hydrazones followed by heating with RCOOH leads to $RCON(CH_2Ar)NMe_2$.

C–O bond cleavage.[3] Benzylidene acetals open on exposure to $Me_3N{\cdot}BH_3$ and $AlCl_3$. Presence of water improves the reaction.

Asymmetric reduction.[4] The CBS-reduction of ketones using borane complex with *N*-*t*-butyl-*N*-trimethylsilylamine as borane source is said to require only neutral aqueous workup.

[1]Sato, S., Sakamoto, T., Miyazawa, E., Kikugawa, Y. *T* **60**, 7899 (2004).
[2]Perdicchia, D., Licandro, E., Maiorana, S., Baldoli, C., Giannini, C. *T* **59**, 7733 (2003).
[3]Sherman, A.A., Mironov, Y.V., Yudina, O.N., Nifantiev, N.E. *CR* **338**, 697 (2003).
[4]Huertas, R.E., Corella, J.A., Soderquist, J.A. *TL* **44**, 4435 (2003).

Borane-tetrahydrofuran. 22, 70

Cleavage of cyclobutenes. 1,2-Disubstituted cyclobutenes undergo rearrangement during hydroboration to afford borolanes which on standard workup give rise to *anti*-1,4-diols.[1]

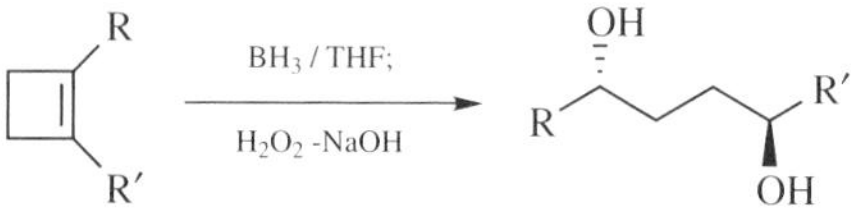

Reduction.[2] A practical asymmetric reduction of ketones uses borane and chiral lactam alcohols such as **1**.

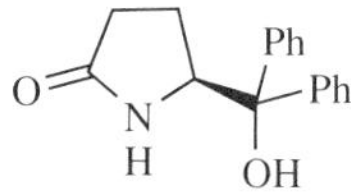

1

[1]Knapp, K.M., Goldfuss, B., Knochel, P. *CEJ* **9**, 5259 (2003).
[2]Kawanami, Y., Murao, S., Ohga, T., Kobayashi, N. *T* **54**, 8411 (2003).

Boric acid.

Esterification. Selective esterification of α-hydroxy carboxylic acids takes place using H_3BO_3 as catalyst because the derived cyclic boronates are reactive.[1] Other carboxyl groups including that of β-hydroxy carboxylic acids are not affected.

Dihydropyrimidinones. The 3-component condensation of β-dicarbonyl compounds, aldehydes and urea is known as the Biginelli reaction. Boric acid is an effective catalyst.[2]

[1]Houston, T.A., Wilkinson, B.L., Blanchfield, J.T. *OL* **6**, 679 (2004).
[2]Tu, S., Fang, F., Miao, C., Jiang, H., Feng, Y., Shi, D., Wang, X. *TL* **44**, 6153 (2003).

Boron tribromide. 13, 43; **14**, 53–54; **18**, 59; **19**, 45; **20**, 54; **21**, 65; **22**, 70

Dealkylation. Besides the well-known application to ether cleavage, BBr_3 is capable of *N*-debenzylation.[1] Aryl *t*-butyl sulfides can be converted to ArSAc using BBr_3 and AcCl thereby rendering such derivatives of ArSH more versatile synthetically.[2]

Rearrangements. *N*-Allyl-α-amino amides form chelates with BBr_3 and a subsequent base treatment effects [2,3]sigmatropic rearrangement to deliver α-allylglycinamides.[3] An assembly of (*Z*)-1-halo-1,4-enynes[4] from 1-alkynes and propargylic alcohols is mediated by BBr_3 or BCl_3.

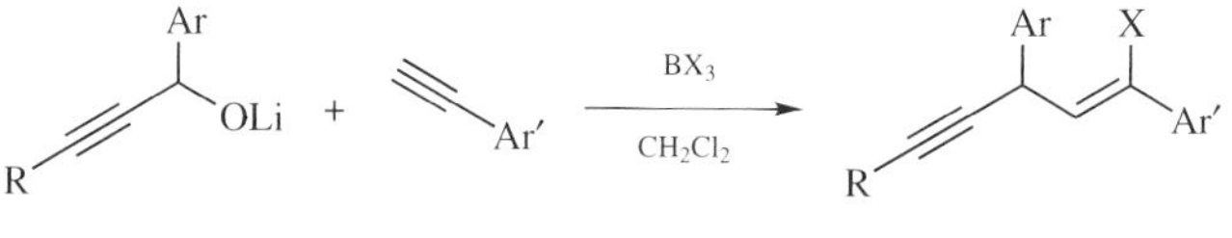

[1]Paliakov, E., Strekowski, L. *TL* **45**, 4093 (2004).
[2]Stuhr-Hansen, N. *SC* **33**, 641 (2003).
[3]Blid, J., Brandt, P., Somfai, P. *JOC* **69**, 3043 (2004).
[4]Kabalka, G.W., Wu, Z., Ju, Y. *OL* **6**, 3929 (2004).

Boron trichloride. **13**, 43; **14**, 54; **15**, 44; **18**, 59–60; **19**, 45–46; **20**, 54–55; **21**, 66; **22**, 70–71

Deprotection. Efficient and selective removal of the *t*-butyl group from protected arenesulfonamides is by treatment with BCl_3.[1]

[1]Wan, Y., Wu, X., Kannan, M.A., Alterman, M. *TL* **44**, 4523 (2003).

Boron trifluoride etherate. **13**, 43–47; **14**, 54–56; **15**, 45–47; **16**, 44–47; **17**, 52–53; **18**, 60–63; **19**, 46–48; **20**, 55–57; **21**, 66–70; **22**, 71–75

Functional group transformations. β-Keto amides are prepared from the esters via dioxaborinanes (formation on adding 1 equiv. of BF_3 $.OEt_2$).[1] Heating nitriles with ROH and BF_3 $.OEt_2$ leads to esters.[2] *N*-Alkoxymethylation of secondary amides is accomplished with $CH_2(OR)_2$ in the presence of $BF_3 \cdot OEt_2$.[3] On a catalyzed reaction of sugars that contain acetoxy groups a C-1 and C-2 with thiourea substitution at C-1 occurs.[4] Subsequent *S*-alkylation gives thioglycosides.

Dehydration of bromohydrins which contain a tertiary hydroxyl group gives allylic bromides.[5] A method for the preparation of benzylic, allylic and propargylic iodides from the corresponding alcohols involves stirring with CsI and $BF_3 \cdot OEt_2$ in MeCN.[6]

Opening of 3-membered heterocycles. In the catalyzed cleavage of 2-(1-aminoalkyl)aziridines there is regiochemical difference in using alcohols and carboxylic acids as nucleophiles.[7]

1
i-PrOH; $NaHCO_3$, H_2O — 83%
BF_3-OEt_2
HOAc, MeCN; $NaHCO_3$, H_2O — 63%

A properly situated π-nucleophile participates in ring opening of an aziridine when the latter is activated.[8] Monocyclic, fused and bridged bicyclic structures can be formed.

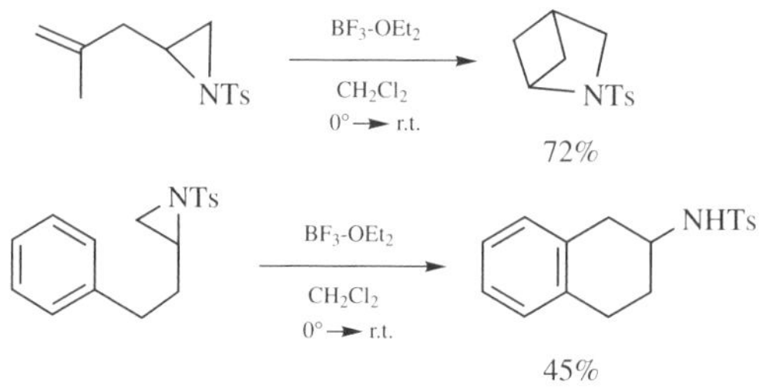

BF3-OEt2, CH2Cl2, 0° → r.t.; 72%

BF3-OEt2, CH2Cl2, 0° → r.t.; 45%

The configuration of C-2 in chiral 1,2-epoxy-3-aminoalkanes is retained on their transformation into acetonides of 3-amino-1,2-alkanediols on the catalyzed reaction with acetone.[9]

Upon treatment with $BF_3{\cdot}OEt_2$, epoxides bearing an allylsilane sidechain cyclize.[10] A catalyzed rearrangement to aldehydes intervenes prior to the desilylative ring closure.

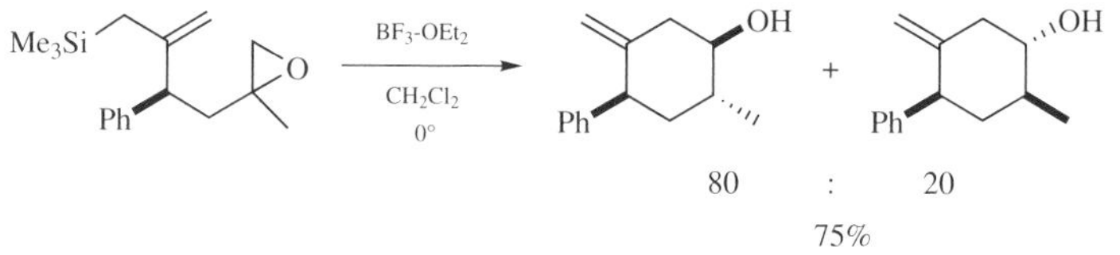

BF3-OEt2, CH2Cl2, 0°; 80 : 20; 75%

Baylis-Hillman reaction. 2-Alkylthiophenyl vinyl ketones undergo α-alkoxyalkylation with $RCH(OMe)_2$ [R=Ph, OMe] that is promoted by $BF_3{\cdot}OEt_2$.[11] The *S*-atom activates such ketones while the Lewis acid makes the acetals electrophilic.

Alkylations. Using alkyl fluorides as electrophiles reaction readily occurs with silyl enolates, allylsilanes, and hydrosilanes in the presence of $BF_3{\cdot}OEt_2$.[12]

Alkynyltrifluoroborate reagents generated from alkynyllithiums and BF_3 $.OEt_2$ attack lactones to give ω-hydroxyalkyl alkynyl ketones.[13] Homoallylamines are obtained when allyltrifluoroborate and *N*-tosylimines are treated with $BF_3{\cdot}OEt_2$. With crotyltrifluoroborates the reaction proceeds with high diastereoselectivity.[14]

A vinylogous Mannich-type reaction between *N*-arylimines and vinyl epoxides is promoted by $BF_3{\cdot}OEt_2$.[15] It involves rearrangement of the epoxides to dienol species to attack the polarized imines. For activation and stabilization of the imines the aryl substituent is installed with an *o*-CF_3 group.

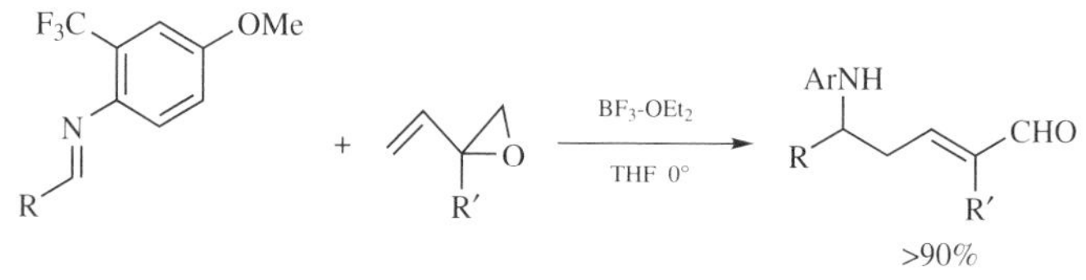

BF3-OEt2, THF 0°; >90%

N-Alkylation involving paraformaldehyde and potassium organotrifluoroborates[16] is complementary to the Mannich reaction.

1-Aryltetralins with structures related to lignans are elaborated by allylation of aldehydes with cyclic allylsiloxanes and intramolecular Friedel-Crafts alkylation in tandem, both reactions are catalyzed by $BF_3 \cdot OEt_2$.[17] However, success is limited to reaction partners bearing hydroxyl and/or alkoxy substiuents in the aromatic rings.

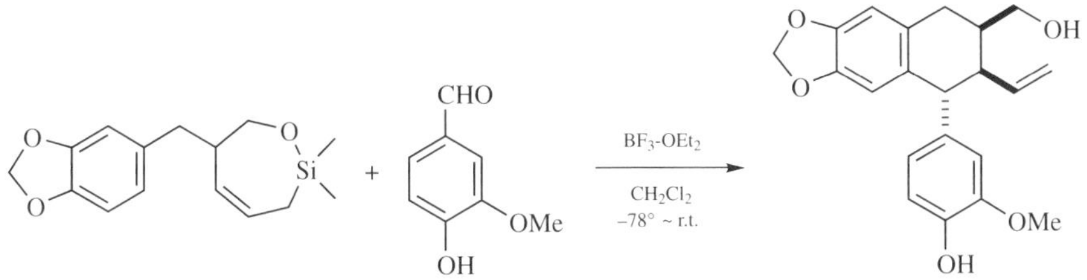

When catalyzed by $BF_3 \cdot OEt_2$ the allylation of phenacyl bromides with allystannanes proceeds with rearrangement.[18]

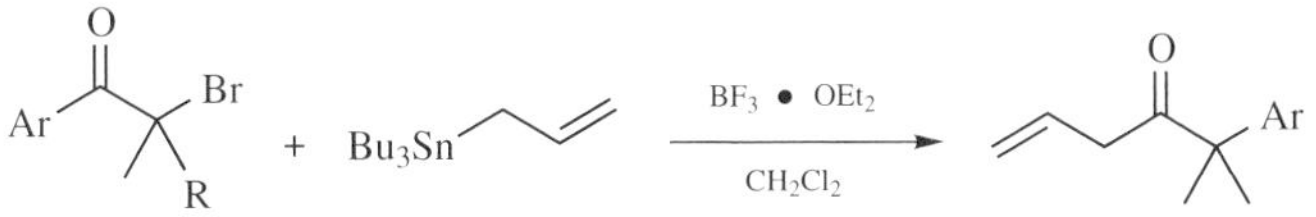

Ring formation. A tandem reaction involving 3,4-disilyl-1-butene and aldehydes gives rise to trisubstituted tetrahydrofurans,[19] whereas 3-hydroxyprolines are formed by a catalyzed reaction of benzyl diazoacetate with *N*-protected β-amino aldehydes,[20] and aziridines with a perfluoroalkyl group at C-2 from ArNHCH(OH)R^f [aziridination from a chiral diazoacetate shows very high diastereoselectivity.][21]

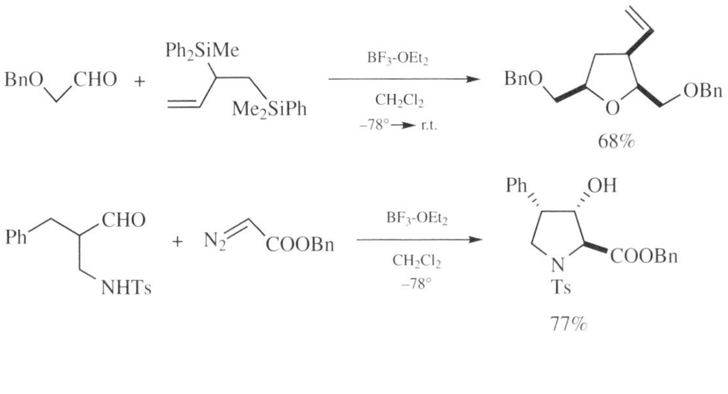

Cyclization of unsaturated *N*-chloramines in the presence of $BF_3 \cdot OEt_2$ proceeds via chloronium ions.[22] 2-Isocyanostyrenes react with aldehydes to afford substituted quinolines.[23]

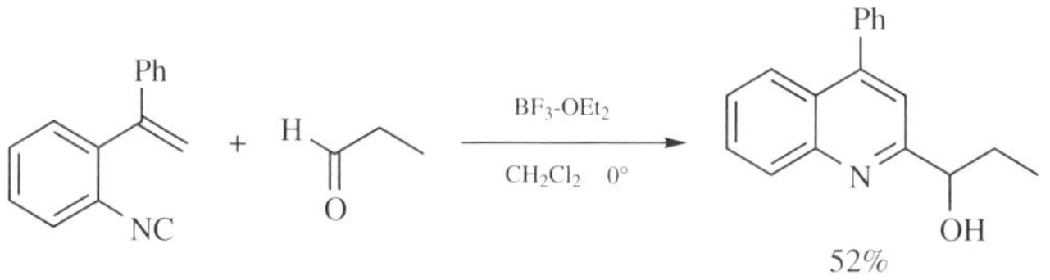

BF_3-OEt_2, CH_2Cl_2 0°

52%

Hydrazones and alkenes including cyclopentadiene undergo [3 + 2]cycloaddition leading to pyrazolidines.[24]

A synthesis of (-)-elisapterosin-B is completed on treatment of a benzoquinone precursor with $BF_3 \cdot OEt_2$. Intramolecular Diels-Alder reaction to give the methyl ether of (-)colombiasin-A occurs on heating the same substrate without any additive.[25]

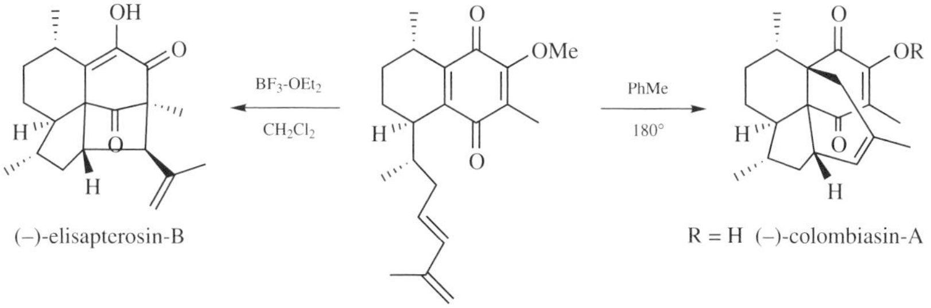

(–)-elisapterosin-B

R = H (–)-colombiasin-A

Bicyclic products are obtained from a Diels-Alder/ene reaction tandem of 1,3,7-alkatrienes and enals. With $BF_3 \cdot OEt_2$ as catalyst, it is completed at room temperature.[26]

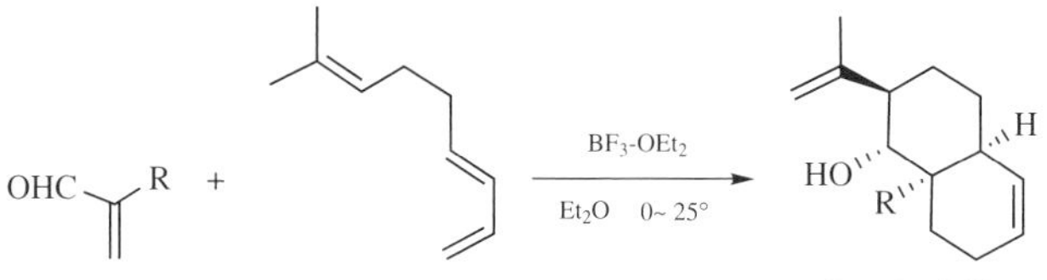

BF_3-OEt_2, Et_2O 0~ 25°

R = Br 77%

Coupling reactions. Suzuki coupling of $ArB(OH)_2$ with 1-aryltriazines to provide biaryls under Pd-catalysis is induced by $BF_3 \cdot OEt_2$.[27] Preparation of α-amino amides can be accomplished by carbamoylation of aldimines,[28] with $BF_3.OEt_2$ serving a dual role: activiating the aldimines and furnishing $[F^-]$ to remove the silyl group from the reagents $Me_3SiCONR_2$.

Si-*Deanisylation.* For replacement of the aromatic residue in a *p*-anisyldimethylsilyl group by a fluorine atom, $BF_3 \cdot OEt_2$ and 2 equiv. of HOAc can be used instead of the more expensive $BF_3{\cdot}2HOAc$ complex.[29] The exchange is precedent to transformation of the silyl moiety into a hydroxyl function.

[1]Stefane, B., Polanc, S. *SL* 698 (2004).
[2]Jayachitra, G., Yasmeen, N., Rao, K.S., Ralte, S.L., Srinivasan, R., Singh, A.K. *SC* **33**, 3461 (2003).
[3]Szmigielski, R., Danikiewicz, W. *SL* 372 (2003).
[4]Ibatullin, F.M., Shabalin, K.A., Janis, J.V., Shavva, A.G. *TL* **44**, 7961 (2003).
[5]Bettadaiah, B.K., Srinivas P. *SC* **33**, 3615 (2003).
[6]Hayat, S., Atta-ur-Rahman, Khan, K.M., Choudhary, M.I., Maharvi, G.M., Zia-Ullah, Bayer, E. *SC* **33**, 2531 (2003).
[7]Concellon, J.M., Riego, E., Suarez, J.R. *JOC* **68**, 9242 (2003).
[8]Bergmeier, S.C., Katz, S.J., Huang, J., McPherson, H., Donoghue, P.J., Reed, D.D. *TL* **45**, 5011 (2004).
[9]Concellon, J.M., Suarez, J.R., Garcia-Granda, S., Diaz, M.R. *OL* **7**, 247 (2005).
[10]Barbero, A., Castreno, P., Pulido, F.J. *OL* **5**, 4045 (2003).
[11]Kinoshita, H., Osamura, T., Kinoshita, S., Iwamura, T., Watanabe, S., Kataoka, T., Tanabe, G., Muraoka, O. *JOC* **68**, 7532 (2003).
[12]Hirano, K., Fujita, K., Yorimitsu, H., Shinokubo, H., Oshima, K. *TL* **45**, 2555 (2004).
[13]Doubsky, J., Streinz, L., Leseticky, L., Koutek, B. *SL* 937 (2003).
[14]Li, S.-W., Ratey, R.A. *CC* 1382 (2004).
[15]Lautens, M., Tayama, E., Nguyen, D. *TL* **45**, 5131 (2004).
[16]Tremblay-Morin, J.-P., Raeppel, S., Gaudette, F. *TL* **45**, 3471 (2004).
[17]Miles, S.M., Marsden, S.P., Leatherbarrow, R.J., Coates, W.J. *CC* 2292 (2004).
[18]Miyake, H., Hirai, R., Nakajima, Y., Sasaki, M. *CL* 164 (2003).
[19]Sarkar, T.K., Haque, S.A., Basak, A. *ACIEE* **43**, 1417 (2004).
[20]Angle, S.R., Belanger, D.S. *JOC* **69**, 4361 (2004).
[21]Akiyama, T., Ogi, S., Fuchibe, K. *TL* **44**, 4011 (2003).
[22]Noack, M., Kalsow, S., Gottlich, R. *SL* 1110 (2004).
[23]Kobayashi, K., Takagoshi, K., Kondo, S., Morikawa, O., Konishi, H. *BCSJ* **77**, 553 (2004).
[24]Kobayashi, S., Hirabayashi, R., Shimizu, H., Ishitani, H., Yamashita, Y. *TL* **44**, 3351 (2003).
[25]Kim, A.I., Rychnovsky, S.D. *ACIEE* **42**, 1267 (2003).
[26]Kraus, G.A., Kim, J. *OL* **6**, 3115 (2004).
[27]Saeki, T., Son, E.-C., Tamao, K. *OL* **6**, 617 (2004).
[28]Chen, J., Cunico, R.F. *TL* **44**, 8025 (2003).
[29]Clive, D.L.J., Cheng, H., Gangopadhyay, P., Huang, X., Prabhudas, B. *T* **60**, 4205 (2004).

Bromine. 13, 47; **14**, 56–57; **15**, 47; **18**, 64; **19**, 48; **20**, 57–58; **21**, 70; **22**, 76

De-t-butylation.[1] *t*-Butyl sulfides are cleaved by bromine and the presence of AcCl converts them into RSAc.

Bromination. On treatment with Br_2, silylated 1,3-butadienes lose the silyl groups and are thereby converted into 1,1,4,4-tetrabromo-2-butenes.[2] Brominative fragmentation of Reformatsky reaction products of cyclic ketones occurs readily.[3]

OH COOEt → (Br_2 - K_2CO_3, $CHCl_3$ 0°) Br–(CH₂)₄–CO–CBr₂–COOEt

87%

A technique for selective bromination at the less reactive site of an unsymmetrical ketone (e.g., acetophenone) is by a transfer reaction to acetone.[4] In other words, dibrominated products submit the bromine atom from the doubly activated position to acetone.

Oxidation. Primary and secondary alcohols are oxidized to carbonyl compounds in excellent yields by bromine using *N-t*-butyl(*o*-nitrobenzenesulfenamide) as catalyst.[5]

[1]Blaszczyk, A., Elbing, M., Mayor, M. *OBC* **2**, 2722 (2004).
[2]Xi, Z., Liu, X., Lu, J., Bao, F., Fan, H., Li, Z., Takahashi, T. *JOC* **69**, 8547 (2004).
[3]Zhao, M.-X., Wang, M.-X., Yu, C.-Y., Huang, Z.-T., Fleet, G.W.J. *JOC* **69**, 997 (2004).
[4]Choi, H.Y., Chi, D.Y. *OL* **5**, 617 (2003).
[5]Matsuo, J., Kawana, A., Yamanaka, H., Mukaiyama, T. *CL* 182 (2003).

Bromine-dimethyl sulfide.

Protection-deprotection. Cleavage of TBS ethers[1] and THP ethers[2] in MeOH is catalyzed by Br_2-SMe_2. Acetalization and thioacetalization are also readily achieved.[3]

[1]Rani, S., Babu, J.L., Vankar, Y.D. *SC* **33**, 4043 (2003).
[2]Khan, A.T., Mondal, E., Borah, B.M., Ghosh, S. *EJOC* 4113 (2003).
[3]Khan, A.T., Mondal, E., Ghosh, S., Islam, S. *EJOC* 2002 (2004).

Bromine trifluoride. 19, 48; **21**, 71; **22**, 76

Bromodesulfurization. α,α-Difluoroalkanoic acid derivatives are readily prepared[1] from 2-alkyl-2-alkoxycarbonyl-1,3-dithianes on reaction with BrF_3. If the corresponding carboxylic acids are submitted to the reaction 1,1,1-trifluoroalkanes are produced.[2] 2-Substituted 1,3-dithianes give 1,1-difluoroalkanes.[3]

[1]Hagooly, A., Sasson, R., Rozen, S. *JOC* **68**, 8287 (2003).
[2]Sasson, R., Rozen, S. *T* **61**, 1083 (2005).
[3]Sasson, R., Hagooly, A., Rozen, S. *OL* **5**, 769 (2003).

β-Bromo-α-ethylthiocinnamonitriles.

3-Cyanoflavones.[1] The reagents are obtained feom condensation of ArCHO with $EtSCH_2CN$ followed by bromination. They lose the bromine atom on reaction with *o*-hydroxyaroic esters to give products which cyclize by treatment with $AlCl_3$.

[1]Lassagne, F., Pochat, F. *TL* **44**, 9283 (2003).

Bromopentacarbonylrhenium.

Alkenyl carboxylates.[1] 1-Alkynes add carboxylic acids in the anti-Markovnikov manner. The homogeneous catalyst is recoverable.

Cyclic carbonates.[2] Insertion of CO_2 into epoxides is accomplished by heating the components under pressure with $Re(CO)_5Br$ in the neat.

[1]Hua, R., Tian, X. *JOC* **69**, 5782 (2004).
[2]Jiang, J.-L., Gao, F., Hua, R., Qiu, X. *JOC* **70**, 381 (2005).

***N*-Bromosuccinimide.** **13**, 49; **14**, 57–58; **15**, 50–51; **16**, 49; **18**, 65–67; **19**, 50–51; **20**, 58–59; **22**, 77

Bromination. NH_4OAc catalyzes bromination of ketones[1] with NBS. Halogenation of deactivated arenes[2] by NXS (X=Cl, Br, I) in the presence of NH_4NO_3 has also been reported. A potent brominating system is NBS-BF_3-H_2O that converts pentafluorobenzene into bromopentafluorobenzene in 96% yield.[3] Analogously, NCS for chlorination is also effective (e.g., *m*-chloronitrobenzene, 69%),

Access to RBr from ROH via *O*-alkylisoureas by treatment with NBS avoids use of phosphorus reagents.[4]

Oxidations. Derivatization of oligonucleotides by oxidation of the corresponding phosphites can be carried out rapidly with NBS-DMSO.[5] Conditions are nonaqueous and nonbasic. Various kinds of glycols (including tartaric acid) are converted into 1,2-diketones by NBS in hot CCl_4.[6]

Oxidative deoximation[7] and conversion of *N*-tosyl-2-arylaziridines into α-tosylamino ketones[8] have been carried out with NBS in water in the presence of β-cyclodextrin.

Activation. A method for replacing the sulfenyl group of thioglycosides while exposing a C-2 hydroxyl consists of treating 2-acyloxythioglycosides with NBS.[9] The reaction conditions are so mild that TBS and benzylidene groups survive.

Aziridines and oxazolidin-2-ones are formed[10] when 2-aminoalkyl selenides and their *N*-Boc derivatives, respectively, are treated with NBS and then a base.

(*Z*)-Alkenes with substituents such as COOR and Ph undergo isomerization under free radical conditions (NBS/AIBN or Bz_2O_2).[11]

Heterocycles. A few interesting examples of bromoheterocyclization initiated by NBS pertain to formation of bicyclic aziridines,[12] pyrroles,[13] and pyridines.[14] The two latter types of products arise from intramolecular reaction of δ-amino dienyl carbonyl compounds.

HN Ph O —NBS / DME-H_2O→ Br N H Ph O

76%

Addition. Aminobromination of alkenes with NBS and $TsNH_2$ can give rise to different regioisomers, depending on what metal salts serve as catalysts.[15]

Ph —$TsNH_2$ / NBS / CH_2Cl_2 r.t.→ TsNH Ph Br + Br Ph TsNH

catalyst:		
CuI	>99	: 1
(salen)Mn(¢ó)	<1	: 99

Spirocycles are formed[16] by rearrangement immediately following the attack of Br^+ from NBS. The relief of ring strain provides the driving force.

[1]Tanemura, K., Suzuki, T., Nishida, Y., Satsumabayashi, K., Horaguchi, T. *CC* 470 (2004).
[2]Tanemura, K., Suzuki, T., Nishida, Y., Satsumabayashi, K., Horaguchi, T. *CL* 932 (2003).
[3]Prakash, G.K.S., Mathew, T., Hoole, D., Esteves, P.M., Wang, Q., Rasul, G., Olah, G.A. *JACS* **126**,15770 (2004).
[4]Li, Z., Crosignani, S., Linclau, B. *TL* **44**, 8143 (2003).
[5]Uzagare, M.C., Padiya, K.J., Salunkhe, M.M., Sanghvi, Y.S. *BMCL* **13**, 3537 (2003).
[6]Khurana, J.M., Kandpal, B.M. *TL* **44**, 4909 (2003).
[7]Reddy, M.S., Narender, M., Rao, K.R. *SC* **34**, 3875 (2004).
[8]Reddy, M.S., Narender, M., Rao, K.R. *TLC* **46**, 1299 (2005).
[9]Yu, H., Ensley, H.E. *TL* **44**, 9363 (2003).
[10]Miniejew, C., Outurquin, F., Pannecoucke, X. *OBC* **2**, 1575 (2004).
[11]Baag, M.M., Kar, A., Argade, N.P. *T* **59**, 6489 (2003).
[12]Sasaki, M., Yudin, A.K. *JACS* **125**, 14242 (2003).
[13]Agami, C., Dechoux, L., Hamon, L., Hebbe, S. *S* 859 (2003).
[14]Bagley, M.C., Glover, C., Merritt, E.A., Xiong, X. *SL* 811 (2003).
[15]Thakur, V.V., Talluri, S.K., Sudalai, A. *OL* **5**, 861 (2003).
[16]Hurley, P.B., Dake, G.R. *SL* 2131 (2003).

t-Butoxydiphenylsilylacetonitrile.

(Z)-2-Alkenitriles.[1] Highly (*Z*)-selective Peterson reaction is realized with this reagent.

[1]Kojima, S., Fukuzaki, T., Yamakawa, A., Murai, Y. *OL* **6**, 3917 (2004).

4-(*t*-Butyldimethylsiloxy)benzylidene dimethylacetal.

Diol protection.[1] An exchange reaction renders 1,2-diols to cyclic acetals. To cleave these derivatives K_2CO_3, hydroxylamine, or CsF can be employed.

[1]Kaburagi, Y., Osajima, H., Shimada, K., Tokuyama, H., Fukuyama, T. *TL* **45**, 3817 (2004).

(*t*-Butyldimethylsilyl)diphenylphosphine.

Silylation.[1] On activation with DEAD and PPTS the title reagent becomes a useful silylating agent for alcohols.

[1]Hayashi, M., Matsuura, Y., Watanabe, Y. *TL* **45**, 1409 (2004).

***t*-Butyl hydroperoxide-metal salts.** **19**, 51–53; **20**, 61–62; **21**, 76; **22**, 80–81

Oxidation. Allylic alcohols undergo either oxidation or epoxidation by *t*-BuOOH in the presence of silica modified by cobalt or vanadium phosphonate salts, respectively.[1] The former system is also highly effective in converting alkenes to conjugated ketones.[2] Oxidation catalyzed by a water-soluble Cu complex in the presence of Bu_4NCl can be performed without using organic solvents.[3]

The double bond of *o*-hydroxystyrenes is affected by reaction with *t*-BuOOH-$VO(acac)_2$. Benzyl ketones are formed which are readily converted into benzo[*b*]furans on contact with CF_3COOH.[4]

OH → OH O
$VO(acac)_2$
t-BuOOH
CH_2Cl_2 , 40¡ã
56%

2-Alkoxyglycals arise when glycosyl selenides are treated with *t*-BuOOH-$(i\text{-PrO})_4Ti$. Rapid elimination after oxidation to the selenoxides ensues.[5] Benzylic oxidation of *N*-heteroaromatic compounds by SeO_2 to aldehydes is improved by adding *t*-BuOOH (less carboxylic acids are produced).[6]

Coupling.[7] A preparation of cinnamic esters from Ar_3Sb and acrylic esters involves treatment with *t*-BuOOH (or H_2O_2), HOAc and Li_2PdCl_4. The oxidant is used to convert Ar_3Sb into $Ar_3Sb(OAc)_2$.

[1]Jurado-Gonzalez, M., Sullivan, A.C., Wilson, J.R.H. *TL* **45**, 4465 (2004).
[2]Jurado-Gonzalez, M., Sullivan, A.C., Wilson, J.R.H. *TL* **44**, 4283 (2003).
[3]Ferguson, G., Ajjou, A.N. *TL* **44**, 9139 (2003).
[4]Lattanzi, A., Senatore, A., Massa, A., Scettri, A. *JOC* **68**, 3691 (2003).
[5]Chambers, D.J., Evans, G.R., Fairbanks, A.J. *T* **60**, 8411 (2004).
[6]Tagawa, Y., Yamashita, K., Higuchi, Y., Goto, Y. *H* **60**, 953 (2003).
[7]Moiseev, D.V., Morugova, V.A., Gushchin, A.V. *TL* **44**, 3155 (2003).

Butyllithium. **13**, 56; **14**, 63–68; **15**, 59–61; **17**, 59–60; **18**, 74–77; **19**, 54–59; **20**, 62–65; **21**, 77–7; **22**, 81–84

Lithiation. A route to cyclohexenones by ring expansion of cyclobutenones involves reaction with lithiated conjugated sulfones.[1]

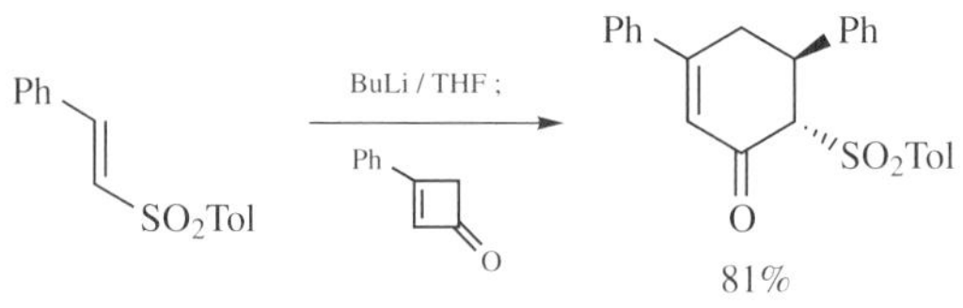

Ph, SO$_2$Tol; BuLi / THF ; Ph (cyclobutenone); Ph, Ph, SO$_2$Tol, O; 81%

1-Bromo-1-lithioethene is a practical reagent for the preparation of 2-bromo-1-alken-3-ols. It is generated by lithiation of bromoethene at $-110°$ with BuLi.[2] The presence of LiBr during metallation increases the yields of the alkylation products. Strict temperature control is critical.

Oligopyridines in which the heterocycles are bridged by methylene or methine groups are obtained readily.[3]

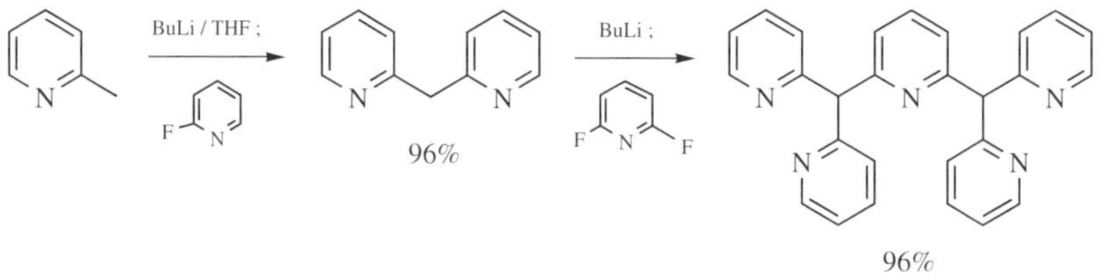

N; BuLi / THF ; F, N; N, N; 96%; BuLi ; F, N, F; N, N, N, N, N; 96%

Cyclization of 4-pentenamines to pyrrolidines is catalyzed by BuLi.[4]

2-benzenesulfonyl-1,3-cyclohexadienes are obtained by sequential alkylation of lithiated allyl phenyl sulfone with conjugated carbonyl compounds, acetylation, elimination, and electrocyclic reaction.[5]

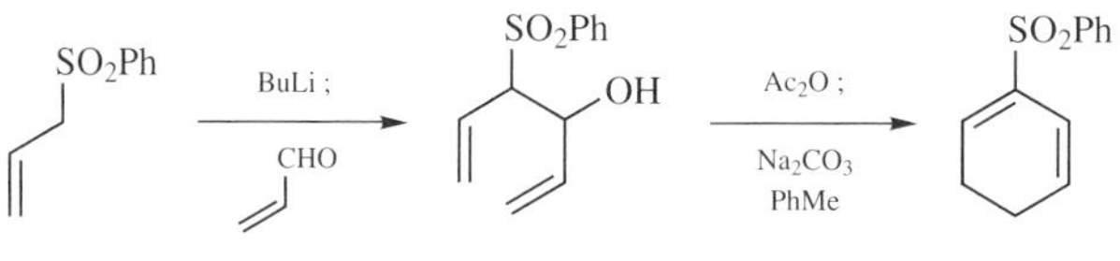

SO$_2$Ph; BuLi ; CHO; SO$_2$Ph, OH; Ac$_2$O ; Na$_2$CO$_3$, PhMe; SO$_2$Ph

A key step in an access to 5-allyl-4-aminofuran-2(5*H*)-ones, analogues of tetronic acid, is by the BuLi-induced [2,3] Wittig rearrangement.[6]

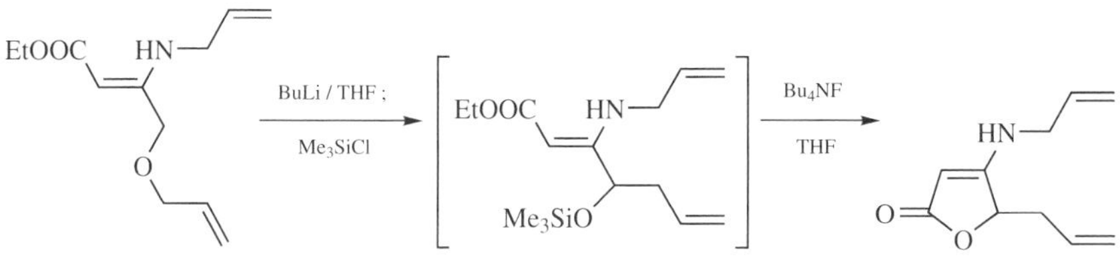

EtOOC, HN, O; BuLi / THF ; Me$_3$SiCl; [EtOOC, HN, Me$_3$SiO]; Bu$_4$NF, THF; HN, O, O

The deprotonated species of 3-(benzotriazol-2-ylmethyl)pyrrole derivatives react with enones to give adducts which are converted into indoles on treatment with sulfuric acid.[7]

Reaction of silanes. Deprotonation of α-silyl selenoamides and reaction with aldehydes furnishes conjugated selenoamides.[8] A Peterson reaction follows the alkylation.

Unstable silenes are generated by the elimination method.[9] Trapping by conjugated dienes results in silacyclohexenes which are useful precursors of 1,5-diols.

HO, Ph–Si(SiMe$_3$)–SiMe$_3$ + diene —BuLi / Et$_2$O ; LiBr→ Si–SiMe$_3$, Ph 50%

Dehalogenation. 3,5-Dibromo-4-hydroxypyridine loses one bromine atom through Br/Li exchange and protonation.[10] Benzyne formation from 3-fluoro-2-iodo-1-chlorobenzene shows strong solvent dependence. After I/Li exchange the elimination of LiCl is favored in THF but that of LiF, in pentane.[11]

3-Alkylindoles.[12] Butyllithiums (*n*-,*s*-, *t*-) add to *N*-Boc 2-vinylanilines in the presence of TMEDA to furnish benzylic organolithium compounds. Addition of DMF at this point before acid treatment leads to indoles.

[1]Magomedov, N.A., Ruggiero, P.L., Tang, Y. *JACS* **126**, 1624 (2004).
[2]Novikov, Y.Y., Sampson, P. *OL* **5**, 2263 (2003).
[3]Dyker, G., Muth, O. *EJOC* 4319 (2004).
[4]Ates, A., Quinet, C. *EJOC* 1623 (2003).
[5]Brandange, S., Leijonmarck, H. *CC* 292 (2004).
[6]Pevet, I., Meyer, C., Cossy, J. *SL* 663 (2003).
[7]Katritzky, A.R., Ledoux, S., Nair, S.K. *JOC* **68**, 5728 (2003).
[8]Murai, T., Fujishima, A., Iwamoto, C., Kato, S. *JOC* **68**, 7979 (2003).
[9]Berry, M.B., Griffiths, R.J., Sanganee, M.J., Steel, P.G., Whelligan, D.K. *TL* **44**, 9135 (2003).
[10]Meana, A., Rodriguez, J.F., Sanz-Tejedor, M.A., Garcia-Ruano, J.L. *SL* 1678 (2003).
[11]Coe, J.W., Wirtz, M.C., Bashore, C.G., Candler, J. *OL* **6**, 1589 (2004).
[12]Kessler, A., Coleman, C.M., Charoenying, P., O'Shea, D.F. *JOC* **69**, 7836 (2004).

Butyllithium – ***N,N,N'N'*-tetramethylethylenediamine (TMEDA). 19**, 60; **20**, 66–67; **21**, 80–81; **22**, vvv

Hydroamination. Cyclization of aminoethyltetralins involves formation of allyllithium species. Particularly interesting is the stereoselectivity attendant the reaction.

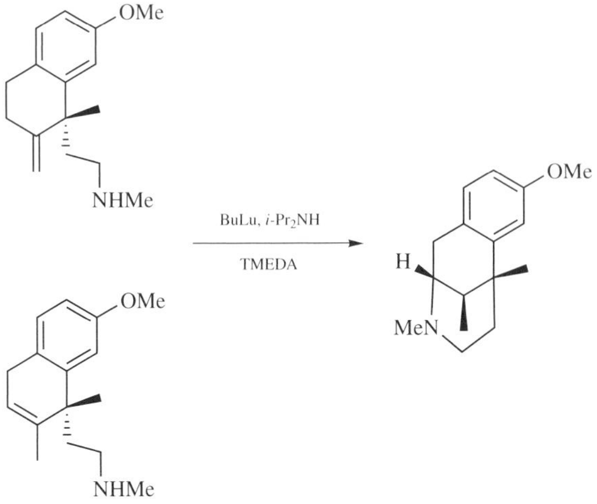

OMe

NHMe

BuLu, i-Pr_2NH

TMEDA

OMe

H

MeN

OMe

NHMe

[1]Trost, B.M., Tang, W. *JACS* **125**, 8744 (2003).

***s*-Butyllithium.** **14**, 69; **16**, 56; **18**, 77–79; **19**, 60–61; **20**, 67; **21**, 81–82; **22**, 85–86

Lithiation.[1] Organocuprates are generated by lithiation of propargylic amines with *s*-BuLi and subsequent metal exchange. These participate in a highly diastereoselective reaction with aldehydes.

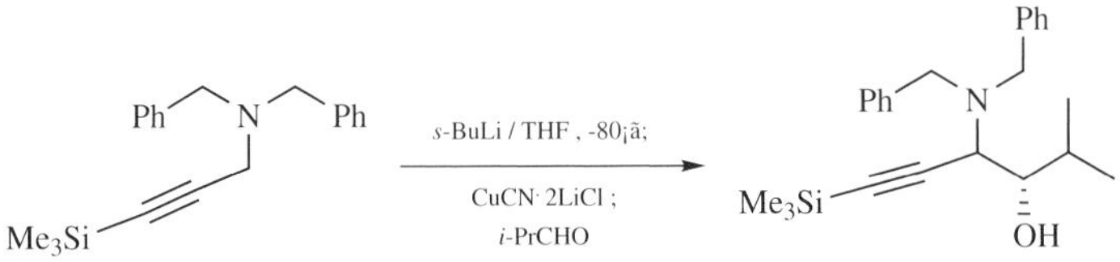

Ph N Ph

Me_3Si

s-BuLi / THF , -80¡ã;

CuCN· 2LiCl ;

i-PrCHO

Ph

Ph N

Me_3Si

OH

Isomerization.[2] Tetrasubstituted allene ethers are obtained from propargyl ethers via a deprotonation-alkylation sequence. The products are useful for the preparation of cyclopentenones.

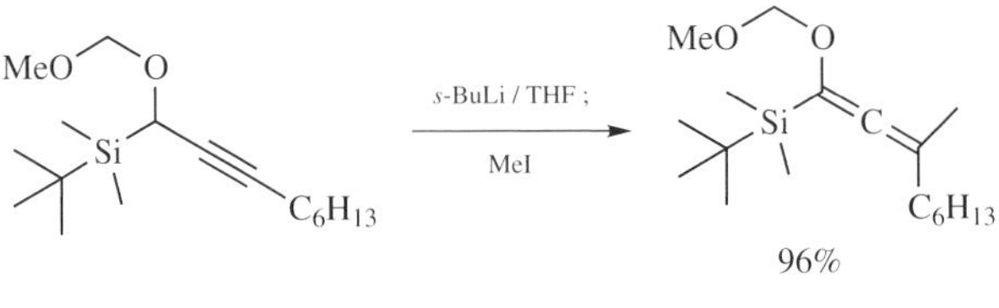

MeO O

Si

C_6H_{13}

s-BuLi / THF ;

MeI

MeO O

Si C

C_6H_{13}

96%

Peterson reaction. Stabilized silylcarbanions are generated by deprotonation with *s*-BuLi. Such anions are involved in the condensation of dimethylamino-

(trimethylsilyl)acetonitrile with carbonyl compounds to give α-cyanoenamines.[3] The method is superior to the Horner-Emmons method because of broader scope and higher yields.

In an analogous manner, conjugated esters containing a trimethylsilylmethyl substituent at the α-position are prepared from ethyl 2,3-bis(trimethylsilyl)propanoate.[4]

4-Aminoalkyl-1,2,3-cyclobutanetriones.[5] The reaction of dianions from *t*-butyl esters of *N*-Boc amino acids (generated with *s*-BuLi) with 3,4-diisopropoxy-3-cyclobutene-1,2-dione leads to replacement of one isopropoxy group. On treatment with conc. HCl the products lose the *t*-butoxycarbonyl and the isopropyl groups.

[1]Bernaud, F., Vrancken, E., Mangeney, P. *OL* **5**, 2567 (2003).
[2]Tokeshi, B.K., Tius, M.A. *S* 786 (2004).
[3]Adam, W., Ortega-Shulte, C.M. *SL* 414 (2003).
[4]Suzuki, H., Ohta, S., Kuroda, C. *SC* **34**, 1383 (2004).
[5]Ishida, T., Shinada, T., Ohfune, Y. *TL* **46**, 311 (2005).

s-Butyllithium-amine.

Benz[1,4]oxathiin-2-ones.[1] These substances can be made from 2-methylthiophenylcarbamates via *O->C* group transfer that is performed on treatment with *s*-BuLi-TMEDA. The products undergo cyclization on heating with HOAc.

Alkylation of epoxides.[2] Terminal epoxides are substituted via lithiation and reaction with electrophiles. The best additive is *N,N'*-dibutylbispidine that somehow suppresses formation of side products.

[1]Pradhan, T.K., Mukherjee, C., Kamila, S., De, A. *T* **60**, 5215 (2004).
[2]Hodgson, D.M., Reynolds, N.J., Coote, S.J. *OL* **6**, 4187 (2004).

s-Butyllithium -(-)-sparteine. **20**, 67; **22**, 86

α-Silyl amines.[1] *N*-Boc-*N*-silyl amines that can give (*S*)-α-silyl amines in excellent ee are those containing an *N*-allyl or *N*-propargyl group.

$RSiPh_2$ / Boc–N (allyl) → s-BuLi ; (-)-sparteine → NHBoc, $SiPh_2R$

>90% ee

Chiral 1,2-diamines. As 1-Boc-pyrrolidine forms chiral cuprates that can be alkenylated and allenylated at C-2,[2] it follows that imidazolidines with only one *N*-Boc substituent are alkylated at carbon next to that nitrogen atom.[3] After ring cleavage such products release chiral 1,2-diamines. The hexahydropyrimidine analogues do not behave in the same manner.

[1]Sieburth, S.M., O'Hare, H.K., Xu, J., Chen, Y., Liu, G. *OL* **5**, 1859 (2003).
[2]Dieter, R.K., Oba, G., Chandupatla, K.R., Topping, C.M., Lu, K., Watson, R.T. *JOC* **69**, 3076 (2004).
[3]Askweek, N.J., Coldham, I., Haxell, T.F.N., Howard, S. *OBC* **1**, 1532 (2003).

***t*-Butyllithium.** **13**, 58; **15**, 64–65; **16**, 56–57; **18**, 79–81; **19**, 61–63; **20**, 68–69; **21**, 82–84; **22**, 86–87

Lithiation. 2-Chloro-6-arylpyridines are deprotonated at an *ortho*-position of the aryl group with *t*-BuLi in ether.[1] On the other hand, *n*-BuLi or LiTMP in THF lithiates these pyridines at C-3. The lithiated species are readily trapped with various electrophiles.

For synthesis of 5-allyl- and 5-propargyloxazoles, the sequence of lithiation of 2-butylthiooxazole followed by conversion into an organocuprate, addition of the proper electrophiles, and desulfurization constitutes a useful routine.[2]

2-(*t*-Butyldimethylsilyl)-1,3-dithiane forms the 2-lithio species with *t*-BuLi. Successive addition of two different electrophiles leads to unsymmetrically 2,2-disubstituted 1,3-dithianes.[3] This is a convenient method for accessing such compounds.

2,5-Dimethyl-3(2*H*)-furanone is alkylated at C-2. Subsequent deprotonation and reaction with electrophiles result in functionalization or chain-elongation at the C-5 methyl group.[4]

A route to vinylidene phosphanes from $(R_2P)_2CHSiMe_3$ starts from deprotonation using *t*-BuLi and the lithiated species are brought into contact with paraformaldehyde.[5]

Ynolates.[6] *t*-BuLi is useful for generating lithium ynolates from 2,2-dibromoalkanoic esters. The ynolates undergo diastereoselective 1,3-dipolar cycloadditions, e.g., with nitrones. Thus, precursors of β-amino acids and γ-lactones are readily prepared.

t-BuLi/THF ; −78¡ã → 0¡ã [MeC≡COLi] t-BuOH/ THF 91%

Cyclization. Br/Li exchange of *N*-allyl-2-bromoanilines is followed by cyclization. Addition of an electrophile E^+ gives indolines bearing a CH_2E group at C-3.[7] Next, the *N*-atom can be substituted (in the same flask).

Trifluoromethylalkynes containing a bulky residue at the other *sp*-carbon atom are susceptible to addition and defluorination.[8] This reaction leading to 3,3-difluorocyclopropenes is quite different from the carbene cycloaddition.

Ph_3X—≡—CF_3 X=C , Si; *t*-BuLi, Et_2O; 55–90%

[1]Fort, Y., Rodriguez, A.L. *JOC* **68**, 4918 (2003).
[2]Marino, J.P., Nguyen, H.N. *TL* **44**, 7395 (2003).
[3]Smith III, A.B., Pitram, S.M., Boldi, A.M., Gaunt, M.J., Sfouggatakis, C., Moser, W.H. *JACS* **125**, 14435 (2003).
[4]Caine, D.S., Arant, M.E. *SL* 2081 (2004).
[5]Izod, K., McFarlane, W., Tyson, B.V. *EJOC* 1043 (2004).
[6]Shindo, M., Itoh, K., Ohtsuki, K., Tsuchiya, C., Shishido, K. *S* 1441 (2003).
[7]Bailey, W.F., Luderer, M.R., Mealy, M.J. *TL* **44**, 5303 (2003).
[8]Brisdon, A.K., Crossley, I.R., Flower, K.R., Pritchard, R.G., Warren, J.E. *ACIEE* **42**, 2399 (2003).

C

N-Carbamoylimidazolium salts.

Tertiary amides. The reagents are prepared by *N*-alkylation. They are reactive toward various nucleophiles and convert carboxylic acids into amides.

[1]Grzyb, J.A., Batey, R.A. *TL* **44**, 7485 (2003).

Carbon, activated.

Dehydrogenation. Activated carbon promotes aromatization of Hantzsch 1,4-dihydropyridines and 1,3,5-trisubstituted pyrazolines using O_2.[1] Condensation of 2-hydroxylanilines with araldehydes by heating with activated carbon in xylene provides 2-arylbenzoxazoles directly.[2]

[1]Nakamichi, N., Kawashita, Y., Hayashi, M. *S* 1015 (2004).
[2]Kawashita, Y., Nakamichi, N., Kawabata, H., Hayashi, M. *OL* **5**, 3713 (2003).

Carbon dioxide. 22, 93–95

Carbonates, carbamates and ureas. Mixed carbonates, thiocarbonates and carbamates are synthesized by DBU-mediated carboxylation of alcohols with subsequent formation of mixed methanesulfonyl anhydrides on treatment with Ms_2O and then reaction with the second nucleophiles.[1] Alkoxycarbonylation of amines to produce carbamates carried out electrochemically in MeCN is said to involve $[NCCH_2]$ anion to deprotonate the amines.[2] Formation of carbazates from hydrazines as accomplished by a carboxylation - alkylation sequence (also dithiocarbazates using CS_2 instead of CO_2) employs Cs_2CO_3 – Bu_4NI in DMF.[3]

Cyclocarbonylation of 2-amino alcohols using the Mitsunobu system proceeds via phosphonium carboxylates $[RR'NCOOPBu_3]^+$. Subsequent steps depend on whether such intermediates can decompose into isocyanate units that are trapped by the hydroxyl group.[4] Those intermediates derived from secondary amino alcohols must undergo *P*-group transfer before ring closure. The consequence is configuration retention or inversion at the carbinolic center.

Rapid transformation of aziridines into 2-oxazolidinones is achieved in a supercritical CO_2 – ionic liquid system (turnover of 500–850 per hour), with insertion occurring mainly at the less substituted C-N bond.[5] Synthetic flexibility in the transformation is offered by changing reaction conditions, as using a catalyst system from a (salen)-Cr(III) complex and DMAP the products are 5-substituted 2-oxazolidinones.[6]

Fiesers' Reagents for Organic Synthesis, Volume 23. Edited by Tse-Lok Ho

Robinson annulation. Using supercritical CO_2 with gradual manipulation of pressure and temperature to conduct the annulation is reported.[7]

[1]Bratt, M.O., Taylor, P.C. *JOC* **68**, 5439 (2003).
[2]Feroci, M., Casadei, M.A., Orsini, M., Palombi, L., Inesi, A. *JOC* **68**, 1548 (2003).
[3]Fox, D.L., Ruxer, J.T., Oliver, J.M., Alford, K.L., Salvatore, R.N. *TL* **45**, 401 (2004).
[4]Dinsmore, C.J., Mercer, S.P. *OL* **6**, 2885 (2004).
[5]Kawanami, H., Matsumoto, H., Ikushima, Y. *CL* **34**, 60 (2005).
[6]Miller, A.W., Nguyen, S.T. *OL* **6**, 2301 (2004).
[7]Kawanami, H., Ikushima, Y. *TL* **45**, 5147 (2004).

Carbonyl(chloro)bis(triphenylphosphine)rhodium.

Oxidation.[1] Secondary alcohols undergo hydrogen transfer to methyl acrylate when irradiated with microwaves in the presence of $(Ph_3P)_2Rh(CO)Cl$. [Primary alcohols are converted into aldehydes using MVK as hydrogen acceptor and $(Ph_3P)_2RuCl_2$ as catalyst.]

Addition reactions. Thioformylation of 1-alkynes under CO and in the presence of thiols affords enals with a β-sulfenyl group.[2] Regioselective addition of alkyl chloroformates to 1,2-alkadienes results in 3-chloro-3-alkenoic esters.[3]

Pauson-Khand reaction. Cyclocarbonylation of 1,3-alkadien-8-ynes catalyzed by the $(Ph_3P)_2Rh(CO)Cl$ – $AgSbF_6$ combination under CO gives bicyclic dienones having one double bond in conjugation.[4] Only the internal double bond of the acyclic diene unit in the substrates is engaged in the reaction.

MeOOC
MeOOC
$(Ph_3P)_2Rh(CO)Cl$
$AgSbF_6$
CO, DCE
MeOOC
MeOOC
O
H
96%

[1]Takahashi, M., Oshima, K., Matsubara, S. *TL* **44**, 9201 (2003).
[2]Kawakami, J., Takeba, M., Kamiya, I., Sonoda, N., Ogawa, A. *T* **59**, 6559 (2003).
[3]Hua, R., Tanaka, M. *TL* **45**, 2367 (2004).
[4]Wender, P.A., Deschamps, N.M., Gamber, G.G. *ACIEE* **42**, 1853 (2003).

Carbonyldihydridotris(triphenylphosphine)ruthenium. 19, 65–66; **21**, 89–90; **22**, 97

Arylation.[1] Aromatic ketones are activated at the *o*-positions by the Ru complex therefore in the presence of $ArB(OR)_2$ an aryl group enters into any unsubstituted position.

Dehydrogenation.[2] Catalytic dehydrogenation of secondary alcohols without solvent or hydrogen acceptor is accomplished by heating (at 130°) with a catalyst prepared from $(Ph_3P)_3Ru(CO)H_2$ and CF_3COOH.

[1]Kakiuchi, F., Kan, S., Igi, K., Chatani, N., Murai, S. *JACS* **125**, 1698 (2003).
[2]Ligthart, G.B.W.L., Meijer, R.H., Donners, M.P.J., Meuldijk, J., Vekemans, J.A.J.M., Hulshof, L.A. *TL* **44**, 1507 (2003).

Catecholborane.

Reductive condensation.[1] *cis*-2-Hydroxycycloalkyl ketones (containing 5- and 6-membered rings) are formed when the proper diketones with one conjugated double bond are treated with catecholborane.

[1]Huddleston, R.R., Cauble, D.F., Krische, M.J. *JOC* **68**, 11 (2003).

Cerium(IV) ammonium nitrate. **13**, 67–68; **14**, 74–75; **15**, 70–72; **16**, 66; **17**, 68; **18**, 85–87; **19**, 67–69; **20**, 73–75; **21**, 90–92; **22**, 98–101

Esterification. Transformation of certain carboxylic acids[1] and *N*-Boc amino acids[2] into methyl esters can be performed with CAN in MeOH at room temperature. On heating the *N*-Boc group is disengaged. The esterification protocol is of questionable synthetic utility in view of the requirement of at least stoichiometic quantities of a high molecular weight salt and there exist so many other methods.

Cyclization. CAN oxidation of electron-rich diarymethyl moiety instigates CC bond scission which is assisted by an alkoxy group.[3,4] Cyclization involving a well distanced activated alkene unit (enol acetate, allylsilane, allenylsilane, etc.) is observed.

OAc; OC_8H_{17}; $CHAr_2$ — CAN/ $NaHCO_3$; 4A-MS; MeCN, DCE; 45° → O; OC_8H_{17}; 94%

$SiMe_3$; C; OC_8H_{17}; $CHAr_2$ — CAN/NaOAc; MeCN, DCE; 40° → OC_8H_{17}; 91%

CAN promotes tetrahydropyran formation from glycidyl cinnamyl ethers by electron abstraction from the epoxide.[5]

Ar; Ph; O; O — CAN / t-BuOH → Ar; Ph; HO; O; O

Oxidative addition and cycloaddition. Oxidation of β-dicarbonyl compounds with CAN in the presence of alkenyl sulfides leads to dihydrofuranyl sulfides which are easily converted into furan derivatives.[6] Capture of the oxidation products with enamino esters gives substituted pyrroles.[7] Using CTAN [cerium(IV) tetrabutylammonium nitrate] to oxidize methyl acetoacetate and allyltrimethylsilane as trapping agent, solvent plays an important role in product formation. In MeCN methyl 2-allylacetoacetate is obtained, but in CH_2Cl_2, a dihydrofuran that contains a trimethylsilylmethyl substituent.[8]

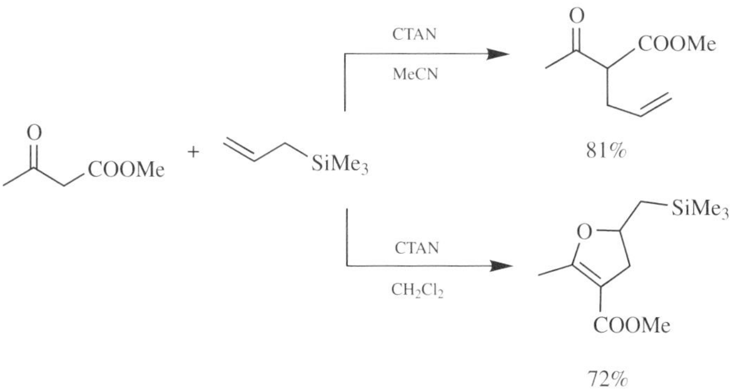

CTAN, MeCN → 81%; CTAN, CH_2Cl_2 → 72%

Methyl ketones (acetone, acetophenone) are converted by CAN into acylnitrile oxides via α-nitromethyl ketones. Reaction of such reactive species in situ with alkenes and alkynes affords 3-acylisoxazolines and 3-acylisoxazoles, respectively.[9]

o-Quinone methides can be trapped to form chromanes[10] when liberated from their osmium complexes by CAN oxidation.

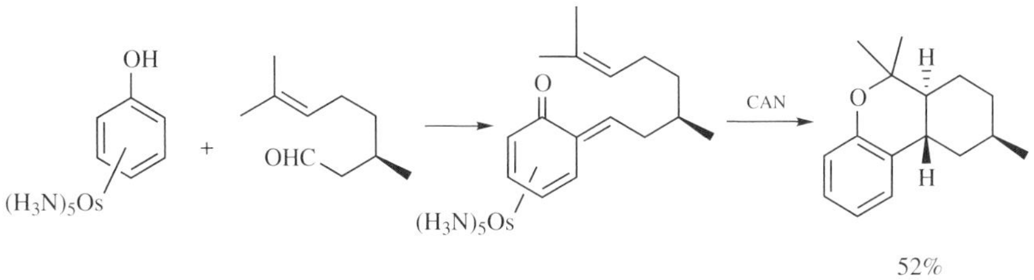

52%

Ring cleavage. Arylcyclobutanes and 2-aryloxetanes are cleaved by CAN in MeOH to afford benzyl methyl ethers.[11]

CAN / MeOH → 69%

CAN / MeOH → 91%

[1]Pan, W.-B., Chang, F.-R., Wei, L.-M., Wu, M.-J., Wu, Y.-C. *TL* **44**, 331 (2003).
[2]Kuttan, A., Nowshudin, S., Rao, M.N.A. *TL* **45**, 2663 (2004).
[3]Seiders II, J.R., Wang, L., Floreancig, P.E. *JACS* **125**, 2406 (2003).
[4]Wang, L., Seiders II, J.R., Floreancig, P.E. *JACS* **126**, 12596 (2004).
[5]Nair, V., Balagopal, L., Rajan, R., Deepthi, A., Mohanan, K., Rath, N.P. *TL* **45**, 2413 (2004).
[6]Lee, Y.R., Kang, L.Y., Lee, G.J., Lee, W.K. *S* 1977 (2003).
[7]Chuang, C.-P., Wu, Y.-L. *T* **60**, 1841 (2004).
[8]Zhang, Y., Raines, A.J., Flowers II, R.A. *OL* **5**, 2363 (2003).
[9]Itoh, K., Horiuchi, C.A. *T* **60**, 1671 (2004).
[10]Stokes Jr, S.M., Ding, F., Smith, P.L., Keane, J.M., Kopach, M.E., Jervis, R., Sabat, M., Harman, W.D. *OM* **22**, 4170 (2003).
[11]Nair, V., Rajan, R., Mohanan, K., Sheeba, V. *TL* **44**, 4585 (2003).

Cerium(III) bromate.

Oxidations.[1] The title reagent is prepared from $Ce_2(SO_4)_3$ and $Ba(BrO_3)_2$. It oxidizes alcohols to carbonyl compounds, alkylarenes to aryl ketones and sulfides to sulfoxides.

[1]Shaabani, A., Lee, D.G. *SC* **33**, 1845 (2003).

Cerium(III) chloride heptahydrate. **14**, 75–77; **15**, 72–73; **16**, 67–68; **18**, 87; **20**, 75; **21**, 92–93; **22**, 101

Chlorination.[1] Allylic chlorination of terminal alkenes using the $CeCl_3.7H_2O$ – NaOCl system is safe and efficient.

[1]Moreno-Dorado, F.J., Guerra, F.M., Manzano, F.L., Aladro, F.J., Jorge, Z.D., Massanet, G.M. *TL* **44**, 6691 (2003).

Cerium(III) chloride heptahydrate – sodium iodide. **21**, 93; **22**, 101–102

Michael reactions. Solvent-free conditions are established for the addition of amines to electron-defient alkenes by catalysis of $CeCl_3.7H_2O$ – NaI supported on alumina.[1]

The silica-supported reagents promote reaction of indoles with enones,[2] but the same reaction is also performed with CAN with ultrasound irradiation.[3]

Allylation. Good yields of homoallylic alcohols are obtained by allylation of aldehydes with allylstannanes.[4]

[1]Bartoli, G., Bartolacci, M., Giuliani, E., Marcantoni, E., Masaccesi, M., Torregiani, E. *JOC* **70**, 169 (2005).
[2]Bartoli, G., Bartolacci, M., Bosco, M., Foglia, G., Giuliani, E., Marcantoni, E., Sambri, L., Torregiani, E. *JOC* **68**, 4594(2003).
[3]Ji, S.-J., Wang, S.-Y. *SL* 2074 (2003).
[4]Bartoli, G., Bosco, M., Giuliani, E., Marcantoni, E., Palmieri, A., Petrini, M., Sambri, L. *JOC* **69**, 1290 (2004).

Cerium(III) nitrate.

4-Methylenetetrahydropyrans.[1] Intramolecular allylation of 1,3-dioxanes bearing a trimethylsilylisobut-2-enyl chain at C-4 results releases one of the alkoxy units to form the tetrahydropyrans. In the capacity of a Lewis acid $Ce(NO_3)_3$ catalyzes the transformation that can be carried out in water in the presence of a surfactant.

Me3Si; H; O; H; O; $Ce(NO_3)_3 \cdot 6H_2O$; H_2O; H; H; O; HO; 80%

[1]Aubele, D.L., Lee, C.A., Floreancig, P.E. *OL* **5**, 4521 (2003).

Cerium(IV) triflate. 20, 75; 21, 94; 22, 102

Deprotection. For cleavage of PMB ethers[1] $Ce(OTf)_4$ in $MeNO_2$ is more effective than the $CeCl_3$ – NaI system and it requires substoichimetric amounts of the catalyst while providing better yields.

Acetylation. $Ce(OTf)_4$ is said to be a useful catalyst for *O*-acetylation.[2]

[1]Bartoli, G., Dalpozzo, R., De Nino, A., Maiuolo, L., Nardi, M., Procopio, A., Tagarelli, A. *EJOC* 2176 (2004).
[2]Dalpozzo, R., De Nino, A., Maiuolo, L., Procopio, A., Nardi, M., Bartoli, G., Romeo, R. *TL* **44**, 5621 (2003).

Cesium carbonate.

Deprotection. Silyl ethers of the type ArOTBS are cleaved by Cs_2CO_3 in aq. DMF at room temperature.[1] For removing trichloroacetyl group from those amides heating with Cs_2CO_3 in DMSO at 100° is recommended because the conditions are compatible with benzoates, acetals, and silyl ethers.[2]

Substitution reactions. As a base Cs_2CO_3 is useful in *N*-alkylation of indoles[3] and synthesis of some diaryl ethers from activated ArX using ArOTBS.[4]

Epoxidation. Methylene-transfer to aldehydes is achieved by decomposing carboxymethylsulfonium betaines with Cs_2CO_3.[5]

Sulfinimines. Using Cs_2CO_3 for activation and dehydration is expedient in the condensation of sulfinamides with aldehydes.[6]

[1]Jiang, Z.-Y., Wang, Y.-G. *TL* **44**, 3859 (2003).
[2]Urabe, D., Sugino, K., Nishikawa, T., Isobe, M. *TL* **45**, 9405 (2004).
[3]Fink, D.M. *SL* 2394 (2004).
[4]Cui, S.-L., Jiang, Z.-Y., Wang, Y.-G. *SL* 1829 (2004).
[5]Forbes, D.C., Standen, M.C., Lewis, D.L *OL* **5**, 2283 (2003).
[6]Higashibayashi, S., Tohmiya, T., Mori, T., Hashimoto, K., Nakata, M. *SL* 457 (2004).

Cesium fluoride. 13, 68; **14**, 79; **15**, 75–76; **16**, 69–70; **17**, 68; **18**, 88–89; **19**, 70–72; **20**, 77–78; **21**, 95–96; **22**, 103

Substitutions. Nucleophilic substitution of 2- and 4-cyanopyridines is achieved using lithium dialkylamides[1] in the presence of CsF (1.5 equiv.). Treatment of *o*-trimethylsilylaryl triflates with CsF in the presence of phenols leads to diaryl ethers (and with carboxylic acids to aryl esters).[2]

Celite-supported CsF is useful as base for alkylation and acylation of thiols.[3]

Stille coupling. A synergistic effect is manifested between CuI and CsF for enhancing the Stille coupling[4] owing to removal of Bu_3SnX from equilibrium as the polymeric Bu_3SnF so that formation of RCuX from $RSnBu_3$ is favored. Of course, the organocopper intermediates are further converted into R-Pd-R′ prior to reductive elimination to complete one reaction cycle.

[1]Penney, J.M. *TL* **45**, 2667 (2004).
[2]Liu, Z., Larock, R.C. *OL* **6**, 99 (2004).
[3]Shah, S.T.A., Khan, K.M., Heinrich, A.M., Voelter, W. *TL* **43**, 8281 (2002).
[4]Mee, S.P.H., Lee, V., Baldwin, J.E. *ACIEE* **43**, 1132 (2004).

Cesium hydroxide.

Substitutions. A reagent system containg $CsOH \cdot H_2O$, CS_2, Bu_4NI, and MeI is used in derivatizing one of the functional groups of diols and diamines, i.e, into xanthates and dithiocarbamates, respectively. Selectivity for the XH compounds is primary > secondary, and for amino alcohols reaction at N takes precedence.[1] Utility for $CsOH{\cdot}H_2O$ is also found in *N*-arylation of secondary cyclic amines with activated ArX.[2]

β-Hydroxyalkylphosphines are formed by the CsOH-mediated opening of epoxides with R_2PH.[3]

[1]Nagle, A.S., Salvatore, R.N., Cross, R.M., Kapshiu, E.A., Sahab, S., Yoon, C.H., Jung, K.W. *TL* **44**, 5695 (2003).
[2]Varala, R., Ramu, E., Alam, M.M., Adapa, S.R. *SL* 1747 (2004).
[3]Fox, D.L., Robinson, A.A., Frank, J.B., Salvatore, R.N. *TL* **44**, 7579 (2003).

Chiral auxiliaries and catalysts. 18, 89–97; **19**, 72–93; **20**, 78–103; **21**, 97–125; **22**, 104–138

Resolutions. Acylation as a means of kinetically resolving alcohols and amines is well recognized. Chemoenzymatic method involving aminocyclopentadienyl ruthenium complexes and Novozym 435 produces chiral acetates,[1] a modified lipase catalyzes enantioselective allyloxycarbonylation of 1-methyl-1,2,3,4-tetrahydroisoquinolines.[2]

2-Arylpropanoic acids can be resolved via reaction of the corresponding acyl chloride with chiral *N*-trimethylsilyl-2-oxazolidinones.[3] Derivatization of racemizable α-substituted carbonyl compounds into imines and crystallization of one diastereomer set induce dynamic resolution.[4] 2-Substituted 2-cyclopropenecarboxylic acids are converted into diastereomeric *N*-(2-cyclopropenecarbonyl)-2-oxazolidinones and thereby resolved.[5]

Synthesizing di- and tripeptide analogues via displacement of α-bromo amides with amines is subject to dynamic kinetic resolution and the presence of stoichiometric Bu_4NI and Et_3N is most important.[6] The situation has been analyzed theoretically.[7] Kinetic resolution of α-amino acids is accomplished via alcoholysis of the azlactones with allyl alcohol in the presence of urea-based bifunctional catalysts.[8] Based on the Pd-catalyzed displacement reaction allylic alcohols are resolved or transformed into chiral derivatives via the racemic esters/carbonates. In aqueous media chiral alcohols are obtained,[9] whereas lithium *t*-butanesulfinate converts one enantiomeric series into allylic sulfones.[10] By the same technique chiral allylic sulfones are also accessible via 1,3-rearrangement of racemic allylic sulfenates (without adding any nucleophile).[11] Using chiral [2.2.2]bicyclooctadiene-iridium complexes to catalyze transformation of allylic carbonates into phenoxides is another option.[12]

Substitution of secondary/tertiary benzylic alcohols, ethers and silyl ethers by an allyl group in the presence of **1** creates a center of chirality.[13]

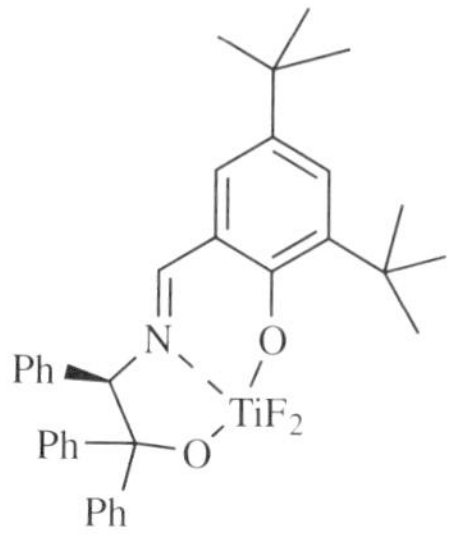

1

Enantioselective oxidation of secondary alcohols is achieved aerobically in the presence of (-)-sparteine and $PdCl_2$,[14,15] or with $PhI(OAc)_2$ and a chiral (salen)-Mn complex.[16] Vanadium-catalyzed asymmetric aerobic oxidation of α-hydroxy esters is also practical.[17]

One enantiomeric series of racemic epoxides is transformed into allylic alcohols on treatment with chiral 3-*exo*-pyrrolidinylmethyl-2-azabicyclo[2.2.1]heptane and LDA and the method serves to kinetically resolve epoxides.[18] Another method involves reaction with indoles catalyzed by a chiral (salen)-Cr complex to remove epoxides of one enantiomeric series.[19]

Unwanted enantiomers remained from resolution finds utility on racemization. To deal with chiral alcohols, catalysis at room temperature by $(C_5Ph_5)Ru(CO)_2Cl$ under basic conditions is effective.[20]

[1]Choi, J.H., Choi, Y.K., Kim, Y.H., Park, E.S., Kim, E.J., Kim, M.-J., Park, J. *JOC* **69**, 1972 (2004).
[2]Breen, G.F. *TA* **15**, 1427 (2004).
[3]Fukuzawa, S., Chino, Y., Yokoyama, T. *TA* **13**, 1645 (2002).
[4]Kosmrlj, J., Weigel, L.O., Evans, D.A., Downey, C.W., Wu, J. *JACS* **125**, 3208 (2003).
[5]Liao, L., Zhang, F., Yan, N., Golen, J.A., Fox, J.M. *T* **60**, 1803 (2004).
[6]Nam, J., Chang, J.-Y., Hahm, K.-S., Park, Y.S. *TL* **44**, 7727 (2003).
[7]Santos, A.G., Pereira, J., Afonso, C.A.M., Frenking, G. *CEJ* **11**, 330 (2005).
[8]Berkessel, A., Cleemann, F., Mukherjee, S., Müller, T.N., Lex, J. *ACIEE* **44**, 807 (2005).
[9]Lussem, B.J., Gais, H.-J. *JACS* **125**, 6066 (2003).
[10]Gais, H.-J., Jagusch, T., Spalthoff, N., Gerhards, F., Frank, M., Raabe, G. *CEJ* **9**, 4202 (2003).
[11]Jagusch, T., Gais, H.-J. *JOC* **69**, 2731 (2004).
[12]Fischer, C., Defieber, C., Suzuki, T., Carreira, E.M. *JACS* **126**, 1628 (2004).
[13]Braun, M., Kotter, W. *ACIEE* **43**, 514 (2004).
[14]Mandal, S.K., Sigman, M.S. *JOC* **68**, 7535 (2003).
[15]Jensen, D.R., Sigman, M.S. *OL* **5**, 63 (2003).
[16]Sun, W., Wang, H., Xia, C., Li, J., Zhao, P. *ACIEE* **42**, 1042 (2003).
[17]Radosevich, A.T., Musich, C., Toste, F.D. *JACS* **127**, 1090 (2005).
[18]Gayet, A., Bertilsson, S., Andersson, P.G. *OL* **4**, 3777 (2002).
[19]Bandini, M., Cozzi, P.G., Melchiorre, P., Umani-Ronchi, A. *ACIEE* **43**, 84 (2004).
[20]Csjernyik, G., Bogar, K., Backvall, J.-E. *TL* **45**, 6799 (2004).

Halogenation. Fluorination of β-keto esters with *N*-fluorobenzenesulfonimide in the presence of $Cu(OTf)_2$ and chiral bis(oxazoline) ligand give products with moderate to good ee.[1,2] In fluorination of *t*-butyl 1-indanone-2-carboxylate in the presence of Ph-BOX much better enantioselectivity (with opposite sense of chirality) is observed by changing the metal salt to $Ni(ClO_4)_2$, probably because the bidentate substrate is fixed on a plane by Cu but its aromatic ring is parallel to one of the ligand and a different face is exposed to attack by the fluorinating agent.[2] 1,3-Transpositional fluorination of allylsilanes with Selectfluor and cinchona alkaloid-derived DHQ ligands is effective in the access to chiral tertiary fluorides.[3]

The Cu-bis(oxazoline)-catalyzed reaction is also useful to replace the active hydrogen atom of a β-keto ester with chlorine or bromine atom by using NXS as halogen source.[4] Asymmetric chlorination is also performed with Ti-TADDOLate in combination with $ArICl_2$.[5] For chlorination of simple aldehydes (*S*)-prolinamide,[6] (*R,R*)-2,5-diphenylpyrrolidine,[6] or (*S*)-2,2,3-trimethyl-5-benzylimidazolidin-4-one[7] is used as the chiral catalyst. Chiral *trans*-4,5-diphenylimidazolidine shows good activities in promoting α-chlorination of ketones (up to 98% ee).[8]

[1]Ma, J.-A., Cahard, D. *TA* **15**, 1007 (2004).
[2]Shibata, N., Ishimaru, T., Nagai, T., Kohno, J., Toru, T. *SL* 1703 (2004).
[3]Greedy, B., Paris, J.-M., Vidal, T., Gouverneur, V. *ACIEE* **42**, 3291 (2003).

[4]Marigo, M., Kumaragurubaran, N., Jorgensen, K.A. *CEJ* **10**, 2133 (2004).
[5]Ibrahim, H., Kleinbeck, F., Togni, A. *HCA* **87**, 605 (2004).
[6]Halland, N., Braunton, A., Bachmann, S., Marigo, M., Jorgensen, K.A. *JACS* **126**, 4790 (2004).
[7]Brochu, M.P., Brown, S.P., MacMillan, D.W.C. *JACS* **126**, 4108 (2004).
[8]Marigo, M., Bachmann, S., Halland, N., Braunton, A., Jorgensen, K.A. *ACIEE* **43**, 5507 (2004).

Alkylations and acylations. Cyclic β-keto esters undergo asymmetric allylation in the presence of the diaminophosphine oxide **2**.[1] For Michael addition to 1-alkyn-3-ones catalyzed by a DHQ ligand the presence of Bu_3P for (Z)-to-(*E*) isomerization of the double bond in the adducts is critical.[2]

To present the allylating agent in a chiral environment for reaction with ketone enolates reaction conditions similar to allylic substitutions are applicable.[3] Pd-catalyzed decomposition of allyl enol carbonates is rendered asymmetrically by adding a proper chiral ligand.[4]

A synthesis of chiral α-branched aldehydes involves alkylation of phenylalanine-derived *N*-acyl-4-benzyl-5,5-dimethyl-2-oxazolidinones and Dibal-H reduction of the products.[5] Both enantioselective and diastereoselective issues pertaining to alkylation of α-hydroxy acids are addressed via Mo-catalyzed allylation of the derived 5*H*-2-phenyloxazol-4-ones.[6]

Asymmetric synthesis of α-amino acids attracts a lot of attention. *N*-Diphenylmethyleneglycine *t*-butyl ester is alkylated effectively under phase-transfer conditions, with quaternized cinchona alkaloid derivatives[7,8,9] or **3**[10] serving dual purposes. Such a process has been extended to enantioselective formation of quaternary α-amino acid derivatives from alanine *t*-butyl ester via the 2-naphthylaldimine.[11] [For other examples, see Phase-transfer catalysts section.]

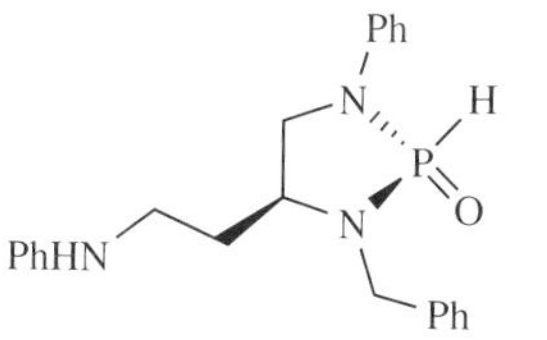

Ph H PhHN N P O N Ph

2

CF3 OMe CF3 Br⊖ ⊕N H MeO CF3 F3C

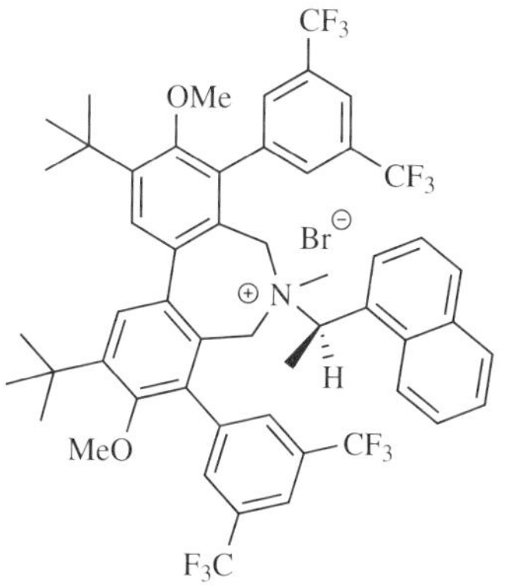

3

Deprotonation at C-3 of 1-alkenyl carbamates by BuLi in the presence of (-)-sparteine creates homoenolates that undergo asymmetric alkylation.[12] With a similar base system (*s*-BuLi) diarylmethanes that are *o*-substituted with a MOMO group also deliver chiral alkylates.[13]

Alkylative trapping of pre-quenched 1,4-adducts from Bn_2NLi and (*S*)-*N*-propenoyl-4-isopropyl-5,5-dimethyl-2-oxazolidinone is highly diastereoselective and enantioselective.[14] After conversion of 1,3-dihydroxyacetone into 2,2-dimethyl-1,3-dioxan-5-one SAMP-hydrazone, differentiation of the two methylene groups is possible. Alkylation furnishes products with a (*R*)-configuration to satisfy many synthetic purposes.[15] Imines generated ketones and chiral 2-methoxyethylamines are easily converted into chelated zinc enamides on consecutive treatment with $ZnCl_2$ and MeLi. Incorporation of 1-alkenes is stereocontrolled (1,4-induction).[16]

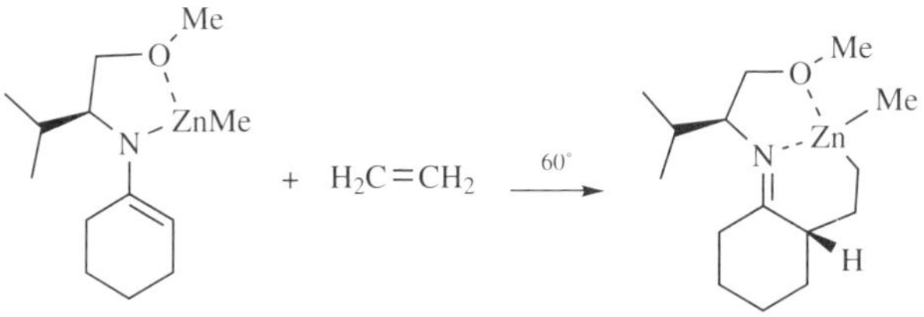

Me O ZnMe N + $H_2C{=}CH_2$ 60° → Me O Me N Zn H

Carboxylic acid derivatives such as α-sulfinyl thioacetamide[17] and *N*-*t*-butanesulfinyl amidines[18] are subjected to asymmetric induction during alkylation by the chiral sulfur atom.

Substrate-directed carbon chain elongation by Mitsunobu reaction of chiral benzylic alcohols with malonic esters[19] should be useful in view of the alcohols are readily available (e.g., from CBS reduction of the ketones).

Lithiated *t*-butylthiomethyl 2-pyridyl sulfide reacts asymmetrically with aldehydes in the presence of chiral bis(oxazoline) ligands to afford protected α-hydroxy aldehydes. For each of two diastereomeric products ee ranges from 55% to 88%.[20]

Enantiopure sulfoxides are synthesized in three steps from *N*-tosylnorephedrine by consecutive reactions with $SOCl_2$, RMgX, and R′MgX.[21] The first step creates a chiral heterocycle with (*R*)-configuration at the sulfur atom, the first Grignard reagent releases the tosylamido group and the final step involves cleavage of the sulfinate ester.

Silyl ketene acetals[22] and silyl ketene imines[23] undergo asymmetric *C*-acylation with ferrocene-fused DMAP analogue **4A** as catalyst. Some *meso*-diols (1,3- and 1,4-) are monoacylated with assistance by a dinuclear zinc complex (**5A**).[24]

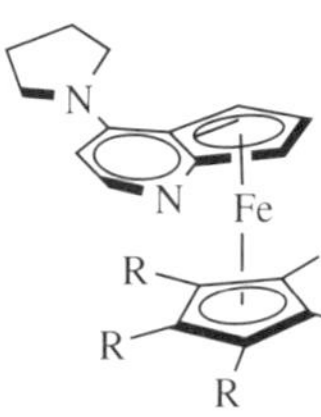

N N Fe R R R R R

4A R = Ph
4B R = Me

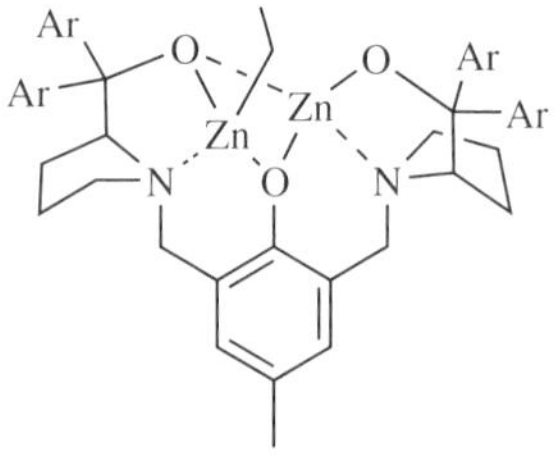

Ar Ar O Zn Zn O Ar Ar N O N

5A Ar = 4-Ph-C_6H_4
5B Ar = α-Np
5C Ar = Ph

A ferrocenyldiphosphine is employed to create a chiral Pd catalyst to desymmetrize *meso*-succinic anhydrides by reaction with R_2Zn.[25]

[1]Nemoto, T., Matsumoto, T., Masuda, T., Hitomi, T., Hatano, K., Hamada, Y. *JACS* **126**, 3690 (2004).
[2]Bella, M., Jorgensen, K.A. *JACS* **126**, 5672 (2004).
[3]Trost, B.M., Schroeder, G.M. *CEJ* **11**, 174 (2005).
[4]Behenna, D., Stoltz, M. *JACS* **126**, 15044 (2004).
[5]Bull, S.D., Davies, S.G., Nicholson, R.L., Sanganee, H.J., Smith, A.D. *OBC* **1**, 2886 (2003).
[6]Trost, B.M., Dogra, K., Franzini, M. *JACS* **126**, 1944 (2004).
[7]Park, H.-G., Jeong, B.-S., Yoo, M.-S., Lee, J.-H., Park, B.-S., Kim, M.G., Jew, S.-S. *TL* **44**, 3497 (2003).
[8]Yu, H., Koshima, H. *TL* **44**, 9209 (2003).
[9]Elango, S., Venugopal, M., Suresh, P.S., Eni. *T* **61**, 1443 (2005).
[10]Lygo, B., Allbutt, B. *SL* 326 (2004).
[11]Jew, S., Jeong, B.-S., Lee, J.-H., Yoo, M.-S., Lee, Y.-J., Park, B., Kim, M.G., Park, H. *JOC* **68**, 4514 (2003).
[12]Seppi, M., Kalkofen, R., Reupohl, J., Frohlich, R., Hoppe, D. *ACIEE* **43**, 1423 (2004).
[13]Wilkinson, J.A., Rossington, S.B., Leonard, J., Hussain, N. *TL* **45**, 5481 (2004).
[14]Beddow, J.E., Davies, S.G., Smith, A.D., Russell, A.J. *CC* 2778 (2004).
[15]Enders, D., Müller-Hüwen, A. *EJOC* 1732 (2004).
[16]Nakamura, M., Hatakeyama, T., Hara, K., Nakamura, E. *JACS* **125**, 6362 (2003).
[17]Nowaczyk, S., Alayrac, C., Metzner, P., Averbuch-Pouchot, M.-T. *JOC* **67**, 6852 (2002).
[18]Kochi, T., Ellman, J.A. *JACS* **126**, 15652 (2004).
[19]Hillier, M.C., Desrosiers, J.-N., Marcoux, J.-F., Grabowski, E.J.J. *OL* **6**, 573 (2004).
[20]Nakamura, S., Ito, Y., Wang, L., Toru, T. *JOC* **69**, 1581 (2004).
[21]Han, Z., Krishnamurthy, D., Grover, P., Wilkinson, H.S., Fang, Q.K., Su, X., Lu, Z.-H., Magiera, D., Senanayake, C.H. *ACIEE* **42**, 2032 (2003).
[22]Mermerian, A.H., Fu, G.C. *JACS* **125**, 4050 (2003).
[23]Mermerian, A.H., Fu, G.C. *ACIEE* **44**, 949 (2005).
[24]Trost, B.M., Mino, T. *JACS* **125**, 2410 (2003).
[25]Bercot, E.A., Rovis, T. *JACS* **126**, 10248 (2004).

Allylic substitutions. New ligands developed for the Pd-catalyzed allylic substitution include fluorous bis(oxazolines),[1] *N*-(2-diphenylphosphino-6-methoxy)-prolinol,[2] butylthiomethylaziridine **6**,[3] (1*R*,2*S*)-1-phenylphospholane-2-carboxylic acid (**7**),[4] and ferrocenylphosphines that contain a SAMP-hydrazone at ortho position (**8**).[5] However, the Trost ligand **9** and its analogues remain reliable workhorses, an excellent example being the preparation of a chiral bis(allylic) ether from 2-iodoresorcinol and a racemic Baylis-Hillman adduct for the synthesis of the furaquinocins.[6] An interesting observation pertains to chroman synthesis by an intramolecular substitution that addition of 1 equivalent of HOAc dramatically increases enantioselectivity (and produces products of opposite absolute configuration!).[7]

BnS N Cl

6

COOH P Ph

7

Azlactones (prepared from α-amino acids) are pronucleophiles that add to allenes asymmetrically in the Pd-catalyzed reaction using ligand **9**. The reaction involves π-allylpalladium complexes derived from the allenes and it proceeds with very high diastereoselectivity in the case of unsymmetrical allenes (with CC bond formation with butoxyallene at the substituted carbon atom and the nature).[8]

Copper-catalyzed S_N2' reaction with organozinc reagents on allylic phosphates in the presence of peptidyl ligands can be used to establish chirality at the α-carbon of an ester,[9] and the method is general other allylic substrates.[10]

EtO, EtO–P=O ... COOMe + Et_2Zn → ($Cu(OTf)_2 \cdot C_6H_6$; THF –50°) → H, COOMe

NHBu, OH, N, H N, O, O

93% (95% ee)

[1]Bayardon, J., Sinou, D. *JOC* **69**, 3121 (2004).
[2]Mino, T., Tanaka, Y., Akita, Sakamoto, M., Fujita, T. *H* **60**, 9 (2003).
[3]Braga, A.L., Paixao, M.W., Miliani, P., Silveira, C.C., Rodrigues, O.E.D., Alves, E.F. *SL* 1297 (2004).
[4]Sun, X.-M., Manabe, K., Lam, W.W.-L., Shiraishi, N., Kobayashi, J., Shiro, M., Utsumi, H., Kobayashi, S. *CEJ* **11**, 361 (2005).
[5]Mino, T., Ogawa, T., Yamashita, M. *JOMC* **665**, 122 (2003).
[6]Trost, B.M., Thiel, O.R., Tsui, H.-C. *JACS* **125**, 13155 (2003).
[7]Trost, B.M., Shen, H.C., Dong, L., Surivat, J.-P. *JACS* **125**, 9276 (2003).
[8]Trost, B.M., Jäkel, C., Plietker, B. *JACS* **125**, 4438 (2003).
[9]Murphy, K.E., Hoveyda, A.H. *JACS* **125**, 4690 (2003).
[10]Kacprzynski, M.A., Hoveyda, A.H. *JACS* **126**, 10676 (2004).

Addition to C=O bond. Numerous bidentate chiral ligands have been tested for their effectiveness to achieve asymmetric addition of R_2Zn to carbonyl compounds (mostly aldehydes). Fluorous (*S*)-*N*-alkyl-α,α-diarylprolinols,[1] (2*S*,4*R*)-*N*-benzyl-α,α-diphenylprolinol,[2] the dimeric **10**,[3] chiral 2-aminoethylchalcogen derivatives,[4–6] and unsymmetrical 1,2-diamine derivatives such as **11**,[7] **12**,[8] and **13**,[9] as well as the *N*-benzylamide of phenylalanine (nickel complex)[10] have been reported. Besides those, bis(oxazoline) incorporated with a dibenzo[*a,c*]cycloheptadiene unit (**14**) has also been developed.[11] Using a chiral diamine containing both mono- and disubstituted N-atoms to form magnesium amide reagents with R_2Mg, reaction with aldehydes is enentioselective.[12] The organozinc reaction is also catalyzed by other metal salts. With titanium the assistance by C_2-symmetrical diamine ligands **15**[13,14] and **16**[15] is typical. The TADDOL-analogue **17** is a more novel ligand.[16]

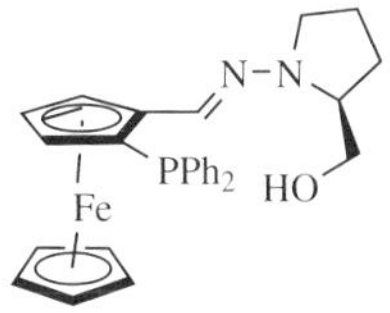

8 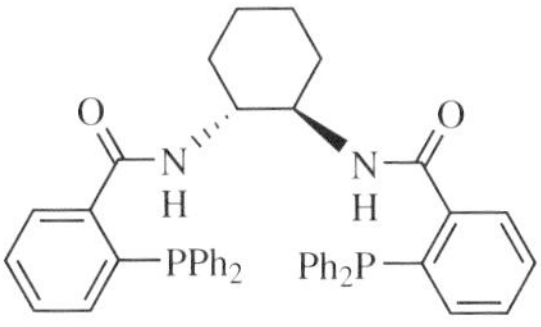 **9** 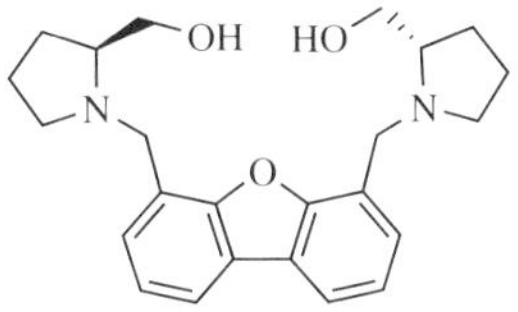 **10**

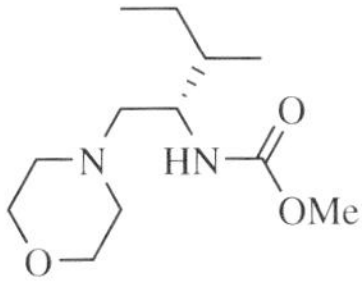

11 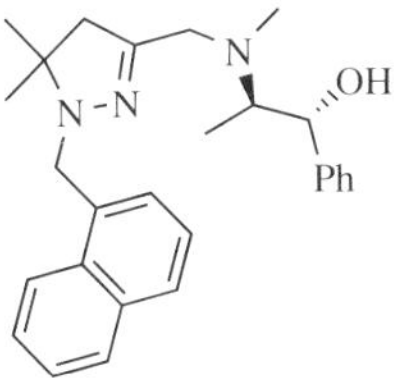 **12** 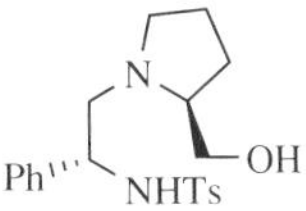 **13**

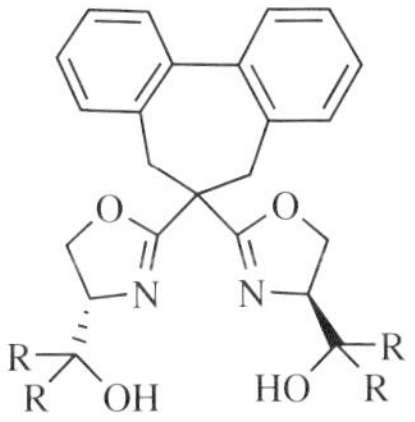

14 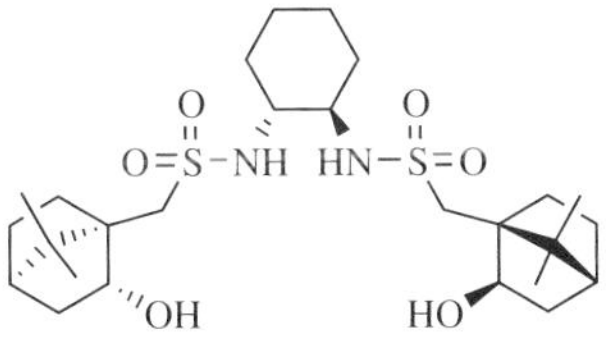 **15** 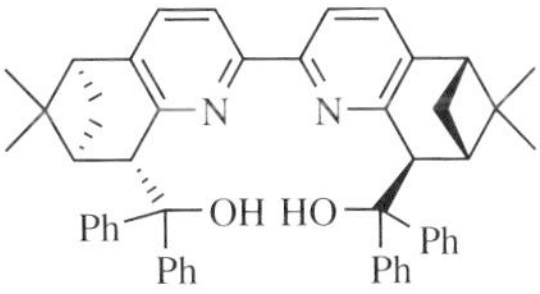 **16**

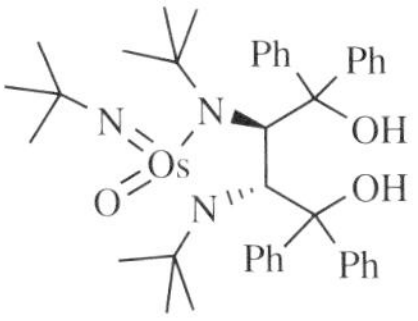

17

Reformatsky reaction with aldehydes reaches 90% ee in some cases when it is carried out in the presence of (*S*)-1,1,1-trifluoro-3-piperidinyl-2-propanol.[17] Chiral epoxides together with asymmetric autocatalysis induce enantioselective R_2Zn addition to aldehydes.[18]

Addition with allylmetals may involve chiral boron and silicon reagents or chiral catalysts. There are several readily available monoterpene-derived allylboranes and boronates that continuing expansion of their use is expected.[19,20] Allylsilane **18** is an excellent reagent for practical enantioselective synthesis of homoallylic alcohols.[21]

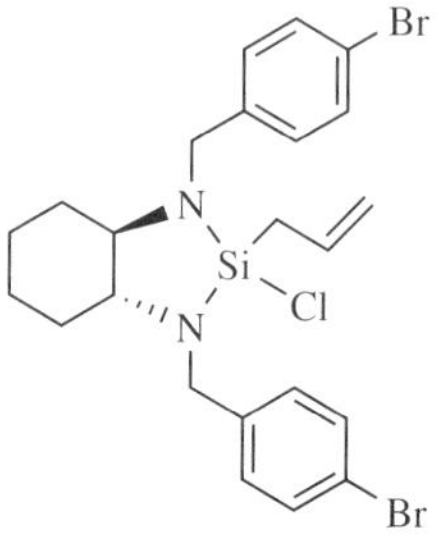

18

1-(2-Methoxynaphth-1-yl)isoquinoline *N*-oxide is a BINOL-analogue useful for mediating the reaction of allyltrichlorosilane with aldehydes.[22]

Allyl group transfer from allyltributylstannane to aldehydes is a much studied reaction, and the chiral catalysts used include different metals and ligands. A polymer-supported pincer bis(oxazolinyl)-linked Rh complex is a heterogeneous catalyst that needs further elaboration.[23] While (salen)-Cr complexes are well known to catalyze enantioselective reactions (new allylation applications[24,25]), the utility of indium(III) complexes [e.g., with PYBOX] emerges only recently.[26]

Bimetallic systems (Fe/Cr and Co/Cr) employing ligand **19** are shown to mediate 2-haloallylation of aldehydes.[27,28] A propargylation is based on $CrCl_2$-Mn and the carbazole **20**, and there are dramatic changes due to varying size of the alkyl pendants on the oxazoline units.[29] The mono-*N*-oxide of a bis(tetrahydroquinoline) analogous to **16** is a useful catalyst to deliver an allyl group from allyltrichlorosilane to aldehydes because it is a Lewis base.[30]

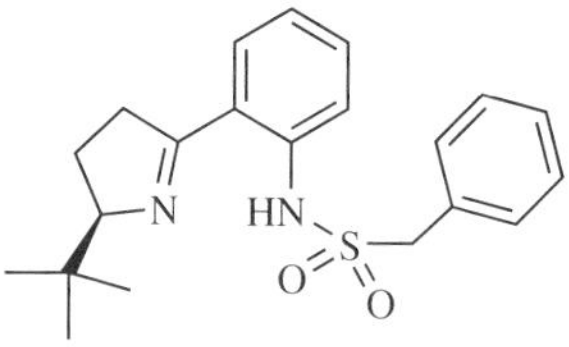

19

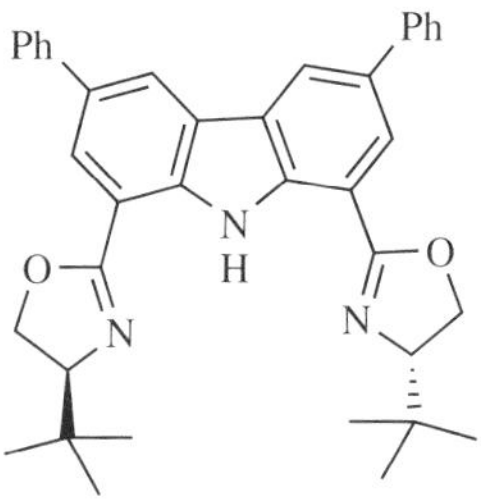

20

Enantioselective allylation of ketones is apparently more difficult. Copper-catalyzed reaction of *B*-allyl(pinacolatoboron) with $(i\text{-PrO})_3La$ additive shows promise.[31]

By means of (*Z*)-crotyl transfer from 2-*endo*-(1-buten-3-yl)isoborneol to aldehydes chiral (*Z*)-2-alken-5-ols are synthesized.[32] The reaction involves a [3,3]sigmatropic rearrangement of oxonium ion intermediates.

Asymmetric alkynylation usually employs zinc derivatives that are generated in situ,[33] therefore zinc-binding *N,O*-ligands are expected to exert great influences to the reaction. The bifunctional (2-pyridylmethyl)aminoethanol **21** performs even better,[34] and the C_2-symmetric SALEN-type ligands[35] and those prepared from BINOL-3-aldehyde also shows good results.[36] With titanium-based catalyst systems several *N,O*-ligands that exhibit reasonably good results are 2-sulfonamino alcohols[37,38] and cinchonidine.[39] For ketone alkynylation Cu complexes as catalysts function well.[40]

Alkenylation often is carried out by insitu generation of alkenylmetals such as hydroboration of 1-alkynes followed by metal exchange. 1-Alkenylzinc reagents are found to add asymmetrically to aldehydes using ligand **22** which is prepared from 2-naphthol.[41] Internal alkynes are transformed into alkenylating agents by $Ni(cod)_2$ and Et_3B, and (+)-(neomenthyl)diphenylphosphine provides conditions for formation of chiral products.[42]

Ph Ph NH OH N Ph Ph N OH

21 **22**

[1]Park, J.K., Lee, H.G., Bolm, C., Kim, B.M. *CEJ* **11**, 945 (2005).
[2]Funabashi, K., Jachmann, M., Kanai, M., Shibasaki, M. *ACIEE* **42**, 5489 (2003).
[3]Ooi, T., Saito, A., Maruoka, K. *CL* 1108 (2001).
[4]Tseng, S.-L., Yang, T.-K. *TA* **15**, 3375 (2004).
[5]Braga, A.L., Miliani, P., Paixao, M.W., Zeni, G., Rodrigues, O.E.D., Alves, E.F. *CC* 2488 (2004).
[6]Braga, A.L., Paixao, M.W., Ludtke, D.S., Silveira, C.C., Rodrigues, O.E.D. *OL* **5**, 2635 (2003).
[7]Richmond, M.L., Seto, C.T. *JOC* **68**, 7505 (2003).
[8]Sibi, M.P., Stanley, L.M. *TA* **15**, 3353 (2004).
[9]Mao, J., Wan, B., Wang, R., Wu, F., Lu, S. *JOC* **69**, 9123 (2004).
[10]Burguete, M.I., Collado, M., Escorihuela, J., Galindo, F., Garcia-Verdugo, E., Luis, S.V., Vicent, M.J. *TL* **44**, 6891 (2003).
[11]Fu, B., Du, D.-M., Wang, J. *TA* **15**, 119 (2004).
[12]Yong, K.H., Taylor, N.J., Chong, J.M. *OL* **4**, 3553 (2002).
[13]Prieto, O., Ramon, D.J., Yus, M. *TA* **14**, 1955 (2003).
[14]Garcia, C., Walsh, P.J. *OL* **5**, 3641 (2003).
[15]Chen, Y-J., Lin, R.-X., Chen, C. *TA* **15**, 3561 (2004).
[16]Muniz, K. *TL* **44**, 3547 (2003).
[17]Fujiwara, Y., Katagiri, T., Uneyama, K. *TL* **44**, 6161 (2003).
[18]Kawasaki, T., Shimizu, M., Suzuki, K., Sato, I., Soai, K. *TA* **15**, 3699 (2004).
[19]Ramachandran, P.V., Padiya, K.J., Rauniyar, V., Reddy, M.V.R., Brown, H.C. *TL* **45**, 1015 (2004).

[20]Lachance, H., Lu, X., Gravel, M., Hall, D.G *JACS* **125**, 10160 (2003).
[21]Kubota, K., Leighton, J.L. *ACIEE* **42**, 946 (2003).
[22]Malkov, A.V., Dufkova, L., Farrugia, L., Kocovsky, P. *ACIEE* **42**, 3674 (2003).
[23]Weissberg, A., Portnoy, M. *CC* 1538 (2003).
[24]Lombardo, M., Licciulli, S., Morganti, S., Trombini, C. *CC* 1762 (2003).
[25]Kwiatkowski, P., Chaladaj, W., Jurczak, J. *TL* **45**, 5343 (2004).
[26]Lu, J., Ji, S.-J., Teo, Y.-C., Loh, T.-P. *OL* **7**, 159 (2005).
[27]Kurosu, M., Lin, M.-H., Kishi, Y. *JACS* **126**, 12248 (2004).
[28]Namba, K., Kishi, Y. *OL* **6**, 5031 (2004).
[29]Inoue, M., Nakada, M. *OL* **6**, 2977 (2004).
[30]Malkov, A.V., Bell, M., Orsini, M., Pernazza, D., Massa, A., Herrmann, P., Kocovsky, P. *JOC* **68**, 9659 (2003).
[31]Wada, R., Oisaki, K., Kanai, M., Shibasaki, M. *JACS* **126**, 8910 (2004).
[32]Lee, C.-L.K., Lee, C.-H.A., Tan, K.-T., Loh, T.-P. *OL* **6**, 1281 (2004).
[33]Pu, L. *T* **59**, 9873 (2003).
[34]Kang, Y.-F., Liu, L., Wang, R., Yan, W.-J., Zhou, Y.-F. *TA* **15**, 3155 (2004).
[35]Cozzi, P.G. *ACIEE* **42**, 2895 (2003).
[36]Li, Z.B., Pu, L. *OL* **6**, 1065 (2004).
[37]Xu, Z., Wang, R., Xu, J., Da, C., Yan, W., Chen, C. *ACIEE* **42**, 5747 (2003).
[38]Xu, Z., Chen, C., Xu, J., Miao, M., Yan, W., Wang, R. *OL* **6**, 1193 (2004).
[39]Kamble, R.M., Singh, V.K. *TL* **44**, 5347 (2003).
[40]Lu, G., Li, X., Jia, X., Chan, W.L., Chan, A.S.C. *ACIEE* **42**, 5057 (2003).
[41]Ji, J.-X., Qiu, L.-Q., Yip, C.W., Chan, A.S.C. *JOC* **68**, 1589 (2003).
[42]Miller, K.M., Huang, W.-S., Jamison, T.F *JACS* **125**, 3442 (2003).

Cyanohydrination of chiral β-keto imides with Et_2AlCN is diastereoselective due to presence of internal stereocontrollers (1,2- + 1,5-stereoinduction).[1] Trapping of the chiral cyanohydrins in situ by Ac_2O in reaction catalyzed by polymeric SALEN - Ti(IV) or V(V) complex [2] and also by ethyl cyanoformate in the reaction with monomeric (salen) TiO_2 are quite successful.

Catalytic silylcyanation performed by Ti-[4–6] and Gd-complexes[7] with chiral ligands, as well as a metal-free process using a bis(dihydroquinine)anthraquinone[8] as catalyst show the maturity of such a reaction. A novel synthesis of α-siloxycyanoacetic esters in chiral form from acylsilanes and cyanoformic esters is accomplished by (salen*)-aluminum isopropoxide.[9]

[1]Flores-Morales, V., Fernandez-Zertuche, M., Ordonez, M. *TA* **14**, 2693 (2003).
[2]Huang, W., Song, Y., Wang, J., Cao, G., Zheng, Z. *T* **60**, 10469 (2004).
[3]Belokon, Y.N., Blacker, A.J., Clutterbuck, L.A., North, M. *OL* **5**, 4505 (2003).
[4]He, K., Zhou, Z., Wang, L., Li, K., Zhao, G., Zhou, Q., Tang, C. *T* **60**, 10505 (2004).
[5]Chen, F., Feng, X., Qin, B., Zhang, G., Jiang, Y. *OL* **5**, 949 (2003).
[6]Fujii, K., Maki, K., Kanai, M., Shibasaki, M. *OL* **5**, 733 (2003).
[7]Masumoto, S., Suzuki, M., Kanai, M., Shibasaki, M. *T* **60**, 10497 (2004).
[8]Tian, S.-K., Hong, R., Deng, L *JACS* **125**, 9900 (2003).
[9]Nicewicz, D.A., Yates, C.M., Johnson, J.S. *JOC* **69**, 6548 (2004).

In recent years the utility of α-amino acids or their derivatives as catalysts for cross-aldol reactions have been scrutinized. The reward of great success with (*S*)-proline further stimulates chemists' efforts. Co-catalysis by (*S*)-valine has been reported.[1] Useful proline-derived catalysts include (*S*)-2-(1-pyrrolidinyl)methylpyrrolidine,[2] (*S*)-5-(pyrrolidin-2-yl)tetrazole,[3] as well as *N*-substituted prolinamides.[4] The zinc complexes **5B, 5C** that contain two diphenylprolinol units efficiently direct asymmetric condensation between alkynyl methyl ketones with aldehydes.[5] Due to formation of new catalytic species from the aldol products, incubation of the Zn complex with achiral alcohols prior to conducting the condensation tends to enhance stereoselectivity.

1,5-Asymmetric induction in boron-mediated aldol reaction of β-alkoxyalkyl methyl ketones leads to products having predominantly the *anti*-configuration.[6] Actually, 4-substituted oxazolidine-based aldol reactions depend on 1,5-asymmetric induction and the principle applies to those involving *N*-acetylthiazolidine-2-thiones.[7,8] *syn*-Aldol reaction of *N*-acylthiazolidine-2-thiones based on (box)-Ni catalysis is performed in the presence of Me_3SiOTf.[9]

For biomimetic decarboxylative condensation to give β-hydroxy thiolesters in a Cu(II)-catalyzed reaction of malonic acid hemithioesters with aldehydes a tartaric acid-derived benzimidazole **23** exerts some influences.[10]

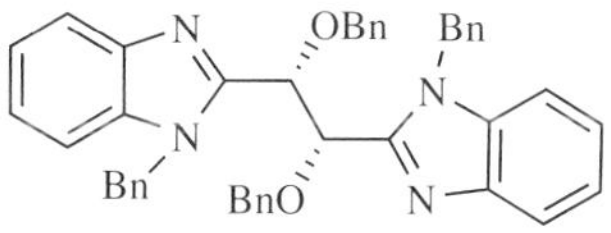

N OBn Bn N N Bn BnO N

23

The Mukaiyama version of aldol reaction using $Sc(OTf)_3$ and ligand **24** with formalin has been performed.[11] *S*-Chirogenic sulfoximines serve well as ligands in the Cu(I)-catalyzed reaction.[12] Silyl dienol ethers[13] and 2-(trimethylsiloxy)furan[14] undergo asymmetric aldol condensation catalyzed by a (salen)-Cr complex. In the latter case two diastereomers are produced, but both in high ee. Trichlorosilyl enol ethers derived from diethyl ketone afford *syn*-aldols by stereoinduction of **25**.[15]

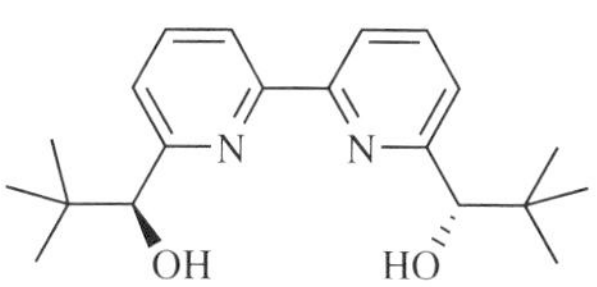

N N OH HO

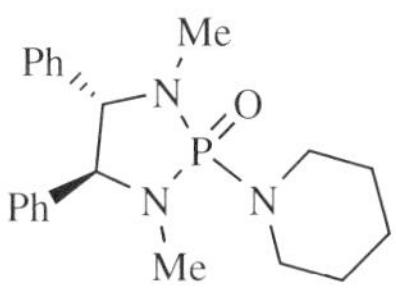

Ph Me N O P Ph N N Me

24 **25**

Aldol reaction employing proline derivatives as catalyst probably involves enamines of donor reactants, and the case of (*S*)-5-(pyrrolidin-2-yl)tetrazole[16] must be included. Aldols are produced from enamine and imines after reaction with carbonyl compounds and subsequent hydrolysis. Very high diastereo- and enantioselective condensation of *N*-alkenyl carbamates with aldehydes[17] and ethyl glyoxylate[18] is effected by Cu(I) ion complexed with a C_2-symmetric bisimine. Excellent 1,4-asymmetric induction is observed in reaction of chloral with imines bearing a chirality center adjacent to the nitrogen atom.[19] The chirality information can come from the *S*-chirogen of sulfoximines.[20]

Chiral enamines are involved in the aldol reaction of two aldehydes using an imidazolidinone catalyst derived from phenylalanine and pivalaldehyde.[21]

The Baylis-Hillman reaction of methyl acrylate with aldehydes in the presence of $TsNH_2$ and a quinine derivative gives α-methylene-β-tosylamino esters in moderate ee.[22]

(*S*)-1,1,1-Trifluoro-3-dimethylamino-2-propanol effectively promotes the asymmetric Reformatsky reaction.[23]

The nitro-aldol reaction (Henry reaction) is also subject to stereoinduction. Metal catalysts include $Cu(OAc)_2$ in conjunction with ligand **26**,[24] **27**,[25] (salen)-Co complexes[26] and analogues.[27] When chiral substrates are readily available (e.g., *N,N*-dibenzyl-α-amino aldehydes[28] and chiral esters of glyoxylic acid[29]) self-directed asymmetric synthesis is most convenient.

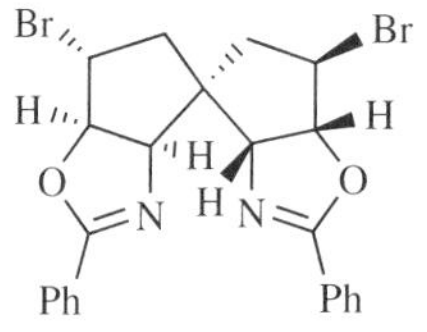

26

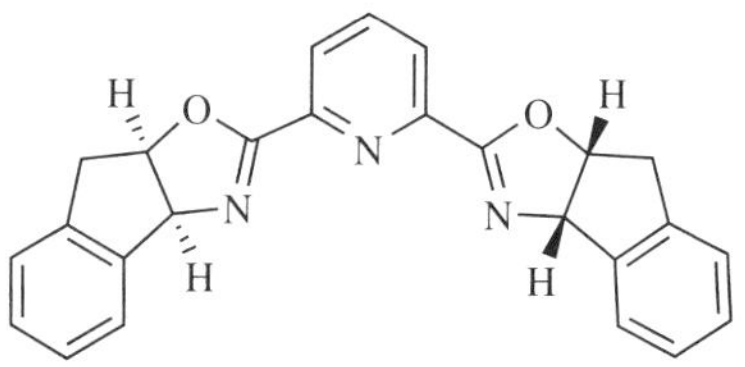

27

The Passerini reaction for combining an aldehyde, an isonitrile and a carboxylic acid to yield an α-acyloxy carboxamide is under stereocontrol by the Cu complex of **27**.[30]

By reaction between ketenes ($RCH_2COCl + Et_3N$ in situ) and triphenylphosphoranylacetate of (-)-10-benzenesulfonylisoborneol a series of chiral allenic esters are obtained.[31]

[1]Gao, M.Z., Gao, J., Lane, B.S., Zingaro, R.A. *CL* **32**, 524 (2003).
[2]Mase, N., Tanaka, F., Barbas III, C.F. *ACIEE* **43**, 2420 (2004).
[3]Torii, H., Nakadai, M., Ishihara, K., Saito, S., Yamamoto, H. *ACIEE* **43**, 1983 (2004).
[4]Tang, Z., Jiang, F., Yu, L.-T., Cui, X., Gong, L.-Z., Mi, A.-Q., Jiang, Y.-Z., Wu, Y.-D. *JACS* **125**, 5262 (2003).
[5]Trost, B.M., Fettes, A., Shireman, B.T. *JACS* **126**, 2660 (2004).
[6]Evans, D.,A., Cote, B., Coleman, P.J., Connell, B.T. *JACS* **125**, 10893 (2003).
[7]Zhang, Y., Phillips, A.J., Sammakia, T. *OL* **6**, 23 (2004).
[8]Zhang, Y., Sammakia, T. *OL* **6**, 3139 (2004).
[9]Evans, D.,A., Downey, C.W., Hubbs, J.L. *JACS* **125**, 8706 (2003).
[10]Orlandi, S., Benaglia, M., Cozzi, F. *TL* **45**, 1747 (2004).
[11]Ishikawa, S., Hamada, T., Manabe, K., Kobayashi, S. *JACS* **126**, 12236 (2004).
[12]Langner, M., Bolm, C. *ACIEE* **43**, 5984 (2004).

[13]Shimada, Y., Matsuoka, Y., Irie, R., Katsuki, T. *SL* 579 (2004).
[14]Matsuoka, Y., Irie, R., Katsuki, T. *CL* **32**, 584 (2003).
[15]Denmark, S.E., Pham, S.M. *JOC* **68**, 5045 (2003).
[16]Torii, H., Nakadai, M., Ishihara, K., Saito, S., Yamamoto, H. *ACIEE* **43**, 1983 (2004).
[17]Matsubara, R., Vital, P., Nakamura, Y., Kiyohara, H., Kobayashi, S. *T* **60**, 9769 (2004).
[18]Matsubara, Nakamura, Y., Kobayashi, S. *ACIEE* **43**, 3258 (2004).
[19]Funabiki, K., Honma, N., Hashimoto, W., Matsui, M. *OL* **5**, 2059 (2003).
[20]Kochi, T., Tang, T.P., Ellman, J.A. *JACS* **125**, 11276 (2003).
[21]Mangion, I.K., Northrup, A.B., MacMillan, D.W.C. *ACIEE* **43**, 6722 (2004).
[22]Balan, D., Adolfsson, H. *TL* **44**, 2521 (2003).
[23]Fujiwara, Y., Katagiri, T., Uneyama, K. *TL* **44**, 6161 (2003).
[24]Kato, T., Marubayashi, K., Takizawa, S., Sasai, H. *TA* **15**, 3693 (2004).
[25]Evans, D.,A., Siedel, D., Rueping, M., Lam, H.W., Shaw, J.T., Downey, C.W. *JACS* **125**, 12692 (2003).
[26]Kogami, Y., Nakajima, T., Ikeno, T., Yamada, T. *S* 1947 (2004).
[27]Kogami, Y., Nakajima, T., Ashizawa, T., Kezuka, S., Ikeno, T., Yamada, T. *CL* **33**, 614 (2004).
[28]Misumi, Y., Matsumoto, K. *ACIEE* **41**, 1031 (2002).
[29]Kudyba, I., Raczko, J., Jurczak, J. *TL* **44**, 8681 (2003).
[30]Andreana, P.R., Liu, C.C., Schreiber, S.L. *OL* **6**, 4231 (2004).
[31]Pinho e Melo, T.M.V.D., Cardoso, A.L., Gonsalves, A.M.d'A.R., Pessoa, J.C., Paixao, J.A., Beja, A.M. *EJOC* 4830 (2004).

Addition to C=N bond. By virtue of 1,5-asymmetric induction, imines formally derived from the C_2-symmetrical *trans*-4,5-diphenyl-2-amino-1,3-dioxolane add organolithium compounds selectively.[1] Excellent enantioselectivity is observed in the addition of R_2Zn to *N*-phosphinoyl arylimines in the presence of (1*R*,2*S*)-*N*-benzyl-1,2-diphenyl-2-aminoethanol.[2] For those reactions catalyzed by $Cu(OTf)_2$ chiral ligands represented by **28**[3] and **29**[4] are highly efficacious.

28

29

Chiral *t*-butanesulfinimines are sources of α-aminoorganostannanes owing to their facile reaction with R_3SnLi.[5] Rh-catalyzed addition of $ArB(OH)_2$ also proceeds well.[6] On the other hand, *N*-arenesulfonamides also give chiral diarylmethylamine derivatives on reaction with either $ArB(OH)_2$ or $(i\text{-}PrO)_3TiAr$, in the presence of proper ligands.[7,8]

Synthesis of homoallylic amines adapting established protocols while adding asymmetric inducers is relatively straightforward. Employing tetraallylsilane as pronucleophile, activation by Bu_4NF is required and a chiral bis-π-allylpalladium complex supplies stereochemical information.[9] In the addition to *N*-acylhydrazones by $CH_2{=}CHCH_2SiX_3$[10] or 1,2-diamine-ZnF_2 complex[11] chiral induction is proven. Of course, chiral information imbedded in the substrate **30** is well transferred during the reaction.[12]

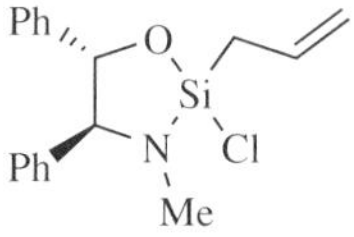

30

A route to chiral α-allylglycine involves formation of allylindium(I) reagents via Pd/In exchange and addition to the *N*-benzyloxyiminoacetyl derivative of Oppolzer's sultam.[13] Equilibration of the adducts is allowed only when water is absent therefore formation of the branched products is due to rapid trapping by water.[14]

The three-component synthesis of propargylic amines is readily rendered asymmetric. Using organozinc addition to imines generated in situ from aldehydes and primary amines, Zr catalyst and a peptide ligand are present.[15] The copper-catalyzed reaction is favored by Quinap (**31**).[16,17] Catalyzed by chloramphenicol derivatives the addition of alkynylzincs to cyclic *N*-acyl ketimines affords products with extremely high ee.[18]

Access to chiral amines via addition of carbon radicals to *N*-acylhydrazones can be accomplished with a (box)-Cu catalyst[19] or in a substrate-directed manner.[20]

Enantioselective Strecker reaction with Me_3SiCN to provide the cyano group has been extended to ketimines catalyzed by $(i\text{-}PrO)_3Gd$ and the dideoxyglucose derivative **32**.[21,22] Addition to *N*-acylhydrazones is catalyzed by (pybox)Ln complexes.[23]

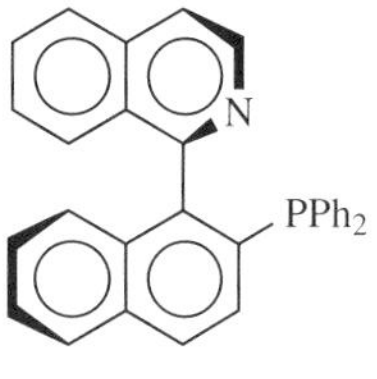

31 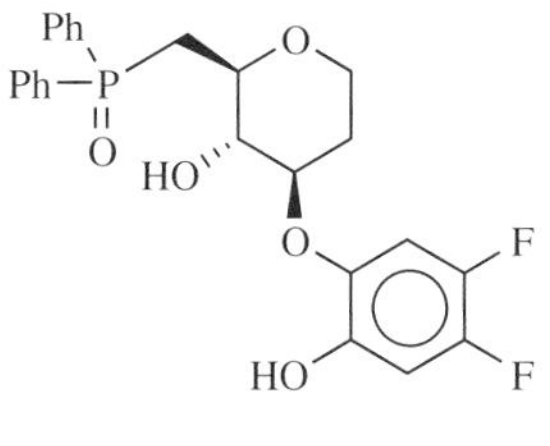 **32**

Mannich reaction using imines is better controlled, hence it is a basis for development of catalytic enantioselective methods. Proline and its many modifications such as (*S*)-5-(pyrrolidin-2-yl)tetrazole,[24] 2-(pyrrolidin-1-ylmethyl)pyrrolidine,[25] and 2-(triflylaminomethyl)pyrrolidine[26] are catalysts of choice at present. Use of **5** enables a direct access to *syn*-2-hydroxy-3-amino ketones from reaction of α-hydroxy ketones.[27]

Further modification of the Mannich reaction involves silyl enol ethers as nucleophiles and the scope of catalytic systems is broadened. α-Amino acid derivatives in conjunction with a silver salt[28] and C_2-symmetric 1,2-diamines with $Cu(OTf)_2$ [29,30] or ZnF_2[31] are applicable. *N*-Benzylidene derivatives of α-amino esters undergo substrate-directed enantioselective addition; interestingly, the $Zn(OTf)_2$-catalyzed reaction with 2-trimethylsiloxy-1,3-dienes is promoted by water.[32]

N-Acylthiazolidine-2-thiones and *O*-methylaldoximes afford *trans*-3,4-disubstituted azetines.[33]

Alkenylcarbamates have similar reactivities to enamines and enol ethers and their reaction with imines is also subject to asymmetric catalysis.[34]

2,3-Diamino acid derivatives are obtained by a Cu(I)-catalyzed enantioselective Mannich reaction of a *N*-diphenylmethyleneglycine ester and aldimines in the presence of **33**.[35] In the presence of β-isocupreidine (cyclic ether derived from quinidine) Baylis-Hillman reaction with imines results in moderate asymmetric induction.[36]

Nitro-Mannich reaction of *N,N*-dibenzyl-α-amino aldehydes under high pressure is substrate-controlled, autocatalytic, and *anti*-selective.[37] Of course such a reaction is influenced by chiral catalysts such as the substituted thiourea **34**.[38] (α-Amino phosphonic acids are similarly acquired via addition of phosphonic acids to aldimines.[39])

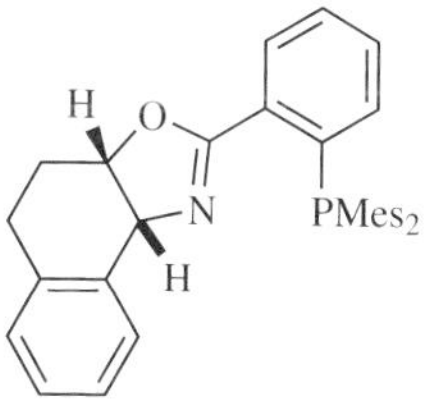

33 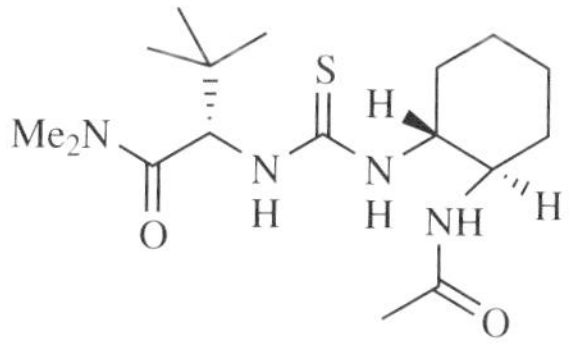 **34**

A related formation of 1,2-diamines involves reductive cross-coupling between nitrones and chiral sulfoximines by SmI_2.[40]

[1]Boezio, A.A., Solberghe, G., Lauzon, C., Charette, A.B. *JOC* **68**, 3241 (2003).
[2]Zhang, H.-L., Jiang, F., Zhang, X.-M., Cui, X., Gong, L.-Z., Mi, A.-Q., Jiang, Y.-Z., Wu, Y.-D. *CEJ* **10**, 1481 (2004).
[3]Boezio, A.A., Pytkowicz, J., Cote, A., Charette, A.B. *JACS* **125**, 14260 (2003).
[4]Soeta, T., Nagai, K., Fujihara, H., Kuriyama, M., Tomioka, K. *JOC* **68**, 9723 (2003).
[5]Kells, K.W., Chong, J.M. *OL* **5**, 4215 (2003).
[6]Weix, D.J., Shi, Y., Ellman, J.A. *JACS* **127**, 1092 (2005).
[7]Tokunaga, N., Otomaru, Y., Okamoto, K., Ueyama, K., Shintani, R., Hayashi, T. *JACS* **126**, 13584 (2004).
[8]Hayashi, T., Kawai, M., Tokunaga, N. *ACIEE* **43**, 6125 (2004).
[9]Fernandes, A., Yamamoto, Y. *JOC* **69**, 735 (2004).
[10]Kobayashi, S., Ogawa, C., Konishi, H., Sugiura, M. *JACS* **125**, 6610 (2003).
[11]Hamada, T., Manabe, K., Kobayashi, S. *ACIEE* **42**, 3927 (2003).
[12]Berger, R., Duff, K., Leighton, J.L. *JACS* **126**, 5686 (2004).

[13]Miyabe, H., Yamaoka, Y., Naito, T., Takemoto, Y. *JOC* **69**, 1415 (2004).
[14]Miyabe, H., Yamaoka, Y., Naito, T., Takemoto, Y. *JOC* **68**, 6745 (2003).
[15]Akullian, L.C., Snapper, M.L., Hoveyda, A.H. *ACIEE* **42**, 4244 (2003).
[16]Gommermann, N., Koradin, C., Polborn, K., Knochel, P. *ACIEE* **42**, 5763 (2003).
[17]Gommermann, N., Knochel, P. *CC* 2324 (2004).
[18]Jiang, B., Si, Y.-G. *ACIEE* **43**, 216 (2004).
[19]Friestad, G.K., Shen, Y., Ruggles, E.L. *ACIEE* **42**, 5061 (2003).
[20]Friestad, G.K., Marie, J.-C., Deveau, A.M. *OL* **6**, 3249 (2004).
[21]Masumoto, S., Usada, H., Suzuki, M., Kanai, M., Shibasaki, M. *JACS* **125**, 5634 (2003).
[22]Kato, N., Suzuki, M., Kanai, M., Shibasaki, M. *TL* **45**, 3147, 3153 (2004).
[23]Keith, J.M., Jacobsen, E.N. *OL* **6**, 153 (2004).
[24]Cobb, A.J.A., Shaw, D.M., Longbottom, D.A., Gold, J.B., Ley, S.V. *OBC* **3**, 84 (2005).
[25]Zhuang, W., Saaby, S., Jorgensen, K.A. *ACIEE* **43**, 4476 (2004).
[26]Wang, W., Wang, J., Li, H. *TL* **45**, 7243 (2004).
[27]Trost, B.M., Terrell, L.R. *JACS* **125**, 338, 2410 (2003).
[28]Josephsohn, N.S., Snapper, M.L., Hoveyda, A.H. *JACS* **126**, 3734 (2004).
[29]Kobayashi, S., Matsubara, R., Nakamura, Y., Kitagawa, H., Sugiura, M. *JACS* **125**, 2507 (2003).
[30]Kobayashi, S., Koyohara, H., Nakamura, Y., Matsubara, R. *JACS* **126**, 6558 (2004).
[31]Hamada, T., Manabe, K., Kobayashi, S. *JACS* **126**, 7768 (2004).
[32]Ishimaru, K., Kojima, T. *JOC* **68**, 4959 (2003).
[33]Ambhaikar, N.B., Snyder, J.P., Liotta, D.C. *JACS* **125**, 3690 (2003).
[34]Matsubara, R., Nakamura, Y., Kobayashi, S. *ACIEE* **43**, 1679 (2004).
[35]Barnardi, L., Gothelf, A.S., Hazell, R.G., Jorgensen, K.A. *JOC* **68**, 2583 (2003).
[36]Kawahara, S., Nakano, A., Esumi, T., Iwabuchi, Y., Hatakeyama, S. *OL* **5**, 3103 (2003).
[37]Misumi, Y., Matsumoto, K. *ACIEE* **41**, 1031 (2002).
[38]Yoon, T.P., Jacobsen, E.N. *ACIEE* **44**, 466 (2005).
[39]Joly, G.D., Jacobsen, E.N. *JACS* **126**, 4102 (2004).
[40]Zhong, Y.-W., Xu, M.-H., Lin, G.-Q. *OL* **6**, 3953 (2004).

Conjugate additions. There are three different ways to achieve enantioselective conjugate addition with lithium amides. The catalytic method employs a chiral ligand such as (1*S*,2*S*)-1,2-dimethoxy-1,2-diphenylethane,[1] reagent-controlled approach depends on the accessibility of suitable chiral amines,[2] and substrate-directed route requires preparation of amides or esters from available chiral units (e.g., pseudoephedrine[3]).

Synthesis of chiral β-amino acids based on conjugated addition to α,β-unsaturated carboxylic derivatives is often thwarted with difficulty. Better stereoinduction is observed by engaging 2-hydroxy-2-methyl-4-alken-3-ones to a Cu-catalyzed 1,4-addition involving carbamate esters, taking advantage of the bidentate nature of the Michael acceptors. A bis(oxazoline) ligand provides asymmetric environment for adduct formation.[4] Oxidative cleavage of the α-hydroxy ketone unit of the products delivers the N-protected β-amino acids.

A method for asymmetric hydration of conjugated imides involves two steps using salicylaldehyde oxime as nucleophile and a (salen)-Al catalyst.[5] The adducts are hydrogenolyzed. A relay 1,3/1,3-stereoinduction apparently operates during addition of (6*S*)-methyltetrahydro-2-pyranol to alkylidenemalonic esters[6] (and conjugated nitroalkenes[7]). Addition of thiols is easier to achieve on chiral substrates (e.g., esters of protected inositols).[8] An intramolecular sulfur atom-transfer is involved in the conversion of *N*-alkenoyloxazolidine-2-thiones.[9]

Many different combinations of metal ions and chiral ligands are found effective for promoting Michael reaction of active *C*-nucleophiles. Ruthenium complexes,[10,11] $Sc(OTf)_3$ with **35**,[12] (salen)-Al complex,[13,14] Ni complex of **36**[15] and oxazaborolidinone **37**[16] are representative. On exposure to imidazolidinecarboxylic acid **38** β-keto esters and enones undergo tandem Michael-aldol reactions in very high ee.[17]

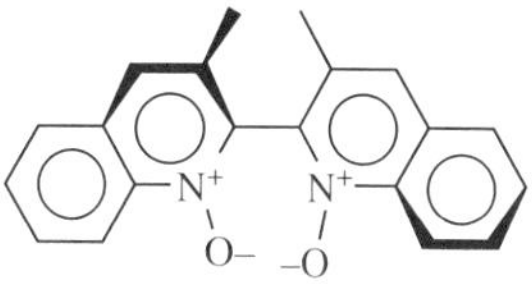

35

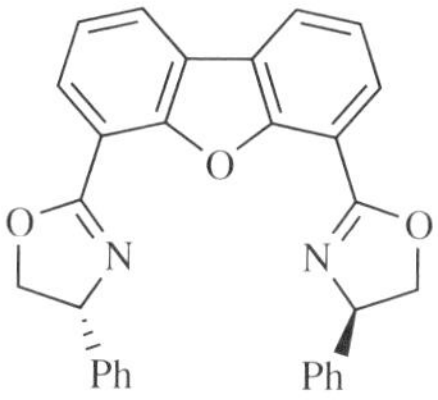

36

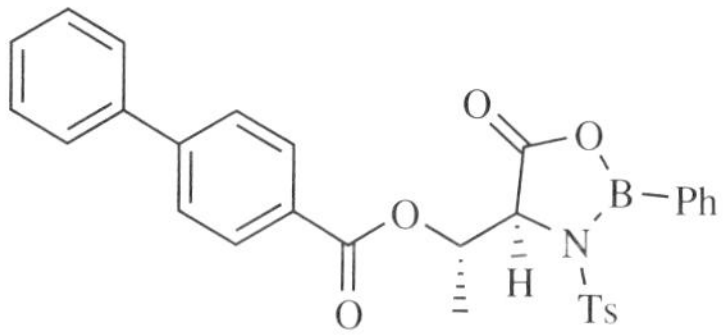

37

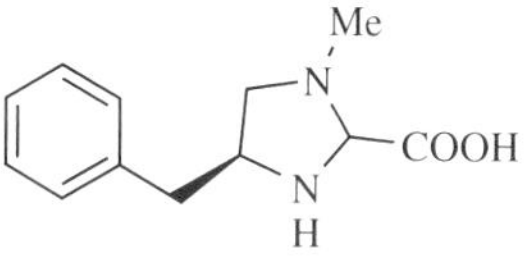

38

Good conversion is observed in Michael addition of aldehydes to enones using (*S*)-2-[bis(3,5-dimethylphenyl)methyl]pyrrolidine as catalyst and a mixture of THF and hexafluoroisopropanol as solvent.[18] With tautomerizable imines of cyclic ketones reaction with electron-deficient alkenes can go through the aza-ene reaction mechanism to provide the Michael adducts.[19]

Copper(II) triflate and bis(oxazoline) ligands form excellent chiral catalysts for β-alkylation (Michael reaction) of indoles.[20,21] The Sc complex of PYBOX **27**[22] and (salen)-Al complexes[23] are equally effective.

2-Trimethylsiloxyfurans react with Michael acceptors at C-5 of the furan ring to afford butenolides. A salt of the imidazolidinone prepared from (*S*)-alanine *N*-methylamide and pivalaldehyde is an excellent catalyst for the reaction.[24]

An asymmetric synthesis of succinic semialdehyde derivatives starts from a Michael addition between (*S*)-2-(diphenylmethoxymethyl)-1-methyleneaminopyrrolidine and alkylidenmalonic esters which is catalyzed by MgI_2.[25]

Conjugated nitroalkenes have similar capacity to conjugated carbonyl compounds as Michael acceptors, therefore catalyst systems are closely related. Cinchona alkaloids,[26] proline derivatives including 5-(pyrrolidin-2-yl)tetrazole,[27] (*S*)-2-(2-pyrrolidinylmethyl)-4-dimethylaminopyridine,[28] and 2-(triflylaminomethyl)pyrrolidine,[29] bipyrrolidine

39[30] as well as a 1,2-diamine-Ru complex[31] have been scrutinized, mostly with satisfactory results. Phosphoramidite **40** serves in the Cu-catalyzed addition of Et_2Zn.[32]

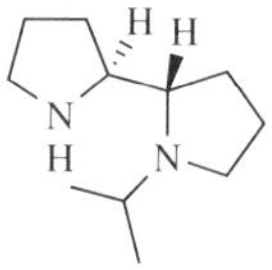

39 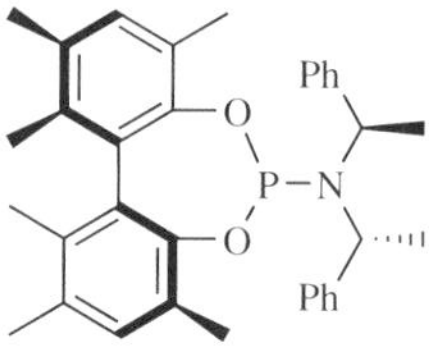 **40**

Asymmetric hydrazination using azodicarboxylic esters as Michael acceptors has received attention due to the fact that the adducts are readily converted into amines. Excellent catalysts for the reaction with β-keto esters and cyanoacetatic esters are $Cu(OTf)_2$ with Ph-BOX[33] and modified cinchona alkaloids.[34]

By an intramolecular Stetter reaction in the presence of **41** which contains a pentafluorophenyl group, carbocyclic and heterocyclic 5- and 6-membered ketones are obtained in much better yields than before.[35] Dimerization of enones catalyzed by *N*-(9-anthracenylmethyl)dihydrocinchonidinium bromide is highly enantioselective, and the products are useful precursors of γ-keto acids bearing asymmetry at the α-carbon.[36]

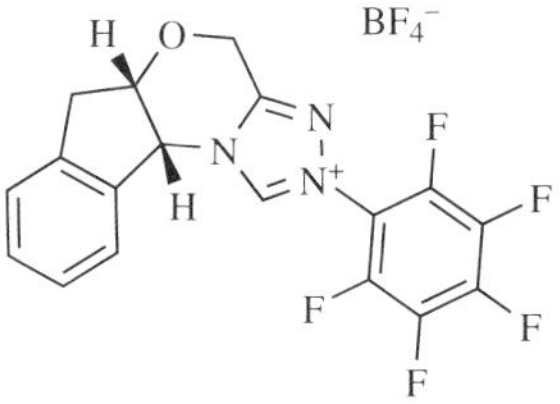

41

Cyanide group introduction to the β-position of conjugated imides is subject to exalted asymmetric catalysis by a (salen)-Al complex and (Py-box)-$ErCl_3$.[37]

In hydroarylation of enones (and other conjugated systems) with $ArB(OH)_2$, Rh complexes formed in situ with (bisphosphino)bipyridyl **42**[38] and the bicycle[2.2.2]octadiene **43**, [39] which is prepared from (-)-carvone, show very good catalytic activities. Using Ph_3Bi as donor the reaction is mediated by Pd(II) and Cu(II) salts with chiral ligands for the Pd ion.[40]

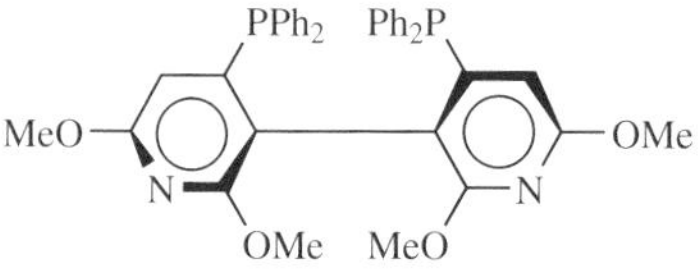

42 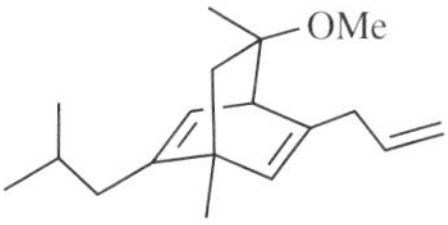 **43**

Alkynyldimethylaluminum reagents submit their unsaturated residues to enones asymmetrically in the presence of nickel catalysts such as **44**,[41] whereas by copper-catalysis together with peptidyl phosphines[42,43] or simpler *P,N*-ligands[44] alkylzinc reagents also accomplish the addition. To acyclic enones the copper-catalyzed addition of Grignard reagents is enantioselective due to ferrocenyl ligand **45**.[45] In the oxazolidinone-controlled asymmetric addition of lithium silylcuprates catalyzed by CuI, the use of dimethyl sulfide as solvent is particularly beneficial to sterically hindered β,β-disubstituted substrates.[46]

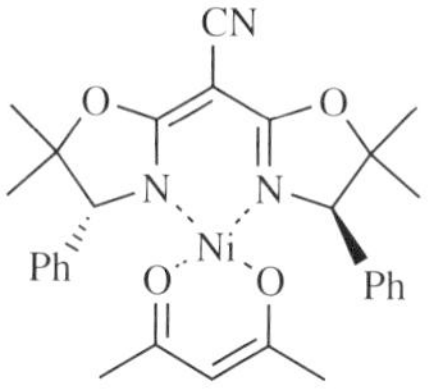

44 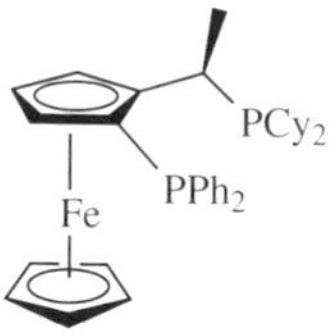 **45**

BOX-type ligands are useful for promoting free radical 1,4-additions in an enantioselective manner.[47–49]

[1]Doi, H., Sakai, T., Iguchi, M., Yamada, K., Tomioka, K. *JACS* **125**, 2886 (2003).
[2]Davies, S.G., Epstein, S.W., Garner, A.C., Ichihara, O., Smith, A.D. *TA* **13**, 1555 (2002).
[3]Etxebarria, J., Vicario, J.L., Badia, D., Carillo, L. *JOC* **69**, 2588 (2004).
[4]Palomo, C., Oiarbide, M., Rajkumar, H., Kelso, M., Gomez-Bengoa, E., Garcia, J.M. *JACS* **126**, 9188 (2004).
[5]Vanderwal, C.D., Jacobsen, E.N. *JACS* **126**, 14724 (2004).
[6]Buchanan, D.J., Dixon, D.J., Hernandez-Juan, F.A. *OL* **6**, 1357 (2004).
[7]Adderley, N.J., Buchanan, D.J., Dixon, D.J., Laine, D.I. *ACIEE* **42**, 4241 (2003).
[8]Cousins, G., Falshaw, A., Hoberg, J.O. *OBC* **2**, 2272 (2004).
[9]Palomo, C., Oiarbide, M., Dias, F., Lopez, R., Linden, A. *ACIEE* **43**, 3307 (2004).
[10]Ikariya, T., Wang, H., Watanabe, M., Murata, K. *JOMC* **689**, 1377 (2004).
[11]Guo, R., Morris, R.H., Song, D. *JACS* **127**, 516 (2005).
[12]Nakajima, M., Yamamoto, S., Yamaguchi, Y., Nakamura, S., Hashimoto, S. *T* **59**, 7307 (2003).
[13]Taylor, M.S., Jacobsen, E.N. *JACS* **125**, 11204 (2003).
[14]Taylor, M.S., Zalatan, D.N., Lerchner, A.M., Jacobsen, E.N. *JACS* **127**, 1313 (2005).
[15]Itoh, K., Oderaotoshi, Y., Kanemasa, S. *TA* **14**, 635 (2003).
[16]Wang, X., Adachi, S., Iwai, H., Takatsuki, H., Fujita, K., Kubo, M., Oku, A., Harada, T. *JOC* **68**, 10046 (2003).
[17]Halland, N., Aburel, P.S., Jorgensen, K.A. *ACIEE* **43**, 1272 (2004).
[18]Melchiorre, P., Jorgensen, K.A. *JOC* **68**, 4151 (2003).
[19]Keller, L., Dumas, F., d'Angelo, J. *EJOC* 2488 (2003).
[20]Jorgensen, K.A. *S* 1117 (2003).
[21]Zhou, J., Tang, Y. *CC* 432 (2004).
[22]Evans, D.A., Scheidt, K.A., Fandrick, K.R., Lam, H.W., Wu, J. *JACS* **125**, 10780 (2003).
[23]Bandini, M., Fagioli, M., Melchiorre, P., Melloni, A., Umani-Ronchi, A. *TL* **44**, 5843 (2003).
[24]Brown, S.P., Goodwin, N.C., MacMillan, D.W.C. *JACS* **125**, 1192 (2003).

[25]Lassaletta, J.M., Vazquez, J., Prieto, A., Fernandez, R., Raabe, G., Enders, D. *JOC* **68**, 2698 (2003).
[26]Li, H., Wang, Y., Tang, L., Deng, L. *JACS* **126**, 9906 (2004).
[27]Cobb, A.J.A., Longbottom, D.A., Shaw, D.M., Ley, S.V. *CC* 1808 (2004).
[28]Ishii, T., Fujioka, S., Sekiguchi, Y., Kotsuki, H. *JACS* **126**, 9558 (2004).
[29]Wang, W., Wang, J., Li, H. *ACIEE* **44**, 1369 (2005).
[30]Andrey, O., Alexakis, A., Bernardinelli, G. *OL* **5**, 2559 (2003).
[31]Watanabe, M., Ikagawa, A., Wang, H., Murata, K., Ikariya, T. *JACS* **126**, 11148 (2004).
[32]Choi, H., Hua, Z., Ojima, I. *OL* **6**, 2689 (2004).
[33]Marigo, M., Juhl, K., Jorgensen, K.A. *ACIEE* **42**, 1367 (2003).
[34]Liu, X., Li, H., Deng, L. *OL* **7**, 167 (2005).
[35]Kerr, M.S., Rovis, T. *JACS* **126**, 8876 (2004).
[36]Zhang, F.-Y., Corey, E.J. *OL* **6**, 3397 (2004).
[37]Sammis, G.M., Danjo, H., Jacobsen, E.N. *JACS* **126**, 9928 (2004).
[38]Shi, Q., Xu, L., Li, X., Jia, X., Wang, R., Au-Yeung, T.T.-L., Chan, A.S.C., Hayashi, T., Cao, R., Hong, M. *TL* **44**, 6505(2003).
[39]Defieber, C., Paquin, J.-F., Serna, S., Carreira, E.M. *OL* **6**, 3873 (2004).
[40]Nishikata, T., Yamamoto, Y., Miyaura, N. *CC* 1822 (2004).
[41]Kwak, Y.-S., Corey, E.J. *OL* **6**, 3385 (2004).
[42]Breit, B., Laungani, A.C. *TA* **14**, 3823 (2003).
[43]Mampreian, D.M., Hoveyda, A.H. *OL* **5**, 2829 (2003).
[44]Krauss, I.J., Leighton, J.L. *OL* **5**, 3201 (2003).
[45]Lopez, F., Harutyunyan, S.R., Minnaard, A.J., Feringa, B.L. *JACS* **126**, 12784 (2004).
[46]Dambacher, J., Bergdahl, M. *JOC* **70**, 580 (2005).
[47]Sibi, M.P., Zimmerman, J., Rheault, T. *ACIEE* **42**, 4521 (2003).
[48]Sibi, M.P., Patil, K. *ACIEE* **43**, 1235 (2004).
[49]Sibi, M.P., Petrovic, G., Zimmerman, J. *JACS* **127**, 2390 (2005).

Addition to C=C bond. Carboalumination of 1-alkenes followed by oxidation gives primary alcohols while introducing an asymmetric center at C-2. The Zr-catalyzed reaction can be chirally manipulated by choosing bis(1-neomenthylindenyl)zirconium dichloride as catalyst.[1]

Addition of ethene to styrenes in the presence of bis(allylbromonickel) gives rise to the (*R*)-3-arylpropenes if **46** is present.[2]

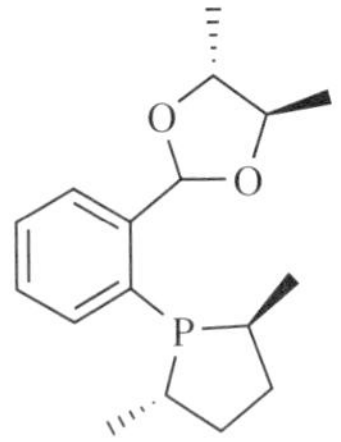

46

[1]Zeng, X., Zeng, F., Negishi, E. *OL* **6**, 3245 (2004).
[2]Zhang, A., RajanBabu, T.V. *OL* **6**, 3159 (2004).

Cycloadditions. Ammonium ylides formed in situ from modified cinchona alkaloids and *t*-butyl bromoacetate react with electron-deficient alkenes to give chiral cyclopropanecarboxylic esters.[1] Bicyclo[n.1.0]alkanones bearing an acyl side chain at the cyclopropane ring are synthesized in an analogous manner from 1-chloro-2-alkanones with an enone unit along the carbon chain.[2] Copper-catalyzed decomposition of diazoalkanes for cyclopropanation and the assistance by bis(oxazoline) ligands to render the reaction enantioselective are well-established, bipyridyl **47** is another useful ligand.[4] Intramolecular reaction of α-diazo-β-keto sulfones is a further example.[3] Chiral porphyrin-cobalt complexes in which the ligands carry in *meso*-positions 2,6-bis(acylamino)phenyl groups are capable of catalyzing enantioselective and diastereoselective cyclopropanation with diazoacetic esters.[5] As for rhodium-catalyzed cyclopropanation, a new type of ligands is represented by **48**.[6]

High enantioselectivity is associated with epoxide synthesis from aldehydes utilizing sulfonium ylides **49**.[7] Efficient aziridination is carried out by decomposition of arenesulfonyl azides in the presence of a C_2-symmetrical (salen)-Ru(CO) complex.[8] A chiral nosyloxy carbamate based on the bornane skeleton is useful for aziridination of electron-deficient alkenes.[9]

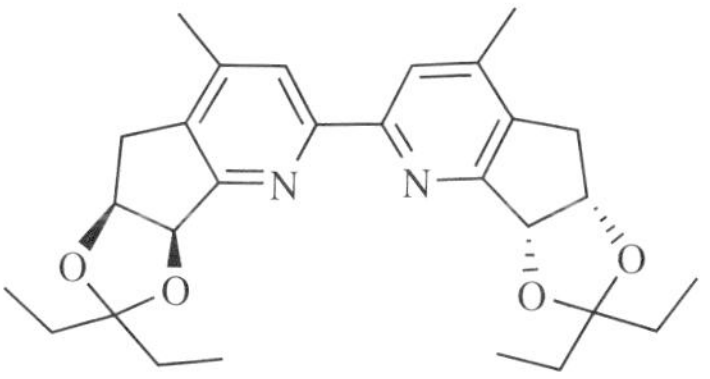

47

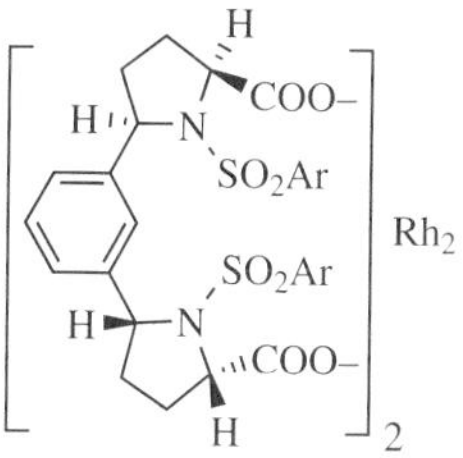

48 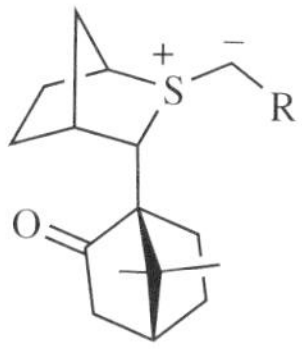 **49**

Relatively few examples of asymmetric synthesis of cyclobutanes are known. Based on the [2 + 2]photocycloaddition (-)-8-(4-nitrophenyl)menthyl 3-oxo-1-cyclohexenecarboxylate forms bicycle[4.2.0]octanone derivative with ethene in a highly diastereoselective fashion.[10] Ru-catalyzed cycloaddition of *N*-alkynoyl derivatives of Oppolzer's sultam and chiral oxazolidinones to norbornenes provides cyclobuteno-fused norbornane analogues.[11]

Chiral catalysts of several different classes are found to mediate asymmetric formation of β-lactones from ketenes and aldehydes. These include *O*-trimethylsilylquinidine,[12] tridentate aminobidsulfonamide-aluminum complexes (**50**)[13] and ferrocenylpyridine **4B**.[14]

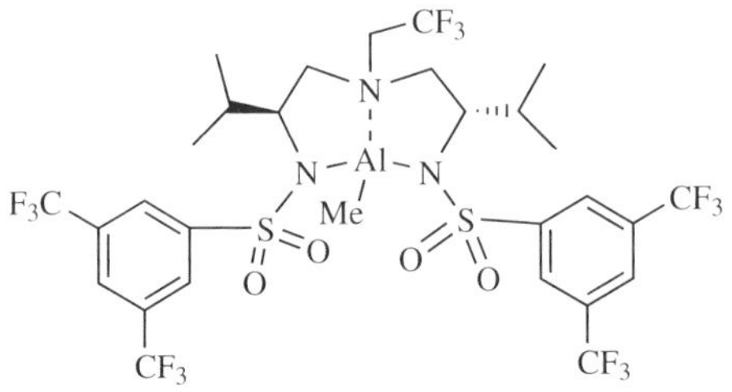

50

Access to β-lactams is perhaps nore attractive to synthetic chemists. A catalytic route using Lewis acid [In(OTf)$_3$] in conjuction with *O*-benzoylquinine to achieve reaction of acyl chlorides with imines provides excellent results.[15] A phosphaferrocene is used to induce chrality in the Cu-catalyzed intramolcular Kinugasa reaction to generate the β-lactam system.[16] The same reaction employing air- and water-tolerant $Cu(ClO_4)_2 \cdot 6H_2O$ and tris(oxazoline) **51** shows adequate levels of enantioselectivity and diastereoselectivity.[17] [2 + 2]Cycloaddition between ketenes and SAMP-hydrazone of aldehydes is also accomplished.[18]

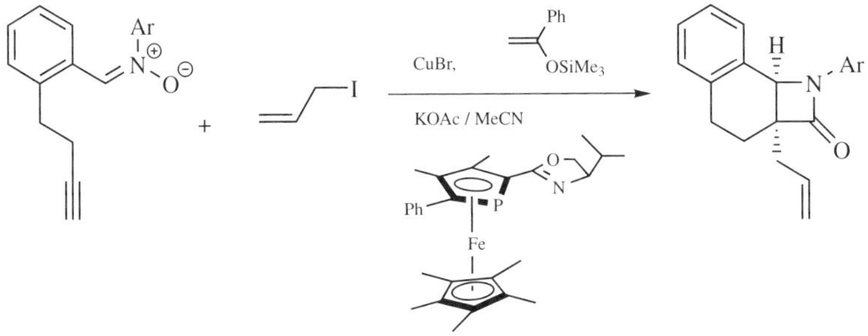

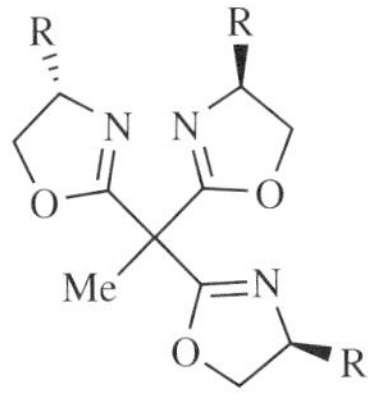

51

Formation of five-membered azacycles in chiral modifications by [3 + 2]cycloaddition is of interest because those compounds often exhibit physiological properties or serve as intermediates for pharmaceutics. Addition of azomethine imines to 1-alkynes is copper-catalyzed and it becomes enantioselective in the presence of a phosphaferrocene (**52**).[19] A pyrrolidine synthesis from Ag-catalyzed azomethine ylide addition involves the C_2-symmetrical **53**.[20]

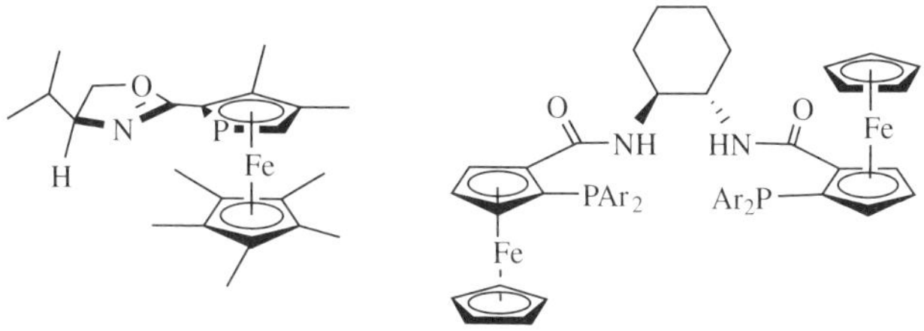

52 **53**

Successful asymmetric synthesis of isoxazolines[21] and isoxazolidines[22] from nitrile oxides and nitrones, respectively, has also been described. Benzo[*c*]pyrylium-3-oxides generated from 2-diazoacetylbenzoic esters are trapped by aldehydes and the [3 + 2]cycloaddition affords chiral adducts in the presence of a (pybox)-Sc$(OTf)_3$ complex.[23]The insertion of isocyanates into a ring bond of *N*-benzyl-2-vinylaziridines in the presence of the naphthalene analogue of **9** manifests dynamic kinetic asymmetric phenomenon in cycloaddition.[24]

A more general catalytic system for asymmetric Diels-Alder reaction involves triflimide activation of chiral oxazaborolidine[25] and protonated species.[26] Addition of another Lewis acid (e.g., $SnCl_4$) to coordinate with the nitrogen atom also provides activation and creates a moisture-tolerant catalyst.[27] Enantioselective Diels-Alder reaction is another application of tris(oxazoline) **51**,[28] whereas a new catalyst for extremely effective reactions of 2-substituted enals is **54**.[29]

By means of a dynamic kinetic resolution using lipase and dimeric (hexamethylbenzene)$RuCl_2$, the acylation of 3-vinyl-2-cyclohexenol and subsequent intramolecular Diels-Alder reaction are accomplished in asymmetric manner.[30]

Asymmetric Diels-Alder reaction with photoenols can be induced by Kemp's acid derivative **55**.[31]

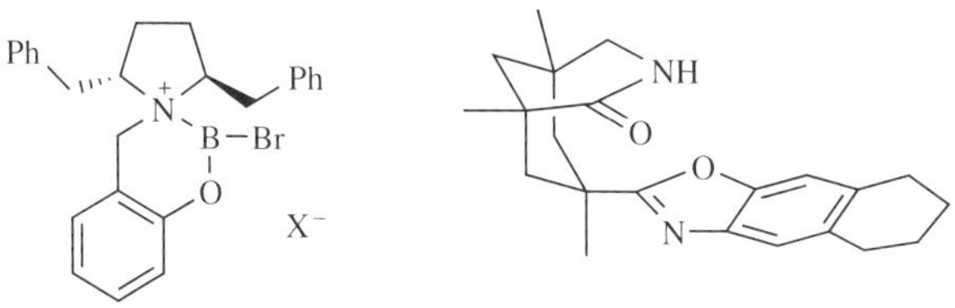

54 **55**

Tetrahydropyrans and dihydro-γ-pyridones with a specified absolute configuration at C-6 are accessible by hetero-Diels-Alder reactions between activated dienes and glyoxylic esters and imines, respectively. Since these reactions are subject to catalysis by Lewis acids, opportunity for asymmetric induction presents itself. New results indicate the usefulness of Cr-complex **56**,[32,33] AgOAc with **57**,[34] and $AgClO_4$ with a *P,S*-substituted ferrocene-CuBr complex.[35] With 2-nitrosopyridines as dienophile, asymmetric hetero-Diels-Alder reaction is catalyzed by **58**.[36]

56

57

58

[4 + 3]Cycloadditions of furans and 4-siloxy-2,4-alkadienals are subject to asymmetric induction by an imidazolidinone derived from phenylalanine.[37]

[1]Papageorgiou, C.D., de Dios, M.A.C., Ley, S.V., Gaunt, M.J. *ACIEE* **43**, 4641 (2004).
[2]Bremeyer, N., Smith, S.C., Ley, S.V., Gaunt, M.J. *ACIEE* **43**, 2681 (2004).
[3]Honma, M., Sawada, T., Fujisawa, Y., Utsugi, M., Watanabe, H., Umino, A., Matsumura, T., Hagihara, T., Takano, M., Nakada,M. *JACS* **125**, 2860 (2003).
[4]Lyle, M.P.A., Wilson, P.D. *OL* **6**, 855 (2004).
[5]Chen, Y., Fields, K.B., Zhang, X.P. *JACS* **126**, 14718 (2004).
[6]Davies, H.M.L., Venkataramani, C. *OL* **5**, 1403 (2003).
[7]Aggarwal, V.K., Bae, I., Lee, H.-Y., Richardson, J., Williams, D.T. *ACIEE* **42**, 3274 (2003).
[8]Omura, K., Murakami, M., Uchida, T., Irie, R., Katsuki, T. *CL* **32**, 354 (2003).
[9]Fioravanti, S., Morreale, A., Pellacani, L., Tardella, P.A. *TL* **44**, 3031 (2003).
[10]Furutani, A., Tsutsumi, K., Nakano, H., Morimoto, T., Kakiuchi, K. *TL* **45**, 7621 (2004).
[11]Villeneuve, K., Tam, W. *ACIEE* **43**, 610 (2004).
[12]Zhu,C., Shen, X., Nelson, S.G. *JACS* **126**, 5352 (2004).
[13]Nelson, S.G., Zhu,C., Shen, X. *JACS* **126**, 14 (2004).
[14]Wilson, J.E., Fu, G.C. *ACIEE* **43**, 6358 (2004).
[15]France, S., Shah, M.H., Weatherwax, A., Wack, H., Roth, J.P., Lectka, T. *JACS* **127**, 1206 (2005).
[16]Shintani, R., Fu, G.C. *ACIEE* **42**, 4082 (2003).
[17]Ye, M.-C., Zhou, J., Huang, Z.-Z., Tang, Y. *CC* 2554 (2003).

[18]Martin-Zamora, E., Ferrete, A., Llera, J.M., Munoz, J.M., Pappalardo, R.R., Fernandez, R., Lassaletta, J.M. *CEJ* **10**, 6111(2004).
[19]Shintani, R., Fu, G.C. *JACS* **125**, 10778 (2003).
[20]Longmire, J.M., Wang, B., Zhang, X. *JACS* **124**, 13400 (2002).
[21]Sibi, M.P., Itoh, K., Jasperse, C.P. *JACS* **126**, 5366 (2004).
[22]Kezuka, S., Ohtsuki, N., Mita, T., Kogami, Y., Ashizawa, T., Ikeno, T., Yamada, T. *BCSJ* **76**, 2197 (2003).
[23]Suga, H., Inoue, K., Inoue, S., Kakehi, A., Shiro, M. *JOC* **70**, 47 (2005).
[24]Trost, B.M., Fandrick, D.R. *JACS* **125**, 11836 (2003).
[25]Ryu, D.H., Corey, E.J. *JACS* **125**, 6388 (2003).
[26]Zhou, G., Hu, Q.-Y., Corey, E.J. *OL* **5**, 3979 (2003).
[27]Futatsugi, K., Yamamoto, H. *ACIEE* **44**, 1484 (2005).
[28]Zhou, J., Tang, Y. *OBC* **2**, 429 (2004).
[29]Sprott, K.T., Corey, E.J. *OL* **5**, 2465 (2003).
[30]Akai, S., Tanimoto, K., Kita, Y. *ACIEE* **43**, 1407 (2004).
[31]Grosch, B., Orlebar, C.N., Herdtweck, E., Massa, W. *ACIEE* **42**, 3693 (2003).
[32]Gao, X., Hall, D.G. *JACS* **125**, 9308 (2003).
[33]Kwiatkowski, P., Asztemborska, M., Jurczak, J. *SL* 1755 (2004).
[34]Josephsohn, N.S., Snapper, M.L., Hoveyda, A.H. *JACS* **125**, 4018 (2003).
[35]Mancheno, O.G., Arrayas, R.G., Carretero, J.C. *JACS* **126**, 456 (2004).
[36]Yamamoto, Y., Yamamoto, H. *JACS* **126**, 4128 (2004).
[37]Harmata, M., Ghosh, S.K., Hong, X., Wacharasindhu, S., Kirchhoefer, P. *JACS* **125**, 2058 (2003).

Coupling reactions. Pinacol coupling of ArCHO proceeds to give mainly the *dl*-isomers in moderate to good ee when catalyzed by titanium complex of the C_2-symmetrical bisimine **59** with Mn as reducing agent in the presence of Me_3SiCl.[1] After lithiation, TADDOL-phosphites are used as catalysts for directed cross-silyl benzoin condensation ($ArCOSiMe_3 + Ar'CHO$).[2]

Ph Ph N N N N

59

Coupling of chiral Grignard reagents with Pd(0) or Ni(0) catalyst proceeds with retention of configuration.[3] With the employment of a chiral 2-(oxazolinylphenyl)diphenylphosphine ligand to form a Ni complex for catalyzing Grignard coupling, ring opening of dinaphthothiophene affords optically active 2'-substituted 1,1'-binaphthalene-2-thiols.[4]

While the asymmetric Heck reaction has undergone development for several years, it still attracts attention. The *P,N*-ligand **60** for Pd-catalysts is superior to BINAP in the synthesis of 3-spiroannulated indolenines.[5] In a synthesis of the furaquinocins the employment of a reductive Heck reaction to form the dihydrobenzofuran unit from a (*R,R*)-ether is the key step.[6]

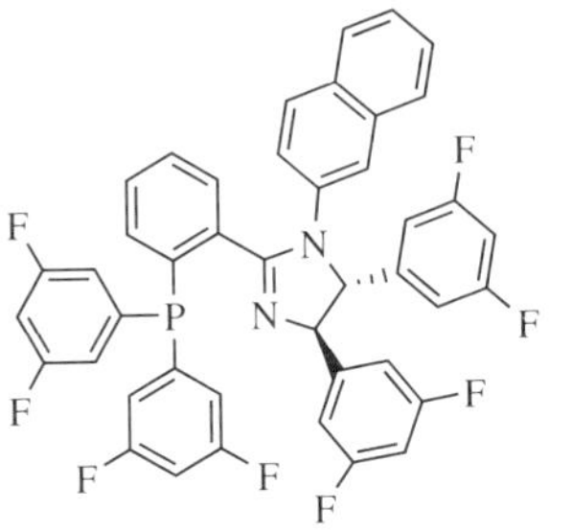

60

Cyclization by an intramolecular Heck reaction of symmetrical bis(homoallylic alcohols) probably involves direction by the hydroxyl group.[7] To achieve chelation-controlled Heck reaction for synthesizing 2-aryl-2-methylcyclopentanones the ketones are converted into enol ethers of (*S*)-*N*-methylprolinol.[8]

In the normal sense Suzuki coupling cannot be used to synthesize chiral compounds (coupling is between two sp^2-carbon atoms), but a desymmetrization strategy involving enantioselective reaction at one site of a *meso*-bistriflate is successful.[9]

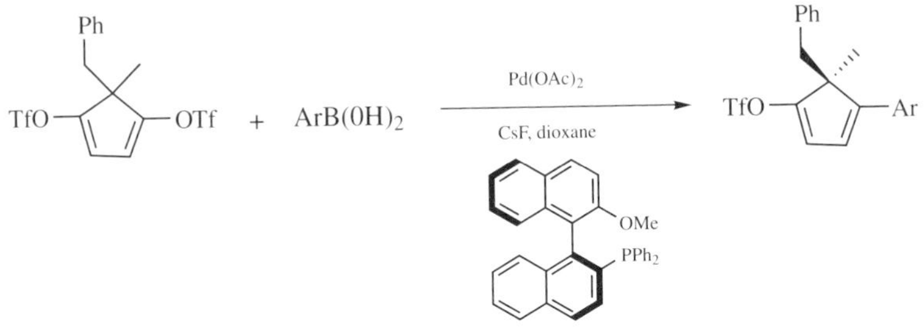

[1]Li, Y.-G., Tian, Q.-S., Zhao, J., Feng, Y., Li, M.-J., You, T.-P. *TA* **15**, 1707 (2004).
[2]Linghu, X., Potnick, J.R., Johnson, J.S. *JACS* **126**, 3070 (2004).
[3]Holzer, B., Hoffmann, R.W. *CC* 732 (2003).
[4]Shimada, T., Cho, Y.-H., Hayashi, T. *JACS* **124**, 13396 (2002).
[5]Busacca, C.A., Grossbach, D., So, R.C., O'Brien, E.M., Spinelli, E.M. *OL* **5**, 595 (2003).
[6]Trost, B.M., Thiel, O.R., Tsui, H.-C. *JACS* **125**, 13155 (2003).
[7]Oestreich, M., Sempere-Culler, F., Machotta, A.B. *ACIEE* **44**, 149 (2005).
[8]Nilsson, P., Larhed, M., Hallberg, A. *JACS* **125**, 3430 (2003).
[9]Willis, M.C., Powell, L.H.W., Claverie, C.K., Watson, S.J. *ACIEE* **43**, 1249 (2004).

Epoxidation and dihydroxylation reactions. An improved synthesis of the ketone catalyst (**61**) for asymmetric epoxidation of alkenes is achieved from D-glucose in 6 steps.[1] Investigation on *N*-aryl analogues of **61** has shown a beneficial substituent effect.[2] Newly introduced catalysts include **62**[3] and **63**.[4] Using tetraphenylphosphonium monopersulfate together with **64** to react with alkenes, epoxides are obtained in moderate ee.[5]

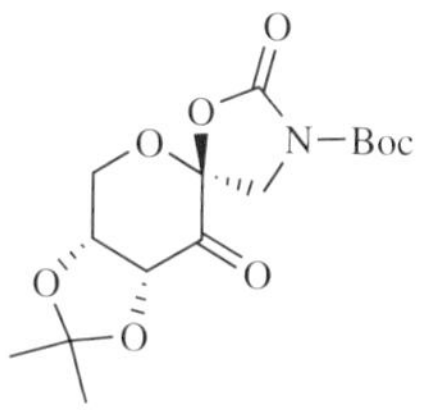

61 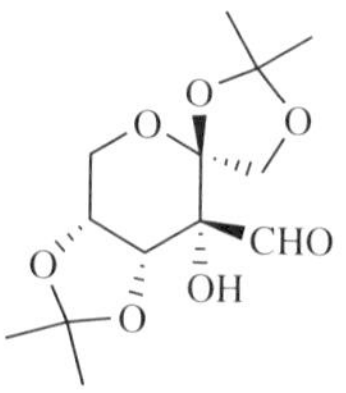 **62** 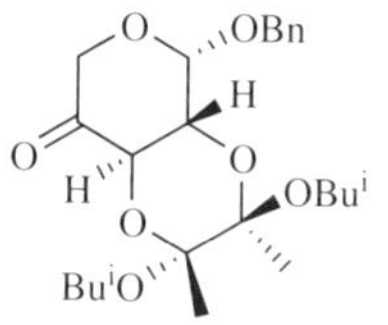 **63** 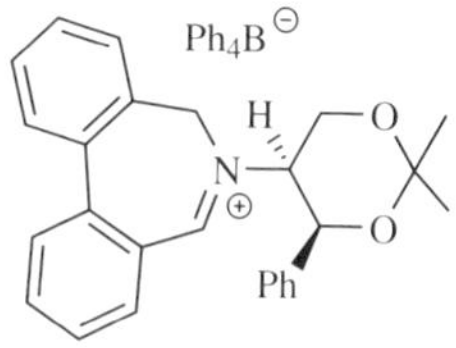 **64**

Addition of hydrogen increase the rate of the (pybox)-Ru -catalyzed asymmetric epoxidation of alkenes by two orders of magnitude.[6] Epoxidation of homoallylic alcohols catalyzed by $(t\text{-BuO})_4Zr$ is chirally assisted by dibenzyl L-tartramide.[7] With chiral TADDOL-derived hydroperoxide and sandwich-type oxovanadium-substituted polyoxometalates for epoxidation of allylic alcohols, enantioselectivity is controlled.[8]

A tandem reaction of aldehydes with alkenylzinc reagents followed by asymmetric epoxidation has been demonstrated.[9] (-)-3-*exo*-Morpholinoisoborneol is the ligand for $(i\text{-PrO})_4Ti$ and oxygen as oxidant for the second-staged transformation. In the presence of the chiral hydroxamic acid **65** satisfactory results are obtained from vanadium-based asymmetric epoxidation of homoallylic alcohols.[10]

Several quaternary ammonium salts derived from cinchona alkaloids serve as catalysts for epoxidation of enones under basic conditions, with H_2O_2,[11] NaOCl,[12] or isocyanuric chloride[13] as activator. Poly-(*S*)-leucine under phase-transfer conditions also shows reasonably good results[14] and moderate success is seen with the amino ether **66**.[15]

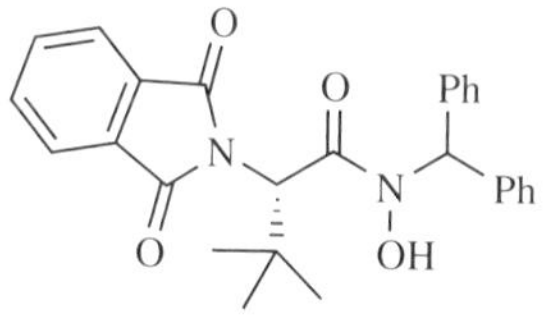

65 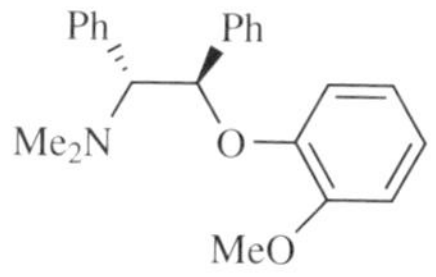 **66**

An excellent ligand for OsO_4-mediated asymmetric dihydroxylation is prepared from 3,6-dichloropyridazine, dihydroquinidine, and the amide derived from (*S*)-proline and indoline.[16]

A modification of the OsO_4-based asymmetric dihydroxylation is to introduce $MeReO_3$ in the system.[17] Following the Rh-catalyzed diboration of alkenes in the presence of (*S*)-Quinap[18] or a ferrocenyldiphosphine[19] oxidative workup with H_2O_2 generates chiral 1,2-diols.

[1]Shu, L., Shen, Y.-M., Burke, C., Goeddel, D., Shi, Y. *JOC* **68**, 4963 (2003).
[2]Shu, L., Shi, Y. *TL* **45**, 8115 (2004).

[3]Bez, G., Zhao, C.-G. *TL* **44**, 7403 (2003).
[4]Shing, T.K.M., Leung, G.Y.C., Yeung, K.W. *TL* **44**, 9225 (2003).
[5]Page, P.C.B., Barros, D., Buckley, B.R., Ardakani, A., Marples, B.A. *JOC* **69**, 3595 (2004).
[6]Tse, M.K., Bhor, S., Klawonn, M., Dobler, C., Beller, M. *TL* **44**, 7479 (2003).
[7]Okachi, T., Murai, N., Onaka, M. *OL* **5**, 85 (2003).
[8]Adam, W., Alsters, P.L., Neumann, R., Saha-Moller, C.R., Seebach, D., Beck, A.K., Zhang, R. *JOC* **68**, 8222 (2003).
[9]Lurian, A.E., Maestri, A., Kelly, A.R., Carroll, P.J., Walsh, P.J. *JACS* **126**, 13608 (2004).
[10]Makita, N., Hoshino, Y., Yamamoto, H. *ACIEE* **42**, 941 (2003).
[11]Jew, S., Lee, J., Jeong, B., Yoo, M., Kim, M., Lee, Y., Lee, J., Choi, S., Lee, K., Lah, M., Park, H. *ACIEE* **44**, 1383 (2005).
[12]Kim, D.Y., Choi, Y.J., Park, H.Y., Juong, C.U., Koh, K.O., Mang, J.Y., Jung, K.-Y. *SC* **33**, 435 (2003).
[13]Ye, J., Wang, Y., Liu, R., Zhang, G., Zhang, Q., Chen, J., Liang, X. *CC* 2714 (2003).
[14]Da, C.-S., Wei, J., Dong, S.-L., Xin, Z.-Q., Liu, D.-X., Xu, Z.-Q., Wang, R. *SC* **33**, 2787 (2003).
[15]Tanaka, Y., Nishimura, K., Tomioka, K. *T* **59**, 4549 (2003).
[16]Huang, J., Corey, E.J. *OL* **5**, 3455 (2003).
[17]Jonsson, S.Y., Adolfsson, H., Backvall, J.-E. *CEJ* **9**, 2783 (2003).
[18]Morgan, J.B., Miller, S.P., Morken, J.P. *JACS* **125**, 8702 (2003).
[19]Morgan, J.B., Morken, J.P. *JACS* **126**, 15338 (2004).

Oxidations. Chiral sulfoxides [RS*(O)R′] and thiosulfinates [RSS*(O)R] are obtained on metal-catalyzed oxidation with H_2O_2 in the presence of appropriate ligands. Systems constituting imines derived from disubstituted salicyaldehydes and chiral aminoethanols together with $Fe(acac)_3$ and $VO(acac)_2$ has been scrutinized.[1,2] A salen ligand is used in the oxidation of sulfides by urea-hydrogen peroxide mediated by $(dme)NbCl_3$.[3]

[1]Legros, J., Bolm, C. *ACIEE* **43**, 4225 (2004).
[2]Ma, Y., Wang, X., Wang, Y., Feng, Y., Zang, Y. *SC* **34**, 501 (2004).
[3]Miyazaki, T., Katsuki, T. *SL* 1046 (2003).

Reduction and hydrogenation. Several chiral ligands to pair with iridium for asymmetric hydrogenation of alkenes have been designed, including those with σ-bonding of *O,N*-type,[1] *P,N*-type,[2] and *C,N*-type.[3] Saturation of the double bonds conjugated to phosphonates[4] and of alkylidenemalonitriles[5] and enones[6] are achieved with different degrees of enantioselectivity, be the hydrogen delivery catalytic or transferred.

Rhodium-based asymmetric hydrogenation of itaconic esters with a bis(tetrahelicenol)/menthyl phosphite ligand **67** reveals the interesting phenomena of matched/mismatched helical and axial chirality.[7] For traditional reasons of practicality and ease of comparison enantioselective hydrogenation of dehydro-α-amino acid derivatives still attracts the most attention. Reactions in the presence of immobilized Rh(binol phosphoramidite) catalyst,[8] with *P*-chirogenic phosphonium salts,[9] particularly with **68** as precursorial ligands [10] and catalyst **69**,[11] both belonging to the diphosphinoethane family, are interesting. In the silica-supported urea ligand terminated with a (2*S*,4*S*)-2-diphenylphosphinomethyl-4-diphenylphosphinopyrrolidine unit the two phosphorus atoms are farther apart.[12] *P,S*-Ligands such as **70** are highly effective to render hydrogenation enantioselective.[13] A relatively simple phosphoramidite ligand **71**[14] for Rh and the air-stable, conformationally rigid complex **72**[15] may also be mentioned. The latter is quite similar to **73**[16]. An *O,N*-bis(diphenylphosphino) derivative of 3-hydroxymethyl-2-azabicyclo[2.2.1]heptane is a useful ligand also.[17] A study comparing homogeneous catalysts and the anchored variants shows that with the latter superior enantioselectivities are achieved.[18]

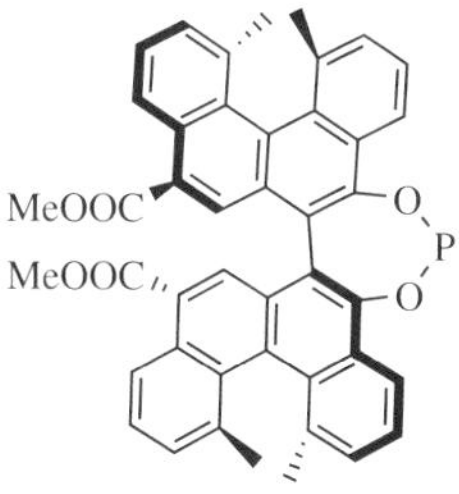

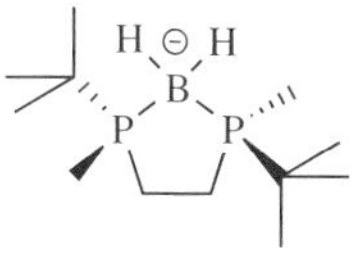

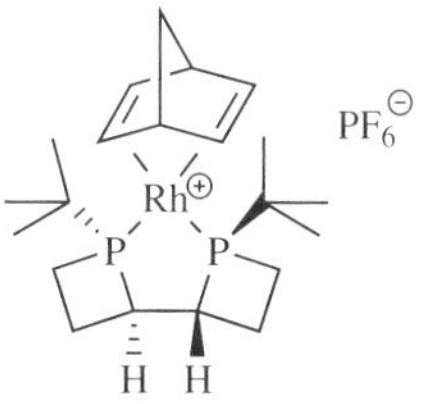

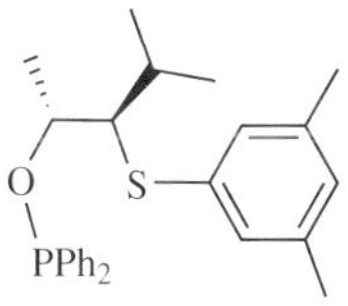

α,β-Unsaturated β-amino esters with an unprotected amino group undergo asymmetric hydrogenation in the presence of a diphosphinoferrocene-ligated Rh catalyst (93–97% ee).[19] A Rh catalyst having only one unhindered quadrant exerts due influence on asymmetric hydrogenation.[20] In a combinatorial approach the use of mixtures of chiral monodentate *P*-ligands is reported.[21]

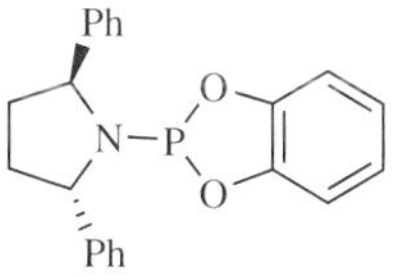

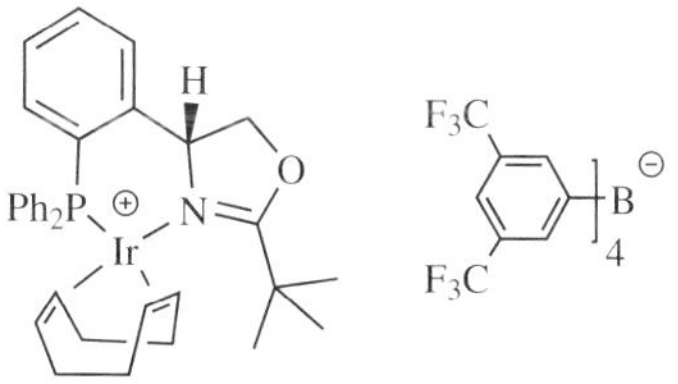

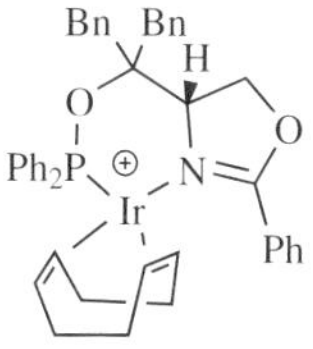

Cyclic β-(acylamino)acrylates are found to be enantioselectively hydrogenated with a Ru complex and a chiral phosphine ligand in good conversion and generally high ee.[22]

α-Acetamino-β-alkoxystyrenes give chiral *N*-acetylbenzylamines, also in a Rh-catalyzed hydrogenation (ligand **74**).[23] Hybrid phosphoramidites of diphenylphosphinoferrocenes and BINOL constitute a new class of practical ligands for Rh-catalyzed asymmetric hydrogenation.[24]

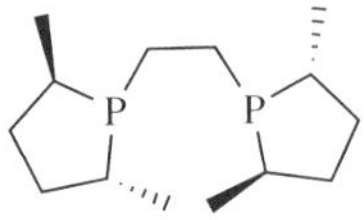

74

Transfer hydrogenation of unsymmetrical ketones (usually aryl methyl ketones for demonstration) catalyzed by Ru catalysts is well developed. It is just different combinations of the metal complexes and chiral ligands are being scrutinized. For example, chiral 1,2-diphenylethylenediamine monotosylate ligands are modifiable by anchoring onto solid supports for reuse using formic acid salts as reducing agent.[25–27] Peptide ligands[28] and Ph-pybox ligands[29] with the Ru complexes are also highly efficient in systems containing NaOH and isopropanol. Spirodiphosphine **75** is a new ligand for Ru-catalyzed hydrogenation.[30] Rhodium- and iridium-centered chiral catalysts[31,32] are also useful in the transfer reduction, although less popular than the Ru counterparts.

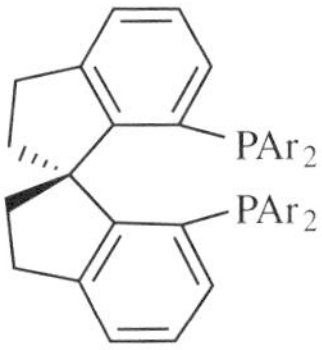

PAr_2
PAr_2

75

Because of its simplicity and high efficiency the CBS-reduction deservedly remains the most popular method for reducing prochiral ketones in research laboratories. New applications include the reduction of allenyl ketones[33] and the preparation of (*R*)-4,4,5-oxazolidinones starting from reduction of α-nitro ketones.[34] Operational modifications such as using the carbonate salt of (*S*)-α,α-diphenylprolinol[32] or *N*-diphenylphosphonyl-α,α-diphenylprolinol[35] in conjuction with $BH_3.SMe_2$ work well. Reduction of diacetylbiphenyls is affected by substitution pattern; more meso contaminants are formed in the cases of *meta*-substituted ketones and a change of experimental conditions ameliorates the situation.[36]

A bidentate 1,6-diaza-1,3,5-triene ligand (**76**) in which one of the imino units is part of a chiral oxazoline binds Zn well and the complex actively promotes enantioselective reduction of ketones by catecholborane.[37] Chiral 1,2-diamines are ligands for asymmetric reduction of ketones by the reducing system of Et_2Zn – PMHS.[38,39] Reductive silylation is also achieved by CuCl-PMHS in the presence of a chiral diphosphinobiphenyl.[40]

Ru-catalyzed asymmetric hydrogenation of α-functionalized ketones has the choice of different diphosphine ligands.[41,42] Using 1,2-diphenylethylenediamine montosylate reduction of α-keto esters with HCOOH proceeds well.[43]

Biphenyl **77** [SYNPHOS] is employed to assist the Ru-catalyzed hydrogenation of β-keto esters.[44] From α-amino-β-keto esters excellent yields of the hydroxy esters are obtained in high enantioselectivity and *anti*-2,3-diastereoselectivity.[45]

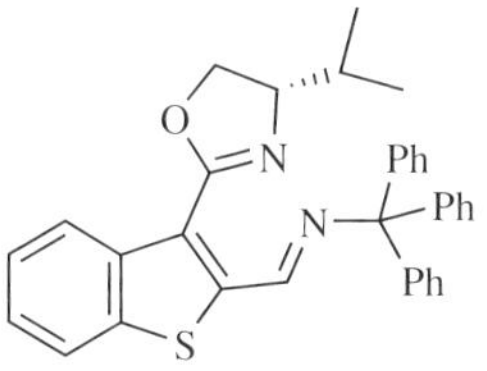

76 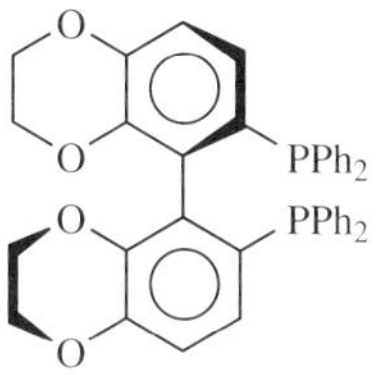 **77**

Acyclic *N*-aryl ketimines are hydrogenated asymmetrically by an iridium complex (**78**),[46] whereas C_2-symmetrical secondary amines can be obtained from chiral primary amines and the corresponding prochiral ketones via Schiff base formation in the presence of $(i\text{-PrO})_4$Ti and then hydrogenation with Pd/C. The whole operation can be carried out without solvent.[47] Borane adducts with chiral oxazaborolidines reduce *N*-alkylketimines and ketoxime ethers.[48]

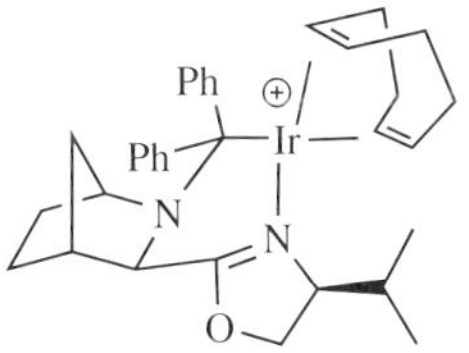

78

2-Substituted quinolines undergo hydrogenation to afford 1,2,3,4-tetrahydro derivatives by Ir-catalysis. The chiral 6,′-dimethoxy-2,2′-bis(diphenylphosphino)biphenyl promotes enantioselective results.[49]

[1]Kallstrom, K., Hedberg, C., Brandt, P., Bayer, A., Andersson, P.G. *JACS* **126**, 14308 (2004).
[2]Tang, W., Wang, W., Zhang, X. *ACIEE* **42**, 943 (2003).
[3]Cui, X., Ogle, J.W., Burgess, K. *CC* 672 (2005).
[4]Rubio, M., Suarez, A., Alvarez, E., Pizzano, A. *CC* 628 (2005).
[5]Chen, Y.-C., Xue, D., Deng, J.-G., Cui, X., Zhu, J., Jiang, Y.-Z. *TL* **45**, 1555 (2004).
[6]Lipshutz, B.H., Servesko, J.M., Petersen, T.B., Papa, P.P., Lover, A.A. *OL* **6**, 1273 (2004).
[7]Nakano, D., Yamaguchi, M. *TL* **44**, 4969 (2003).
[8]Simons, C., Hanefeld, U., Arends, I.W.C.E., Minnaard, A.J., Maschmeyer, T., Sheldon, R.A. *CC* 2830 (2004).
[9]Danjo, H., Sasaki, W., Miyazaki, T., Imamoto, T. *TL* **44**, 3467 (2003).
[10]Miyazaki, T., Sugawara, M., Danjo, H., Imamoto, T. *TL* **45**, 9341 (2004).
[11]Imamoto, T., Oohara, N., Takahashi, H. *S* 1353 (2004).
[12]Aoki, K., Shimada, T., Hayashi, T. *TA* **15**, 1771 (2004).
[13]Evans, D.A., Michael, F.E., Tedrow, J.S., Campos, K.R. *JACS* **125**, 3534 (2003).
[14]Hoen, R., van den Berg, M., Bernsmann, H., Minnaard, A.J., de Vries, J.G., Feringa, B.L. *OL* **6**, 1433 (2004).
[15]Liu, D., Tang, W., Zhang, X. *OL* **6**, 513 (2004).

[16]Smidt, S.P., Menges, F., Pfaltz, A. *OL* **6**, 2023 (2004).
[17]Dubrovina, N.V., Tararov, V.I., Kadyrova, Z., Monsees, A., Borner, A. *S* 2047 (2004).
[18]Jones, M.D., Raja, R., Thomas, J.M., Johnson, B.F.G., Lewis, D.W., Rouzaud, J., Harris, K.D.M. *ACIEE* **42**, 4326 (2003).
[19]Hsiao, Y., Rivera, N.R., Rosner, T., Krska, S.W., Njolito, E., Wang, F., Sun, Y., Armstrong III, J.D., Grabowski, E.J.J., Tillyer,R.D., Spindler, F., Malan, C. *JACS* **126**, 9918 (2004).
[19]Wu, H., Hoge, G. *OL* **6**, 3645 (2004).
[20]Reetz, M.T., Li, X. *T* **60**, 9709 (2004).
[21]Tang, W., Wu, S., Zhang, X. *JACS* **125**, 9570 (2003).
[22]Cong-Dung Le, J., Pagenkopf, B.L. *JOC* **69**, 4177 (2004).
[23]Hu, X., Zheng, Z. *OL* **6**, 3585 (2004).
[24]Liu, P.N., Gu, P.M., Wang, R., Tu, Y.Q. *OL* **6**, 169 (2004).
[25]Liu, P.N., Deng, J.G., Tu, Y.Q., Tu, Y.Q., Wang, S.H. *CC* 2070 (2004).
[26]Li, X., Wu, X., Chen, W., Hancock, F.E., King, F., Xiao, J. *OL* **6**, 3321 (2004).
[27]Bogevig, A., Pastor, I.M., Adolfsson, H. *CEJ* **10**, 294 (2004).
[28]Cuervo, D., Gamasa, M.P., Gimeno, J. *CEJ* **10**, 425 (2004).
[29]Xie, J.-H., Wang, L.-X., Fu, Y., Zhu, S.-F., Fan, B.-M., Duan, H.-F., Zhou, Q.-L. *JACS* **125**, 4404 (2003).
[30]Gade, L.H., Cesar, V., Bellemin-Loponnaz, S. *ACIEE* **43**, 1014 (2004).
[31]Chen, J., Li, Y., Dong, Z., Li, B., Gao, J. *TL* **45**, 8415 (2004).
[32]Yu, C.-M., Kim, C., Kweon, J.-H. *CC* 2494 (2004).
[33]Crich, D., Ranganathan, K., Rumathao, S., Shirai, M. *JOC* **68**, 2034 (2003).
[34]Yanagi, T., Kikuchi, K., Takeuchi, H., Ishikawa, T., Nishimura, T., Kutota, M., Yamamoto, I. *CPB* **51**, 221 (2003).
[35]Li, K., Zhou, Z., Wang, L., Chen, Q., Zhao, G., Zhou, Q., Tang, C. *TA* **14**, 95 (2003).
[36]Delogu, G., Dettori, M.A., Patti, A., Pedotti, S. *T* **60**, 10305 (2004).
[37]Locatelli, M., Cozzi, P.G. *ACIEE* **42**, 4928 (2003).
[38]Bette, V., Mortreux, A., Ferioli, F., Martelli, G., Savoia, D., Carpentier, J.-F. *EJOC* 3040 (2004).
[39]Mastranzo, V.M., Quintero, L., de Parrodi, C.A., Juaristi, E., Walsh, P.J. *T* **60**, 1781 (2004).
[40]Lipshutz, B.H., Noson, K., Chrisman, W., Lower, A. *JACS* **125**, 8779 (2003).
[41]Genov, D.G., Ager, D.J. *ACIEE* **43**, 2816 (2004).
[42]Lei, A., Wu, S., He, M., Zhang, X. *JACS* **126**, 1626 (2004).
[43]Sterk, D., Stephan, M.S., Mohar, B *TL* **45**, 535 (2004).
[44]de Paule, S.D., Jeulin, S., Ratovelomanana-Vidal, V., Genet, J.-P., Champion, N., Dellis, P. *EJOC* 1931 (2003).
[45]Mordant, C., Dunkelmann, P., Ratovelomanana-Vidal, V., Genet, J.-P. *CC* 1296 (2004).
[46]Trifonova, A., Diesen, J., Chapman, C., Andersson, P. *OL* **6**, 3825 (2004).
[47]Alexakis, A., Gille, S., Prian, F., Rosset, S., Ditrich, K. *TL* **45**, 1449 (2004).
[48]Krzeminski, M.P., Zaidlewicz, M. *TA* **14**, 1463 (2003).
[49]Wang, W.-B., Lu, S.-M., Yang, P.-Y., Han, X.-W., Zhou, Y.-G. *JACS* **125**, 10536 (2003).

Other enantioselective reactions. Enantioselective protonation to generate chiral compounds from silyl enol ethers is very effective using Lewis acid-assisted chiral Bronsted acids.[1] Certain chiral amides furnish protons asymmetrically to lithium enolates of α-amino acid derivatives.[2] De-racemization of α-branched carboxylic acids is accomplished on addition of a chiral alcohol to ketenes generated in situ.[3]

Allylic organolithium species internally coordinated to a carbamate group (from lithiation of the enol carbamates with BuLi – (-)-sparteine) are chiral homoenolate ions and their reaction with aldehydes is highly diastereoselective and enantioselective.[4]

In a synthetic approach to the erythrinan alkaloids desymmetrization of a succinimide by selective silylation also renders a subsequent Grignard reaction regioselctive by virtue of steric effects.[5]

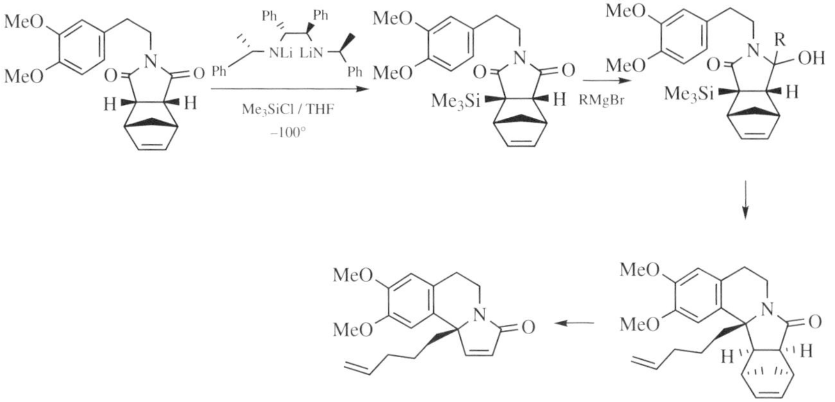

Acylation of *meso*-1,2-diols is catalyzed by phosphate derivatives of cinchona alkaloids to the effect that chiral monoesters are obtained readily.[6] CuCl-catalyzed conversion of dihydrosilanes to siloxanes is asymmetric in the presence of a chiral C_2-symmetric 1,3-dimethyl-DPPP derivative. Chiral allylsiloxanes derived from homopropargylic alcohols serve to extend the carbon chain at the alkyne terminus to provide unsaturated *syn*-1,5-diols.[7]

1,3-Acyl migration of indol-2-yl and benzofuran-2-yl carbonates to afford chiral 3-acyloxindoles and 3-acylbenzofuran-2-ones is promoted by an analogue of **4**.[8] Synthesis of α-keto esters bearing an asymmetric β-carbon atom is readily realized with Claisen rearrangement of allyl alkenyl ethers prepared from allylic alcohols and β-keto esters. The chiral catalyst is a (box)-Cu complex.[9] Rearrangement of allylic trichloroacetimidates delivers the 1,3-transposed *N*-trichloroacetyl allylic amines in chiral form using the cobalt complexes **73**, **74**.[10,11] The products are useful precursors of α-amino acids.

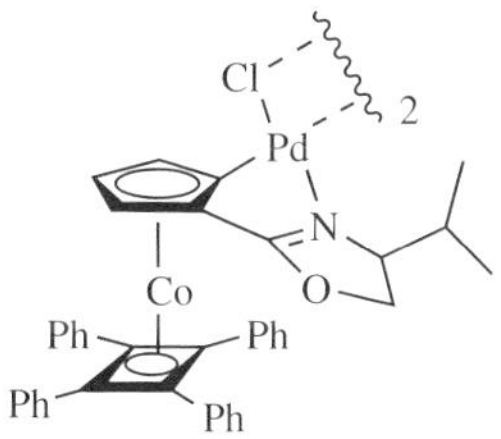

73 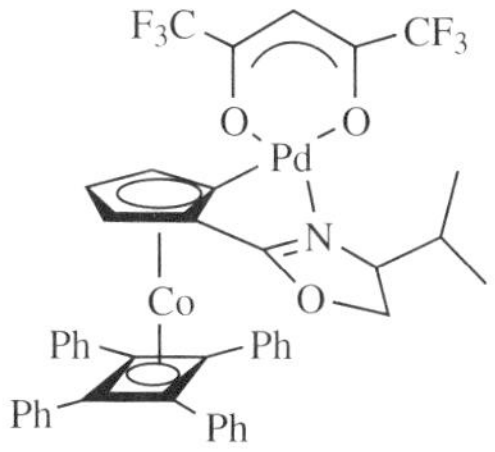 **74**

Ene reaction with a camphor-derived glyoxylamide in the presence of Lewis acid has been studied.[12] A zeolite-immobilized (bisoxazoline)-Cu complex is found in various cases comparable or superior to homogeneous catalysts.[13] Protected chiral aldols are synthesized by a hetero-ene reaction of trimethylsilyl enol ethers and aldehydes using a Cr complex of salicylaldimine derived from (1*S*)-amino-(2*R*)-indanol.[14]

Bis(oxazoline)-Cu complexes are Lewis acids and their employment together with a silver salt in catalyzing asymmetric Nazarov cyclization has gained certain success.[15]

[1]Ishihara, K., Nakashima, D., Hiraiwa, Y., Yamamoto, H. *JACS* **125**, 25 (2003).
[2]Futatsugi, K., Yanagisawa, A., Yamamoto, H. *CC* 566 (2003).
[3]Chen, C.-Y., Dagneau, P., Grabowski, E.J.J., Oballa, R., O'Shea, P., Prasit, P., Robichaud, J., Tillyer, R., Wang, X. *JACS* **125**,2633 (2003).
[4]Seppi, M., Kalkofen, R., Reupohl, J., Frohlich, R., Hoppe, D. *ACIEE* **43**, 1423 (2004).
[5]Gill, C., Greenhalgh, D.A., Simpkins, N.S. *TL* **44**, 7803 (2003).
[6]Mizuta, S., Sadamori, M., Fujimoto, T., Yamamoto, H. *ACIEE* **42**, 3383 (2003).
[7]Schmidt, D.R., O'Malley, S.J., Leighton, J.L. *JACS* **125**, 1190 (2003).
[8]Hills, I.D., Fu, G.C. *ACIEE* **42**, 3921 (2003).
[9]Abraham, L., Korner, M., Hiersmann, M. *TL* **45**, 3647 (2004).
[10]Anderson, C.E., Overman, L.E. *JACS* **125**, 12412 (2003).
[11]Kirsch, S.F., Overman, L.E. *JOC* **69**, 8101 (2004).
[12]Pan, J., Venkatesham, U., Chen, K. *TL* **45**, 9345 (2004).
[13]Caplan, N.A., Hancock, F.E., Page, P.C.B., Hutchings, G.J. *ACIEE* **43**, 1685 (2004).
[14]Ruck, R.T., Jacobsen, E.N. *ACIEE* **42**, 4771 (2003).
[15]Aggarwal, V.K., Belfield, A.J. *OL* **5**, 5075 (2003).

Chloral. 22, 138

Beckmann rearrangement.[1] Ketoximes undergo rearrangement (aldoximes, dehydration) on heating with chloral.

[1]Chandrasekhar, S., Gopalaiah, K. *TL* **44**, 755 (2003).

Chloramine.

N-Amination.[1] The heterocyclic N-H group of pyrroles and indoles is readily converted into N-NH_2 with NH_2Cl after deprotonation with NaH in DMF.

[1]Hynes Jr, J., Doubleday, W.W., Dyckman, A.J., Godfrey Jr, J.D., Grosso, J.A., Kiau, S., Leftheris, K. *JOC* **69**, 1368 (2004).

Chloramine-T. 20, 103; 21, 125; 22, 138

Group exchange. The transformation of $[RBF_3]K$ (in which the R group is aryl or alkenyl) to RX is accomplished by NaX and chloramines-T in aq. THF at room temperature.[1,2]

Cycloadditions. Conversion of alkenes into *N*-tosylaziridines by the chloroamine T-iodine system may involve different sets of conditions: in the presence of silica gel and water,[3] and phase transfer catalyst and phosphomolybdic acid.[4]

A pyrazoline synthesis on soluble polymer support uses chloramines-T to oxidize aryl-hydrazones that are formed in situ.[5]

Aziridines. Chloramine-T and $AgNO_3$ form tosylnitrene that can be trapped by alkenes.[6]

[1]Kabalka, G.W., Mereddy, A.R. *OM* **23**, 4519 (2004).

[2]Kabalka, G.W., Mereddy, A.R. *TL* **45**, 343, 1417 (2004).
[3]Minakata, S., Kano, D., Oderaotoshi, Y., Komatsu, M. *ACIEE* **43**, 79 (2004).
[4]Kumar, G.D.K., Baskaran, S. *CC* 1026 (2004).
[5]Wang, Y.-G., Zhang, J., Lin, X.-F., Ding, H.-F. *SL* 1467 (2003).
[6]Minakata, S., Kano, D., Fukuoka, R., Oderaotoshi, Y., Komatsu, M. *H* **60**, 289 (2003).

Chlorine.

2,2-Dichloroalkanals.[1] Either aldehydes or primary alcohols react with chlorine in the presence of Et_4NCl to afford the dichlorinated aldehydes.

Iminochlorides.[2] The adduct of $(PhO)_3P$ and chlorine transforms RNHCOR′ into iminochlorides which on hydrolysis liberate the amnies. Of course Bischler-Napieralski reaction can be performed by the reagent.

[1]Bellesia, F., De Buyck, L., Ghelfi, F., Pagnoni, U.M., Parsons, A.F., Pinetti, A. *S* 2173 (2003).
[2]Spaggiari, A., Blaszczak, L.C., Prati, F. *OL* **6**, 3885 (2004).

2-Chloroalkoxymethyl(triphenyl)phosphonium chlorides.

Dialkenyl ethers.[1] Wittig reaction with the title reagents gives the twofold unsaturated ethers as a result of dehydrochlorination in situ.

[1]Kulkarni, M.G., Doke, A.K., Davawala, S.I., Doke, A.V. *TL* **44**, 4913 (2003).

Chlorbis(diethylamino)borane.

β-Amino carbonyl compounds.[1] The reagent transforms lithium enolates into boron enolates that in reaction with aldehydes the amino group is transferred besides CC bond formation.

[1]Suginome, M., Uehlin, L., Yamamoto, A., Murakami, M. *OL* **6**, 1167 (2004).

4-Chloro-5-methoxy-2-nitropyridazin-3-one.

N-Nitration. Nitro group transfer to secondary amines is readily achieved with the title reagent at room temperature.

[1]Parl, Y.-D., Kim, H.-K., Kim, J.-J., Cho, S.-D., Kim, S.-K., Shiro, M., Yoon, Y.-J. *JOC* **68**, 9113 (2003).

1-Chloromethyl-4-fluoro-1,4-diazoniabicyclo[2.2.2]octane bis(tetrafluoroborate), Selectfluor®. 22, 195

Fluorination. A preparation of trifluoroacetylsilanes employs the title reagent in the reaction with $CF_2{=}C(OSiMe_3)SiR_3$.[1] For conversion of dithioacetals of diaryl ketones to the *gem*-difluoro compounds[2] a combination of the title reagent with $py(HF)_n$ may be used.

Selectfluor converts sulfilimines into fluoro-λ^6-sulfanenitriles [$Ar_2S(=N)F$] in high yields.[3] [*note = should be 3 lines*]

α-Iodo ketones. Iodine is activated by the title reagent to iodinate ketones. 3-Alkenones are iodinated at C-1.[4]

[1]Chung, W.J., Welch, J.T. *JFC* **125**, 543 (2004).
[2]Reddy, V.P., Alleti, R., Perambuduru, M.K., Welz-Biermann, U., Buchholz, H., Prakash, G.K.S. *CC* 654 (2005).
[3]Fujii, T., Asai, S., Okada, T., Hao, W., Morita, H., Yoshimura, T. *TL* **44**, 6203 (2003).
[4]Stavber, S., Jereb, M., Zupan, M. *S* 853 (2003).

Chloromethyl methyl ether.

Preparation.[1] This valuable reagent (for MOM etherification) is conveniently prepared in 74% yield from methylal and PhCOCl at 60° in the presence of catalytic amount of H_2SO_4.

[1]Reggelin, M., Doerr, S. *SL* 1117 (2004).

Chloromethyl methyl sulfone.

Allylic alcohols.[1] The reagent reacts with aldehydes under basic conditions (*t*-BuOK as base) to deliver 1-alken-3-ols. The Darzens-type reaction is followed by a Ramberg-Bäcklund reaction. Adducts from ketones do not undergo the second reaction.

RCHO + Cl–CH$_2$–SO$_2$Me —(*t*-BuOK, THF, −78° ~ 0°)→ [epoxide (R, O) SO$_2\bar{C}H_2$ → R, $^-$O, episulfone SO$_2$] → R, HO (1-alken-3-ol)

[1]Makosza, M., Urbanska, N., Chesnokov, A.A. *TL* **44**, 1473 (2003).

Chloromethyl vinyl sulfone.

Diels-Alder reaction.[1] The title reagent is a reactive dienophile. On performing a Ramberg-Bäcklund reaction on the cycloadducts 4-methylenecyclohexenes are produced.

Chloromethyl vinyl sulfone + 1,2-dimethylenecyclohexane —(Δ)→ cycloadduct (SO$_2$CH$_2$Cl), 72% —(*t*-BuOK)→ methylene product, 79%

[1]Block, E., Jeon, H.R., Zhang, S.-Z., Dikarev, E.V. *OL* **6**, 437 (2004).

***m*-Chloroperoxybenzoic acid, MCPBA. 13**, 76–79; **14**, 84–87; **15**, 86; **16**, 80–83; **17**, 76; **18**, 101; **19**, 94–95; **20**, 106–108; **21**, 130–131; **22**, 142–143

Oxidative elimination. Use of MCPBA to oxidive α-iodoketones[1] and β-benzylthio nitroalkanes[2] leads to the respective conjugated compounds. However, 2-phenylseleno-1,3-alkanediols derived from allylic alcohols are transformed into epoxy carbinols.[3]

PhSeCl; mCPBA, KOH; 74%

Baeyer-Villiger oxidation. Oxidation of carbonyl compounds by MCPBA is said to be catalyzed by $Bi(OTf)_3$.[4] A practical route to α-fluorinated esters is by BV oxidation of the ketones.[5]

Oxidation of imines. In the presence of $BF_3.OEt_2$ the oxidation leads to amides.[6–8]

[1]Horiuchi, C.A., Ji, S.-J., Matsushita, M., Chai, W. *S* 202 (2004).
[2]Jang, Y.-J., Lin, W.-W., Shih, Y.-K., Liu, J.-T., Hwang, M.-H., Yao, C.-F. *T* **59**, 4979 (2003).
[3]Cooper, M.A., Ward, A.D. *T* **60**, 7963 (2004).
[4]Alam, M.M., Varala, R., Adapa, S.R. *SC* **33**, 3035 (2003).
[5]Kobayashi, S., Tanaka, H., Amii, H., Uneyama, K. *T* **59**, 1547 (2003).
[6]An, G., Rhee, H. *SL* 876 (2003).
[7]An, G., Kim, M., Kim, J.Y., Rhee, H. *TL* **44**, 2183 (2003).
[8]Kim, S.Y., An, G., Rhee, H. *SL* 112 (2003).

***N*-Chlorosuccinimide, NCS. 13**, 79–80; **15**, 86–88; **18**, 101–102; **19**, 95–96; **20**, 108; **21**, 131–132; **22**, 143

Chlorination.[1] A combination of NCS with catalytic PhSeCl is effective for allylic chlorination at room temperature, an addition-elimination sequence is involved. For example, 3-butenoic acid affords 4-chloro-2-butenoic acid in 82% yield.

Diamination. Imidazolines are formed from alkenes, NCS and $TsNH_2$ in MeCN, via chloroamination and solvolysis. The products are further chlorinated at the methyl group.[2] An analogous process uses *N*-chlorosaccharin alone but the methyl group of MeCN is preserved.[3]

+ $TsNH_2$; NCS, MeCN; 72%

[1]Tunge, J.A., Mellegaard, S.R. *OL* **6**, 1205 (2004).
[2]Timmons, C., Chen, D., Xu, X., Li, G. *EJOC* 3850 (2003).
[3]Booker-Milburn, K.I., Guly, D.J., Cox, B., Procopiou, P.A. *OL* **5**, 3313 (2003).

Chlorosulfonic acid.

Aromatization.[1] 2,5-Cyclohexadiene-1-carboxylic acids form acylium ions on brief treatment with $ClSO_3H$ in CH_2Cl_2 at 0°. Rapid loss of proton and CO ensues. The process can be employed in the conversion of benzoic acid to alkylbenzenes via Birch reduction and alkylation.

Rearrangement and dehydration.[2] Nearly quantitative conversion and yields of Beckmann rearrangement products from ketoximes and nitriles from aldoximes are obtained on treatment with $ClSO_3H$. For example, salicylaldoxime gives 2-hydroxybenzonitrile in 99% yield.

[1]Vorndran, K., Linker, T. *ACIEE* **42**, 2489 (2003).
[2]Li, D., Shi, F., Guo, S., Deng, Y. *TL* **46**, 671 (2005).

Chlorosulfonyl isocyanate, CSI. 13, 80–81; **18**, 102, **21**, 132–133; **22**, 143

N-Chlorosulfonyl carbamates.[1] These compounds are readily produced by addition of alcohols to CSI and they can be further transformed into various $RNHSO_2NHCOOR'$.

Allylic amination. Allyl methyl ethers ionize on exposure to CSI and internal return of the ion-pairs lead to C—N bond formation.[2]

83%

[1]Masui, Y., Watanabe, H., Masui, T. *TL* **45**, 1853 (2004).
[2]Kim, J.D., Zee, O.P., Jung, Y.H. *JOC* **68**, 3721 (2003).

Chlorotris(triphenylphosphine)rhodium(I). 19, 96–98; **20**, 108–109; **21**, 133–135; **22**, 143–146

Isomerization. Aldoximes are transformed into amides[1] by heating with $(Ph_3P)_3RhCl$. Cycloisomerization of 1,6-enynes accompanied by halogen shift[2] seems useful because of skeletal and functional group changes can be exploited. 1,2,7-Dienynes cyclize to form trienes composed of two cross-conjugated dienes.[3]

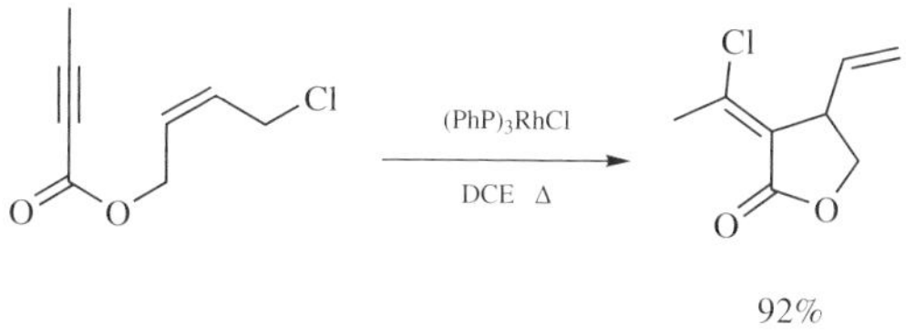

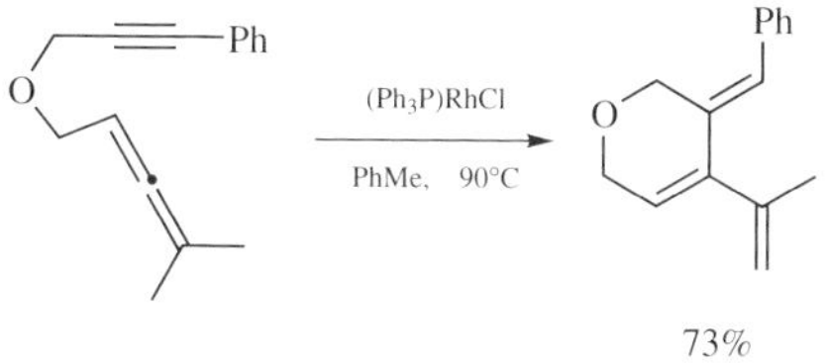

Cl
(PhP)3RhCl
DCE Δ
Cl
O
O
O
O
92%

Ph
O
(Ph3P)RhCl
PhMe, 90°C
Ph
O
73%

Cyclopropylmethyl trimethylsilyl ether is converted into methallyl trimethylsilyl ether on heating with $(Ph_3P)_3RhCl$.[4]

Allylation. The Rh complex catalyzes allylic displacement of allyl carbonates by copper(I) enolates with high regioselectivity (at the more highly substituted carbon atom) and diastereoselectivity.[5,6]

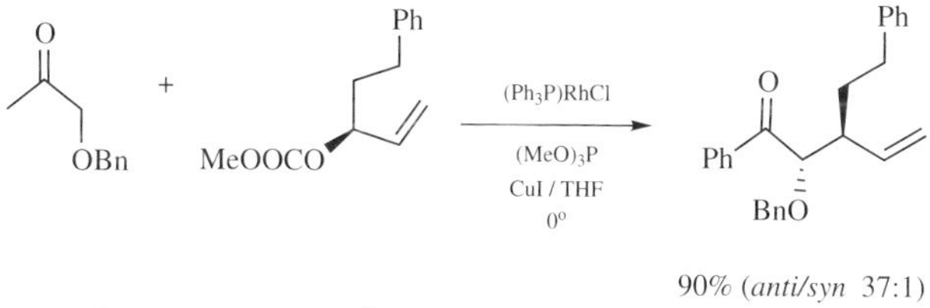

O
OBn
+
Ph
MeOOCO
(Ph3P)RhCl
(MeO)3P
CuI / THF
0°
Ph
O
Ph
BnO
90% (anti/syn 37:1)

Reductive trifluoromethylation.[7] A method for synthesizing α-trifluoromethyl ketones employs Et_2Zn to reduce $(Ph_3P)_3RhCl$ in situ for reduction of conjugated ketones, and the enolate-[Rh]-CF_3 complexes that are subsequently formed on reaction with CF_3I collapse to give the products.

o-Alkylation. Introduction of an alkyl chain into an *o*-position of aromatic ketimines by alkenes can be carried out by microwave heating without solvent.[8] With the successful use of functionalized alkenes (e.g., conjugated esters...)[9] the value of this synthetic method is magnified.

Phenols undergo Rh-catalyzed *o*-arylation with ArBr using arylphosphite cocatalyst,[10] while salicyladehydes add to 1,4- and 1,5-alkadienes to form 2-hydroxyaryl ketones with regiochemistry favoring the α-branched type.[11]

OH
CHO
+
(Ph3P)RhCl
CH2Cl2
OH O
+
OH O
(4 : 1)
100%

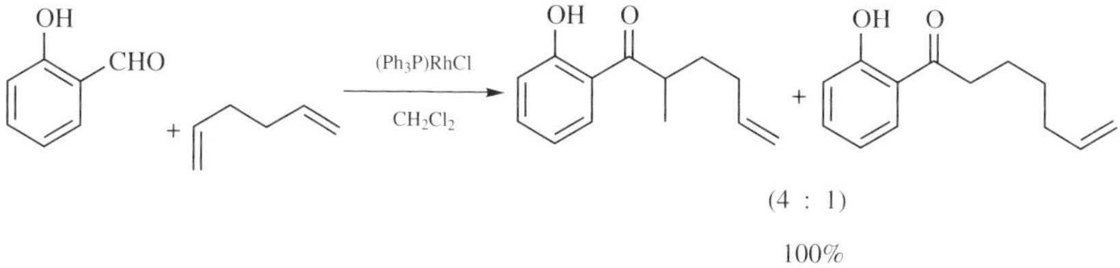

Methylenation.[12] Aldehydes are methylenated by Me_3SiCHN_2 after brief treatment with catalytic $(Ph_3P)_3RhCl$ and stoichiometric Ph_3P and *i*-PrOH. The Wittig reagent is formed by group exchange from the rhodium carbenoid.

[1]Park, S., Choi, Y., Han, H., Yang, S.H., Chang, S. *CC* 1936 (2003).
[2]Tong, X., Zhang, Z., Zhang, X. *JACS* **125**, 6370 (2003).
[3]Shibata, T., Takesue, Y., Kadowaki, S., Takagi, K. *SL* 268 (2003).
[4]Bart, S.C., Chirik, P.J. *JACS* **125**, 886 (2003).
[5]Evans, P.A., Leahy, D.K. *JACS* **125**, 8974 (2003).
[6]Evans, P.A., Lawler, M.J. *JACS* **126**, 8642 (2004).
[7]Sato, K., Omote, M., Ando, A., Kumadaki, I. *OL* **6**, 4359 (2004).
[8]Vo-Thanh, G., Lahrache, H., Loupy, A., Kim, I.-J., Chang, D.-H., Jun, C.-H. *T* **60**, 5539 (2004).
[9]Lim, S.-G., Ahn, J.-A., Jun, C.-H. *OL* **6**, 4687 (2004).
[10]Bedeford, R.B., Limmert, M.E. *JOC* **68**, 8669 (2003).
[11]Tanaka, M., Imai, M., Yamamoto, Y., Tanaka, K., Shimowatari, M., Nagumo, S., Kawahara, S., Suemune, H. *OL* **5**, 1365(2003).
[12]Lebel, H., Paquet, V. *JACS* **126**, 320 (2004).

Chlorotris(triphenylphosphine)rhodium(I)– 2-Amino-3-picoline. 21, 135; **22**, 146–148

Fragmentative hydroacylation. Hydroacylation of alkynes with aldehydes gives ketones that incorporate half of the original alkyne molecules.[1] To be synthetically useful symmetrical alkynes must be employed. The reaction also requires 2-amino-3-picoline and $AlCl_3$.

R
R
+ OHC R′
$(Ph_3P)RhCl$
N NH_2
N N
R R′
R
$CyNH_2$,$AlCl_3$;
H_3O^+
O
R′
R

[1]Lee, D.-Y., Hong, B.-S., Cho, E.-G., Lee, H., Jun, C.-H. *JACS* **125**, 6372 (2003).

Chromium – carbene complexes. 13, 82–83; **14**, 91–93; **15**, 93–95; **16**, 88–92; **17**, 80–84; **18**, 103–104; **19**, 98–101; **20**, 110–111; **21**, 136–138; **22**, 148–150

Cycloadditions. Heavily substituted cyclopentanes or cyclohexanes are obtained by a diastereoselective assemblage of Fischer carbene complexes and nucleophiles such as lithium enolates and allylmagnesium bromide.[10]

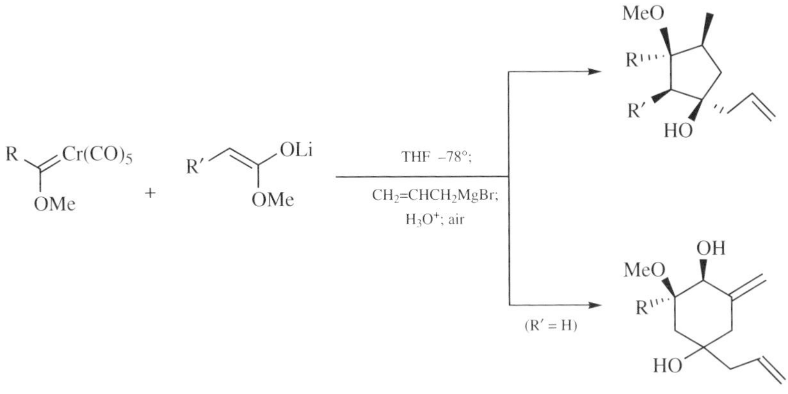

Cyclopentadienes are formed from reaction of β-amino-substituted unsaturated Fischer carbene complexes with internal alkynes.[2] The reaction products appear to be different from those from 1-alkynes.

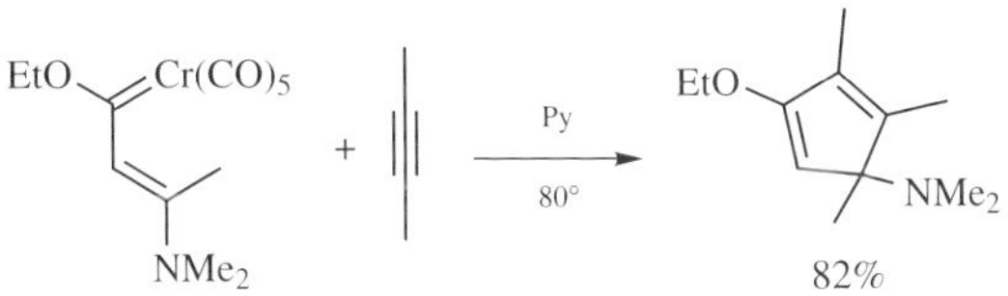

The amine-catalyzed photochemical reaction of $Ph(OMe)=Cr(CO)_5$ with aldehydes under CO affords β-lactones via cycloaddition of Cr-coordinated ketenes, but adducts from some aldehydes decarboxylate in situ to give $Ph(OMe)=CHR$.[3]

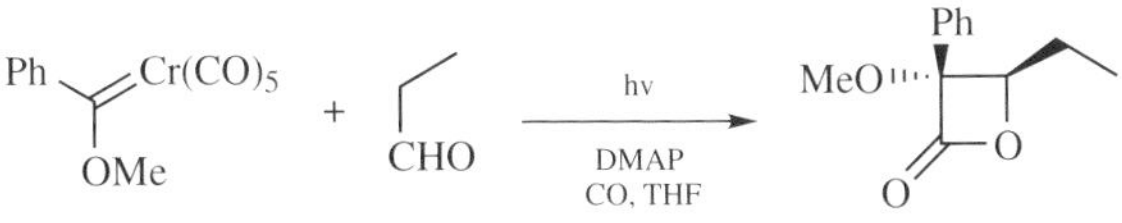

A [3 + 2]cycloaddition involving unsaturated Fischer carbene complexes and alkynes is catalyzed by a cationic Rh(I) complex and it leads to 2-cyclopentenones.[4] The reaction is thought to involve formation of Rh-carbenoids and rhodacyclohexadienes.

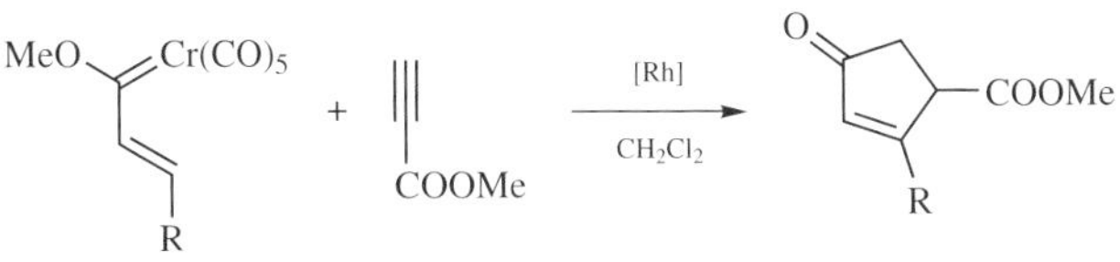

Bridged ring ketones are formed when enamines of cyclic ketones and unsaturated Fischer carbene complexes are brought together.[5]

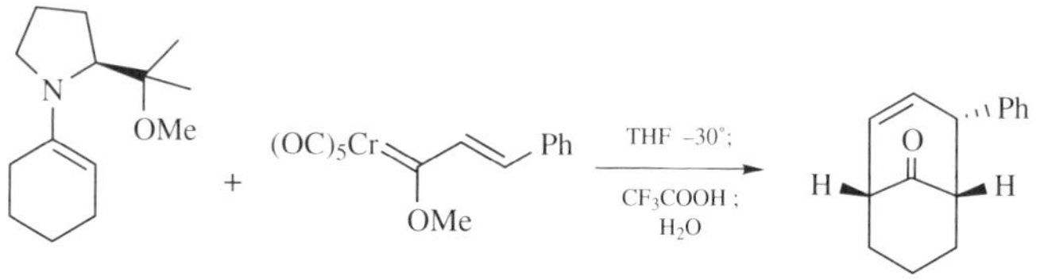

OMe
$(OC)_5Cr$
Ph
OMe
THF –30°;
CF_3COOH ;
H_2O
O
Ph
H
H

Cyclization. *N*-(2-Vinylphenyl)aminocarbene chromium complexes undergo thermal decomposition with extrusion of the metal moiety. Indoles and quinolines are formed but the product distribution is highly dependent on the substitution pattern and electronic properties of the substrates and on the solvent.[6]

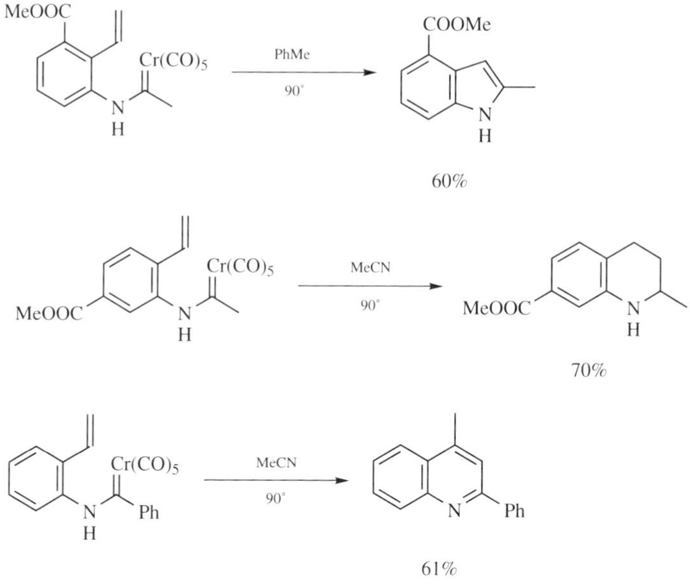

MeOOC
$Cr(CO)_5$
N
H
PhMe
90°
COOMe
N
H
60%

$Cr(CO)_5$
MeOOC
N
H
MeCN
90°
MeOOC
N
H
70%

$Cr(CO)_5$
N
H
Ph
MeCN
90°
N
Ph
61%

2,3-Dihydrofurans. 3-Furylcarbenoids are susceptible to conjugate addition by RLi, the adducts can be trapped by electrophiles. Oxidative demetallation results in 2,3-dihydro-3-furylcarboxylic esters.[7]

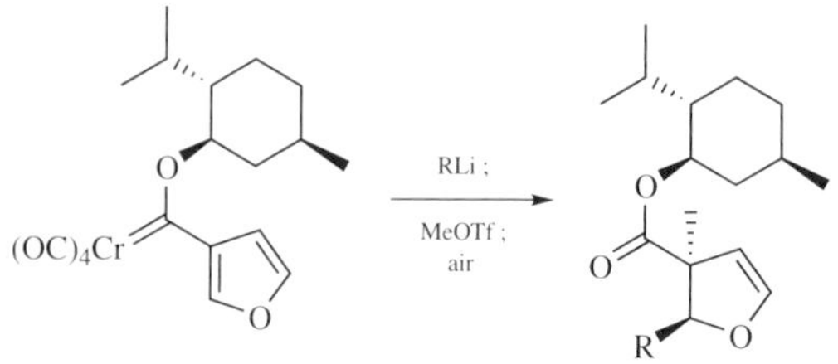

O
$(OC)_4Cr$
O
RLi ;
MeOTf ;
air
O
O
R
O

[1]Barluenga, J., Perez-Sanchez, I., Rubio, E., Florez, J. *ACIEE* **42**, 5860 (2003).
[2]Wu, Y.-T., Flynn, B., Müller, S., Labahn, T., Nötzel, M., de Meijere, A. *EJOC* 724 (2004).
[3]Merlic, C.A., Doroh, B.C. *JOC* **68**, 6056 (2003).
[4]Barluenga, J., Vicente, R., Lopez, L.A., Rubio, E., Florez, J., Tomas, M., Alvarez-Rua, C. *JACS* **126**, 470 (2004).
[5]Barluenga, J., Ballesteros, A., de la Rua, R.B., Santamaria, J., Rubio, E., Tomas, M. *JACS* **125**, 1712 (2003).
[6]Soderberg, B.C.G., Shriver, J.A., Cooper, S.H., Shrout, T.L., Helton, E.S., Austin, L.R., Odens, H.H., Hearn, B.R., Jones, P.C.,Kouadio, T.N., Ngi, T.H., Baswell, R., Caprara, H.J., Meritt, M.D., Mai, T.T. *T* **59**, 8775 (2003).
[7]Barluenga, J., Nandy, S.K., Laxmi, Y.R.S., Suarez, J.R., Merino, I., Florez, J., Garcia-Granda, S., Montejo-Bernardo, J. *CEJ* **9**, 5726 (2003).

Chromium(II) chloride. 13, 84; **14**, 94–97; **15**, 95–96; **16**, 93–94; **17**, 84–85; **18**, 104; **19**, 101; **20**, 111–113; **21**, 138–140; **22**, 150–151

Elimination. As diasteromeric mixtures, 2-Chloro-3-hydroxyalkanoic esters give 2-alkenoic esters in the (*E*)-configuration on heating with $CrCl_2$ in THF.[1]

(Z)-Chloroalkenes. Functionalized chloroalkenes such as α-chloro enones and conjugated esters are synthesized by $CrCl_2$-mediated condensation of aldehydes with 1,1,1-trichloro-2-alkanones[2] and trichloroacetic esters,[3] respectively, in one step. The reaction occurs at room temperature. α-Trihaloalkyl carboxylates rearrange and are partially dehalogenated.[4]

$CrCl_2$ / THF, Δ

R = TBS 84%

Halomethylenation. Homologation of aldehydes to alkenyl halides is easily achieved with CHX_3 - $CrCl_2$. The reaction conditions do not disturb a chiral center adjacent to the aldehyde.[5]

Iodocyclopropanation of alkenes is accomplished in a reaction with CHI_3 - $CrCl_2$ and *N,N,N′,N′*-tetraethylethylenediamine, probably involving a chromium-carbenoid intermediate.[6]

Conjugated esters. By condensation of aldehydes with dichloroacetic esters under the influence of $CrCl_2$ 2-alkenoic esters are readily synthesized.[7]

[1]Concellon, J.M., Rodriguez-Solla, H., Mejica, C. *TL* **45**, 2977 (2004).
[2]Falck, J.R., Bandyopadhyay, A., Barma, D.K., Shin, D.-S., Kundu, A., Kishore, R.V.K. *TL* **45**, 3039 (2004).
[3]Barma, D.K., Kundu, A., Zhang, H., Mioskowski, C., Falck, J.R. *JACS* **125**, 3218 (2003).
[4]Bejot, R., Tisserand, S., Reddy, L., Barma, D.K., Baati, R., Falck, J.R., Mioskowski, C. *ACIEE* **44**, 2008 (2005).
[5]Concellon, J.M., Bernad, P.L., Mejica, C. *TL* **46**, 569 (2005).

[6]Takai, K., Toshikawa, S., Inoue, A., Kokumai, R. *JACS* **125**, 12990 (2003).
[7]Barma, D.K., Kundu, A., Bandyopadhyay, A., Kundu, A., Sangras, B., Briot, A., Mioskowski, C., Falck, J.R. *TL* **45**, 5917 (2004).

Chromium(II) chloride – nickel(II) halide. **14**, 97–98; **15**, 96–97; **17**, 86; **18**, 105; **19**, 102; **20**, 113–114; **21**, 140; **22**, 152

Allylic alcohols. Reductive coupling of alkynes and aldehydes mediated by $CrCl_2$-$NiCl_2$ with Ph_3P and H_2O as additives leads to 2,3-disubstituted 2-propen-1-ols.[1] Formation of 1-alken-2-ylnickel(II) species and Ni/Cr exchange to the chromium(III) species are involved.

[1]Takai, K., Sakamoto, S., Isshiki, T. *OL* **5**, 653 (2003).

Chromium(II) chloride – trialkylsilyl chloride. **22**, 152–153

4-Hydroxyalkanals. Promotion by the reagent, 3-chloro-1-acetoxy-1-propene and aldehydes react to provide doubly protected 4-hydroxyalkanals.

[1]Lombardo, M., Morganti, S., Licciulli, S., Trombini, C. *SL* 43 (2003).

Chromium(III) chloride – zinc.

1-Iodoalkenes.[1] A more convenient iodoalkenation procedure employs $CrCl_3.6H_2O$ with Zn and NaI in THF.

[1]Auge, J., Boucard, V., Gil, R., Lubin-Germain, N., Picard, J., Uziel, J. *SC* **33**, 3733 (2003).

Chromium pentacarbonyl complexes.

Chromium carbenoids. Generation of Cr-carbenoids from diazo compounds and their reaction with furan derivatives lead to dienes.[1] On the other hand, 2-furylcarbene complexes are formed from (*Z*)-1-acylbut-1-en-3-ynes with $(thf)Cr(CO)_5$.[2]

OMe + Ph, COOEt, N_2 — $(coe)Cr(CO)_5$, CH_2Cl_2 → Ph, EtOOC, COOMe

98%

Ph, O + O — $(thf)Cr(CO)_5$ → Ph, O, O

63% (*cis/trans* 76:24)

[1]Hahn, N.D., Nieger, M., Dotz, K.H. *JOMC* **689**, 2662 (2004).
[2]Miki, K., Yokoi, T., Nishino, F., Kato, Y., Washitake, Y., Ohe, K., Uemura, S. *JOC* **69**, 1557 (2004).

Chromium tetraphenylporphyrin salts.

Rearrangement. Epoxides are transformed into aldehydes[1] stereoselectively by the high-valent Cr(tpp)OTf. Thus its catalytic activity differs dramatically from that of Fe(tpp)X [X = OTf, ClO_4] for promoting alkyl migration to furnish ketones.[2,3]

β-Lactones. The tetracarbonylcobaltate salt of the Cr-porphyrin catalyzes carbonylation of epoxides to furnish β-lactones.[4]

[1]Suda, K., Kikkawa, T., Nakajima, S., Takanami, T. *JACS* **126**, 9554 (2004).
[2]Suda, K., Baba, K., Nakajima, S., Takanami, T. *CC* 2570 (2002).
[3]Takanami, T., Hirabe, R., Ueno, M., Hino, F., Suda, K. *CL* 1031 (1996).
[4]Schmidt, J.A.R., Mahadevan, V., Getzler, Y.D.Y.L., Coates, G.W. *OL* **6**, 373 (2004).

Cobalt/carbon.

Cyclodimerization. 1,6-Diynes undergo cyclocarbonylation and dimerization by heating with Co/C under CO (30 atm). Product formation is temperature dependent.[1]

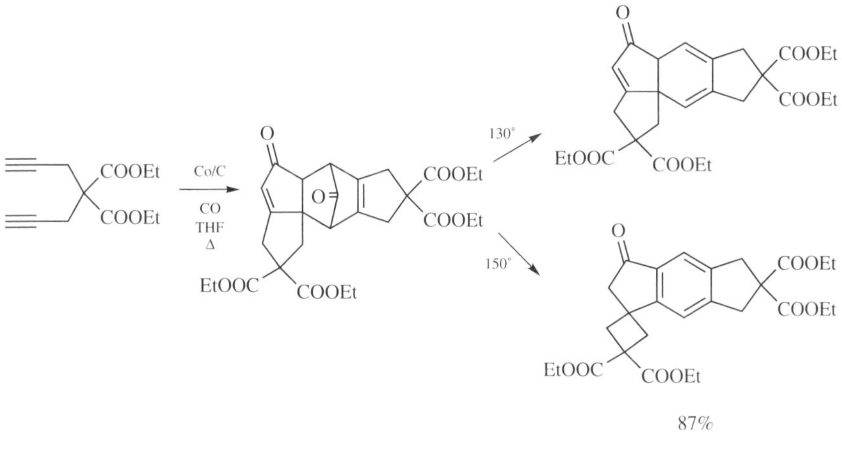

[1]Lee, S.I., Son, S.U., Choi, M.R., Chung, Y.K., Lee, S.-G. *TL* **44**, 4705 (2003).

Cobalt-ruthenium/carbon.

Pauson-Khand reaction. The Co-Ru nanoparticles immobilized on charcoal catalyze formation of cyclopentenones from enynes in the presence of picolyl formate.

[1]Park, K.H., Son, S.U., Chung, Y.K. *CC* 1898 (2003).

Cobalt(II) acetate.

Arylation. Azoles undergo arylation by catalysis with $Co(OAc)_2$. Regioselectivity of the process depends on the presence or absence of a Cu(I) salt.[1] In the case of thiazole the cobalt-catalyzed reaction is superior to that using Pd catalysts.

$Co(OAc)_2$, Cs_2CO_3 [carbene] DMF Δ — 64%

$Co(OAc)_2$, Cs_2CO_3 [carbene] CuI, dioxane Δ — 84%

[1]Sezen, B., Sames, D. *OL* **5**, 3607 (2003).

Cobalt(II) bromide – zinc.

Arylzinc reagents. Functionalized ArZnBr are readily prepared from ArBr using zinc dust and catalytic $CoBr_2$.[1] The preparation expedites the synthesis of aromatic ketones and arylstannanes by further reaction with carboxylic anhydrides[2] and R_3SnCl,[3] respectively. Organozinc reagents do not react with acid chlorides or anhydrides without a catalyst, therefore cobalt species has reactivity enhancing effects.

[1]Fillon, H., Gosmini, C., Perichon, J. *JACS* **125**, 3867 (2003).
[2]Kazmierski, I., Bastienne, M., Gosmini, C., Paris, J.-M., Perichon, J. *JOC* **69**, 936 (2004).
[3]Gosmini, C., Perichon, J. *OBC* **3**, 216 (2005).

Cobalt(II) bromide phosphine – zinc iodide. 22, 154–155

Diels-Alder reaction. With a dppe-complexed $CoBr_2$ as catalyst and Zn-ZnI_2 additive, alkynylboronic esters[1] and alkynyl sulfides,[2] and enynes[3] undergo cyloaddition with conjugated dienes at room temperature. 1-Alkenyl-1,4-cyclohexadienes are obtained in high yields from the last class of reactions.

(dppe)$CoBr_2$ Zn, ZnI_2 CH_2Cl_2 — 96%

[1]Hilt, G., Smolko, K.I. *ACIEE* **42**, 2795 (2003).
[2]Hilt, G., Luer, S., Harms, K. *JOC* **69**, 624 (2004).
[3]Hilt, G., Luer, S., Schmidt, F. *S* 634 (2004).

Cobalt(II) chloride.

Functional group transformations. Tosylation[1] and acetylation[2] with TsOH and HOAc, respectively, are said to be catalyzed by $CoCl_2.6H_2O$, as well as acetalization[3] and dithioacetalization.[4] Catalyzed opening of styrene oxides with anilines[5] is of little synthetic value, while the cleavage of thiiranes with AcCl[6] is somewhat more interesting.

Thionation. Acylsilanes and their latent modifications react with $(Me_3Si)_2S$ in the presence of $CoCl_2$ to replace oxygen atom(s) with sulfur atom. Thiacyclic compounds (5-to 7-membered) are formed from bis(acylsilanes).[7] Very reactive $CH_2{=}CHC({=}S)MR_3$ are generated from acetals composed of an α-triorganosilylallenyl[8] or α-triorganostannylallenyl unit.[9]

$CoCl_2\ 6H_2O$, $(Me_3Si)_2S$, MeCN

R_3Si = TBS 63%

Mannich reaction. Catalyzed by $CoCl_2$ active methylene compounds condense with aldehydes in MeCN and the presence of AcCl. α-Acetamidoalkyl derivatives are obtained.[10]

[1]Veluzamy, S., Kumar, J.S.K., Punniyamurthy, T. *TL* **45**, 203 (2004).
[2]Veluzamy, S., Borpuzari, S., Punniyamurthy, T. *T* **61**, 2011 (2005).
[3]Veluzamy, S., Punniyamurthy, T. *TL* **45**, 4917 (2004).
[4]De, S.K. *TL* **45**, 1035 (2004).
[5]Sundararajan, G., Vijayakrishna, K., Varghese, B. *TL* **45**, 8253 (2004).
[6]Iranpoor, N., Firouzabadi, H., Jafari, A.A. *SC* **33**, 2321 (2003).
[7]Bouillon, J.-P., Capperucci, A., Portella, C., Degl'Innocenti, A. *TL* **45**, 87 (2004).
[8]Capperucci, A., Degl'Innocenti, A., Biondi, S., Nocentini, T., Rinaudo, G. *TL* **44**, 2831 (2003).
[9]Degl'Innocenti, A., Capperucci, A., Nocentini, T., Biondi, S., Fratini, V., Castagnoli, G., Malesci, I. *SL* 2159 (2004).
[10]Rao, I.N., Prabhakaran, E.N., Das, S.K., Iqbal, J. *JOC* **68**, 4079 (2003).

Cobalt(II) iodide phosphine - zinc.

Indenols. *o*-Iodoaryl carbonyl compounds and alkynes react to deliver indenols in good yields on exposure to the title reagents.[1]

Cyclotrimerization. Norbornene and 1,6-diynes combine to give dihydroindanes.[2]

[1]Chang, K.-J., Rayabarapu, D.K., Cheng, C.-H. *OL* **5**, 3963 (2003).
[2]Wu, M.-S., Rayabarapu, D.K., Cheng, C.-H. *T* **60**, 10005 (2004).

Cobalt(II) nitrate.

Oxidation.[1] The cobalt salt supported on silica gel is used to oxidize various primary and secondary alcohols to carbonyl compounds (microwave irradiation, no solvent).

[1]Kiasat, A.R., Kazemi, F., Rafati, M. *SC* **33**, 601 (2003).

Cobalt porphyrins.

Alkenation and cyclopropanation. The cobalt(II) complexes catalyzes decomposition of diazoacetic esters. Trapping with carbonyl compounds gives conjugated esters[1] while cyclopropanes are produced in the presence of alkenes.[2,3]

1,3-Dioxolan-2-ones. Insertion of CO_2 into epoxides catalyzed by the Co(tpp)Cl/DMAP system produces good yields of the cyclic carbonates.[4] The reaction requires high pressure and temperature, however.

[1]Lee, M.-Y., Chen, Y., Zhang, X.P. *OM* **22**, 4905 (2003).
[2]Huang, L., Chen, Y., Gao, G.-Y., Zhang, X.P. *JOC* **68**, 8179 (2003).
[3]Penoni, A., Wanke, R., Tollari, S., Gallo, E., Musella, D., Ragaini, F., Demartin, F., Cenini, S. *EJIC* 1452 (2003).
[4]Paddock, R.L., Hiyama, Y., McKay, J.M., Nguyen, S.T. *TL* **45**, 2023 (2004).

Copper. 22, 155

Conjugate additions. The 1,4-addition of aliphatic amines to conjugated carbonyl compounds, acrylonitrile and phenyl vinyl sulfone in water is efficiently catalyzed by Cu.[1] Michael addend is generated from ethyl bromodifluoromethylacetate by copper powder, and TMEDA improves the reaction.[2]

Aromatic substitution. Copper mediates arylation of amines in *N,N*-dimethylethanol.[3,4] Ultrasound assists the substitution of 2-chlorobenzoic acids by phenols in water using a reagent system consisting of Cu – CuI, pyridine and K_2CO_3.[5]

1,1-Dibromocyclopropanes.[6] Copper and iron activate CBr_4 in MeCN to generate dibromocarbene which adds to alkenes present.

[1]Xu, L.-W., Li, J.-W., Xia, C.-G., Zhou, S.-L., Hu, X.-X. *SL* 2425 (2003).
[2]Sato, K., Nakazato, S., Enko, H., Tsujita, H., Fujita, K., Yamamoto, T., Omote, M., Ando, A., Kumadaki, I. *JFC* **121**, 105 (2003).
[3]Lu, Z., Twieg, R.J., Huang, S.D. *TL* **44**, 6289 (2003).
[4]Lu, Z., Twieg, R.J. *T* **61**, 903 (2005).
[5]Comdom, R.F.P., Palacios, M.L.D. *SC* **33**, 921 (2003).
[6]Leonel, E., Lejaye, M., Oudeyer, S., Paugam, J.P., Nedelec, J.-Y. *TL* **45**, 2635 (2004).

Copper(II) acetate. 18, 109–110; **19**, 106; **20**, 117; **21**, 142–143; **22**, 155–156

Hydrolysis. Peracetyl glycopyranoses are deacetylated at the anomeric position regioselectively in aqueous MeOH using copper(II) acetate as catalyst.[1]

Coupling reactions. Arylation of phenols,[2] aliphatic alcohols,[3] various N-H compounds (amines, amides, imides, sulfonamides),[4,5,6] and sodium sulfinates (to give aryl sulfones)[7] is all based on reactions with arylboron compounds mediated by $Cu(OAc)_2$. Naturally, *N*-alkenylation is a similar process.[8]

The homocoupling protocol for $ArB(OH)_2$ in DMF at 100° can be used to prepare halogenated biaryls.[9]

Addition reactions. Conjugate addition of 1-alkynes to alkylidene Meldrum's acid is realized in aqueous *t*-BuOH (10:1) in the presence of catalytic $Cu(OAc)_2$ and sodium ascorbate.[10] Intramolecular hydroamination of alkynes forms pyrrolines.[11]

N,N'-Diacylhydrazines. Deaminative dimerization of *N*-acylhydrazines is promoted by $Cu(OAc)_2$ under microwave irradiation.[12]

[1]Bhaumik, K., Salgaonkar, P.D., Akamanchi, K.G. *AJC* **56**, 909 (2003).
[2]Sagar, A.D., Tale, R.H., Adude, R.N. *TL* **44**, 7061 (2003).
[3]Quach, T.D., Batey, R.A. *OL* **5**, 1381 (2003).
[4]Lan, J.-B., Zhang, G.-L., Yu, X.-Q., You, J.-S., Chen, L., Yan, M., Xie, R.-G. *SL* 1095 (2004).
[5]Chan, D.M.T., Monaco, K.L., Li, R., Bonne, D., Clark, C.G., Lam, P.Y.S. *TL* **44**, 3863 (2003).
[6]Quach, T.D., Batey, R.A. *OL* **5**, 4397 (2003).
[7]Beaulieu, C., Guay, D., Wang, Z., Evans, D.A. *TL* **45**, 3233 (2004).
[8]Lam, P.Y.S., Vincent, G., Bonne, D., Clark, C.G. *TL* **44**, 4927 (2003).
[9]Demir, A.S., Reis, O., Emrullahoglu, M. *JOC* **68**, 10130 (2003).
[10]Knöpfel, T.F., Carreira, E.M. *JACS* **125**, 6054 (2003).
[11]Knight, D.W., Sharland, C.M. *SL* 119 (2004).
[12]Mogilaiah, K., Prashanthi, M., Reddy, G.R. *SC* **33**, 3741 (2003).

Copper(II) acetylacetonate. 18, 110; 20, 117–118; 22, 156

Benzyl ethers.[1] Alcohols are benzylated by BnX on refluxing with $Cu(acac)_2$.

Aziridination.[2] The transfer of [TsN] group from PhI=NTs to alkenes is catalyzed by $Cu(acac)_2$ microencapsulated in polystyrene. The catalyst is easily recovered and reused.

[1]Sirkecioglu, O., Karliga, B., Talini, N. *TL* **44**, 8483 (2003).
[2]Kantam, M.L., Kavita, B., Neeraja, V., Haritha, Y., Chaudhuri, M.K., Dehury, S.K. *TL* **44**, 9029 (2003).

Copper(I) bromide. 21, 143; 22, 157

Alkylation. Allylation of acetals with allylsilanes in the presence of CuBr is assisted by microwaves.[1] Ligated to **1**, CuBr forms an asymmetric catalyst for alkynylation of imines.[2]

HN Ph
N
N
PPh_2

1

Conjugate addition. Addition of diorganozincs to enones is catalyzed by $CuBr \cdot SMe_2$. The reaction provides chiral adducts with a peptidylphosphine ligand.[3] Adducts from the Cu-catalyzed conjugated addition of 3-trimethylsilyl-1-propenylzirconocene to cross-conjugated dienones are valuable precursors of 4-vinylcyclohexanones.[4]

1,3-Enynes. By coupling protocols stereodefined 1,3-enynes are obtained. Using $(bipy)CuBr \cdot PPh_3$ to mediate the reaction of alkenyl iodides with 1-alkynes is one of such methods.[5] On the other hand, hydroboration of alkynes with 9-BBN and converting the resulting alkenylboranes to copper derivatives produce species for reaction with alkynyl(phenyl)iodonium tosylates.[6]

A route to 1,2-dialkynylcycloalkenes involves an intriguing reaction of 1,ω-diynes with manganese-carbyne complexes.[7]

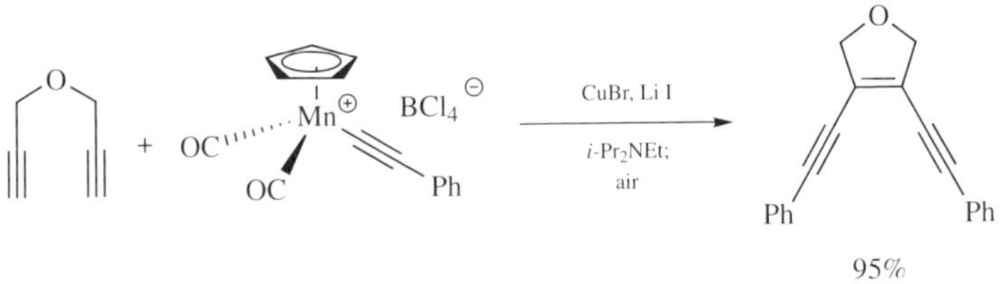

[1]Jung, M.E., Maderna, A. *JOC* **69**, 7755 (2004).
[2]Knöpfel, T.F., Aschwanden, P., Ichikawa, T., Watanabe, T., Carreira, E.M. *ACIEE* **43**, 5971 (2004).
[3]Breit, B., Laungani, A.C. *TA* **14**, 3823 (2003).
[4]Huang, X., Pi, J. *SL* 481 (2003).
[5]Bates, C.G., Saejueng, P., Venkataraman, D. *OL* **6**, 1441 (2004).
[6]Yang, D.-Y., He, J., Miao, S. *SC* **33**, 2695 (2003).
[7]Casey, C.P., Dzwiniel, T.L., Kraft, S., Guzei, I.A. *OM* **22**, 3915 (2003).

Copper(II) bromide. 14, 100; **15**, 100; **18**, 111; **19**, 106; **21**, 143–144; **22**, 157–158

Bromodeacylation.[1] β-Diketones and β-keto esters are converted into α-bromo ketones and α-bromo esters, respectively, by warming with alumina-supported $CuBr_2$ and Na_2CO_3. The individual reagent serves as brominating agent and in deacylation, respectively.

Bromolysis of alkylidenecyclopropanes. The substrates add bromine in a straightforward manner but undergo ring opening with $CuBr_2$ in MeCN to give 1,3-dibromo-3-alkenes.[2] The latter compounds are valued for their versatility for participating in further substitutions and coupling reactions. Cyclopropylideneacetic esters afford 3-bromo-5,6-dihydro-2-pyrones.[3] [Other CuX_2 can replace $CuBr_2$ in these reactions to furnish other halogenated analogues.]

Nazarov reaction. With a chiral complex of Cu as Lewis acid and assisted by $AgSbF_6$, cyclopentenones are synthesized with ee ranging from 44% to 88%.[4]

[1]Aoyama, T., Takido, T., Kodomari, M. *TL* **45**, 1873 (2004).

[2]Zhou, H., Huang, X., Chen, W. *SL* 2080 (2003).
[3]Huang, X., Zhou, H., Chen, W. *JOC* **69**, 839 (2004).
[4]Aggarwal, V.K., Belfield, A.J. *OL* **5**, 5075 (2003).

Copper(I) *t*-butoxide.

Desilylative alkylation. *Ipso*-substitution of the silyl group from a (*Z*)-3-triorganosilyl-2-alkenol by a benzyl[1] or allyl group[2] is performed in the presence of *t*-BuOCu. The reaction proceeds even with the silyl substituent located in an aromatic ring (*o*-position) of benzylic alcohols.[3]

OH SiMe$_3$ Ph + PhCH$_2$Br —(*t*-BuOCu, DMF - THF)→ OH Ph Ph

76%

Me$_3$Si OH + Cl —(*t*-BuOCu, DMF - THF)→ OH

75%

β-Cyanohydrins. Catalytic amounts of *t*-BuOCu and DPPE in DMSO bring about the condensation of alkyl cyanides with aldehydes.[4]

[1]Taguchi, H., Tsubouchi, A., Takeda, T. *TL* **44**, 5205 (2003).
[2]Tsubouchi, A., Itoh, M., Onishi, K., Takeda, T. *S* 1504 (2004).
[3]Taguchi, H., Takami, K., Tsubouchi, A., Takeda, T. *TL* **45**, 429 (2004).
[4]Suto, Y., Kumagai, N., Matsunaga, S., Shibasaki, M. *OL* **5**, 3147 (2003).

Copper(I) chloride. **13**, 85; **15**, 101; **18**, 112–113; **19**, 107–108; **20**, 118–120; **21**, 144–146; **22**, 158–159

Coupling reactions. The CuCl-Cs_2CO_3 system previously used in diaryl ether synthesis can be adapted to preparation of alkenyl aryl ethers.[1] *N*-Arylimidazoles are available from CuCl-catalyzed reaction of imidazole with $ArB(OH)_2$ in MeOH.[2]

2-Alkenoic esters bearing an α-(pinacolatoboryl)methyl substituent are obtained from acetylated Baylis-Hillman adducts by the CuCl-catalyzed coupling with (bispinacolato)-diboron. The products are useful for synthesis of substituted α-methylene-γ-butyrolactone.[3]

Coupling of benzylzirconocenes with allyl halides relies on Cu catalysis.[4] Reactions involving such partners are rare, especially for those containing a vinyl substituent in the aromatic ring of the benzyl residue.

Cyclizations. Aldimines derived from *o*-alkynylanilines cyclize to *N*-(α-alkoxybenzyl)indoles[5] in the presence of CuCl and ROH. *N*-Chloroalkenamides in which the double bond is four bonds from the nitrogen atom undergo cycloisomerization to afford 5-chloromethylpyrrolidones or *N*-acyl-2-chloromethylpyrrolidines.[6]

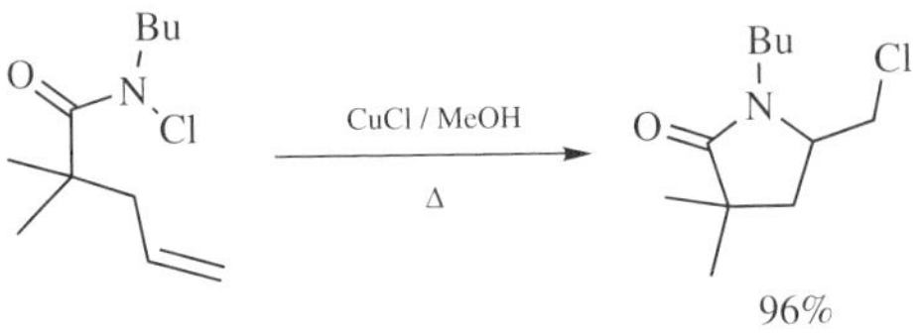
Bu
N
Cl
O
CuCl / MeOH
Δ
Bu
N
Cl
O
96%

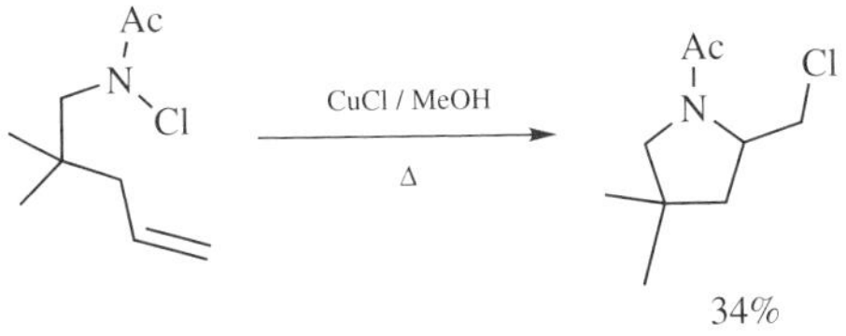
Ac
N
Cl
CuCl / MeOH
Δ
Ac
N
Cl
34%

Alkenylsilanes. α-Haloallylsilanes react with polyhalides through 1,3-transpositional radical substitution.[7] The role of CuCl in halogen abstraction directs to both the polyhalides as well as the haloallylsilanes.

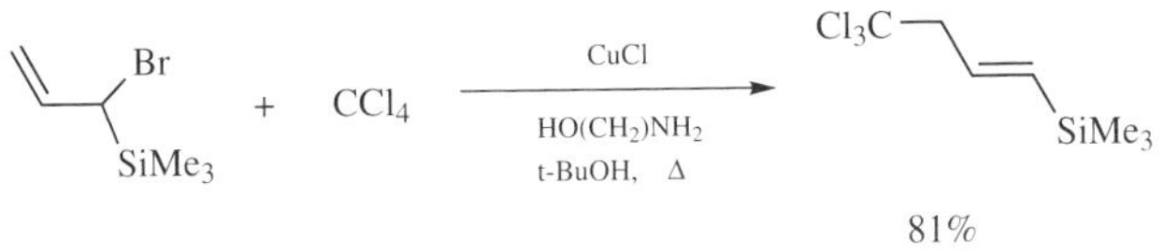
Br
SiMe3
+ CCl4
CuCl
HO(CH2)NH2
t-BuOH, Δ
Cl3C
SiMe3
81%

[1]Wan, Z., Jones, C.D., Koenig, T.M., Pu, Y.J., Mitchell, D. *TL* **44**, 8257 (2003).
[2]Lan, J.B., Chen, L., Yu, X.-Q., You, J.-S., Xie, R.-G. *CC* 188 (2004).
[3]Ramachandran, P.V., Pratihar, D., Biswas, D., Srivastava, A., Reddy, M.V.R. *OL* **6**, 481 (2004).
[4]Ikeuchi, Y., Taguchi, T., Hanzawa, Y. *TL* **45**, 3717 (2004).
[5]Kamijo, S., Sasaki, Y., Yamamoto, Y. *TL* **45**, 35 (2004).
[6]Shulte-Wulwer, I.A., Helaja, J., Gottlich, R. *S* 1886 (2003).
[7]Mitani, M., Masuda, M., Inoue, A. *SL* 1227 (2004).

Copper(II) chloride.

Chlorinations. Hydrosilanes are converted into silyl chlorides by $CuCl_2$ – CuI in Et_2O at room temperature.[1] Admixture of lithium allyloxides and conjugated nitroalkenes followed by treatment with $CuCl_2$ leads to 3-chloromethyl-4-nitrotetrahydrofurans.[2]

***N*-Allylanilines.** *N*-Functionalization of alkenes at the less substituted carbon site is achieved with nitrosobenzene. The reaction is mediated by $CuCl_2$ – Cu and deoxygenation of the adducts (hydroxylamines) is involved.[3]

[1]Kunai, A., Ohshita, J. *JOMC* **686**, 3 (2003).
[2]Jahn, U., Rudakov, D. *SL* 1207 (2004).
[3]Srivastava, R.S. *TL* **44**, 3271 (2003).

Copper(I) cyanide.

Ynamides. *N*-Alkynylation with 1-bromoalkynes catalyzed by CuCN is applicable to synthesis of carbamates, lactams, and chiral oxazolidinones.[1]

[1]Frederick, M.O., Mulder, J.A., Tracey, M.R., Hsung, R.P., Huang, J., Kurtz, K.C.M., Shen, L., Douglas, C.J. *JACS* **125**, 2368(2003).

Copper(II) 2-ethylhexanoate.

α-(3-Indolyl) ketones. Oxidative cross-coupling of ketone enolates and indoles (3-unsubstituted) occurs in the presence of the title compound. This reaction is applicable to a synthesis ofr hapalindole-Q.

hapalindole-Q

[1]Baran, P.S., Richter, J.M. *JACS* **126**, 7450 (2004).

Copper(I) fluoride - triphenylphosphine.

Aldol reaction.[1] With CuF as catalyst silyl ketene acetals condense with various ketones.

[1]Oisaki, K., Suto, Y., Kanai, M., Shibasaki, M. *JACS* **125**, 5644 (2003).

Copper(I) iodide. 16, 98; 18, 114–115; 19, 109–110; 20, 120–121; 21, 147–148; 22, 160–161

Arylations. Refinement and modifications of the CuI-mediated arylation procedures continue. For *N*-arylation of amines, additives include CsOAc,[1] (*S*)-proline,[2] salicylamide,[3] Bu_3P.[4] *N*-Arylation of heteroaromatic compounds such as pyrroles, pyrazoles, imidazoles, triazoles, indazoles is performed with CuI and a 1,2-diamine[5] or (*S*)-proline.[6] The cross-coupling of such heteroarenes with aryl bromide bearing a side chain functionalized with an NH_2 group under microwave irradiation[7] is interesting.

Amides also undergo Cu-catalyzed *N*-arylation, in the presence of glycine[8] and diamine in the case of α-pyridones.[9] Benzannulated 7-to-10 membered diazacycles are prepared from β-lactams and *o*-iodoarylamines and *o*-iodoarylalkylamines.[10] It involves *N*-arylation and intramolecular transacylation.

CuI, $MeNH(CH_2)NHMe$, K_2CO_3, PhMe, 110° — 92%

Urea is transformed into *N,N'*-diarylureas by a twofold arylation using the CuI-diamine method.[11] *N*-Arylsulfoximines are obtained from the parent compounds.[12]

Preparation of azoarenes[13] from bis-Boc arylhydrazines and iodoarenes in hot DMF involves deacylation and oxidation besides the cross-coupling.

Ullmann coupling to form diaryl ethers is performed with microwave heating[14] and in ionic liquids.[15] *N,N*-Dimethylglycine is an effective promoter for the same reaction,[16] whereas monomethylglycine is used to ligate CuI in synthesizing diaryl sulfides.[17] Ring closure of *N*-(*o*-bromoaryl)thioureas is a convenient access to 2-aminobenzothiazoles, especially by the copper route.[18] Conditions for *S*-arylation appear to be compatible with some unusual functional groups (e.g, triazene) present in the aryl halides which are used to synthesize arylthioglycosides.[19]

There was no simple way to prepare unsymmetrical diaryl selenides. Reaction of triorganostannyl arylselenides with ArX ameliorates the situation. The catalyst system consists of $(Ph_3P)CuI$ and phenanthroline.[20]

Direct coupling of secondary phosphines and phosphites with aryl halides (and alkenyl halides)[21,22] provides unsymmetrical tertiary phosphines and phosphites.

Following the arylation experiences *N*-acylenamines[23] and *N*-acylynamines[24] are readily prepared. For *O*-alkenylation of allylic alcohols the products are substituted 4-pentenals as a result of Claisen rearrangement of the alkenyl allyl ethers.[25]

Some technical improvements for the synthesis of 1,2-diarylethynes by the cross-coupling method are application of microwaves[26] and *N,N*-dimethylglycine as additive.[27] 1,4-Bis(tosylamino)-1,3-butadiynes are the homocoupling products from the tosylaminoethynes. The reaction is carried out in the presence of oxygen.[28]

Alkenyl(2-pyridyl)silanes undergo homocoupling while detaching the pyridylsilyl group to provide 1,3-dienes. Besides CuI acting as promoter, CsF is also required to activate the silyl substituent.[29]

Substitution and coupling reactions. 1,4-Diynes are prepared from 1-alkynes and propargylic halides in the presence of CuI and Cs_2CO_3.[30] These two salts are instrumental to formation of copper alkynides.

Conversion of ArX and alkenyl iodides to the corresponding azides is rather facile. The reagent system consists of NaN_3, CuI, (*S*)-proline, NaOH (DMSO as solvent).[31] Access to 2-oxo-3-alkynoic esters is also simple through a copper-mediated acylation of 1-alkynes with oxalyl monochlorides at room temperature.[32]

α-Tributylstannyl alcohols that are available from carbonyl compounds by reaction with $(Bu_3Sn)_2Zn$ undergo coupling with allyl halides after thiocarbamoylation.[33] Formation of alkynyl ketones in water containing a surfactant by a Pd-catalyzed, CuI-promoted reaction of acid chlorides and 1-alkynes is surprisingly effective.[34]

Cleaner reaction, shortened reaction time, and improved yields are the beneficial effects attributed to CuI (0.2 equivalent) in its presence in the 3-component reaction of ArI, allenes, and Ar'CHO promoted by $Pd(OAc)_2$ and indium.[35] The reaction involves formation of (2-arylallyl)palladium complexes and then allylindium reagents.

Propargylic amines. Condensation of 1-alkynes, formaldehyde and secondary amines in the presence of CuI proceeds in good yields.[36] Addition to aldimines using RCOCl and 1-alkynes leads to *N*-acyl progargylic amines.[37]

Rearrangement. When exposed to CuI or a Pd complex, cyclopropenyl ketones pursue different reaction pathways toward transformation into furans.[38]

CuI, MeCN Δ: R = Bu 96%

$(MeCN)_2PdCl_2$, CH_2Cl_2 Δ: R = Bu 88%

[1]Okano, K., Tokuyama, H., Fukuyama, T. *OL* **5**, 4987 (2003).
[2]Ma, D., Cai, Q., Zhang, H. *OL* **5**, 2453 (2003).
[3]Kwong, F.Y., Buchwald, S.L. *OL* **5**, 793 (2003).
[4]Patil, N.M., Kelkar, A.A., Nabi, Z., Chaudhari, R.V. *CC* 2460 (2003).
[5]Antilla, J.C., Baskin, J.M., Barder, T.E., Buchwald, S.L. *JOC* **69**, 5578 (2004).
[6]Ma, D., Cai, Q. *SL* 128 (2004).
[7]Wu, Y.-J., He, H., L'Hereux, A. *TL* **44**, 4217 (2003).
[8]Deng, W., Wang, Y.-F., Zou, Y., Liu, L., Guo, Q.-X. *TL* **45**, 2311 (2004).
[9]Li, C.S., Dixon, D.D. *TL* **45**, 4257 (2004).
[10]Klapars, A., Parris, S., Anderson, K., Buchwald, S.L. *JACS* **126**, 3529 (2004).
[11]Nandakumar, M.V. *TL* **45**, 1989 (2004).
[12]Cho, G.Y., Remy, P., Jansson, J., Moessner, C., Bolm, C. *OL* **6**, 3293 (2004).

[13]Kim, K.-Y., Shin, J.-T., Lee, K.-S., Cho, C.-G. *TL* **45**, 117 (2004).
[14]Wu, Y.-J., He, H. *TL* **44**, 3445 (2003).
[15]Luo, Y., Xin, J., Ren, R.X. *SL* 1734 (2003).
[16]Ma, D., Cai, Q. *OL* **5**, 3799 (2003).
[17]Deng, W., Zou, Y., Wang, Y.-F., Liu, L., Guo, Q.-X. *SL* 1254 (2004).
[18]Joyce, L.L., Evindar, G., Batey, R.A. *CC* 446 (2004).
[19]Naus, P., Leseticky, L., Smrcek, S., Tislerova, I., Sticha, M. *SL* 2117 (2003).
[20]Beletskaya, I.P., Sigeev, A.S., Peregudov, A.S., Petrovskii, P.V. *TL* **44**, 7039 (2003).
[21]Gelman, D., Jiang, L., Buchwald, S.L. *OL* **5**, 2315 (2003).
[22]Allen, D.V., Venkataraman, D. *JOC* **68**, 4590 (2003).
[23]Jiang, L., Job, G.E., Klapars, A., Buchwald, S.L. *OL* **5**, 3667 (2003).
[24]Dunetz, J.R., Danheiser, R.L. *OL* **5**, 4011 (2003).
[25]Nordmann, G., Buchwald, S.L. *JACS* **125**, 4978 (2003).
[26]He, H., Wu, Y.-J. *TL* **45**, 3237 (2004).
[27]Ma, D., Liu, F. *CC* 1934 (2004).
[28]Rodriguez, D., Castedo, L., Saa, C. *SL* 377 (2004).
[29]Itami, K., Ushiogi, Y., Nokami, T., Ohashi, Y., Yoshida, J. *OL* **6**, 3695 (2004).
[30]Caruso, T., Spinella, A. *T* **59**, 7787 (2003).
[31]Zhu, W., Ma, D. *CC* 888 (2004).
[32]Guo, M., Li, D., Zhang, Z. *JOC* **68**, 10172 (2003).
[33]Mohapatra, S., Bandyopadhyay, A., Barma, D.K., Capdevila, J.H., Falck, J.R. *OL* **5**, 4759 (2003).
[34]Chen, L., Li, C.-J. *OL* **6**, 3151 (2004).
[35]Cleghorn, L.A.T., Cooper, I.R., Grigg, R., MacLachlan, W.S., Sridharan, V. *TL* **44**, 7969 (2003).
[36]Bieber, L., Silva, M. *TL* **45**, 8281 (2004).
[37]Black, D.A., Arndtsen, B.A. *OL* **6**, 1107 (2004).
[38]Ma, S., Zhang, J. *JACS* **125**, 12386 (2003).

Copper(I) oxide. 16, 99; 21, 148–149; 22, 161

Arylations. Mild conditions for pyrazole arylation[1] and diaryl ether synthesis[2] involve Cu_2O as catalyst and CsCO as base.

[1]Cristau, H.-J., Cellier, P.P., Spindler, J.-F., Taillefer, M. *EJOC* 695 (2004).
[2]Cristau, H.-J., Cellier, P.P., Hamada, S., Spindler, J.-F., Taillefer, M. *OL* **6**, 913 (2004).

Copper(II) sulfate.

Substitution reactions. Ynamides are formed by *N*-alkynylation in the presence of $CuSO_4$ and 1,10-phenanthroline.[1] Conversion of arylamines to aryl azides on reaction with TfN_3 is achieved by catalysis with aq. $CuSO_4$.[2]

1,2,3-Triazoles. A one-pot synthesis of 1,4-disubstituted 1,2,3-triazoles from ArI, 1-alkynes and NaN_3 is most efficient. $CuSO_4$ plays a dual role in promoting both cycloaddition and *N*-arylation.[3]

[1]Zhang, Y., Hsung, R.P., Tracey, M.R., Kurtz, K.C.M., Vera, E.L. *OL* **6**, 1151 (2004).
[2]Liu, Q., Tor, Y. *OL* **5**, 2571 (2003).
[3]Feldman, A.K., Colasson, B., Fokin, V.V. *OL* **6**, 3897 (2004).

Copper(I) triflate.

Homopropargylic amines. Ring opening of aziridines by lithium alkynides occurs at room temperature in the presence of CuOTf.[1]

Addition reactions. Regioselective addition to enones by $TsNCl_2$ leads to β-chloro-α-amino ketones.[2]

[1]Ding, C.-H., Dai, L.-X., Hou, X.-L. *SL* 1691 (2004).
[2]Chen, D., Timmons, C., Chao, S., Li, G. *EJOC* 3097 (2004).

Copper(II) triflate. 19, 112; 20, 122–123; 21, 149; 22, 162–163

Addition reactions. Stabilized carbonyl compounds such as β-keto esters and malonic esters add to aldimines under catalysis by $Cu(OTf)_2$, and the presence of certain chiral ligands (e.g., *t*-Bu-BOX) the reaction becomes asymmetric.[1] Destannylative addition provides a way to 1,2-amino alcohol derivatives using α-siloxystannanes.[2] The reaction between trimethylsilyl nitronates and *N*-(*o*-methoxybenzyl)aldimines is *anti*-selective and 1,2-diamines of that configuration are readily prepared.[3]

Alkylidenemalonic esters undergo *C,O*-disilylation in a Cu-catalyzed reaction with hexaorganodisilanes.[4] The addends are activated by Lewis bases. Hydration of arylidenecyclopropanes causes ring rupture to provide homoallylic alcohols.[5]

Conjugate addition to electron-deficient alkenes with electrophilic trapping in tandem is synthetically efficient. To render organometallic addends effective it often requires in situ exchange into copper species, therefore catalysis by $Cu(OTf)_2$ has certain appeal.[6]

Ph, O, O, + Et_2Zn, $Cu(OTf)_2$, $(Et_2O)_3P$, CH_2Cl_2, O, H, OH, Ph, 81%

For effecting Nazarov cyclization $Cu(OTf)_2$ is an exceptionally mild catalyst, especially for substrates possessing polar substituents.[7]

O, O, COOMe, Ph, $Cu(OTf)_2$, DCE, 25°, O, O, COOMe, Ph, 99%

Through Cu-catalyzed formation of pyrylium ions to induce [4 + 2]cycloaddition with alkenes, *trans*-1-acyl-1,2-dihydronaphthalenes are produced.[8]

Cu(OTf)$_2$ / THF 80°

Acetylation. Using Cu(OTf)$_2$ as catalyst and Ac_2O in the neat, hexopyranoses are peracetylated[9] and α-hydroxyphosphonates are converted into the corresponding acetates.[10]

[1]Marigo, M., Kjaersgaard, A., Juhl, K., Gathergood, N., Jorgensen, K.A. *CEJ* **9**, 2359 (2003).
[2]Kagoshima, H., Shimada, K. *CL* **32**, 514 (2003).
[3]Anderson, J.C., Blake, A.J., Howell, G.P., Wilson, C. *JOC* **70**, 549 (2005).
[4]Clark, C.T., Lake, J.F., Scheidt, K.A. *JACS* **126**, 84 (2004).
[5]Siriwardana, A.I., Nakamura, I., Yamamoto, Y. *TL* **44**, 4547 (2003).
[6]Agapiou, K., Cauble, D.F., Krische, M.J. *JACS* **126**, 4528 (2004).
[7]He, W., Sun, X., Frontier, A.J. *JACS* **125**, 14278 (2003).
[8]Asao, N., Kasahara, T., Yamamoto, Y. *ACIEE* **42**, 3504 (2003).
[9]Tai, C.-A., Kulkarni, S.S., Hung, S.-C. *JOC* **68**, 8719 (2003).
[10]Firouzabadi, H., Iranpoor, N., Sobhani, S., Amoozgar, Z. *S* 295 (2004).

Cyanomethylenetriorganophosphoranes. 21, 150

Substitutions. Reagents such as $Me_3P{=}CHCN$ are available in two steps from $ClCH_2CN$. They can replace the Mitsunobu system for alkylation.[1]

[1]Sakamoto, I., Kaku, H., Tsunoda, T. *CPB* **51**, 474 (2003).

β-Cyclodextrin. 21, 151; 22, 164

Epoxide and aziridine cleavage. In the presence of β-cyclodextrin in aqueous media epoxides are opened with ArONa.[1] The cleavage of *N*-tosylaziridines by HX is unexceptional.[2]

Aromatic aldehydes are regenerated from their acetals under neutral conditions with aq. solution of β-cyclodextrin.[3]

[1]Surendra, K., Krishnaveni, N.S., Nageswar, Y.V.D., Rao, K.R. *JOC* **68**, 4994 (2003).
[2]Krishnaveni, N.S., Surendra, K., Narender, M., Nageswar, Y.V.D., Rao, K.R. *S* 501 (2004).
[3]Krishnaveni, N.S., Surendra, K., Reddy, M.A., Nageswar, Y.V.D., Rao, K.R. *JOC* **68**, 2018 (2003).

D

Decaborane. 21, 154; **22**, 167

Reduction. Nitroalkenes are reduced to oximes with decaborane in the presence of Pd/C,[1] while azides are converted into amines or their *t*-butyl carbamates directly.[2]

N-Alkylation.[3] Arylamines are alkylated with carbonyl compounds using decaborane as reducing agent. This procedure is just an extension of the previously reported *N*-methylation with HCHO.

[1]Lee, S.H., Park, Y.J., Yoon, C.M. *OBC* **1**, 1099 (2003).
[2]Jung, Y.J., Chang, Y.M., Lee, J.H., Yoon, C.M. *TL* **43**, 8735 (2002).
[3]Jung, Y.J., Bae, J.W., Park, E.S., Chang, Y.M., Yoon, C.M. *T* **59**, 10331 (2003).

Dess-Martin periodinane. 21, 154–155; **22**, 167

Oxidation.[1] Aldehydes are converted into acyl azides by the periodinane and NaN_3 in CH_2Cl_2 at 0°.

Dethioacetalization.[2] Use of the periodinane in a mixture of MeCN, CH_2Cl_2 and H_2O at room temperature regenerates carbonyl compounds from dithioacetals. Some siloxy groups are unaffected.

[1]Bose, D.S., Reddy, A.V.N. *TL* **44**, 3543 (2003).
[2]Langille, N.F., Dakin, L.A., Panek, J.S. *OL* **4**, 575 (2003).

Dialkylaluminum chloride. 20, 126–127; **21**, 155; **22**,

Cyclization. An extraordinary cyclization to form a bicycle[3.1.0]heptene derivative occurs when a conjugated alkatrienoic ester is exposed to Me_2AlCl.[1]

COOEt

Me_2AlCl

CH_2Cl_2

COOEt

H

78%

[1]Miller, A.K., Trauner, D. *ACIEE* **42**, 549 (2003).

Fiesers' Reagents for Organic Synthesis, Volume 23. Edited by Tse-Lok Ho

Dialkyl azodicarboxylates.

Hydrohydrazination.[1] Addition of hydrazine-1,2-dicarboxylic ester to alkenes by a cobalt-catalyzed reaction with di-*t*-butyl azodicarboxylate and phenylsilane proceeds in the Markovnikov sense. It involves hydrocobaltation of the alkenes and nitrogen atom insertion into the C-Co bond.

Debenzylation. Benzylamines react with diisopropyl azodicarboxylate. The debenzylation does not affect *O*-benzyl, *N*-tosyl and azido groups.[2]

[1]Waser, J., Carreira, E.M. *JACS* **126**, 5676 (2004).
[2]Kroutil, J., Trnka, T., Cerny, M. *S* 446 (2004).

Diaminochloroboranes.

Boron enolates.[1] Mannich adducts are formed from boron enolates [prepared from lithium enolates and $ClB(NR_2)_2$] and aldehydes.

$OB(NR_2)_2$ + PhCHO → (O, OH, Ph) + (O, NEt_2, Ph)

R = Me	THF, −78°~20°	63%	-
R = Et	DMF, 50°	-	72%

[1]Suginome, M., Uehlin, L., Yamamoto, A., Murakami, M. *OL* **6**, 1167 (2004).

1,4-Diazabicyclo[2.2.2]octane, DABCO. 13, 92; **15**, 109; **18**, 120; **19**, 116–117; **20**, 128–129; **21**, 157–158; **22**, 170–171

Alkylation. Using DABCO as base, heterocyclic amines such as indole and benzimidazole are alkylated with dialkyl carbonates.[1,2]

Baylis-Hillman reaction. The DABCO-promoted reaction has been extended to alkynals[3] and *N*-tosylimines.[4] *N,N'*- Di[3,5-bis(trifluoromethyl)phenyl]urea is found to accelerate the condensation through H-bonding,[5] and sulfolane serves as a useful solvent.[6] Phosphonium salts are useful cocatalysts.[7]

Renewed condensation of the acetylated adducts leads to 3-substituted 1,4-pentadienes.[7]

NO_2, AcO, COOMe + COOMe —DABCO; THF, H_2O, r.t.→ NO_2, MeOOC, COOMe

51%

The Baylis-Hillman reaction of 3-trimethylsilyl propynoic esters adopts a somewhat different pathway. The initial condensation is followed by generation of alkylidene carbene and the process is terminated by further insertion into the C-H bond of another aldehyde molecule.[8]

Me_3Si—≡—COOMe —(DABCO; RCHO, PhH, Δ)→ R(C=O)CH=C(COOMe)CH(R)$OSiMe_3$

Ring formation. The Baylis-Hillman adducts of many substrates are disposed toward ring closure. For example, 2-formylbenzoic acid and ethyl acrylate give an alkylidenephthalide,[9] 2-aminobenzaldehyde and a nitroalkene furnish a 3-nitro-1,2-dihydroquinoline.[10]

A reflexive Michael addition between *o*-hydroxy-ω-nitrostyrene and a nitroalkene leads to 3-nitrochromane derivative.[11]

Cyclopropanation occurs when α-chloroketones and various alkenes (*t*-butyl acrylate, vinyl phenyl sulfone,...) are exposed to DABCO.[12] Perhaps the intramolecular version[13] is more interesting.

DABCO; Na_2CO_3, DCE, 60°

59

Treatment of imines and 3-nitropropyl mesylate with DABCO results in the formation of 3-nitropyrrolidines.[14]

[1]Shieh, W.-C., Dell, S., Bach, A., Repic, O., Blacklock, T.J. *JOC* **68**, 1954 (2003).
[2]Shieh, W.-C., Lozanov, M., Repic, O. *TL* **44**, 6943 (2003).
[3]Krishna, P.R., Sehar, E.R., Kannan, V. *TL* **44**, 4973 (2003).
[4]Xu, Y.-M., Shi, M. *JOC* **69**, 417 (2004).
[5]Maher, D.J., Connon, S.J. *TL* **45**, 1301 (2004).
[6]Krishna, P.R., Manjuvani, A., Kannan, V., Sharma, G.V.M. *TL* **45**, 1183 (2004).
[7]Johnson, C.L., Donkor, R.E., Nawaz, W., Karodia, N. *TL* **45**, 7359 (2004).
[8]Matsuya, Y., Hayashi, K., Nemoto, H. *JACS* **125**, 646 (2003).
[9]Lee, K.Y., Kim, J.M., Kim, J.N. *SL* 357 (2003).
[10]Yan, M.-C., Tu, Z., Lin, C., Ko, S., Hsu, J., Yao, C.-F. *JOC* **69**, 1565 (2004).
[11]Yao, C.-F., Jang, Y.-J., Yan, M.-C. *TL* **44**, 3813 (2003).
[12]Papageorgiou, C.D., Ley, S.V., Gaunt, M.J. *ACIEE* **42**, 828 (2003).
[13]Bremeyer, N., Smith, S.C., Ley, S.V., Gaunt, M.J. *ACIEE* **43**, 2681 (2004).
[14]Baricordi, N., Benetti, S., Biondini, G., De Risi, C., Pollini, G.P. *TL* **45**, 1373 (2004).

1,8-Diazabicyclo[5.4.0]undec-7-ene, DBU. 13, 92; **14**, 109; **15**, 109–110; **16**, 105–106; **17**, 99–100; **18**, 120–121; **19**, 117; **20**, 129–130; **21**, 158–160; **22**, 171–172

Elimination. Heating of tosylates with NaI and DBU in glyme generates alkenes.[1] 2,3-Dibromopropyl phenyl sulfide gives the useful synthetic intermediate, 2-phenylthioallyl bromide, on dehydrobromination with DBU.[2]

Conducting certain Wittig reactions using DBU as a base[3] is convenient.

Heating 2-sulfonylalkylcyclopropane-1,1-dicarboxylic esters with DBU in benzene generates a conjugate diene unit.[4]

PhSO$_2$ COOEt COOEt R —DBU / PhH, Δ→ R COOEt COOEt

Condensations. The reaction of nitroalkanes with α-chloro nitriles to give conjugated nitriles[5] is useful because the substrates may contain other functional groups. Nitronate anions react with *N,N*-bis(trimethylsiloxy) enamines and β-nitroalkoximes are obtained.[6]

A route to 2-acylindole-3-acetic esters[7] starting from *o*-benzenesulfonaminocinnamic esters involves *N*-alkylation with α-bromo ketones and intramolecular Michael reaction, both capably effected by DBU. Here is a final treatment with NaOH to eliminate PhSO$_2$H from the substituted indolines.

1,4-Dicarbonyl compounds. Thiazolium ylide generation using DBU as base initiates the acyl transfer from acylsilanes to conjugated ketones and esters.[8]

R(C=O)SiMe$_3$ + Ph–CH=CH–C(=O)Ph —(thiazolium salt S, NEt$^+$, Br$^-$, HO; DBU, iPrOH, THF; H$_2$O)→ R–C(=O)–CH(Ph)–CH$_2$–C(=O)Ph

[1]Phukan, P., Bauer, M., Maier, M.E. *S* 1324 (2003).
[2]Masuyama, Y., Sano, T., Oshima, M., Kurusu, Y. *BCSJ* **76**, 1679 (2003).
[3]Okuma, K., Sakai, O., Shioji, K. *BCSJ* **76**, 1675 (2003).
[4]Jeon, H.-S., Koo, S. *TL* **45**, 7023 (2004).
[5]Ballini, R., Fiorini, D., Gil, M.V., Palmieri, A. *TL* **44**, 9033 (2003).
[6]Kunetsky, R.A., Dilman, A.D., Tsvaygboym, K.P., Ioffe, S.L., Strelenko, Y.A., Tartakovsky, V.A. *S* 1339 (2003).
[7]Nakao, K., Murata, Y., Koike, H., Uchida, C., Kawamura, K., Mihara, S., Hayashi, S., Stevens, R.W. *TL* **44**, 7269 (2003).
[8]Mattson, A.E., Bharadwaj, A.R., Scheidt, K.A. *JACS* **126**, 2314 (2004).

Dibenzoyl peroxide.

Radical addition. Styrylation of ethers[1] such as THF by PhCH=$CHNO_2$ on heating with Bz_2O_2 proceeds via radical addition and elimination. Esters such as ethyl acetate generate carbinyl radicals[2] and react similarly allylic esters are formed in moderate yields.

[1]Jang, Y.-J., Shih, Y.-K., Liu, J.-Y., Kuo, W.-Y., Yao, C.-F. *CEJ* **9**, 2123 (2003).
[2]Yan, M.-C., Jang, Y.-J., Wu, J., Lin, Y.-F., Yao, C.-F. *TL* **45**, 3685 (2004).

Dibromodifluoroethane.

Bromination.[1] For the preparation of α,α-dibromoalkanoic esters, which are useful precursors of ynolate anions, α-bromoalkanoic esters are further brominated (LDA, $BrCF_2CF_2Br$, −78°).

[1]Shindo, M., Sato, Y., Koretsune, R., Yoshikawa, T., Matsumoto, K., Itoh, K., Shishido, K. *CPB* **51**, 477 (2003).

Dibromodifluoromethane. 22, 172

Difluoromethylenation.[1] ArCHO are converted into ArCH=CF_2 via their hydrazones on treatment with ethylenediamine-copper chloride and Br_2CF_2.

Ramberg-Bäcklund reaction.[2] A synthesis of *C*-glycosides involves condensation of normal sugars with $BnSO_2CH_2PO(OEt)_2$ and the subjecting sulfones to the R-B reaction using a base in the supported form (KOH/Al_2O_3).

[1]Nanajdenko, V.G., Varseev, G.N., Korotchenko, V.N., Shastin, A.V., Balenkova, E.S. *JFC* **124**, 115 (2003).
[2]McAllister, G.C., Paterson, D.E., Taylor, R.J.K. *ACIEE* **42**, 1387 (2003).

1,3-Dibromo-5,5-dimethylhydantoin.

Bromodeboronation.[1] Arylboronic acids are converted into ArBr by the title reagent (and NaOMe as a base).
[4]Szumigala, R.H., Devine, P.N., Gauthier, D.R., Volante, R.P. *JOC* **69**, 566 (2004).

Dibromomethane.

α-Methylenation.[1] Reaction of enolizable carbonyl compounds with CH_2Br_2/Et_2NH is accelerated by microwave irradiation.

[1]Hon, Y.-S., Hsu, T.-R., Chen, C.-Y., Lin, Y.-H., Chang, F.-J., Hsieh, C.-H., Szu, P.-H. *T* **59**, 1509 (2003).

Dibromonitrosotris(1-pyrazolyl)boratomolybdenum.

Dearomatization.[1] Arenes are η^2-complexed to Mo on treatment with the title reagent $TpMo(NO)Br_2$, Na/Hg and 1-methylimidazole. The complexation enables the arenes to protonate at an "uncomplexed" double bond to form stabilized allyl cations that are reactive toward nucleophiles.

[1]Meiere, S.H., Keane, J.M., Gunnoe, T.B., Sabat, M., Harman, W.D. *JACS* **125**, 2024 (2003).

Dibutylboron triflate. 20, 132, **21**, 161; **22**, 173

Aldol reaction. Aldol reactions involving cyclic acetals are reported (intermolecular[1] and intramolecular[2]).

[1]Li, L.-S., Das, S., Sinha, S.C. *OL* **6**, 127 (2004).
[2]Das, S., Li, L.-S., Sinha, S.C. *OL* **6**, 123 (2004).

Dibutylchlorotin hydride.

Hydrostannylation.[1] Propargylic alcohols afford (*Z*)-allylic alcohols. Regioselectivity of the reaction depends on the substitution pattern of the alkyne linkage.

$ClBu_2Sn$ OH
R OH $\xrightarrow[\text{9-BBN}]{Bu_2SnHCl}$
R=H 65%
$SnBu_2Cl$ OH
R=Me (major)

[1]Thiele, C.M., Mitchell, T.N. *EJOC* 337 (2004).

Dibutyltin methoxide.

Aldol reaction. In a mixture of THF and MeOH the catalytic activity of $Bu_2Sn(OMe)_2$ for promoting aldol reaction of enol trichloroacetates at room temperature is much higher than that of Bu_3SnOMe.[1] Slight variation of conditions leads to enones.[2]

[1]Yanagisawa, A., Sekiguchi, T. *TL* **44**, 7163 (2003).
[2]Yanagisawa, A., Goudu, R., Arai, T. *OL* **6**, 4281 (2004).

Dibutyltin oxide. 13, 95–96; **15**, 116–117; **16**, 112; **18**, 125; **20**, 133–134; **21**, 163; **22**, 174–176

Alkylation. Diols are monobenzylated in the presence of Bu_2SnO, and the method has been critically reevaluated.[1] Allylation of arabinofuranose derivatives via the 1,2-*O*-stannylene acetals is regioselective, allyl glycosides are obtained.[2]

[1]Simas, A.B.C., Pais, K.C., da Silva, A.A.T. *JOC* **68**, 5426 (2003).
[2]Darwish, O.S., Callam, C.S., Hadad, C.M., Lowary, T.L. *JCC* **22**, 963 (2003).

N,N-Dichloroarenesulfonamides.

Diamination. Addition to the enone double bond by one of the title reagents (Ar = Tol, $O_2NC_6H_4$) in MeCN leads to imidazoline derivatives (via aziridines). The methyl group (from MeCN) is chlorinated, but whether to dichloro or trichloro stage is controllable (temperature and addition of molecular sieves).

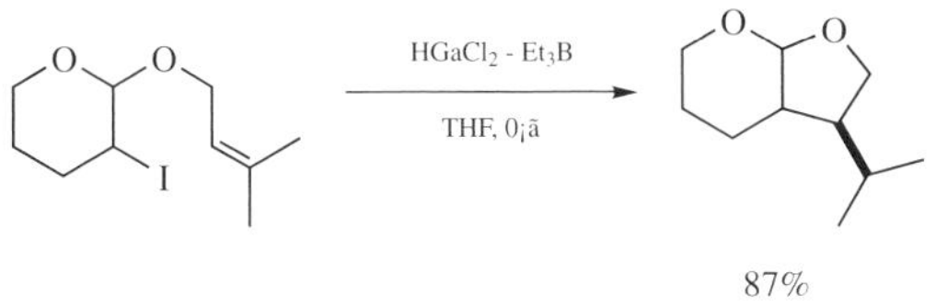

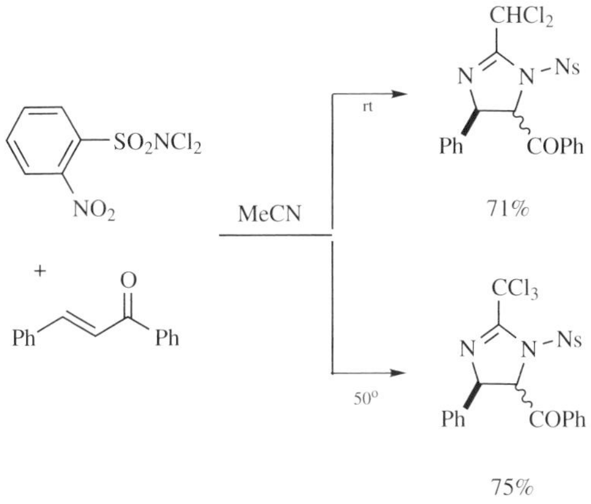

HGaCl2 - Et3B
THF, 0¡ã
87%

SO2NCl2
NO2
+
Ph Ph
O
MeCN
rt
CHCl2
N N-Ns
Ph COPh
71%
50°
CCl3
N N-Ns
Ph COPh
75%

[1]Chen, D., Timmons, C., Wei, H.-X., Li, G. *JOC* **68**, 5742 (2003).
[2]Pei, W., Wei, H.-X., Chen, D., Headley, A.D., Li, G. *JOC* **68**, 8404 (2003).

Dichloroborane.

Hydroboration. The dioxane complex of dichloroborane is useful for the regioselective preparation of (alkyl and alkenyl) boronic acids.[1] These are obtained on aqueous quench of the reaction adducts with alkenes and alkynes.

Reduction. Dichloroborane-dimethyl sulfide is a chemoselective reducing agent for azides (to amines).[2] Thus, this borane is more reactive toward RN_3 than a C=C bond.

[1]Josyula, K.V.B., Gao, P., Hewitt, C. *TL* **44**, 7789 (2003).
[2]Salunkhe, A.M., Ramachandran, P.V., Brown, H.C. *T* **58**, 10059 (2002).

2,3-Dichloro-5,6-dicyano-1,4-benzoquinone, DDQ. **13**, 104–105; **14**, 126–127; **15**, 125–126; **16**, 120; **18**, 130; **19**, 121–122; **20**, 137–138; **21**, 164–165; **22**, 176

Dehydrogenation. 2,2-Disubstituted pyrrolidines give 2*H*-pyrroles on treatment with DDQ.[1] Benzylic oxidation of electron-rich aryl methyl ethers for immediate condensation with silyl enol ethers is accomplished with DDQ-$LiClO_4$.[2]

2-Substituted indoles. For substitution of the indole nucleus at C-2 the benzene ring is subjected to a Birch reduction before reaction of the resulting pyrroles with Michael acceptors (may need $Bi(NO_3)_3$ to catalyze this step). Rearomatization is done by a quinone such as DDQ.[3]

[1]Cheruku, S.R., Padmanilayam, M.P., Vennerstrom, J.L. *TL* **44**, 3701 (2003).
[2]Ying, B.-P., Trogden, B.G., Kohlman, D.T., Liang, S.-X., Xu, Y.-C. *OL* **6**, 1523 (2004).
[3]Cavdar, H., Saracoglu, N. *T* **61**, 2401 (2005).

Dichloromethane.

Bisaryloxymethanes.[1] Good yields of $ArOCH_2OAr$ are obtained from treatment of ArONa with CH_2Cl_2 in NMP.

Methylenation. A methylene-titanium complex is formed from CH_2Cl_2, $TiCl_4$, and Mg. The complex is reactive toward carbonyl compounds[2] including esters[3] (conversion into vinyl ethers).

[1]Liu, W., Szewczyk, J., Waykole, L., Repic, O., Blacklock, T.J. *SC* **33**, 2719 (2003).
[2]Yan, T.-H., Tsai, C.-C., Chien, C.-T., Cho, C.-C., Huang, P.-C. *OL* **6**, 4961 (2004).
[3]Yan, T.-H., Chien, C.-T., Tsai, C.-C., Lin, K.-W., Wu, Y.-H. *OL* **6**, 4965 (2004).

α,α-Dichloromethyl ether.

Homologation.[1] Functionalization of an alkene $RCH{=}CH_2$ via hydroboration (with 9-BBN) and rearrangement to the stable *B*-alkyl-9-oxa-10-borabicyclo[3.3.2]-decane by Me_3NO is terminated by reaction with Cl_2CHOMe and oxidative workup. The product is RCH_2CH_2COOH.

[1]Soderquist, J.A., Martinez, J., Oyola, Y., Kock, I. *TL* **45**, 5541 (2004).

Dichloro(triphos)rhodium(II).

Glycosylation.[1] The complex **1** catalyzes formation of glycosides from sugars and alcohols.

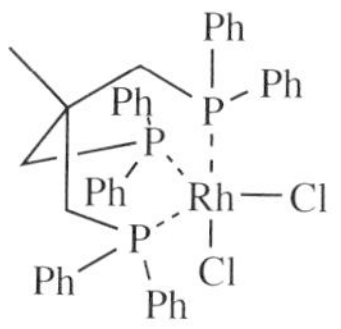

1

[1]Wagner, B., Heneghan, M., Schnabel, G., Ernst, B. *SL* 1303 (2003).

Dichlorotris(triphenylphosphine)ruthenium(II). 22, 178

Isomerization-aldol reaction.[1] Allylic alcohols serve as enol precursors for condensation with aldehydes in the presence of $(Ph_3P)_3RuCl_2$. The reaction can be performed in aq. media.

N-Tosylimines.[2] Imido transfer from $Ph_3P{=}NTs$ to aldehydes is accomplished at room temperature by catalysis of $(Ph_3P)_3RuCl_2$.

Coupling reactions. Benzhydryl halides are coupled to furnish 1,1,2,2-tetraarylethanes on exposure to $(Ph_3P)_3RuCl_2$ under hydrogen.[3] The unusual condensation between secondary and primary alcohols with the latter compounds behaving as alkylating agents[4] is applicable to a quinoline synthesis.[5] Actually both alcohols are dehydrogenated (hydrogen transferred to 1-dodecene) and then undergo aldol reaction and reduction.

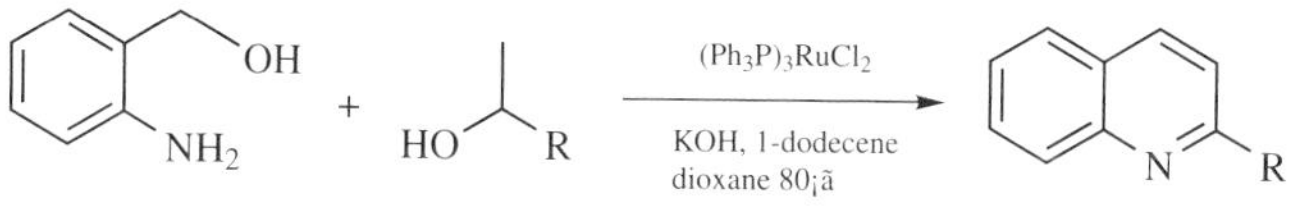

[1]Wang, M., Yang, X.-F., Li, C.-J. *EJOC* 998 (2003).
[2]Jain, S.L., Sharma, V.B., Sain, B. *TL* **45**, 4341 (2004).
[3]Li, Y., Izumi, T. *SC* **33**, 3583 (2003).
[4]Cho, C.S., Kim, B.T., Kim, H.-S., Kim, T.-J., Shim, S.C. *OM* **22**, 3608 (2003).
[5]Cho, C.S., Kim, B.T., Choi, H.-J., Kim, T.-J., Shim, S.C. *T* **59**, 7997 (2003).

Dicobalt octacarbonyl. 13, 99–101; **14**, 117–119; **15**, 117–118; **16**, 113–115; **17**, 102–105; **18**, 132; **19**, 125–126; **20**, 139–141; **21**, 166–167; **22**, 178–181

Reduction. Unsaturated carbonyl compounds are reduced (at the double bond) with $Co_2(CO)_8$-H_2O in DME.[1]

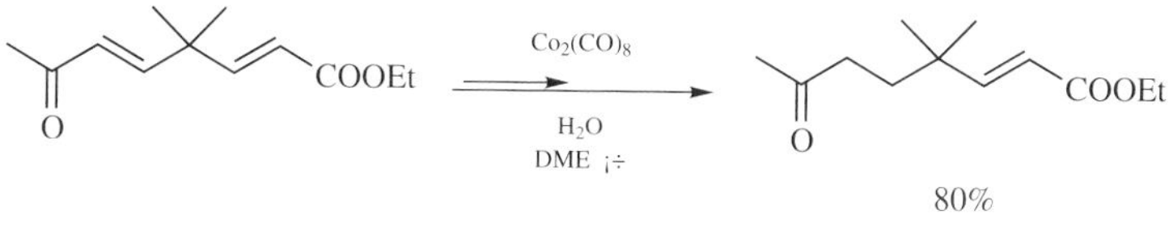

80%

Cycloisomerization. 1,6-Enynes are converted into cyclopentenes.[2] Equimolar $Co_2(CO)_8$ and $(MeO)_3P$ improve the reaction selectivity against mere double bond migration.

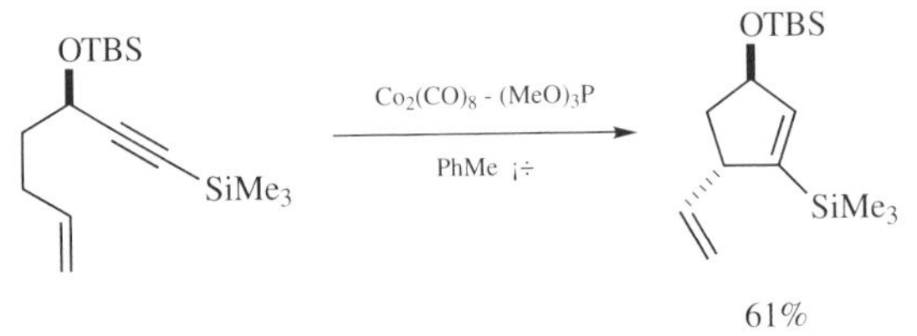

OTBS SiMe3 $\xrightarrow[\text{PhMe } \Delta]{Co_2(CO)_8 \text{ - } (MeO)_3P}$ OTBS SiMe3 61%

Pauson-Khand reaction. Allenamides undergo P-K reaction in the manner shown below.[3] Tandem [5 + 1]/[2 + 2 + 1]cycloaddition processes are involved in the formation of tricyclic δ-lactones from alkynyl epoxides which also contain a remote double bond.[4]

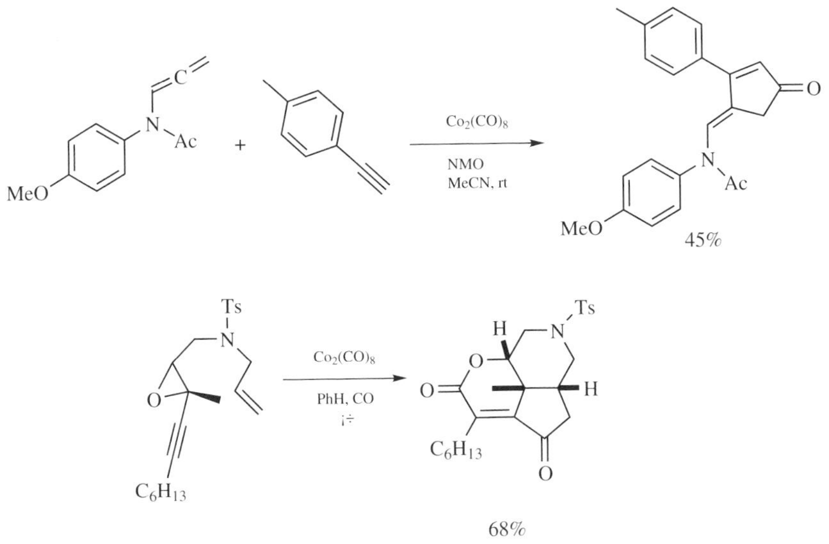

MeO N Ac + $\xrightarrow[\text{NMO, MeCN, rt}]{Co_2(CO)_8}$ O N Ac MeO 45%

Ts N O C_6H_{13} $\xrightarrow[\text{PhH, CO } \Delta]{Co_2(CO)_8}$ H N Ts O O H C_6H_{13} O 68%

Quinolines.[5] *N,N*-Diallylanilines isomerize to quinolines. The presence of imines allows group exchange in an intermediate stage to form 2-arylquinoline derivatives.

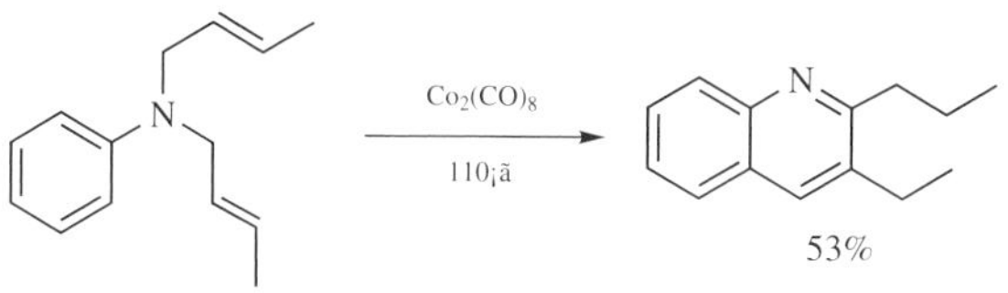

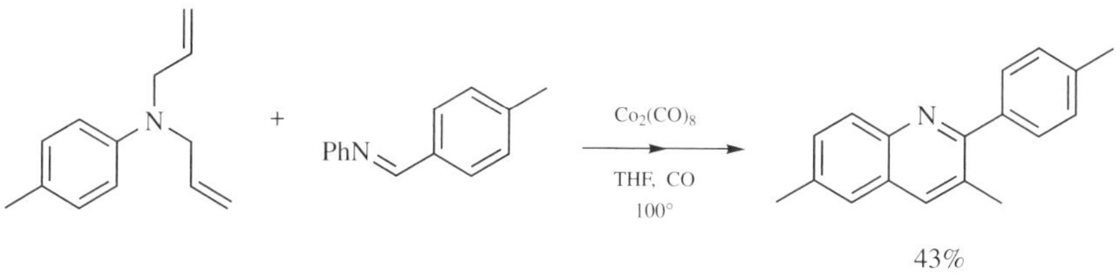

$Co_2(CO)_8$

110¡ã

53%

+

PhN

$Co_2(CO)_8$

THF, CO

100°

43%

Carbonylation. Epoxides are cleaved and homologated to give β-hydroxy carboxamides.[6]

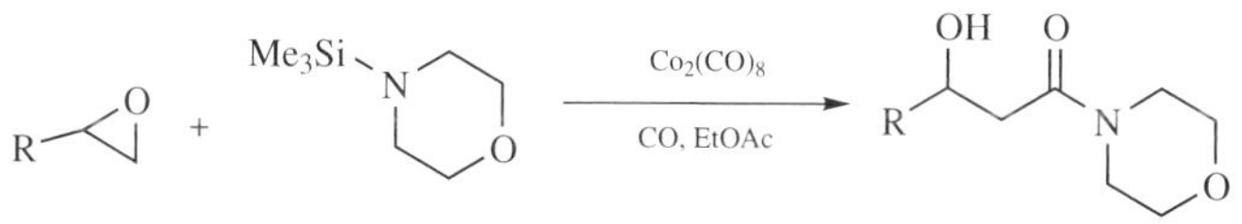

Me_3Si N O R O + $Co_2(CO)_8$ CO, EtOAc OH O R N O

[1]Lee, H.-Y., An, M. *TL* **44**, 2775 (2003).
[2]Ajamian, A., Gleason, J.L. *OL* **5**, 2409 (2003).
[3]Anorbe, L., Poblador, A., Dominguez, G., Perez-Castells, J. *TL* **45**, 4441 (2004).
[4]Odedra, A., Wu, C.-J., Madhushaw, R.J., Wang, S.-L., Liu, R.-S. *JACS* **125**, 9610 (2003).
[5]Jacob, J., Jones, W.D. *JOC* **68**, 3563 (2003).
[6]Goodman, S.N., Jacobsen, E.N. *ACIEE* **41**, 4703 (2002).

N-(Diethoxyphosphoryl)-*O*-benzylhydroxylamine.

N-Alkylation.[1] Various RNHOBn are easily prepared from $(EtO)_2P(=O)NHOBn$ via alkylation and acid hydrolysis.

[1]Blazewska, K., Gajda, T. *T* **59**, 10249 (2003).

4-(Diethoxyphosphorylmethyl)benzenesulfonylhydrazide.

Reduction.[1] The double bond of a 1,1-diiodo-1-alkene is reduced on heating with the title reagent in xylene. This diimide source is valuable because the polar group facilitates removal of byproducts while causing no complication to the reaction.

[1]Cloarec, J.-M., Charette, A.B. *OL* **6**, 4731 (2004).

Diethylaminosulfur trifluoride (DAST).

Fluorination. 1,1,1-Trifluoro-2-alkanols are fluorinated with DAST at −70° and the tetrfluoro derivatives give 1,1,2-trifluoro-1-alkenes on treatment with LiHMDS.[1]

Reaction of arylcyclopropyl silyl ethers with DAST shows dependence on the nature of the aryl substituent. The three-membered ring is retained when the Ar group is located at C-1 and strongly electron-donating or at C-2 and electron-withdrawing. Otherwise, ring cleavage to give allylic fluorides ensues.[2]

Me_3SiO Ar $\xrightarrow[CH_2Cl_2,\ r.t.]{Et_2N\text{-}SF_3}$ Ph F (Ar=Ph, 45%); MeO MeO F (Ar = 2,4-$(MeO)_2C_6H_3$, 65%)

[1]Anilkumar, R., Burton, D.J. *TL* **44**, 6661 (2003).
[2]Kirihara, M., Kakuda, H., Tsunooka, M., Shimajiri, A., Takuwa, T., Hatano, A. *TL* **44**, 8513 (2003).

Diethyl 4-(hydrazinosulfonyl)benzylphosphonate.

1,1-Diiodoalkanes.[1] Convenient access to this class of compounds from 1,1-diiodoalkenes is by heating with the title diimide precursor in xylene.

[1]Cloarec, J.-M., Charette, A.B. *OL* **6**, 4731 (2004).

Difluoromethyl phenyl sulfone.

Difluoromethylation. Alkylation followed by reductive desulfonylation constitutes a method for homologation of RI into $RCHF_2$.[1] When the alkylation products are warmed, elimination to afford 1,1-difluoro-1-alkenes[2] occurs readily.

The reagent also acts as an equivalent of $[CF_2]$-anion. On treatment with a base (e.g., *t*-BuOK) and an aldehyde a twofold condensation occurs (desulfonylation in the second step).[3]

ArCHO + $PhSO_2CHF_2$ $\xrightarrow[DMF,\ -50° \sim rt]{t\text{-BuOK}}$ Ar OH OH Ar F F

anti:syn > 94:6

[1]Prakash, G.K.S., Hu, J., Wang, Y., Olah, G.A. *OL* **6**, 4315 (2004).
[2]Prakash, G.K.S., Hu, J., Wang, Y., Olah, G.A. *ACIEE* **43**, 5203 (2004).

[3]Prakash, G.K.S., Hu, J., Mathew, T., Olah, G.A. *ACIEE* **42**, 5216 (2003).

N,N'-Dihalo-N,N'-ditosylethylenediamine.

Oxidation.[1] Primary and secondary alcohols are oxidized to carbonyl compounds by a combination of the brominated reagent with Me_2S (and subsequent Et_3N treatment). Essentially it is a modified Corey-Kim oxidation.

Iodination. The diiodo reagent efficiently transfers its iodine atoms to arenes.[2]

[1]Ghorbani-Vaghei, R., Khazaei, A. *TL* **44**, 7525 (2003).
[2]Ghorbani-Vaghei, R. *TL* **44**, 7529 (2003).

Diiodomethane. 13, 110–115; 275–276; **16**, 184–185; **17**, 155; **18**, 139–140; **19**, 128; **20**, 143–144; **21**, 171; **22**, 183–184

Cyclopropanation. Cyclopropanation of cinnamyl alcohols in the presence of a chiral diamine derivative reveals moderate degrees of asymmetric induction.[1] Alkene cyclopropanation[2] using indium powder and CH_2I_2 is possible.

A better route to cyclopropylboronic esters involves hydroboration of alkynes with 1,1,2,2-tetraphenyl-1,2-ethanediolatoborane and reaction with Et_2Zn - CH_2I_2.[3]

[1]Imai, N., Nomura, T., Yamamoto, S., Ninomiya, Y., Nokami, J. *TA* **13**, 2433 (2002).
[2]Virender, Jain, S.L., Sain, B. *TL* **46**, 37 (2005).
[3]Pietruszka, J., Witt, A. *SL* 91 (2003).

Diisobutylaluminum hydride, Dibal-H. 13, 115–116; **15**, 137–138; **16**, 134–135; **17**, 123–125; **18**, 140–141; **19**, 128–129; **20**, 144–146; **21**, 171; **22**, 184

Tishchenko reaction.[1] Dibal-H promotes redox condensation of aldehydes.

Hydroalumination.[2] After treatment of 2-alkynoic esters with Dibal-H the addition of aldehydes leads to adducts of the Baylis-Hillman type. The scope is broader because the final products contain a β-alkyl group and such structures are not accessible from the normal Baylis-Hillman reaction.

[1]Hon, Y.-S., Chang, C.-P., Wong, Y.-C. *TL* **45**, 3313 (2004).
[2]Ramachandran, P.V., Rudd, M.T., Burghardt, T.E., Reddy, M.V.R. *JOC* **68**, 9310 (2003).

Diisobutylaluminum isopropoxide.

Reduction.[1] The reagent is prepared from i-Bu_2AlH in PhMe. It shows remarkable solvent dependence for stereoselectivity in the reduction of cyclic ketones. For cyclohexanones, reduction in PhMe favors formation of equatorial alcohols but in CH_2Cl_2, axial alcohols. Norbornanone gives *exo*-alcohol (in PhMe) and *endo*-alcohol (in CH_2Cl_2), respectively.

[1]Bahia, P.S., Jones, M.A., Snaith, J.S. *JOC* **69**, 9289 (2004).

B-Diisopropylaminocatecholborane.

Mannich reaction. The title reagent is a generator of iminium ions from RCHO, thus compatible to conditions for Mannich reaction of silyl enol ethers with aldehydes and amines.

[1]Suginome, M., Uehlin, L., Murakami, M. *JACS* **126**, 13196 (2004).

(2,6-Diisopropylphenyl)imidovanadium trichloride.

Guanidines.[1] Aromatic amines and carbodiimides combine to give guanidines under the influence of the title reagent. Vanadium pentoxide is totally ineffective and $VO(acac)_2$ gives much lower yields.

NH_2 + N=C=N (N)VCl_3 PhMe, 105¡ã NH N H N 84%

[1]Montilla, F., Pastor, A., Galindo, A. *JOMC* **689**, 993 (2004).

Di(isopropylprenyl)borane.

Boronic acids. The reagent is available from hydroboration of 2,5-dimethyl-2,4-hexadiene. On its addition to 1-alkynes or 1-alkenes with subsequent hydrolysis boronic acids are produced.[1]

$NaBH_4$ - Me_2SO_4 diglyme 0¡ã H−B

[1]Kalinin, A.V., Scherer, S., Snieckus, V. *ACIEE* **42**, 3399 (2003).

Dilauroyl peroxide. 21, 172–173; **22**, 185

Addition-elimination. Dilauroyl peroxide not only promotes radical addition to nitroalkenes a subsequent defunctionalization is also achieved. An overall result of either hydrodenitrosation[1] or chain lengthening (by reaction with iodoacetic esters and amides)[2] has been observed.

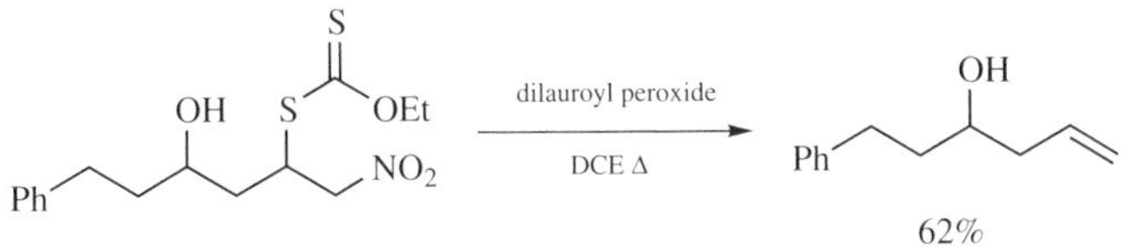

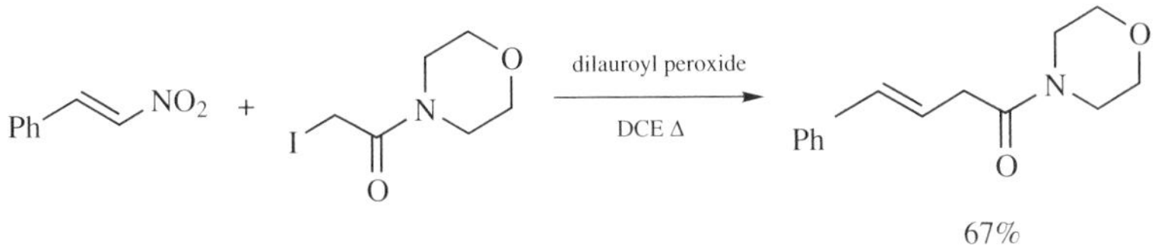

dilauroyl peroxide
DCE Δ

62%

dilauroyl peroxide
DCE Δ

67%

Replacement of a xanthate unit by a ketone chain is easily achieved using an alkene precursor containing a latent functionality. [3]

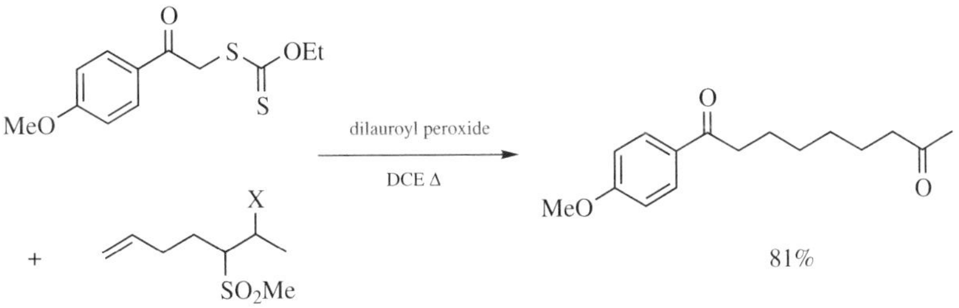

dilauroyl peroxide
DCE Δ

81%

Heterocycle modification. To 2-(*N*-alkenylamino)pyridines the radical addition with a xanthate ester and cyclization are both promoted by dilauroyl peroxide.[4]

Radical addition to heteroaromatic systems (e.g., *N*-methylpyrrole-2-carbaldehyde) and rearomatization serve to access to substituted compounds.[5] Tin reagents are avoided by using the method.

[1]Ouvry, G., Quiclet-Sire, B., Zard, S.Z. *OL* **5**, 2907 (2003).
[2]Garcia-Torres, A., Cruz-Almanza, R., Miranda, L.D. *TL* **45**, 2085 (2004).
[3]Ouvry, G., Zard, S.Z. *CC* 778 (2003).
[4]Bacque, E., El Qacemi, M., Zard, S.Z. *OL* **6**, 3671 (2004).
[5]Osornio, Y.M., Cruz-Almanza, R., Jimenez-Montano, V., Miranda, L.D. *CC* 2315 (2003).

Dimanganese decacarbonyl. 21, 173; **22**, 185

Coupling reactions. Radical addition to 1,6-dienes with ring formation as mediated by $Mn_2(CO)_{10}$ has extended to chlorosulfonylation.

[1]Huther, N., McGrail, P.T., Parsons, A.F. *EJOC* 1740 (2004).

_N,N'_-Dimesitylimidazolium chloride.

Butyrolactones. The ylide generated in situ from the title compound and DBU promotes reaction of enals with aldehydes to provide predominantley *cis*-β,γ-disubstituted γ-butyrolactones. The ylide functions in the same way as the better-known thiazolium ylides.

[1]Sohn, S.S., Rosen, E.L., Bode, J.W. *JACS* **126**, 14370 (2004).

2,2-Dimethoxy-5,5-dimethyl-1,3,4-oxadiazoline.

Dimethoxycarbene. This heterocycle generates dimethoxycarbene on thermolysis. It combines with conjugated ketenes to give cyclopentenones. Sulfur analogues behave in the same manner.

RX = MeO, PrS...

[1]Rigby, J.H., Wang, Z. *OL* **5**, 263 (2003).

4-(4,6-Dimethoxy-1,3,5-triazin-2-yl)-4-methylmorpholinium chloride.

Weinreb amides. The title reagent activates carboxylic acids for reaction with MeNHOMe.[1]

[1]Hioki, K., Kobayashi, H., Ohkihara, R., Tani, S., Kunishima, M. *CPB* **52**, 470 (2004).

Dimethyl acetylenedicarboxylate.

Dehydration. Aldoximes are converted into nitriles at room temperature by this reagent (together with Et_3N).

[1]Coskun, N. *SC* **34**, 1625 (2004).

4-Dimethylaminopyridine, DMAP. 21, 176; **22**, 186–187

Protection-deprotection. The use of Boc_2O and DMAP in esterification of carboxylic acids[1] is convenient, but probably uneconomical. The acylation of oxazolidinones and thiazolidinethiones using RCOOH is mediated by DCC- DMAP.[2] DMAP is an effective catalyst in the detachment (with BnOH) of thiazolidinethione moiety from their amides.[3]

Condensations. The Baylis-Hillman reaction conducted under solvent-free conditions and mediated by DMAP[4] is complete in much shorter time at 76°.

[1]Goossen, L.J., Dohring, A. *SL* 263 (2004).
[2]Andrade, C.K.Z., Rocha, R.O., Vercillo, O.E., Silva, W.A., Matos, R.A.F. *SL* 2351 (2003).
[3]Wu, Y., Sun, Y.-P.,Yang, Y.-Q., Hu, Q., Zhang, Q. *JOC* **69**, 6141 (2004).
[4]Octavio, R., de Souza, M.A., Vasconcellos, M.L.A.A. *SC* **33**, 1383 (2003).

Dimethylammonium dimethyl carbamate.

Arylidenation.[1] Aldehydes and enolizable ketones condense with ArCHO to afford monoarylidene derivatives.

[1]Kreher, U.P., Rosamilia, A.E., Raston, C.L., Scott, J.L., Strauss, C.R. *OL* **5**, 3107 (2003).

2,6-Dimethylbenzoquinone.

Hydroxyl inversion.[1] Derivatization of alcohols (even tertiary) to $ROPPh_2$ followed by treatment with an acid (e.g., PhCOOH) and 2,6-dimethylbenzoquinone leads to carboxylic esters. The reaction proceeds with configuration inversion.

Etherification.[2] After conversion of one alcohol to $ROPPh_2$ the admixture with another alcohol (R′OH) and the quinone gives ROR′. Symmetrical and unsymmetrical ethers including ArOR are readily prepared.

[1]Mukaiyama, T., Shintou, T., Fukumoto, K. *JACS* **125**, 10538 (2003).
[2]Shintou, T., Mukaiyama, T. *JACS* **126**, 7359 (2004).

Dimethyl carbonate.

N-Methylation.[1] Anilines undergo mono-*N*-methylation with $CO(OMe)_2$ over NaY faujasite.

Deacylation. The NaOMe/ $CO(OMe)_2$ system is useful for cleaving *N*-acyloxazolidinones.[2]

[1]Selva, M., Tundo, P., Perosa, A. *JOC* **68**, 7374 (2003).
[2]Kanomata, N., Maruyama, S., Tomono, K., Anada, S. *TL* **44**, 3599 (2003).

Dimethyldioxirane, DMD. **12**, 413; **13**, 120; **14**, 148; **15**, 143–144; **16**, 142–144; **18**, 144–146; **19**, 135–136; **20**, 150–152; **21**, 177–178; **22**, 187–188

Epoxidation. Alkenyltrifluoroborate salts are successfully epoxidized with DMD at room temperature.[1]

Oxidation. Using DMD to convert epoxy alcohols into the corresponding ketones can preserve the acid-sensitive and hydrolytically labile functionality. The more powerful 3-trifluoromethyl-3-trimethyldioxirane is capable of oxyfunctionalizing C-H bond and thereby producing epoxy ketones, e.g., epoxycyclohexane to 3,4-epoxycyclohexanone (86%).[2]

Removal of a *p*-methoxybenzyl group from nitrogen is due to double oxidation (at N and benzylic sites) to set up a fragmentation.[3]

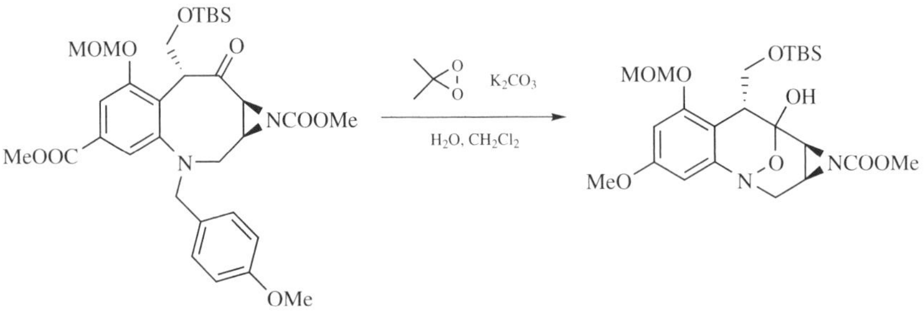

Oxidation of the nitronate anions obtained from reaction of nitroarenes with ester enolates using DMD leads phenols.[4] The nitro group is lost therefore it differs from results with $KMnO_4$ oxidation.

4,5-Disubstituted imidazoles undergo oxidative rearrangement to afford imidazolin-4-ones.[5]

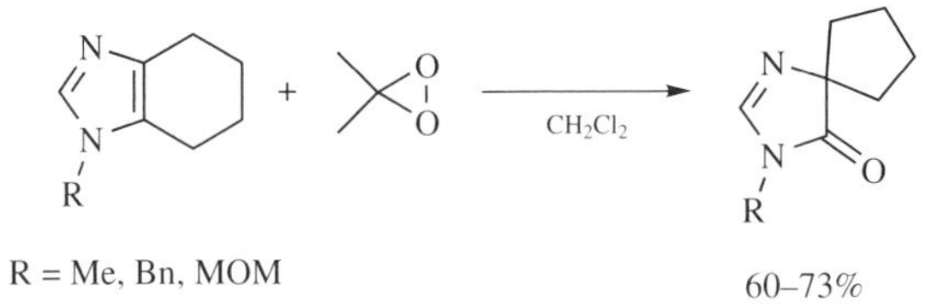

[1]Molander, G.A., Ribagorda, M. *JACS* **125**, 11148 (2003).
[2]D'Accolti, L., Fusco, C., Annese, C., Rella, M.R., Turteltaub, J.S., Willard, P.G., Curci, R. *JOC* **69**, 8510 (2004).
[3]Judd, T.C., Williams, R.M. *ACIEE* **41**, 4683 (2002).
[4]Makosza, M., Surowiec, M. *T* **59**, 6261 (2003).
[5]Lovely, C.J., Du, H., He, Y., Dias, H.V.R. *OL* **6**, 735 (2004).

S,S'-Dimethyl dithiocarbonate

Nitriles.[1] Dehydration of aldoximes is achieved with this reagent and Et_3N in hot dioxane.

[1]Khan, T.A., Peruncheralathan, S., Ila, H., Junjappa, H. *SL* 2019 (2004).

N,N-Dimethylformamide – phosphoryl chloride. **18**, 146; **22**, 188

Heterocycles. A convenient synthesis of certain pyrones and pyridines[1] by reaction of carbonyl compounds with the reagent is established with limited examples. Also acyclic ketene-*S,S*-acetals condense with the Vilsmeier-Haack reagent to provide 2-alkylthio-4-chloropyridines.[2]

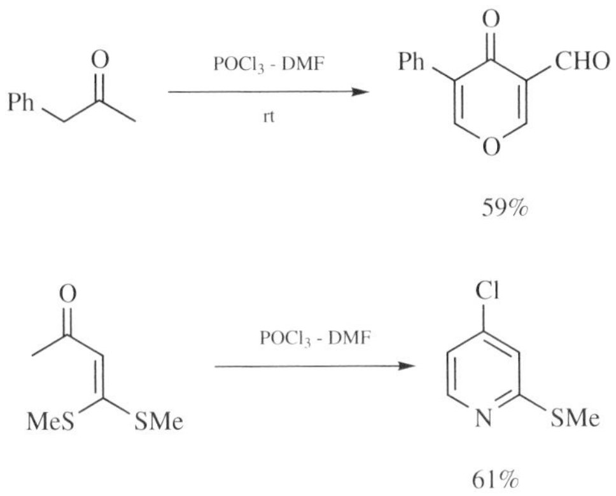

2-Hydroxyacetophenones are converted into 3-cyano-4-benzopyrones in simple operations of treatment with DMF-$POCl_3$ followed by hydroxylamine.[3]

Acyl azides.[4] The Vilsmeier-Haack reagent is useful for activating acids to form $RCON_3$ with NaN_3.

Formic esters.[5] Simple and direct transformation of silyl ethers to formic esters is observed on mixing with DMF-$POCl_3$.

[1]Thomas, A.D., Josemin, Asokan, C.V. *T* **60**, 5069 (2004).
[2]Sun, S., Liu, Y., Liu, Q., Zhao, Y., Dong, D. *SL* 1731 (2004).
[3]Reddy, G.J., Latha, D., Thirupathaiah, C., Rao, K.S. *TL* **45**, 847 (2004).
[4]Sridhar, R., Perumal, P.T. *SC* **33**, 607 (2003).
[5]Lellouche, J.-P., Kotlyar, V. *SL* 564 (2004).

N,N-Dimethylformamide dimethylacetal.

Acylation. Warming ArCCH with the title reagent in DMF gives moderate yields of ArCCCHO. A similar reaction with $MeC(OMe)_2NMe_2$ afford the alkynones.[1]

[1]Lee, K.Y., Lee, M.J., GowriSankar, S., Kim, J.N. *TL* **45**, 5043 (2004).

1,3-Dimethylimidazolidin-2-one oxime *O*-methoxyacetate.

Amination.[1] The reagent **1** reacts with arenes to give guanidine derivatives which are readily converted into $ArNH_2$ by CsOH.

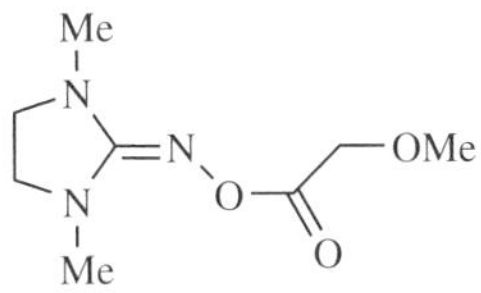

1

[1]Baldovini, N., Kitamura, M., Narasaka, K. *CL* **32**, 548 (2003).

Dimethylmalonyltrialkylphosphoranes.

Esterification.[1] Either inversion or retention of the carbinyl center can be controlled in the acylation of secondary alcohols mediated by $Bu_3P{=}C(COOMe)_2$. 4-Nitrobenzoic esters are formed without disturbing the original configuration of the alcohols, whereas inverted esters of hindered acids such as 2,6-disubstituted benzoic acids (including mesitoic acid) are obtained.

[1]McNulty, J., Capretta, A., Laritchev, V., Dyck, J., Robertson, A.J. *JOC* **68**, 1597 (2003).

Dimethylsulfonium methylide. 18, 149; **20**, 154; **22**, 189

Conjugate addition. A unique access to substituted vinylsilanes and styrenes is via the addition-elimination pathway. The product profile is dependent of the amount of base (cyclopropanation if 1 equiv. of dimsyl sodium is used).[1]

PhMe2Si–CH=C(COOEt)2 + Me2S=CH2 —(DMSO, THF)→ Me2PhSi–C(=CH2)–CH(COOEt)2 66%

Aziridines. Cycloaddition of the ylide to sulfinyl imines affords *N*-sulfinylaziridines and chiral products are accessible analogously.[2]

[1]Ghosh, S.K., Singh, R., Date, S.M. *CC* 636 (2003).
[2]Morton, D., Pearson, D., Field, R.A., Stockman, R.A. *SL* 1985 (2003).

Dimethyl sulfoxide.

Aldehydes. Conversion of $ArCHBr_2$ to ArCHO is accomplished by heating with DMSO at 100.[1] Note aq. Me_2NH can also be used for the same reaction.[2]

N-Methoxyindoles. Application of the demethoxycarbonylation method (NaX-DMSO, Δ) to *o*-nitrophenylmalonic esters leads to *N*-methoxyindoles via *o*-nitrosostyrene intermediates.[3]

R R / NO2 —(NaCl, DMSO, 155;ã)→ R / N / OMe

R=COOEt , CN R=COOEt 55%

(E)-1-Alkoxy-1,3-butadienes.[4] Propargylic ethers react with lithiated DMSO at 80–100° to give the homologated alkoxydienes in moderate yields. The transformation involves isomerization to the allenyl ethers and addition of the carbanion to the unsubstituted terminus, and elimination. The usefulness of the method lies in its simplicity.

[1]Li, W., Li, J., De Vincentis, D., Mansour, T.S. *TL* **45**, 1071 (2004).
[2]Bankston, D. *S* 283 (2004).
[3]Selvakumar, N., Reddy, B.Y., Azhagan, A.M., Khera, M.K., Babu, J.M., Iqbal, J. *TL* **44**, 7065 (2003).
[4]Lysek, R., Wozny, E., Danh, T.T., Chmielewski, M. *TL* **44**, 7541 (2003).

Dimethylsulfoxonium methylide. 14, 152; **15**, 147; **16**, 146; **17**, 126–127; **18**, 148; **19**, 139; **20**, 155–156; **22**, 190

Tetrahydrodibenzofurans. 3,4-Cyclobutanocoumarins bearing an electron-withdrawing group at C-3 are converted by the title reagent to the tricyclic system.

R H CO$_2$Et O O + $H_2C{=}SOMe_2$ DMF → R H H O COOEt OH

Chloromethyl ketones. For large-scale preparation of chloromethyl ketones from carboxylic acids via the acid chlorides using diazomethane is harzardous. It is proposed to employ the method of Corey to react with esters and then treatment wqith LiCl and MsOH.[2]

[1]Yamashita, M., Inaba, T., Shimizu, T., Kawasaki, I., Ohta, S. *SL* 1897 (2004).
[2]Wang, D., Schwinden, M.D., Radesca, L., Patel, B., Kronenthal, D., Huang, M.-H., Nugent, W.A. *JOC* **69**, 1629 (2004).

***N,N*-Dimethylthiocarbamoyl chloride.**

Alcohol protection.[1] The derived thiocarbamates show moderate to high stability toward many reagents and conditions, such as metal hydrides, ylides, boranes, RLi, RMgX, DDQ, PDC, TBAF, $CrCl_2$, common acids and bases. Deprotection uses $NaIO_4$ or H_2O_2.

[1]Barma, D.K., Bandyopadhyay, A., Capdevila, J.H., Falck, J.R. *OL* **5**, 4755 (2003).

Dimethyltitanocene. 21, 181

Indole synthesis. Succeeding addition of amines to 2-alkynylaryl halides in the presence of Cp_2TiMe_2 the introduction of a proper Pd catalyst can turn the products into indoles.[1]

Reduction.[2] The reaction course for Cp_2TiMe_2/Cp_2TiF_2 –catalyzed reductive dimerization of benzamides by a hydrosilane is changed to that leading to *N*-ethylamines from the corresponding acetamides.

[1]Siebeneicher, H., Bytschkov, I., Doye, S. *ACIEE* **42**, 3042 (2003).
[2]Selvakumar, K., Rangareddy, K., Harrod, J. *CJC* **82**, 1244 (2004).

Dinitrogen tetroxide.

Oxidation.[1] Sulfides and disulfides are rapidly converted into sulfoxides and thiosulfonates, respectively, by N_2O_4 impregnated in activated charcoal at room temperature.

[1]Iranpoor, N., Firouzabadi, H., Pourali, A.-R. *SL* 347 (2004).

***O*-(2,4-Dinitrophenyl)hydroxylamine.**

N-Amination.[1] Pyridines, quinolines, and isoquinolines react with the reagent and direct benzoylation of the products furnish *N,N′*-ylides. The reagent is available from reaction of *N*-hydroxyphthalimide and 2,4-dinitrochlorobenzene.

[1]Legault, C., Charette, A.B. *JOC* **68**, 7119 (2003).

Diphenylboron perchlorate.

Aldol reaction. Condensation of aldehydes (self- and cross-) is promoted by Ph_2BClO_4.

[1]Kiyooka, S., Fujimoto, H., Mishima, M., Kobayashi, S., Uddin, K.M., Fujio, M. *TL* **44**, 927(2003).

Diphenyl diselenide.

Functionalization of alkenes. The catalytic role of PhSeSePh in the reaction of unsaturated acids with NBS to provide bromolactones[1] is noted. The diselenide is oxidized by $TolIF_2$ to generate a selenofluorination reagent.[2]

Photoinduced reaction of PhSeSePh with acrylonitrile and acrylic esters leads to cyclopentane derivatives while 1:1:1-adducts are generated with enol ethers and propynoic esters.[3]

BuO + R′OOC —PhSeSePh, hν (>300nm)→ BuO, PhSe, CHSePh, COOR′

R′ = Et 66%

X —PhSeSePh, hν (>300nm)→ X, CH(SePh)$_2$, COOEt, X

X = CN, COOR′

Seleno carbonyl compounds. The terminal triple bond of 3- and 4-alkynols is readily converted into geminal seleno tosylates. Further reaction with PhSeSePh in the presence of DDQ leads to α-selenolactones.[4]

HO–CH₂CH₂–CH=C(SePh)(OTs) + PhSeSePh —(DDQ, MeCN)→ α-(phenylseleno)-γ-butyrolactone

Construction of β-seleno acrylamides from alkynes, sulfenamides, CO and PhSeSePh is Pd-catalyzed.[5]

R–C≡CH + PhSeSePh + PhS-NR′₂ —($(Ph_3P)_4Pd$, CO / PhH)→ R–C(SePh)=CH–C(=O)–NR′₂

Alkynyl selenides. Mixed selenides can be prepared from 1-alkynes and PhSeSePh in the presence of CuI in DMSO.[6] Analogous sulfides and tellurides are also accessible by this method.

[1]Mellegaard, S.R., Tunge, J.A. *JOC* **69**, 8979 (2004).
[2]Panunzi, B., Picardi, A., Tingoli, M. *SL* 2339 (2004).
[3]Tsuchii, K., Doi, M., Hirao, T., Ogawa, A. *ACIEE* **42**, 3490 (2003).
[4]Tiecco, M., Testaferri, L., Temperini, A., Bagnoli, L., Marini, F., Santi, C. *SL* 655 (2003).
[5]Knapton, D.J., Meyer, T.Y. *OL* **6**, 687 (2004).
[6]Bieber, L.W., da Silva, M.F., Menezes, P.H. *TL* **45**, 2735 (2004).

(Diphenylphosphinoethane)rhodium(I) perchlorate. 21, 182; **22**, 191

Hydroacylation.[1] Addition of 3-methylthiopropanal to alkenes (preferably electron-deficient specimens) is catalyzed by (dppe)$RhClO_4$. The MeS group is important for forming a chelate-stabilized acyl-Rh species for the reaction because homologues and the corresponding oxy analogues are totally ineffective addends.

[1]Willis, M.C., McNally, S.J., Beswick, P.J. *ACIEE* **43**, 340 (2004).

Diphenylphosphonyldimethoxymethane.

Homologation. The reagent is a d^1-synthon that on alkylation (with ROTf) and acid hydrolysis affords methyl esters (RCOOMe). α-Amino aldehydes are transformed into β-amino esters.[1]

[1]Brunjes, M., Kujat, C., Monenschein, H., Kirschning, A. *EJOCB* 1149 (2004).

Diphenylphosphoroazidate. 21, 182–183

Azidation. Direct transformation of pyridin-4-one, quinolin-4-one and related compounds into the heteroaryl azides is easily performed with $Ph_2P(=O)N_3$.

[1]Aizikovich, A., Kuznetsov, V., Gorohovsky, S., Levy, A., Meir, S., Byk, G., Gellerman, G. *TL* **45**, 4241 (2004).

Dipyridyliodonium tetrafluoroborate. 22, 191–192

Cyclization. Py_2IBF_4 induces cyclization of *o*-alkynylanilines to afford 3-iodoindoles.[1] Ring closure also occurs with *o*-alkynylaryl aldehydes when a proper nucleophile is added to neutralize the oxonium salts.[2]

CHO; Ph → IPy_2BF_4- HBF_4; CH_2=$CHCH_2SiMe_3$, CH_2Cl_2 → O; Ph; I; 42%

Initiated by iodination at the terminal double bond of a polyene complex ring structures can be assembled in one step.[3] It is not surprising simpler substrates undergo iodinative cyclization to give tetralins[4] (conditions are mild: reaction occurs at −90°).

MeO; Ms; N; OMe → IPy_2BF_4 - HBF_4; CH_2Cl_2 80° → MeO; Ms; N; OMe; I; 41%

[1]Barluenga, J., Trincado, M., Rubio, E., Gonzalez, J.M. *ACIEE* **42**, 2406 (2003).
[2]Barluenga, J., Vazquez-Villa, H., Ballesteros, A., Gonzalez, J.M. *JACS* **125**, 9028 (2003).
[3]Barluenga, J., Trincado, M., Rubio, E., Gonzalez, J.M. *JACS* **126**, 3416 (2004).
[4]Appelbe, R., Casey, M., Dunne, A., Pascarella, E. *TL* **44**, 7641 (2003).

3-(1,3-Dithian-2-ylidene)-2,4-pentanedione.

1,3-Propanedithiol equivalent.[1] The title reagent is odorless and it can be used to dithioacetalize carbonyl compounds.

[1]Yu, H., Liu, Q., Yin, Y., Fang, Q., Zhang, J., Dong, D. *SL* 999 (2004).

Di-2-thienyl carbonate.

Esterification. Various alcohols and carboxylic acids condense to form esters by this reagent.

[1]Mukaiyama, T., Oohashi, Y., Fukumoto, K. *CL* **33**, 552 (2004).

Dysprosium(III) triflate. 21, 184; **22**, 192

Alkylation.[1] Reaction of indole with carbonyl compounds and imines to give di(3-indolyl)alkanes is promoted by $Dy(OTf)_3$ in ionic liquids.

Diels-Alder reaction.[2] Schiff base formation from 5-alkynals and anilines catalyzed by $Dy(OTf)_3$ sets up a molecular system conducive to intramolecular cycloaddition, as demonstrated by an expedient synthesis of luotonin A.

NH_2 + OHC N O N → $Dy(OTf)_3$, MeCN r.t. → 51% luotonin-A

[1]Mi, X., Luo, S., He, J., Cheng, J.-P. *TL* **45**, 4567 (2004).
[2]Twin, H., Batey, R.A. *OL* **6**, 4913 (2004).

E

Erbium(III) triflate.

Deacetalization.[1] $Er(OTf)_3$ is found to be another Lewis acid capable of catalyzing the hydrolysis of acetals.

Esterification.[2] Acylation of alcohols with this catalyst is reported.

[1]Dalpozzo, R., De Nino, A., Maiuolo, L., Nardi, M., Procopio, A., Tagarelli, A. *S* 496 (2004).
[2]Procopio, A., Dalpozzo, R., De Nino, A., Maiuolo, L., Russo, B., Sindona, G. *ASC* **346**, 1465 (2004).

Ethylenebis(triphenylphosphine)platinum.

Boration-allylation.[1] The diene unit of 6-methylene-7-octenals is diborated by bis(catecholato)diboron, but due to the Lewis acidity of the Pt complex intramolecular allylation ensues. Standard oxidation transforms the remaining borane to a primary alcohol.

CHO + B—B; $(Ph_3P)_2Pt(CH_2{=}CH_2)$; PhH, 80°; H_2O_2; OH, OH; 56%

[1]Ballard, C.E., Morken, J.P. *S* 1321 (2004).

Ethyl *N*-nosyloxycarbamate.

Aziridines.[1] Cycloaddition to substituted acrylonitriles to give 2-cyanoaziridines is conducted at room temperature. Yields vary in the 80–90% range.

[1]Fioravanti, S., Morreale, A., Pellacani, L., Tardella, P.A. *SL* 1083 (2004).

Ethyl propynoate.

Peptide synthesis.[1] The ester activates N-protected amino acids by forming enol esters that easily couple with amines.

[1]Iorga, B., Campagne, J.-M. *SL* 1826 (2004).

Fiesers' Reagents for Organic Synthesis, Volume 23. Edited by Tse-Lok Ho

Europium(III) triflimide.

Deacetalization.[1] Eu(NTf_2)$_3$ catalyzes high temperature (e.g., 250°) Friedel-Crafts acylation of activated arenes by carboxylic acids. Benzene gives hexyl phenyl ketone in only 4% yield, therefore the synthetic utility is quite limited.

[1]Kawamura, M., Cui, D.-M., Hayashi, T., Shimada, S. *TL* **44**, 7715 (2003).

F

Ferricenium hexafluorophosphate.

Radical cyclization. The ferricenium salt oxidizes stabilized carbanions. Intramolecular coupling with an allylsilane unit gives cyclic products.

COOEt COOEt SiMe$_3$ $[Cp_2Fe]PF_6$ DME COOEt COOEt

93%

[1]Jahn, U., Hartmann, P., Kaasalainen, E.M. *OL* **6**, 257 (2004).

Ferricenyl methyl sulfide.

Epoxides.[1] Ylide formation from the reagent and RCH_2X and then reaction with an aldehyde provides an epoxide product.

[1]Miniere, S., Reboul, V., Arrayas, R.G., Metzner, P., Carretero, J.C. *S* 2249 (2003).

Fluoranil.

Ethers.[1] Oxidation at phosphorus of $ROPPh_2$ provides a driving force for C-O bond scission in the presence of alcohols. The oxidation can be achieved with fluoranil. (For same effect, see 2,6-dimethyl-1,4-benzoquinone.)

[1]Shintou, T., Mukaiyama, T. *JACS* **126**, 7359 (2004).

Fluorine. **13**, 135; **14**, 167; **15**, 160; **18**, 161; **19**, 146; **20**, 165; **21**, 188–189; **22**, 194

Amine fluorination.[1] Tertiary alkyl amines are perfluorinated, usually in good yields.

a-Fluoro ketones.[2] Silyl enol ethers including β,β-difluoro counterparts are fluorinated.

Oxidation.[3] Continuous flow gas-liquid thin film microreactors are effective for oxidation of alcohols and Baeyer-Villiger oxidation of ketones by elemental fluorine.

[1]Felling, K.W., Lagow, F.J. *JFC* **123**, 233 (2003).
[2]Prakash, G.K.S., Hu, J., Alauddin, M.M., Conti, P.S., Olah, G.A. *JFC* **121**, 239 (2003).
[3]Chambers, R.D., Holling, D., Rees, A.J., Sandford, G. *JFC* **119**, 81 (2003).

Fiesers' Reagents for Organic Synthesis, Volume 23. Edited by Tse-Lok Ho

1-Fluoro-4-hydroxy-1,4-diazoniabicyclo[2.2.2]octane bis(tetrafluoroborate). 19, 146–147; **21**, 190

Fluorination.[1] α-Fluorination of ketones with the title reagent in hot MeOH usually proceeds in >70% yield.

[1]Stavber, S., Jereb, M., Zupan, M. *S* 853 (2003).

2-(α-Fluoroethylsulfonyl)benzthiazole.

Fluoroethylidenation.[1] Julia alkenation using the reagent rapidly furnishes compounds of the RCH=CHF structure.

[1]Chevrie, D., Lequeux, T., Demoute, J.P., Pazenok, S. *TL* **44**, 8127 (2003).

1-Fluoropyridinium triflates.

Glycosylation.[1] Thioglycosides are transformed into *O*-glycosides as well as other glycosyl derivatives such as azides, and sulfoxides when activated by the title reagents.

[1]Tsukamoto, H., Kondo, Y. *TL* **44**, 5247 (2003).

Fluorotetraphenylbismuth.

Phenylation.[1] Ph_4BiF is a new reagent for the introduction of a phenyl group to the α-position of a carbonyl compound (on reaction with silyl enol ether).

[1]Ooi, T., Goto, R., Maruoka, K. *JACS* **125**, 10494 (2003).

Fluorous reagents and ligands. 21, 191–192; **22**, 195–197

Functional group transformations. Heavily fluorinated reagents have been developed for achieving known transformation, taking advantage of the distinctive properties of the fluorous derivatives. A 1,2-diol bearing fluoroalkyl chains form acetals that can be separated and purified by simple fluorous-organic extraction.[1] Cleavage of *o*-nitrobenzenesulfonamides using fluorinated decanethiol[2] and conversion of alkanols to bromides by CBr_4 as mediated by a fluorous phosphine[3] are also new developments.

Glycosylation using thioglycosides in which the arenethiol moiety bears a fluoroacyl-amino group benefits from facile removal of the byproduct.[4] Fluorous benzyloxy-carbonyl derivatives of α-amino acids are readily prepared,[5] and a fluorous support containing a Fmoc-aminobenzhydryl residue is proposed to be used in peptide synthesis.[6]

In Mitsunobu reaction bis(tridecafluorooctyl) azodicarboxylate underperforms DIAD and is not recommended for use. Either the diester containing one methylene group longer on each fluorous chain or unsymmetrical analogues (*t*-butyl ester at one end) should be used.[7] Esterification of sterically hindered carboxylic acids in C_6F_{14} procceds in higher yields than in toluene (by a factor of 1.2 to 5).[8]

Redox reactions. Good results are obtained in the bioreduction of β-keto esters with immobilized baker's yeast in perfluorooctane.[9] The fluorous Rh catalyst used in hydrosilylation of ketones and enones is easy to recycle.[10]

A fluorinated triflamide of tin supported on fluorous silica gel has several synthetic applications. Baeyer-Villiger oxidation of ketones using aq. H_2O_2 and mediated by tin(IV) bis(perfluoroalkanesulfonyl)amide is certainly advantageous.[11,12]

Allylic oxidation of alkenes to enones by a fluorous seleninic acid (together with $PhIO_2$) is a high-yielding reaction.[13]

Amination. A C-H bond at the β-carbon of α-keto esters is inserted by nitrogen when they are treated with fluoroalkanesulfonyl azides.[14]

Coupling reactions. Synthesis of 2-acetamido-1,3-dienes[15] by the Heck reaction using fluorous-tagged DPPP ligands is simplified because product separation is faster.

Me Ac N + TfO → $Pd(OAc)_2$ / DPPP / Et_3N, DMSO → Me Ac N

53%

[1]Huang, Y., Qing, F.-L. *T* **60**, 8341 (2004).
[2]Christensen, C., Clausen, R.P., Begtrup, M., Kristensen, J.L. *TL* **45**, 7991 (2004).
[3]Desmaris, L., Percina, N., Cottier, L., Sinou, D. *TL* **44**, 7589 (2003).
[4]Jing, Y., Huang, X. *TL* **45**, 4615 (2004).
[5]Curran, D.P., Amatore, M., Guthrie, D., Campbell, M., Go, E., Luo, Z. *JOC* **68**, 4643 (2003).
[6]Mizuno, M., Goto, K., Miura, T., Hosaka, D., Inazu, T. *CC* 972 (2003).
[7]Dandapani, S., Curran, D.P. *JOC* **69**, 8751 (2004).
[8]Gacem, B., Jenner, G. *TL* **44**, 1391 (2003).
[9]Yajima, A., Naka, K., Yabuta, G. *TL* **45**, 4577 (2004).
[10]Dinh, L.V., Gladysz, J.A. *NJC* **29**, 173 (2005).
[11]Hao, X., Yamazaki, O., Yoshida, A., Nishikido, J. *TL* **44**, 4977 (2003).
[12]Yamazaki, O., Hao, X., Yoshida, A., Nishikido, J. *TL* **44**, 8791 (2003).
[13]Crich, D., Zou, Y. *OL* **6**, 775 (2004).
[14]Zhu, S., Jin, G., Xu, Y. *T* **59**, 4389 (2003).
[15]Vallin, K.S.A., Zhang, Q., Larhed, M., Curran, D.P., Hallberg, A. *JOC* **68**, 6639 (2003).

Formic acid.

Ritter reaction.[1] *N*-Benzhydryl amides are formed when Ph_2CHOH is heated with RCN in formic acid.

N-Arylformamides.[2] Aryl azides are reduced and formylated in refluxing HCOOH.

[1]Gullickson, G.C., Lewis, D.E. *S* 681 (2003).
[2]Kamal, A., Ramana, A.V., Reddy, K.S., Ramana, K.V., Babu, A.H., Prasad, B.R. *TL* **45**, 8187 (2004).

G

Gadolinium(III) triflate.

β-Amino alcohols. $Gd(OTf)_3$ mediates regioselective aminolysis of steroid 2α,3α-epoxides.

[1]Thibeault, D., Poirier, D. *SL* 1192 (2003).

Gallium. 21, 194; **22**, 198

Allylation. The reaction of allyl halides with imines mediated by Ga under solvent-free conditions is reported.[1] Gallium is better than indium for giving pure amines.

Deoxygenation.[2] Amine oxides are reduced by Ga in refluxing water.

[1]Andrews, P.C., Peatt, A.C., Raston, C.L. *TL* **45**, 243 (2004).
[2]Han, J.H., Choi, K.I., Kim, J.H., Yoo, B.W. *SC* **34**, 3197 (2004).

Gallium(III) halides. 20, 169–170; **21**, 195–196; **22**, 198–199

Vinylation. β-Silylethenylation of ketones catalyzed by $GaCl_3$ may use a pyridine base.[1]

Alkynylation. Silylethynylation at an *o*-position of *N*-trimethylsilylanilines and *N*-benzylanilines[2] proceeds via the dichlorogallium amides. The reaction of silyl enol ethers with 1-chloro-2-trimethylsilylethyne proceeds via the metal exchange, carbometallation, elimination sequence, and a further reaction leads to 2-chloroenynyl ketones.[3]

Me3SiO, Ph, R + Me3Si—≡—Cl → ($GaCl_3$; H_2SO_4, MeOH) O, Ph, R, Cl

Addition. Vicinal disulfides are formed through a $GaCl_3$-mediated reaction of alkenes and alkynes with PhSSPh.[4] The two sulfenyl groups in the adducts with alkynes are (*E*)-disposed.

Fiesers' Reagents for Organic Synthesis, Volume 23. Edited by Tse-Lok Ho

Heterocycles. Insertion of nitriles into cyclopropenes to form pyrroles,[5] of hindered isonitriles into epoxides in a 2:1 ratio to form 2-imino-2,5-dihydrofurans,[6] and the [4 + 1]cycloaddition of isonitriles to enones to give unsaturated imino lactones[7] are all catalyzed by $GaCl_3$.

Cyclization. Ring closure with group migration occurs when 1,6-enynes are exposed to $GaCl_3$.[8]

COOEt COOEt $\xrightarrow[\text{PhMe, 0°}]{GaCl_3}$ COOMe COOMe

77%

1-Formyladamantane. $GaCl_3$-catalyzed carbonylation of adamantane with CO at room temperature give the aldehyde in 84% yield.

[1]Amemiya, R., Nishimura, Y., Yamaguchi, M. *S* 1307 (2004).
[2]Amemiya, R., Fujii, A., Yamaguchi, M. *TL* **45**, 4333 (2004).
[3]Amemiya, R., Fujii, A., Arisawa, M., Yamaguchi, M. *CL* **32**, 298 (2003).
[4]Usugi, S., Yorimitsu, H., Shinokubo, H., Oshima, K. *OL* **6**, 601 (2004).
[5]Araki, S., Tanaka, T., Toumatsu, S., Hirashita, T. *OBC* **1**, 4025 (2003).
[6]Bez, G., Zhao, C.-G. *OL* **5**, 4991 (2003).
[7]Chatani, N., Oshita, M., Tobisu, M., Ishii, Y., Murai, S. *JACS* **125**, 7812 (2003).
[8]Chatani, N., Inoue, H., Kotsuma, T., Murai, S. *JACS* **124**, 10294 (2002).
[9]Oshita, M., Chatani, N. *OL* **6**, 4323 (2004).

Germanium(IV) chloride. 20, 170

Trichlorogermanes. A convenient preparation of $ArGeCl_3$ and R^fGeCl_3 involves heating organostannanes with $GeCl_4$ without solvent.

[1]Kultyshev, R.G., Prakash, G.K.S., Olah, G.A., Faller, J.W., Parr, J. *OM* **23**, 3184 (2004).

Glyoxylic acid.

Deoximation. Facile regeneration of carbonyl compounds from oximes is observed in aq. medium containing glyoxylic acid.

[1]Chavan, S.P., Soni, P. *TL* **45**, 3161 (2004).

Gold(III) bromide.

Propargylic amines. Deoxygenative C-C and N-C bond formation of carbonyl compounds with 1-alkynes and amines leads to propargylic amines. The reaction is carried out in water and catalyzed by $AuBr_3$.[1]

Annulation. Enynals and *o*-alkynylbenzaldehydes condense with either alkynes[2] or aldehydes[3] to provide benzene and naphthalene derivatives, respectively. The reaction is thought to involve Diels-Alder reaction of pyrylium ions.

CHO + Ph —($AuBr_3$, DCE 80°)→ Ph, Ph, O
100%

CHO + Ph CHO —($AuBr_3$, dioxane 100°)→ Ph, Ph, O + Ph
87 : 13
90%

o-Alkynylnitrobenzenes afford two different kinds of heterocycles depending on the terminal *sp*-substituent (alkyl or aryl).[4]

NO_2, Ar —($AuBr_3$)→ O⊖, N⊕, Ar, O

NO_2, R —($AuBr_3$)→ N, O, O, R

[1]Wei, C., Li, C.-J. *JACS* **125**, 9584 (2003).
[2]Asao, N., Nogami, T., Lee, S., Yamamoto, Y. *JACS* **125**, 10921 (2003).
[3]Asao, N., Aikawa, H., Yamamoto, Y. *JACS* **126**, 7458 (2004).
[4]Asao, N., Sato, K., Yamamoto, Y. *TL* **44**, 5675 (2003).

Gold(I) chloride.

Cycloisomerization. Siloxy enynes are transformed, through extensive skeletal reorganization, into cyclohexadienes by AuCl.[1]

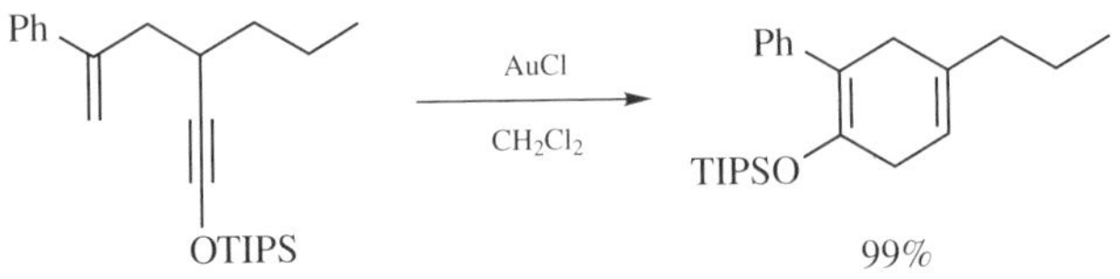

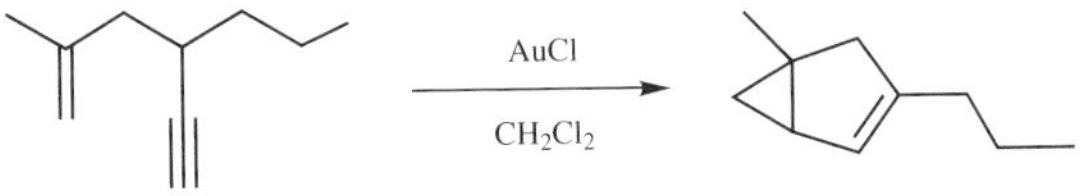

AuCl, CH_2Cl_2 — 99%

AuCl, CH_2Cl_2

[1]Zhang, L., Kozmin, S.A. *JACS* **126**, 11806 (2004).

Gold(I) chloride - triphenylphosphine.

Cyclization. α-(3-Alkynyl)-β-dicarbonyl compounds rapidly cyclize on treatment with $(Ph_3P)AuCl$ and AgOTf, the newly created ring may be part of a fused system or bridged system.[1] The 4-alkynyl homologues are similarly cyclized to provide cyclopentanes containing an exocyclic double bond.[2]

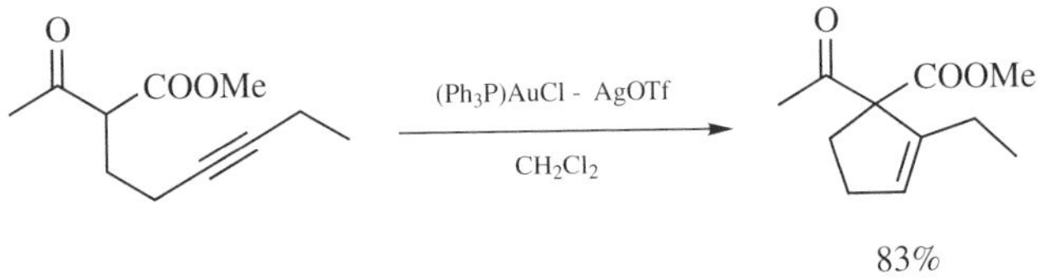

$(Ph_3P)AuCl$ - AgOTf, CH_2Cl_2 — 83%

Cycloisomerization of 1,6-enynes with end-group migration is also effected by a similar catalyst combination. But under acidic conditions and using $(Ph_3P)AuMe$ as catalyst alone (no Ag salt) cyclization with addition of ROH occurs.[3]

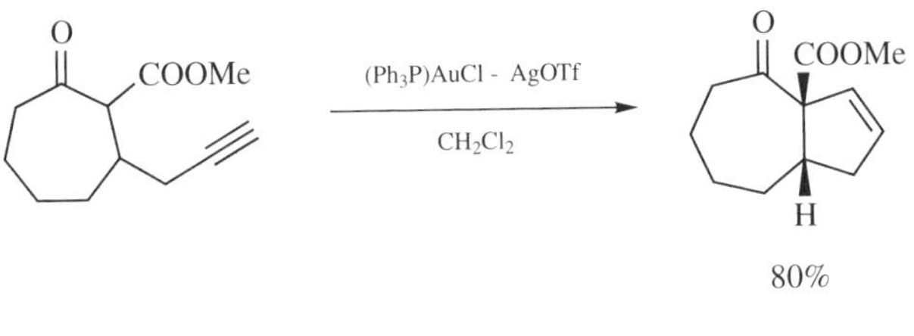

$(Ph_3P)AuCl$ - AgOTf, CH_2Cl_2 — 80%

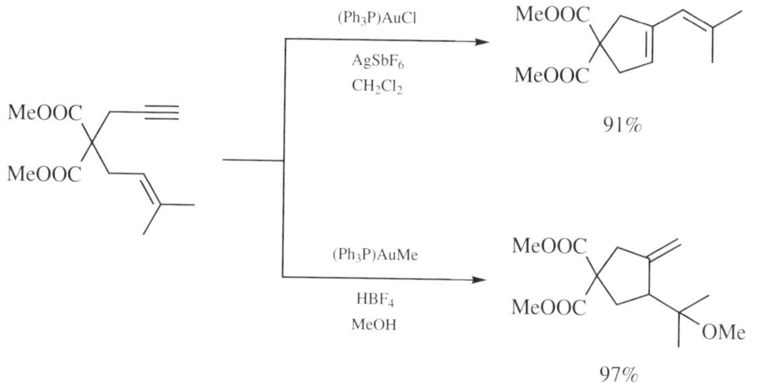

MeOOC
MeOOC
(Ph$_3$P)AuCl
AgSbF$_6$
CH$_2$Cl$_2$
MeOOC
MeOOC
91%
(Ph$_3$P)AuMe
HBF$_4$
MeOH
MeOOC
MeOOC
OMe
97%

[1]Staben, S.T., Kennedy-Smith, J.J., Toste, F.D. *ACIEE* **43**, 5350 (2004).
[2]Kennedy-Smith, J.J., Staben, S.T., Toste, F.D. *JACS* **126**, 4526 (2004).
[3]Nieto-Oberhuber, C., Munoz, M.P., Bunel, E., Nevado, C., Cardenas, D.J., Echavarren, A.M. *ACIEE* **43**, 2402 (2004).

Gold(III) chloride. 21, 196; 22, 200

Cyclization. Cyclization to furans from 2-(1-alkynyl)-2-alkenones is observed in a $AuCl_3$-catalyzed reaction.[1] Butenolides are formed when *t*-butyl 2,3-alkadienoates are treated with the Lewis acidic and soft $AuCl_3$.[2] Similarly, activation of the allene moiety by $AuCl_3$ is the key to transform 2,3-alkadienyl amines into pyrrolines.[3]

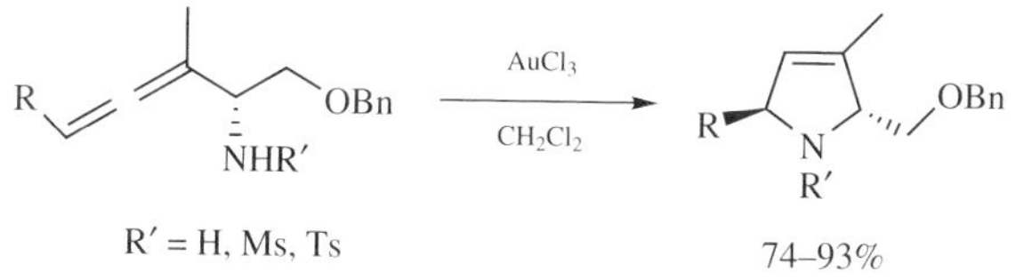

R
OBn
NHR′
AuCl$_3$
CH$_2$Cl$_2$
R
N
R′
OBn
R′ = H, Ms, Ts
74–93%

Cycloaddition. Catalyzed by $AuCl_3$ benzopyrylium ions are formed from 2-alkynylphenyl carbonyl compounds. Various nucleophiles can be added to the reaction media to trap such reactive species.[4]

[1]Yao, T., Zhang, X., Larock, R.C. *JACS* **126**, 11164 (2004).
[2]Kang, J.-E., Lee, E.-S., Park, S.-I., Shin, S. *TL* **46**, 7431 (2005).
[3]Morita, N., Krause, N. *OL* **6**, 4121 (2004).
[4]Dyker, G., Hildebrandt, D., Liu, J., Merz, K. *ACIEE* **42**, 4399 (2003).

Gold(III) chloride – silver triflate.

Alkylation. Activated methylene compounds add to alkenes (most of the demonstrated cases are styrenes).[1] Cyclization of trifloxyalkylarenes[2] and alkenylation[3] as well as aminoalkylation[4] of arenes in the presence of $AuCl_3$-AgOTf are typical Friedel-Crafts type reactions. The Au-catalyzed alkylation process[2] does not has the limitation of the Friedel-Crafts reaction that cannot be used to directly prepare long-chain *n*-alkylarenes. Also the reaction with imines is carried out.[3]

MsO ... =X, O → $AuCl_3$ - AgOTf, DCE, 120° → =X, O

X=H,H 90%
X=O 72%

Hydroarylation.[4] Arenes add to electron-rich alkynes. In reaction with electron-deficient alkynes the use of $(Ph_3P)AuCl$ is preferable. Two types of addition show different regioselectivity.

[1]Yao, X., Li, C.-J. *JACS* **126**, 6884 (2004).
[2]Shi, Z., He, C. *JACS* **126**, 13596 (2004).
[3]Luo, Y., Li, C.-J. *CC* 1930 (2004).
[4]Reetz, M.T., Sommer, K. *EJOC* 3485 (2003).

Graphite. **20**, 170; **21**, 197; **22**, 200

Nitriles.[1] Aldehydes form nitriles on heating with $NH_2OH{\cdot}HCl$ and MsCl on graphite.

Aromatic alkylations. Graphite is a catalyst for arene alkylation with benzyl, secondary and tertiary alkyl halides,[2] perhaps acting as a Lewis base to stabilize carbocation-like intermediates.

[1]Sharghi, H., Sarvari, M.H. *S* 243 (2003).
[2]Sereda, G.A. *TL* **45**, 7265 (2004).

Grignard reagents. **13**, 138–140; **14**, 171–172; **16**, 172–173; **17**, 141–142; **18**, 167–171; **19**, 151–154; **20**, 170–173; **21**, 197–202; **22**, 200–206

Substitution reactions. Alkynyl epoxides are very reactive toward allylmagnesium chloride and the nucleophilic attack occurs at the more highly substituted carbon atom with inversion of configuration.[1] A new method for deoxygenative coupling of alcohols involves Grignard reaction of the derived phosphonium salts $[ROPPh_2Me]X$.[2] 2-Imidazolidinone oxime tosylates are attacked by RMgX at the exocyclic nitrogen atom and the resulting guanidines can be hydrolyzed to give primary amines RNH_2 or subjected to reduction with $LiAlH_4$ to afford RNHMe.[3]

Sulfoxides of very high optical purity are synthesized by two consecutive Grignard reactions on chiral *N*-sulfonyl-1,2,3-oxathiazolidine-2-oxide derivatives.[4] Such strategy has been reported but the chiral substrate is different. The benzenesulfinyl group of α-chloro sulfoxides is removed by *i*-PrMgCl, and further reaction with RCH(Li)-SO_2Ph leads to alkenes. The method is suitable for the preparation of alkylidenecyclopropanes,[5] alkenes,[6] and α-amino acid derivatives.[7]

4-MeOC$_6$H$_4$N(Li)Me; ClCOOEt

PhCH$_2$SO$_2$Ph BuLi,THF

+ i-PrMgCl

93% (*E/Z* 1 : 2)

Dithioacetal monoxides generate Grignard reagents with an α-thio substituent. Those species undergo Cu-catalyzed alkylation.[8]

Phosphonites are obtained from trialkyl phosphates after hydrolysis of the Grignard reaction products.[9] Access to aryltris(trimethylsilyl)silanes from tetrakis(trimethylsilyl)-silane through removal of one silyl group is by arylation using ArMgBr.[10]

A synthesis of cyclopropyl ketones[11] is based on reaction of cyclopropylmagnesium bromide with 1-boryl-1-bromoalkenes. The bromine atom is displaced and oxidation of the products leads to the ketones.

−78°; NaOMe MeOH

NaOAc; H_2O_2

75–78%

Grignard reagents are aminated by reaction with 4,4,5,5-tetramethyl-1,3-dioxolan-2-one *O*-benzenesulfonyloxime followed by acid hydrolysis.[12] Reaction of *N,S*-acetals results in displacement of the RS group, and the reaction can be applied to synthesis of amines from HC(=S)NR_2 by in situ activation and two different organometallic reactions.[13]

Formation of other reactive species. Metallation of α-diazo ketones with MeMgX followed by acylation provides α-diazo β-keto esters.[14] Benzynes are generated from *o*-iodoaryl sulfonates on treatment with *i*-PrMgCl.[15] However, mixed Li/Mg-species (*i*-PrMgCl - LiCl) makes the Br/Mg exchange under mild conditions that 1,2-dibromobenzene is readily converted into the mono-magnesium chloride for the preparation of boronic esters.[16]

The Br/Mg exchange reaction of β,β-dibromoacrylic esters at low temperature leads to Grignard reagents. Since capture with carbonyl compounds gives γ-lactones,[17] selective exchange of the (*Z*)-bromine atom occurs. α-Cyanoalkylmagnesium halides are also readily formed by the exchange reaction.[18]

Addition reactions. Mentionable reactions are: synthesis of chiral α-oxoalkanoic esters from oxalic esters of optically active alcohols (e.g., dimenthyl oxalate),[19] addition-alkylation of γ-hydroxy α,β-unsaturated nitriles,[20] addition to latent nitrones from *N*-benzyl-*N*-glycosylhydroxylamines,[21] (and double addition to *N*-glycosyl nitrones[22]).

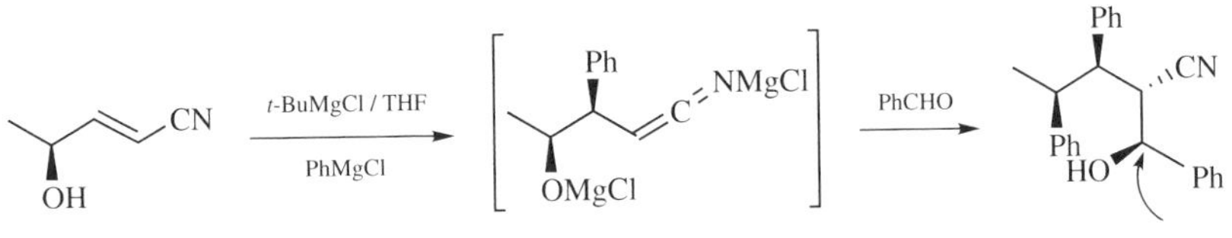

t-BuMgCl / THF; PhMgCl → [intermediate: OMgCl, Ph, C=NMgCl] → PhCHO → 4:1 isomers 64%

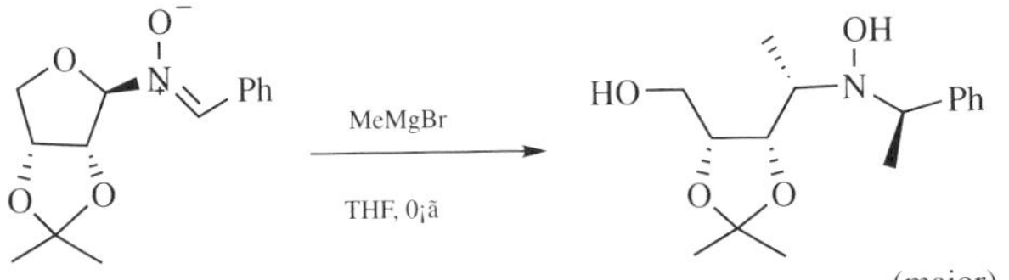

MeMgBr, THF, 0¡ã → 82% (major)

Diarylamines are obtained by the addition of ArMgX to nitrosoarenes,[23] or arylazo tosylates (ArN=NTs which are available from reaction of ArN_2BF_4 with NaTs),[24] besides the previously reported reaction with $ArNO_2$. A reducing agent is required in the workup. Reaction of nitroarenes with vinylmagnesium bromide gives indoles.[25]

To elaborate $ArSO_2NR_2$ it requires only to add to ArMgX in sequence: SO_2, $SOCl_2$, and R_2NH.[26] Alkynylmagnesium bromides and organoazides undergo cycloaddition to afford triazol-4-ylmagnesium derivatives that can be trapped with electrophiles.[27]

cis-2,6-Disubstituted piperidines are obtained from a reaction sequence starting from a Grignard reaction on 2-substituted *N*-benzoyliminopyridinium ylides.[28] Regioselectivity is derived from complexation.

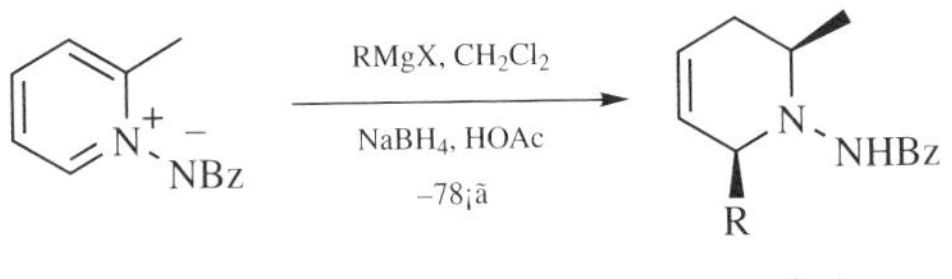

$RMgX, CH_2Cl_2$; $NaBH_4$, HOAc, −78¡ã → NHBz, R (major)

Condensation of *N*-acylenamines with stabilized iminium species followed by Grignard reaction furnishes protected diamines.[29]

Metal carbenoid reactions. On treatment with *i*-PrMgCl the TolSO group of 1-chloro-1-*p*-toluenesulfinyl-1-alkenes is replaced by [MgCl] and the alkylidene-metal carbenoids readily insert into an *o*-position of arylamines.[30]

SOTol, Cl —(i-PrMgCl)→ MgCl, Cl —(X–C6H4–NH2)→ X–C6H3(NH2)–CH=CMe2

X = H, OMe 43%

[1]Taber, D.F., He, Y., Xu, M. *JACS* **126**, 13900 (2004).
[2]Shintou, T., Kikuchi, W., Mukaiyama, T. *CL* **32**, 676 (2003).
[3]Kitamura, M., Chiba, S., Narasaka, K. *BCSJ* **76**, 1063 (2003).
[4]Han, Z., Krishnamurthy, D., Grover, P., Wilkinson, H.S., Fang, Q.K., Su, X., Lu, Z.-H., Magiere, D., Senanayake, C.H. *ACIEE* **42**, 2032 (2003).
[5]Satoh, T., Saito, S. *TL* **45**, 347 (2004).
[6]Satoh, T., Kondo, A., Musashi, J. *T* **60**, 5453 (2004).
[7]Satoh, T., Osawa, A., Kondo, A. *TL* **45**, 6703 (2004).
[8]Satoh, T., Akita, K. *CPB* **51**, 181 (2003).
[9]Petnehazy, I., Jaszay, Z.M., Szabo, A., Everaert, K. *SC* **33**, 1665 (2003).
[10]Sanganee, M.J., Steel, P.G., Whelligan, D.K. *JOC* **68**, 3337 (2003).
[11]Bhat, N.G., Garcia, L., Tamez, Jr, V. *TL* **44**, 7175 (2003).
[12]Kitamura, M., Suga, T., Chiba, S., Narasaka, K. *OL* **6**, 4619 (2004).
[13]Murai, T., Mutoh, Y., Ohta, Y., Murakami, M. *JACS* **126**, 5968 (2004).
[14]Cuevas-Yanez, E., Muchowski, J.M., Cruz-Almanza, R. *TL* **45**, 2417 (2004).
[15]Sapountzis, I., Lin, W., Fischer, M., Knochel, P. *ACIEE* **43**, 4364 (2004).
[16]Krasovskiy, A., Knochel, P. *ACIEE* **43**, 3333 (2004).
[17]Vu, V.A., Marek, I., Knochel, P. *S* 1797 (2003).
[18]Fleming, F.F., Zhang, Z., Knochel, P. *OL* **6**, 501 (2004).
[19]MaGee, D.I., Mallais, T.C., Eic, M. *TA* **14**, 3177 (2003).
[20]Fleming, F.F., Wang, Q., Steward, O.W. *JOC* **68**, 4235 (2003).
[21]Dondoni, A., Perrone, D. *T* **59**, 4261 (2003).
[22]Bonnani, M., Marradi, M., Cicchi, S., Faggi, C., Goti, A. *OL* **7**, 319 (2005).
[23]Kopp, F., Sapountzis, I., Knochel, P. *SL* 885 (2003).
[24]Sapountzis, I., Knochel, P. *ACIEE* **43**, 897 (2004).
[25]Knepper, K., Brase, S. *OL* **5**, 2829 (2003).
[26]Pandya, F., Murashima, T., Tedeschi, L., Barrett, A.G.M. *JOC* **68**, 8274 (2003).
[27]Krasinski, A., Fokin, V.V., Sharpless, K.B. *OL* **6**, 1237 (2004).
[28]Legault, C., Charette, A.B. *JACS* **125**, 63604 (2003).
[29]Suga, S., Nishida, T., Yamada, D., Nagaki, A., Yoshida, J. *JACS* **126**, 14338 (2004).
[30]Satoh, T., Ogino, Y., Nakamura, M. *TL* **45**, 5785 (2004).

Grignard reagents/cobalt(II) salts. **22**, 207

Couplings. Cobalt-catalyzed cross-coupling has been extended to heteroaromatic chlorides (with ArMgX[1] and RMgX[2]). The 3-component coupling[3] that involves alkyl halides, 1,3-dienes, and Me_3SiCH_2MgCl is more significant.

Br + Ph $\xrightarrow[\text{Ph}_2\text{P(CH}_2)_6\text{PPh}_2\ \ \text{Et}_2\text{O rt}]{\text{Me}_3\text{SiCH}_2\text{MgCl - CoCl}_2}$ SiMe3 Ph

87%

A synthesis of allylsilanes consisting of two consecutive Grignard coupling reactions starts from $Co(acac)_3$-catalyzed reaction of (*E*)-1,2-dichloroethene with Me_3SiCH_2MgCl and then a nickel-mediated cross-coupling with other Grignard reagents.[4] Cross-coupling following a radical cyclization can be effected in one flask, after forming the catalyst by mixing $CoCl_2$ with DPPP, the substrate (e.g., an iodoacetaldehyde allyl acetal) and RMgX are added to allow the reaction to proceed.[5]

Allylic ethers react with ArMgX and Me_3SiCH_2MgCl in a S_N2 manner when catalyzed by a $CoCl_2$-phosphine complex.[6]

[1]Korn, T.J., Cahiez, G., Knochel, P. *SL* 1892 (2003).
[2]Ohmiya, H., Yorimitsu, H., Oshima, K. *CL* **33**, 1240 (2004).
[3]Mizutani, K., Shinokubo, H., Oshima, K. *OL* **5**, 3959 (2003).
[4]Kamachi, T., Kuno, A., Matsuno, C., Okamoto, S. *TL* **45**, 4677 (2004).
[5]Ohmiya, H., Tsuji, T., Yorimitsu, H., Oshima, K. *CEJ* **10**, 5640 (2004).
[6]Mizutani, K., Yorimitsu, H., Oshima, K. *CL* **33**, 832 (2004).

Grignard reagents/copper salts. 18, 171–173; **19**, 154–156; **20**, 174–175; **21**, 202–203; **22**, 207–209

Epoxide opening. Grignard reaction of glycidol derivatives is reliably established with the addition of CuI.[1]

Coupling and addition reactions. Alkyl fluorides couple with RMgX (including *t*-BuMgCl) in the presence of $CuCl_2$ and 1,3-butadiene.[2] Highly arylated alkenes are synthesized from 2-alkynylsilylpyridines via carbometallation with ArMgX in the presence of CuI and then Pd-catalyzed coupling. The pyridylsilyl group is replaceable after transforming into a boronate.[3]

Allylic substitution. For the Cu-catalyzed displacement reaction of *cis*-4-cyclopentene-1,3-diol monoacetate, vinyl and arylmagnesium chlorides are better than the corresponding bromides for regioselectivity. Since some of the RMgCl are difficult to prepare, one way to solve the problem is to add LiCl to the reaction media.[4] 1,4-Oxa-1,4-dihydronaphthalenes are opened by RMgX-CuCl-Ph_3P to afford *trans*-1-hydroxy-2-alkyl-1,2-dihydronaphthalenes.[5]

3,3-Disubstituted allyl phosphates in which one of the substituents is a CF_3 group usually undergo S_N2' substitution with RMgX in the presence of CuCN.[6] But ArMgBr favor direct (S_N2) displacement.

Ketone synthesis. Reactivity moderation of substrates and reagents makes the Grignard reaction serviceable to ketone synthesis. For example, presence of $AlCl_3$ together with CuCl allows reaction of acyl chlorides to yield ketone products.[7]

3-Butenyl ketones are formed as the major products when esters react with vinylmagnesium bromide in the presence of $Cu(OAc)_2$.[8] Without the Cu-salt substantial amounts of divinyl carbinols appear.

MeOOC— NSO$_2$Ph → ($H_2C{=}CHMgBr$; $Cu(OAc)_2$; THF, – 45;ã) NSO$_2$Ph

55%

[1]Bonini, C., Chiummiento, L., Lopardo, M.T., Pullez, M., Colobert, F., Solladie, G. *TL* **44**, 2695 (2003).
[2]Terao, J., Ikumi, A., Kuniyasu, H., Kambe, N. *JACS* **125**, 5646 (2003).
[3]Itami, K., Kamei, T., Yoshida, J. *JACS* **125**, 14670 (2003).
[4]Kobayashi, Y., Nakata, K., Ainai, T. *OL* **7**, 183 (2005).
[5]Arrayas, R.G., Cabrera, S., Carretero, J.C. *OL* **5**, 1333 (2003).
[6]Kimura, M., Yamazaki, T., Kitazume, T., Kubota, T. *OL* **6**, 4651 (2004).
[7]Vicha, R., Potacek, M. *T* **61**, 83 (2005).
[8]Hansford, K.A., Dettwiler, J.E., Lubell, W.D. *OL* **5**, 4887 (2003).

Grignard reagents/iron(III) acetylacetonate. 22, 210

Diaryl ketones. Iron-catalyzed reaction between ArMgX and Ar'COCN is useful for the synthesis of ArCOAr'.[1]

Coupling reactions. From ArMgX and RX the formation of Ar-R is readily accomplished in the presence of $FeCl_3$-TMEDA[2] or $Fe(acac)_3$.[3] Generally, undesirable side reactions such as dehydrohalogenation can be suppressed by adding TMEDA.[2] Coupling of RMgX with enol triflates, acyl chlorides, and dichloroarenes has also been successfully carried out in the presence of ester/lactone group in the same molecule.[4] The involvement of only one C-Cl bond in dichloroarenes has good synthetic value. A synthesis of conjugated enynes from chloroenynes is based on the same process.[5] [A (salen)-Fe complex is useful for coupling of ArMgBr with alkyl halides bearing β-hydrogen atoms.[6]]

Alkynyl epoxides undergo ring opening to give allenyl alcohols when exposed to RMgCl-$Fe(acac)_3$.[7]

O H → (RMgCl; $Fe(acac)_3$; PhMe) C R OH

For cross-coupling of alkyl halides with ArMgX application of a complex composing (tetraethene)iron and lithium-TMEDA is effective.[8] Besides direct coupling, reaction relayed by a double bond leading to cyclized products is also observed.

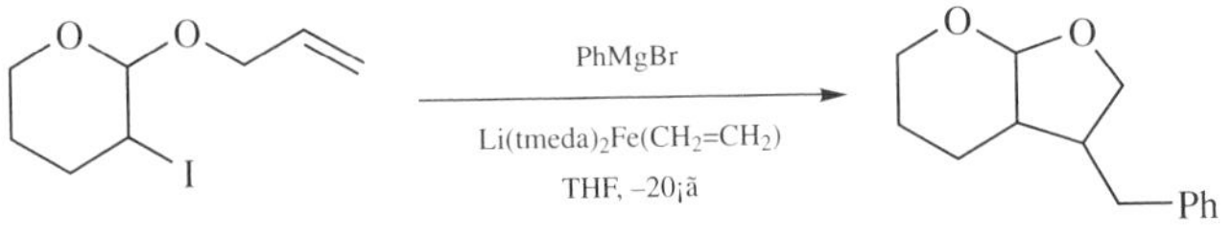

PhMgBr

$Li(tmeda)_2Fe(CH_2{=}CH_2)$

THF, –20¡ã

85%

[1]Duplais, C., Bures, F., Sapountzis, I., Korn, T.J., Cahiez, G., Knochel, P. *ACIEE* **43**, 2968 (2004).
[2]Nakamura, M., Matsuo, K., Ito, S., Nakamura, E. *JACS* **126**, 3686 (2004).
[3]Nagano, T., Hayashi, T. *OL* **6**, 1297 (2004).
[4]Scheiper, B., Bonnekessel, M., Krause, H., Furstner, A. *JOC* **69**, 3943 (2004).
[5]Seck, M., Franck, X., Hocquemiller, R., Figadere, B., Peyrat, J.-F., Provot, O., Brion, J.-D., Alami, M. *TL* **45**, 1881 (2004).
[6]Bedford, R., Bruce, D., Frost, R., Goodby, J., Hird, M. *CC* 2822 (2004).
[7]Furstner, A., Mendez, M. *ACIEE* **42**, 5355 (2003).
[8]Martin, R., Furstner, A. *ACIEE* **43**, 3955 (2004).

Grignard reagents/manganese(II) halide.

Substitution. An alkoxy residue of an acetal becomes a leaving group and is replaced on reaction with RMgX when the reaction medium also contains $MnBr_2$, LiBr, and $BF_3 \cdot OEt_2$.[1] Thus a better route to ethers may involve this reaction instead of the Grignard reaction on carbonyl compounds followed by etherification (which is often inefficient).

Addition to allenes. The organometallic species derived from allylmagnesium chloride and $MnCl_2$ is capable of transferring an allyl group to the more substituted terminus of an allene to form a 1,5-diene. When an excess Grignard reagent is used another allyl group replaces the proximal alkenyl hydrogen atom. Intramolecular trapping of the alkenylmanganese intermediate by another double bond leads to methyl enecycloalkanes.

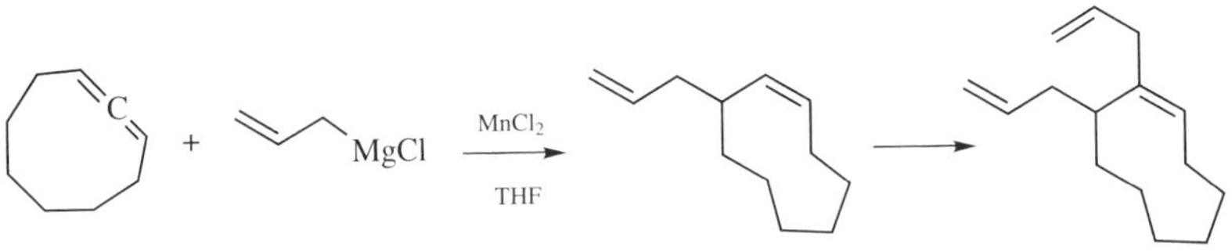

C + MgCl $\xrightarrow[THF]{MnCl_2}$

69% 64%

OMe OMe C + MgCl $\xrightarrow[HMPA - THF]{MnCl_2}$ OMe OMe

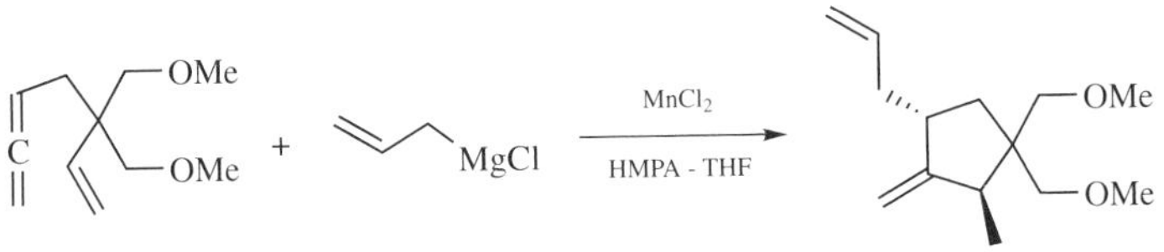

46%

[1]Hojo, M., Ushioda, N., Hosomi, A. *TL* **45**, 4499 (2004).
[2]Nishikawa, T., Shinokubo, H., Oshima, K. *OL* **5**, 4623 (2003).

Grignard reagents/nickel complexes. 18, 173; **19**, 156–157; **20**, 176–177; **21**, 204–205; **22**, 210–211

Cross couplings. It is noteworthy that *S*-activation by nickel makes possible of forming Ar-Ph from $PhSO_3CH_2C(Bn)Me_2$ in reaction with ArMgBr.[1] Sulfonamides[2] are also competent. For the replacement of an alkenyl PhTe group the Ni-catalyzed coupling works well, and since the reaction does not disturb a geminal silyl residue, alkenylsilanes can be prepared by this method.[3]

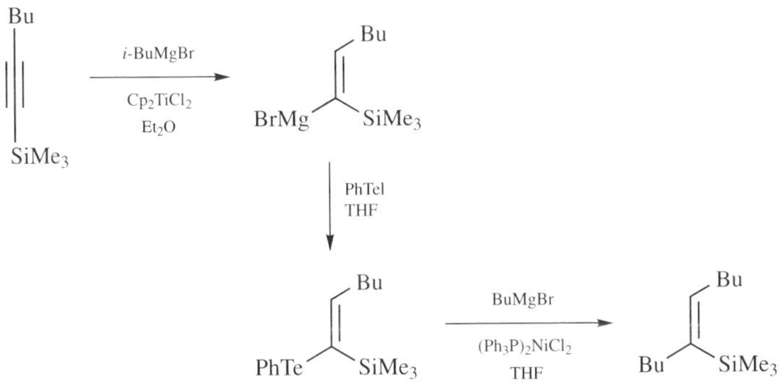

Coupling to replace the polymethylenedithio unit from a cyclic dithioacetal by the trimethylsilylvinylidene group is extendable to more ordinary substrates by using t-Bu_2PMe or t-Bu_3P as an activating ligand.[4] (*Z*)-Alkenes are produced in the Ni-catalyzed Grignard reaction of (*Z*)-1,2-bis(ethylseleno)ethene, and monoarylation is feasible.[5]

Stereoretentive replacement of a (*Z*)-*N*,*N*-diisopropylcarbamoyloxy group on reaction with $CH_2{=}CHMgBr$ represents a valuable diene synthesis.[6] Conjugated dienes (1,3-butadiene and 2-substituted congeners) are dimerized while attaching to both the organic residue of a Grignard reagent and the silyl group of a chlorosilane, when they are mixed with $Ni(acac)_2$.[7]

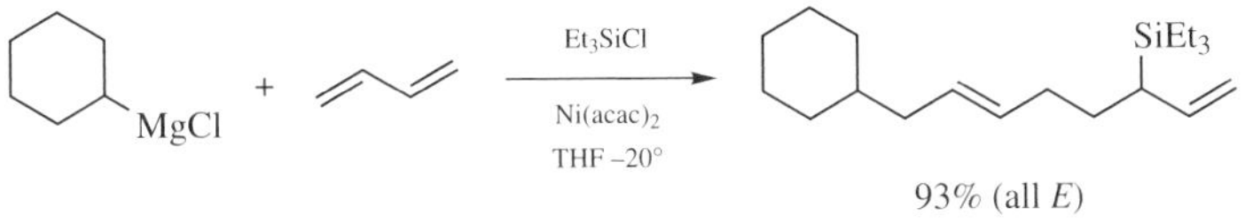

Alkoxyarenes are reactive toward arylnickel species therefore they can participate in biaryl synthesis.[8] Promoted by $NiCl_2$ and in the presence of 1,3-butadiene Wurtz-type coupling of Grignard reagents with primary alkyl fluorides is feasible.[9]

Modified Grignard reagents (RMgSPh) cross-couple with ArCN with Ni catalyst to give ArR.[10]

Desulfonylation. *o*-Anisylsulfonamides of secondary amines are cleaved by *i*-PrMgCl-Ni(acac)$_2$.[11] The reaction serves as a basis for the protection of such amines.

Ketones. *N*-Benzenesulfonyl α-amino acids form ketones from the Ni-catalyzed Grignard reaction.[12]

[1]Cho, C.-H., Yun, H.-S., Park, K. *JOC* **68**, 3017 (2003).
[2]Milburn, R.R., Snieckus, V. *ACIEE* **43**, 888 (2004).
[3]Cai, M., Hao, W., Zhao, H., Xia, J. *JOMC* **689**, 1714 (2004).
[4]Huang, L.-F., Huang, C.-H., Stulgies, B., de Meijere, A., Luh, T.-Y. *OL* **5**, 4489 (2003).
[5]Martynov, A.V., Potapov, V.A., Amosova, S.V., Makhaeva, N.A., Beletskaya, I.P., Hevesi, L. *JOMC* **674**, 101 (2003).
[6]Poree, F.-H., Clavel, A., Betzer, J.-F., Pancrazi, A., Ardisson, J. *TL* **44**, 7553 (2003).
[7]Terao, J., Oda, A., Ikumi, A., Nakamura, A., Kuniyasu, H., Kambe, N. *ACIEE* **42**, 3412 (2003).
[8]Dankwardt, J.W. *ACIEE* **43**, 2428 (2004).
[9]Terao, J., Ikumi, A., Kuniyasu, H., Kambe, N. *JACS* **125**, 5646 (2003).
[10]Miller, J.A., Dankwardt, J.W. *TL* **44**, 1907 (2003).
[11]Milburn, R.R., Snieckus, V. *ACIEE* **43**, 892 (2004).
[12]Sharma, A.K., Hergenrother, P.J. *OL* **5**, 2107 (2003).

Grignard reagents/palladium complexes. 21, 205; 22, 211

Allylic substitution. The long-standing problem of preparing substituted cyclopentenols of definite constitution and stereochemistry from cyclopentediol derivatives has gradually receded due to intensive studies of the displacement reactions. *cis*-2-Alkyl-3-cyclopentenols are easily available by a Pd-catalyzed Grignard reaction.[1] The organic group is introduced from a chelate intermediate.

HO OAc + RMgX —(Ph$_3$P)$_4$Pd, THF, 0¡ã→ HO, R + HO, R

R=Bu 91 : 9 (67%)

Cross couplings. Various combinations are possible in the Pd-catalyzed Grignard reaction: RMgX/ArMgX with RBr/ROTs,[2] ArMgBr with ArOTs,[3] and RCCMgI with RBr.[4]

A formal *trans*-addition of organic groups to the triple bond of propargylic alcohols is accomplished using the Pd-catalyzed coupling technique.[5] Formation of oxamagnesiacyclopentene intermediates is indicated.

OH —RMgCl→ Mg, O, R —ArI, (Ph$_3$P)$_4$Pd→ Ar, R, OH

3,8-Disilyl-1,6-octadienes are formed from the coupling of chlorosilanes with 1,3-butadiene.[6] The Grignard reagent (e.g., PhMgBr) is used for activating the Pd complex and it does not contribute an organic residue to the products.

[1]Hattori, H., Abbas, A.A., Kobayashi, Y. *CC* 884 (2004).
[2]Terao, J., Naitoh, Y., Kuniyasu, H., Kambe, N. *CL* 890 (2003).
[3]Roy, A.H., Hartwig, J.F. *JACS* **125**, 8704 (2003).
[4]Yang, L.-M., Huang, L.-F., Luh, T.-Y. *OL* **6**, 1461 (2004).
[5]Tessier, P.E., Penwell, A.J., Souza, F.E.S., Fallis, A.G. *OL* **5**, 2989 (2003).
[6]Terao, J., Oda, A., Kambe, N. *OL* **6**, 3341 (2004).

Grignard reagents/titanium(IV) compounds. 14, 121–122; **18**, 174; **19**, 158–161; **20**, 177–180; **21**, 205–210; **22**, 211–215

Cyclopropane derivatives. The access to cyclopropylamines from RCN, addition of $BF_3.OEt_2$ before hydrolytic workup is essential for reasonable yields because of its ability to activate the azatitaniacyclopentene adducts to perform the ring-contracting rearrangement.[1] Expedient and almost unique synthesis of a β-amino ester containing two cyclopropane rings is as follows.[2]

CN, COOEt → $(i\text{-PrO})_4Ti$, $EtMgBr / Et_2O$; $BF_3 \cdot OEt_2$ → COOEt, H_2N

49%

Generally, the synthetic process finds greater significance in the further transformations of the products. Oxidative ring cleavage to provide larger cyclic ketones[3] is one such possibility. The ready access to cyclopropylamines from nitriles and amides including *N,N*-dibenzyl-2-tributylstannylcyclopropylamine opens routes to other substituted cyclopropylamines based on Stille coupling.[4] Intramolecular cyclopropanation of unsaturated nitriles affords bicyclo[n.1.0]alkanamines.[5]

Reductive cyclopropanation of cyclic ketones cannot be carried out in THF, and to conserve the three-membered ring the presence of a Lewis acid is very beneficial.[7]

OMe, O → $(i\text{-PrO})_4Ti$, RCH_2CH_2MgBr; $BF_3 \cdot OEt_2 / Et_2O$ → OMe, R

Coupling reactions. The hexacarbonyldicobalt complexes of alkynals are coupled to give *syn*-diols.[6] Reductive head-to-head coupling of homoallylic alcohols furnishes 1,8-diols.[8]

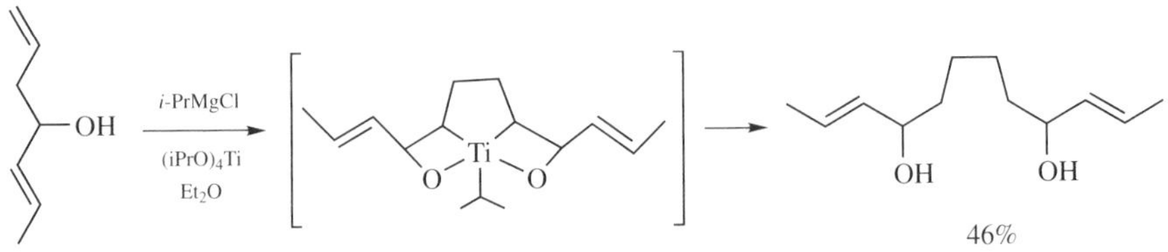

OH
i-PrMgCl
$(iPrO)_4Ti$
Et_2O
Ti O O
OH OH
46%

Subvalent titanium alkoxide species derived from EtMgBr and $(i\text{-PrO})_4Ti$ are reactive as pinacol coupling reagent.[9]

Carbon chain elongation of styrenes is easily achieved by coupling with alkyl halides.[10] Many useful alkenes are synthesized by reductive functionalization of alkynes: carboxylic acids,[11] alkenylsilanes,[12] dienylsilanes,[13] silylated allylic alcohols.[14]

Titanium-alkyne adducts add to *N*-tosylalkynylamines to afford 1-amino-1,3-dienes.[15] Alkenyloxazoline-Ti complexes react with 1-alkynes and aldehydes in sequence, to deliver highly functionalized carbon chains.[16]

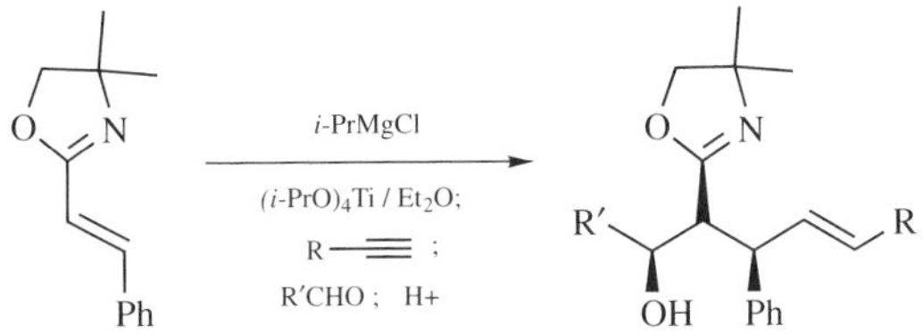

O N
Ph
i-PrMgCl
$(i\text{-PrO})_4Ti$ / Et_2O;
R—≡ ;
R′CHO ; H+
O N
R′ R
OH Ph

Cycloadditions. A synthesis of 2-indanols by combining 1,6-diynes with alkynyl sulfones[17] while mediated by *i*-PrMgBr/$(i\text{-PrO})_4Ti$ is quite practical.

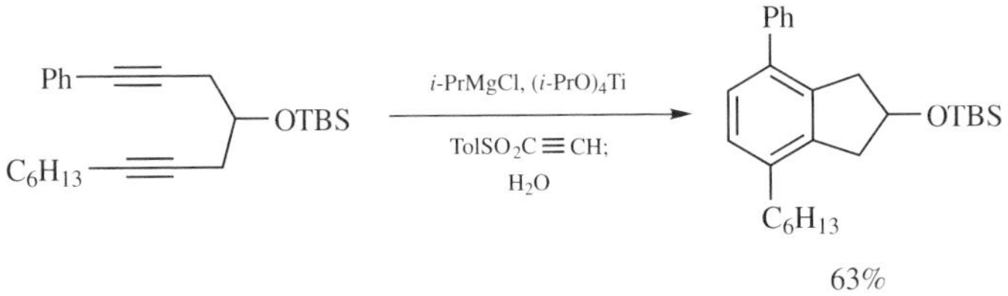

Ph
OTBS
C_6H_{13}
i-PrMgCl, $(i\text{-PrO})_4Ti$
$TolSO_2C{\equiv}CH$;
H_2O
Ph
OTBS
C_6H_{13}
63%

1-Acyloxyalken-6-ynes undergo a new type of cyclization.[18]

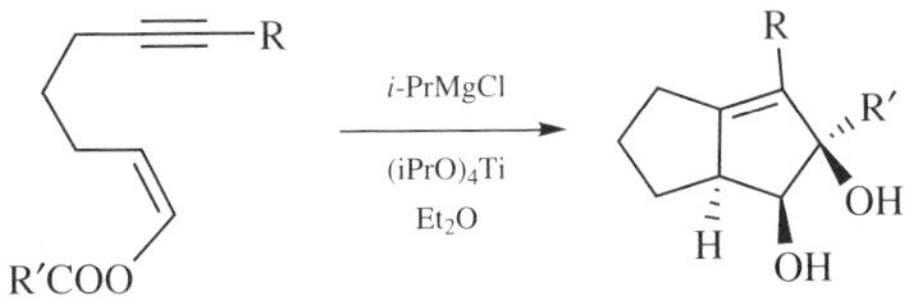

R
R′COO
i-PrMgCl
$(iPrO)_4Ti$
Et_2O
R
R′
OH
H OH

Reduction. Isoxazoles and isoxazolines are cleaved reductively with EtMgBr/$(i\text{-PrO})_4Ti$ to give β-amino enones and aldols, respectively.[19]

2-Butene-1,4-dianion. A titanacyclopentene complex is formed by admixing 3-buten-1-ylmagnesium chloride with $(i\text{-PrO})_4Ti$. It reacts with various electrophiles.[20]

R, R′ C=O + MgCl
i-PrMgCl
$(i\text{-PrO})_4Ti$
Et_2O −50°
R′, R, HO, R′, R, OH
Z/E >95 : 5

[1] Bertus, P., Szymoniak, J. *JOC* **68**, 7133 (2003).
[2] Bertus, P., Szymoniak, J. *SL* 265 (2003).
[3] Lecormue, F., Ollivier, J. *OBC* **1**, 3600 (2003).
[4] Wiedemann, S., Rauch, K., Savchenko, A., Marek, I., de Meijere, A. *EJOC* 631 (2004).
[5] Laroche, G., Bertus, P., Szymoniak, J. *TL* **44**, 2485 (2003).
[6] Lake, K., Dorrell, M., Blackman, N., Khan, M.A., Nicholas, K.M. *OM* **44**, 4260 (2003).
[7] Masalov, N., Feng, W., Cha, J.K. *OL* **6**, 2365 (2004).
[8] Isakov, V.E., Kulinkovich, O.G. *SL* 967 (2003).
[9] Matiushenkov, E.A., Sokolov, N.A., Kulinkovich, O.G. *SL* 77 (2004).
[10] Terao, J., Watabe, H., Miyamoto, M., Kambe, N. *BCSJ* **76**, 2209 (2003).
[11] Six, Y. *EJOC* 1157 (2003).
[12] Cai, M.-Z., Hao, W., Zhao, H., Song, C. *JOMC* **679**, 14 (2003).
[13] Zhao, H., Cai, M.-Z. *SC* **33**, 1643 (2003).
[14] Hirano, S., Tanaka, R., Urabe, H., Sato, F. *OL* **6**, 727 (2004).
[15] Tanaka, R., Hirano, S., Urabe, H., Sato, F. *OL* **5**, 67 (2003).
[16] Mitsui, K., Sato, T., Urabe, H., Sato, F. *ACIEE* **43**, 490 (2004).
[17] Hanazawa, T., Sasaki, K., Takayama, Y., Sato, F. *JOC* **68**, 4980 (2003).
[18] Urabe, H., Suzuki, D., Sasaki, M., Sato, F. *JACS* **125**, 4036 (2003).
[19] Churykau, D.H., Zinovich, V.G., Kulinkovich, O.G. *SL* 1949 (2004).
[20] Goeke, A., Mertl, D., Jork, S. *CC* 166 (2004).

Grignard reagents/zinc halide. 22, 215–216

Negishi coupling. Conjugated enynes are synthesized from chloroenynes by the regioselective and stereoselective coupling.[1] Grignard reagents are transformed into organozincates for the reaction.

Cl, C_5H_{11} + BrMg–C_6H_4–OMe
$(dppf)PdCl_2$
$ZnCl_2$
THF Δ
MeO, C_5H_{11}
83%

[1]Peyrat, J.-F., Thomas, E., L'Hermite, N., Alami, M., Brion, J.-D. *TL* **44**, 6703 (2003).

Grignard reagents/zirconium compounds. 18, 174; **19**, 161; **20**, 180–181; **21**, 210–211; **22**, 216–217

Allylation. Allylzirconium species derived from $CH_2=CHMgX$ and Cp_2ZrMe_2 has been used to couple with haloacetic esters in the presence of Et_3B to furnish 4-pentenoic esters.[1] When allylzirconium reagents prepared from zirconocene-alkene complexes (i.e., from Cp_2ZrCl_2 and Grignard reagents) are treated with diisopropyl ketone (to reduce the Zr-H species) they are ready for allyl transfer to carbonyl compounds.[2] 1,4-Diketones are allylated at one carbonyl site while reduced at the other.[3]

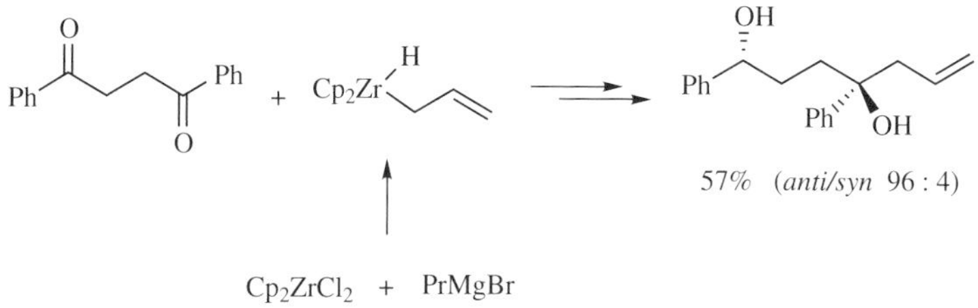

57% (*anti/syn* 96 : 4)

Cp_2ZrCl_2 + PrMgBr

The redox reaction in the above reaction is also responsible for the ketone formation from reaction of Cp_2ZrEt_2-EtMgBr with alkynes and then aldehydes.[4] Apparently, the 7-membered oxazirconacycle intermediates break down with hydride transfer to the unreacted aldehyde molecules.

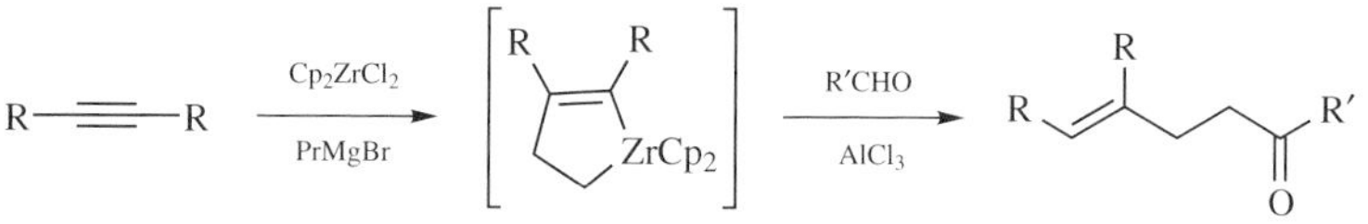

[1]Hirano, K., Fujita, K., Shinokubo, H., Oshima, K. *OL* **6**, 593 (2004).
[2]Fujita, K., Yorimitsu, H., Shinokubo, H., Oshima, K. *JOC* **69**, 3302 (2004).
[3]Fujita, K., Shinokubo, H., Oshima, K. *ACIEE* **42**, 2550 (2003).
[4]Zhao, C., Yan, J., Xi, Z. *JOC* **68**, 4355 (2003).

Guanidinium nitrate.

Nitration.[1] A potent nitrating agent (possibly nitronium ion) is generated from the title reagent in 85% H_2SO_4. With which nitrobenzene gives *m*-dinitrobenzene in 88% yield.

[1]Ramana, M.M.V., Malik, S.S., Parihar, J.A. *TL* **45**, 8681 (2004).

H

Hafnium(IV) triflate. 22, 218–219

Esterification.[1] Carboxylic acids activated in the form of mixed anhydrides with 2-thienylcarboxylic acid are attacked by alcohols in the presence of $Hf(OTf)_4$.

Glycosides.[2] Glycosyl carbonates are easy to make, thus their decarboxylation on treatment with $Hf(OTf)_4$ is a valuable for synthesis of glycosides. From 2α-O-pivaloyl derivatives β-glycosides are formed.

[1]Oohashi, Y., Fukumoto, K., Mukaiyama, T. *CL* **34**, 190 (2005).
[2]Azumaya, I., Kotani, M., Ikegami, S. *SL* 959 (2004).

Hexaalkylditins. 13, 142; **14**, 173–174; **16**, 174; **17**, 143–144; **18**, 175–176; **19**, 162–163; **20**, 182–184; **21**, 213; **22**, 219–220

Preparation.[1] For preparing hexabutylditin in practically quantitative yield it requires only heating stoichiometric amounts of Bu_3SnH with $Bu_3SnOSnBu_3$.

Cyclization.[2] Acylphosphonates participate in cyclization with a remote (4 or 5 bonds away) radical. Cyclic ketones can be generated.

(EtO)$_2$P(O) O COOEt COOEt EtOOC COOEt I

$Me_3SnSnMe_3$ / PhH, *hv*

O COOEt COOEt EtOOC EtOOC

79%

Hydrodehalogenation.[3] The $Bu_3SnSnBu_3$ -TBAF combination reduces ArBr and ArI. It permits regioselective deuteration. Another advantage of the method is that sulfur substituents are not affected.

[1]Darwish, A., Chong, J.M. *SC* **34**, 1885 (2004).
[2]Kim, S., Cho, C.H., Lim, C.J. *JACS* **125**, 9574 (2003).
[3]Harrowven, D.C., Guy, I.L., Nunn, M.I.T. *CC* 1966 (2004).

2,4-Hexadienyl 4-nitrophenyl carbonate.

Amine protection.[1] Transformation of amines into *O*-2,4-hexadienyl carbamates render them stable to certain reaction conditions. While the derivatives are cleaved

Fiesers' Reagents for Organic Synthesis, Volume 23. Edited by Tse-Lok Ho

with 1% TFA in CH_2Cl_2 their stability to heat (100°), light, 10% HOAc, 2*N*-NaOH, DBU, N_2H_4, TBAF, $NaBH_4$, and Pd(0) reagents has been established.

[1]Lingard, I., Bhalay, G., Bradley, M. *SL* 1791 (2003).

Hexafluoroacetone.

2,2-Bis(trifluoromethyl)oxazolidin-5-ones. These compounds are prepared from α-amino acids and hexafluoroacetone in DMSO at room temperature. Peptide bond formation takes place on mixing them with amino acid esters. Accordingly, hexafluoroacetone serves as protecting and activating agent.

[1]Burger, K., Lange, T., Rudolph, M. *H* **59**, 189 (2003).

1,1,1,3,3,3-Hexafluoro-2-propanol. **21**, 213–214; **22**, 221

Hydrogenation.[1] For hydrogenation of arenes containing functional groups such as OH, COOR, CF_3 to cyclohexane derivatives, use of $RuCl_3$ in this fluorinated solvent is very advantageous.

[1]Fache, F., Piva, O. *SL* 1294 (2004).

η^4-Hexamethylbenzene(pentamethylcyclopentadienyl)rhodium. **22**, 222

Borylation. The methyl group of an alkyl chain that contains a heteroatom (O, N, F) is borylated by the Rh complex at high temperature.

Cp*Rh(¦Ç4-C6Me6)
KHF2
MeOH 150°
86%

[1]Lawrence, J.D., Takahashi, M., Bae, C., Hartwig, J.F. *JACS* **126**, 15334 (2004).

Hexamethyldisilazane, HMDS. **13**, 141; **18**, 177–178; **19**, 163–164; **20**, 184–185; **21**, 214; **22**, 222

Triarylimidazolines. Heating ArCHO with HMDS leads to *cis*-2,4,5-triarylimidazolines.[1] At lower temperatures bis(imino)arylmethanes are detected. Thermal cyclization then follows the disrotatory mode. Microwave assistance in each step reduces the preparation time.[2]

[1]Uchida, H., Shimizu, T., Reddy, P.Y., Nakamura, S., Toru, T. *S* 1236 (2003).
[2]Uchida, H., Tanikoshi, H., Nakamura, S., Reddy, P.Y., Toru, T. *SL* 1117 (2003).

Hexamethylenetetramine.

Baylis-Hillman reaction. [1] This inexpensive amine is an effective catalyst.

[1]Krishna, P.R., Sekhar, E.P., Kannan, V. *S* 857 (2004).

Hydrazine hydrate.

Reductions. Azoarenes are reduced to *N,N'*-diarylhydrazines by hydrazine.[1] Reductive cleavage of α,β-epoxy ketones to aldols proceeds well if the temperature is controlled (usually 0° to −20° to minimize formation of allylic alcohols).[2]

Deacetylation. Regioselective removal of a 2*O*-acetyl group from protected glycosides[3] by hydrazine, albeit in moderate yields, is accomplished. Interestingly, thioglycosides behave differently.

[1]Zhang, C.-R., Wang, Y.-L. *SC* **33**, 4205 (2003).
[2]Salvador, J.A.R., Leitaio, A.J.L., Sa e Melo, M.L., Hanson, J.R. *TL* **46**, 1067 (2005).
[3]Li, J., Wang, Y. *SC* **34**, 211 (2004).

Hydridobiruthenium complex. 22, 223

Dehydrogenation. Complex **1**, after heterogenization by a sol-gel process with $(MeO)_4Si$, converts alcohols to carbonyl compounds and liberates hydrogen without any additive. The stable and reusable catalyst is recovered by filtration.[1]

1

[1]Choi, J.H., Kim, N., Shin, Y.J., Park, J.H., Park, J. *TL* **45**, 4607 (2004).

Hydridotetrakis(triphenylphosphine)rhodium. 22, 223

Disulfide exchange. In the presence of $(Ph_3P)_4RhH$, TfOH, and *p*-Tol_3P, mixed disulfides are formed from two homodisulfides.[1]

Addition to 1-alkynes.[2] The Rh catalyst promotes reaction of 1-alkynes with RSSR, and RSeSeR in a stereoselective and regioselective fashion, (*Z*)-1-organoseleno-2-organothio-1-alkenes are obtained due to preferential formation of 2-organothioalkenylrhodium intermediates.

X = S, Se
R′ = Ph, Bu

[1]Arisawa, M., Yamaguchi, M. *JACS* **125**, 6624 (2003).
[2]Arisawa, M., Kozuki, Y., Yamaguchi, M. *JOC* **68**, 8964 (2003).

Hydrido(tetraphenylporphyrin)rhodium(I).

Hydrometallation.[1] 1-Alkenes form adducts with the reagent in which the Rh atom is attached to the terminal position. Demetallative heterocyclization follows on addition of Me_4NOH in DMSO if the substrate contains a heteroatom at a viable distance.

[1]Sanford, M.S., Groves, J.T. *ACIEE* **43**, 588 (2004).

[Hydridotris(3,5-dimethylpyrazolyl)borato]copper(I).

O-H insertion. Etherification of alcohols by way of O-H bond insertion by Cu-carbenoids derived from diazoacetic esters is accomplished with the catalyst. Allyl alcohol and propargyl alcohol give ethers, formation of 3-membered ring (cycloaddition) does not occur.

[1]Morilla, M.E., Molina, M.J., Mar Diaz-Requejo, M., Belderrain, T.R., Nicasio, M.C., Trofimenko, S., Perez, P.J. *OM* **22**, 2914 (2003).

[Hydridotris(3,5-dimethylpyrazolyl)borato]trioxorhenium.

Deoxygenation. Epoxides are converted into alkenes with reaction conditions that tolerate many functional groups. *cis*-Epoxides are more reactive and it is further favored by the presence of an electron-withdrawing substituent.[1]

[1]Gable, K.P., Brown, E.C. *SL* 2243 (2003).

Hydrogen fluoride - amine. **16**, 286–287; **18**, 181; **19**, 164–165; **20**, 185; **21**, 215–216; **22**, 224

Halogen exchange. α,α-Dichloro sulfides are converted into the corresponding α,α-difluoro congeners by $3HF\text{-}Et_3N$ in the presence of $ZnBr_2$.[1] The mild Lewis acid facilitates the second stage exchange. Sulfenyl group of the substrates remains intact.

Alkyl fluorides.[2] Alcohols are transformed into fluorides by perfluorobutanesulfonyl fluoride and HF- Et_3N (additional Et_3N also added).

Trifluoromethylation.[3] A CF_3 group is introduced into the benzene nucleus of indole, indoline, and related compounds by treatment with $HF\text{-}SbF_5\text{-}CCl_4$, and then HF-pyridine. The first stage involves generation of $[CCl_3]^+$ and its electrophilic substitution. Cl/F exchange completes the process.

HF, SbF_5, CCl_4; HF Py, 0°

F_3C 83%

HF, SbF_5, CCl_4; HF Py, 0°

F_3C 60%

[1]Gouault, S., Guerin, C., Lemoucheux, L., Lequeux, T., Pommelet, J.-C. *TL* **44**, 5061 (2003).
[2]Yin, J., Zarkowsky, D.S., Thomas, D.W., Zhao, M.M., Huffman, M.A. *OL* **6**, 1465 (2004).
[3]Debarge, S., Kassou, K., Carreyre, H., Violeau, B., Jouannetaud, M.-P., Jacquesy, J.-C. *TL* **45**, 21 (2004).

Hydrogen peroxide. 22,

Sulfoxides. Oxidation of sulfides to sulfoxides by H_2O_2 usually requires an additive. The effectiveness of 3-fluorophenol[1] and a flavin derivative[2] has been noted.

Amides. Conversion of aldehydes to primary amides is accomplished by treatment with iodine and NH_4OH in THF, and after some time, aqueous H_2O_2.[3]

[1]Xu, W.L., Li, Y.Z., Zhang, Q.S., Zhu, H.S. *SC* **34**, 231 (2004).
[2]Linden, A.A., Kruger, L., Backvall, J.-E. *JOC* **68**, 5890 (2003).
[3]Shie, J.-J., Fang, J.-M. *JOC* **68**, 1158 (2003).

Hydrogen peroxide, acidic. **14**, 176; **15**, 167–168; **16**, 177–178; **17**, 145; **18**, 182–183; **19**, 166; **20**, 187; **21**, 216–217; **22**, 225–226

Oxidations. An economical and clean route for dihydroxylation of alkenes involves reaction with 30% H_2O_2 in the presence of Nafion beads.[1] Alkene epoxidation is also promoted by polyfluoroalkanesulfonyl fluoride.[2]

Phenol is a useful additive in the oxidation of sulfides to sulfoxides.[3] Its effect is due to H-bonding to H_2O_2.

gem-Bishydroperoxides.[4] Acetals react with H_2O_2 in the presence of BF_3 complexes (ether or MeOH) at room temperature to afford $R_2C(OOH)_2$.

[1]Usui, Y., Sato, K., Tanaka, M. *ACIEE* **42**, 5623 (2003).
[2]Yan, Z., Tian, W. *TL* **45**, 2211 (2004).
[3]Xu, W.L., Li, Y.Z., Zhang, Q.S., Zhu, H.S. *S* 227 (2004).
[4]Terent'ev, A.O., Kutkin, A.V., Platonov, M.M., Ogibin, Y.N., Nikishin, G.I. *TL* **44**, 7359 (2003).

Hydrogen peroxide, metal catalysts. **13**, 145; **14**, 177; **15**, 294; **17**, 146–148; **18**, 184–185; **19**, 166–167; **20**, 188; **21**, 217–218; **22**, 226–227

Epoxidation. A number of minerals and their modified forms continue to attract attention as catalysts for reactions involving H_2O_2 as oxidant. Effectiveness and chemoselectivity of various systems must be evaluated individually. Epoxidation of alkenes including enones over hydrotalcites is assisted by ultrasound.[1] A network consisting of phosphotungstate and a copolymer is an oxidation catalyst.[2] Sandwich-type polyoxometalates also are effective in promoting epoxidation of allylic alcohols.[3]

A Ru-complex shows activity for asymmetric epoxidation, although ee values of the products are $<85\%$.[4]

Oxidations. The effectiveness of a (tetrahydrosalen)-Co catalyst for oxidation of alcohols to acids is found.[5] Sulfated tin(IV) oxide with zeolite oxidizes ArCHO to esters (media contain aliphatic alcohol).[6] Solvent-free protocol for $FeBr_3$-catalyzed oxidation indicates faster rates for secondary alcohols (vs. primary alcohols).[7]

Stronger oxidizing systems compose H_2O_2 and the peroxi-Mo(VI) complex of 3,5-dimethylpyrazole[8] or a (tetrahydrosalen)-Cu complex,[9] both can oxidize a benzylic C-H bond.

Heteroatom oxidation such as sulfide to sulfoxide conversion catalyzed by $Sc(OTf)_3$[10] or a (porphyrin)-Fe complex,[11] to sulfone by a Mn complex,[12] and arylamine to nitrosoarene[13] by Mo oxide [and further to $ArNO_2$ with $(NH_4)_6Mo_7O_{24} \cdot H_2O$] are rather unexceptional. A ligand derived from diiodosalicylaldehyde and chiral 2-amino-ethanol for Fe achieves variable successes in the preparation of optically active sulfoxide.[14]

Tungstate-exchanged double layer hydroxides catalyze the oxidation of phenols to quinols and quinol ethers by H_2O_2 that is assisted by bromide ion.[15]

[1]Pillai, U.R., Sahle-Demessie, E., Varma, R.S. *SC* **33**, 2017 (2003).
[2]Yamada, Y.M.A., Tabata, H., Ichinohe, M., Takahashi, H., Ikegami, S. *T* **60**, 4087 (2004).
[3]Adam, W., Alsters, P.L., Neumann, R., Saha-Möller, C.R., Sloboda-Rozner, D., Zhang, R. *JOC* **68**, 1721 (2003).
[4]Tse, M.K., Dobler, C., Bhor, S., Klawonn, M., Magerlein, W., Hugl, H., Beller, M. *ACIEE* **43**, 5255 (2004).
[5]Das, S., Punniyamurthy, T. *TL* **44**, 6033 (2003).
[6]Qian, G., Zhao, R., Ji, D., Lu, G., Qi, Y., Suo, J. *CL* **33**, 834 (2004).
[7]Martin, S.E., Garrone, A. *TL* **44**, 549 (2003).
[8]Das, S., Bhowmick, T., Punniyamurthy, T., Dey, D., Nath, J., Chaudhuri, M.K. *TL* **44**, 4915 (2003).
[9]Velusamy, S., Punniyamurthy, T. *TL* **44**, 8955 (2003).
[10]Matteucei, M., Bhalay, G., Bardley, M. *OL* **5**, 235 (2003).
[11]Baciocchi, E., Gerini, M.F., Lapi, A. *JOC* **69**, 3586 (2004).
[12]Barker, J.E., Ren, T. *TL* **45**, 4681 (2004).
[13]Defoin, A. *S* 706 (2004).
[14]Legros, J., Bolm, C. *ACIEE* **42**, 5487 (2003).
[15]Sels, B., De Vos, D., Jacobs, P. *ACIEE* **44**, 310 (2005).

Hydrosilanes. **19**, 167–169; **20**, 188–192; **21**, 218–222; **22**, 227–233

Reduction. Reactive species generated from both free radical and cationic cleavages of C-O bond are trapped by hydrosilanes. Tertiary trifluoroacetates are reduced by Ph_2SiH_2-(*t*-BuO)$_2$.[1] Conversion of benzylidene acetals to monobenzylated diols by Et_3SiH-$EtAlCl_2$ is highly regioselective and chemoselective.[2] Thus, 2,4-diphenyl-1,3-dioxolane is reduced to 2-benzyloxy-2-phenylethanol as major product. Many functional groups (e.g., OMs, OTBS, *N*-Boc, *N*-Cbz) are tolerated.

Several β-*C*-glycosides have been prepared by deoxygenation with hydrosilane that is assisted by a Lewis acid (e.g., Me_3SiOTf).[3] Phenylsilane $PhSiH_3$ is effective for 1,4-reduction of enones in EtOH when catalyzed by $In(OAc)_3$.[4] Reductive aldol reaction is achieved by the combination (*syn*-selective)[4] or Et_3SiH- $InBr_3$.[5]

A synthesis of *cis*-2,5-disubstituted pyrrolidines from pyroglutamic acid involves Grignard reaction of the proper derivatives and reductive ring closure.[6] The latter step is easily achieved by a hydrosilane with Lewis acid catalysis.

The paucity of methods for direct reduction of carboxylic acids and derivatives to hydrocarbons makes the finding that catalytic $(C_6F_5)_3B$ combined with excess Et_3SiH can effect the transformation is very welcome. ArCOX (X = OH, OMe, Cl, H) are converted to ArMe in the 82–87% range.[7]

While double bond migration of 1-alkenes is observed in the presence of $PdCl_2$-Et_3SiH in EtOH at room temperature,[8] prolonged exposure eventually leads to alkanes.[9] For preparation of $[Ph_3PCuH]_6$ an expedient way involves addition of $PhSiHMe_2$ to the mixture of CuCl, Ph_3P and *t*-BuOK in benzene.[10]

Hydrosilylation. Generation of CuH in situ by hydrosilanes for hydrosilylation of dialkyl ketones is very efficient, because optimization by varying the reagents and stoichiometry is possible.[11] Copper-heterocyclic carbene complexes also serve effectively as catalysts.[12]

A method for elaboration of the *syn*-1,3-diol system from aldol derivatives is via hydrosilylation in the presence of a Lewis acid.[13]

OMe O O O R R Et_3SiH $SnCl_4$ CH_2Cl_2 O O R R

dr > 200:1

Rhodium-catalyzed hydrosilylation of ketones that uses a bowl-shaped phosphine ligand has been described.[14]

Chemoselective reduction by role reversal of the M-O π-bond has been designed (e.g., using a Re(IV)-dioxo complex as catalyst).[15] The [M = O] bond is reduced in situ to generate a metal hydride for the reduction. Usually the metal-oxo species are used in oxidation, now their stability in the air is a great advantage in dealing with reducing events.

2-Aminopyrimidines are reduced to semicyclic guanidines.[16] A conjugated double bond connected to the ring system, when present, is also reduced while disulfide linkage is not affected. Reductive hydrazination of alkenes occurs when the latter compounds are mixed with azodicarboxylic esters, a hydrosilane and a Co-salen type complex.[17]

Hydrosilylation of 1-alkynes leads to alkenylsilanes. Water and air-stable transition metal catalysts,[18] and radical initiators[19] have been used for the purpose. *syn*-Hydrosilylation of internal alkynes is catalyzed by titanocene.[20]

$+ (Me_3Si)_3SiH$ — CO / AIBN / PhH, 85° → $(Me_3Si)_3Si$ … 72%

1-Cyclopropyl-1,6-heptadienes add hydrosilanes at the termini while suffering cleavage of the three-membered ring and bridging the internal sp^2 carbon atoms to form a cyclopentane.[21]

MeOOC, MeOOC — Et_3SiH / (phen)Pd(Me)Cl / $NaB[C_6H_3(CF_3)_2]$ / CH_2Cl_2, 0° → MeOOC, MeOOC, $SiEt_3$ 69%

Many alkenylsilanes do not survive conditions for cleavage of silyl ethers, therefore their presence is in jeopardy if such manipulations are necessary before their turn of transformation (e.g., coupling reactions). Benzyldimethylsilyl group attached to a double bond has no such limitation and it is also stable to mild acid and strong alkali. 2-Benzyldimethylsilyl-1-alkenes are readily prepared from a Ru-catalyzed hydrosilylation.[22]

Reductive etherification. Carbonyl compounds RR'C = O are converted into ethers RR'CHOR" by reaction with $R''OSiMe_3$ and Et_3SiH which is catalyzed by $Cu(OTf)_2$.[23]

Hydrogenolysis. Cleavage of trityl ethers with Et_3SiH in the presence of R_3SiOTf is very rapid at room temperature.[24]

[1]Kim, J.-G., Cho, D.H., Jang, D.O. *TL* **45**, 3031 (2004).
[2]Balakumar, V., Aravind, A., Baskaran, S. *SL* 647 (2004).
[3]Terauchi, M., Abe, H., Matsuda, A., Shuto, S. *OL* **6**, 3751 (2004).
[4]Miura, K., Yamada, Y., Tomita, M., Hosomi, A. *SL* 1985 (2004).
[5]Shibata, I., Kato, H., Ishida, T., Yasuda, M., Baba, A. *ACIEE* **43**, 711 (2004).
[6]Rudolph, A.C., Machauer, R., Martin, S.F. *TL* **45**, 4895 (2004).
[7]Bajracharya, G.B., Nogami, T., Jin, T., Matsuda, K., Gevorgyan, V., Yamamoto, Y. *S* 308 (2004).

[8]Mirza-Aghayan, M., Boukherroub, R., Bolourtchian, M., Hoseini, M., Tabar-Hydar, K. *JOMC* **678**, 1 (2003).
[9]Mirza-Aghayan, M., Boukherroub, R., Bolourtchian, M., Hoseini, M. *TL* **44**, 4579 (2003).
[10]Chiu, P., Li, Z., Fung, K.C.M. *TL* **44**, 455 (2003).
[11]Lipshultz, B.H., Caires, C.C., Kuipers, P., Chrisman, W. *OL* **5**, 3085 (2003).
[12]Kaur, H., Zinn, F.K., Stevens, E.D., Nolan, S.P. *OM* **23**, 1157 (2004).
[13]Cullen, A.J., Sammakia, T. *OL* **6**, 3143 (2004).
[14]Niyomura, O., Tokunaga, M., Obora, Y., Iwasawa, T., Tsuji, Y. *ACIEE* **42**, 1287 (2003).
[15]Kennedy-Smith, J.J., Nolin, K.A., Gunterman, H.P., Toste, F.D. *JACS* **125**, 4056 (2003).
[16]Baskaran, S., Hanan, E., Byun, D., Shen, W. *TL* **45**, 2107 (2004).
[17]Waser, J., Carreira, E.M. *JACS* **126**, 5676 (2004).
[18]Wu, W., Li, C.-J. *CC* 1668 (2003).
[19]Tojino, M., Otsuka, N., Fukuyama, T., Matsubara, H., Schiesser, C.H., Kuriyama, H., Miyazato, H., Minakata, S., Komatsu, M., Ryu, I. *OBC* **1**, 4262 (2003).
[20]Takahashi, T., Bao, F., Gao, G., Ogasawara, M. *OL* **5**, 3479 (2003).
[21]Wang, X., Stankovich, S.Z., Widenhoefer, R.A. *OM* **21**, 901 (2002).
[22]Trost, B.M., Machacek, M.R., Ball, Z.T. *OL* **5**, 1895 (2003).
[23]Yang, W.-C., Lu, X.-A., Kulkarni, S.S., Hung, S.C. *TL* **44**, 7837 (2003).
[24]Imagawa, H., Tsuchihashi, T., Singh, R.K., Yamamoto, H., Sugihara, T., Nishizawa, M. *OL* **5**, 153 (2003).

***N*-Hydroxyphthalimide, NHPI.** **22**, 233–234

Nitrosation.[1] The importance of cycloalkanone oximes in the nylon industry makes the availability of such compounds critical. One way to provide nitrosocyclohexane (and homologues) is by direct functionalization of the cycloalkane with *t*-BuONO in the presence of NHPI.

Hydroxysilylation.[2] Addition of silyl and hydroxy groups to alkenes such as methyl acrylate from Et_3SiH and O_2 under the influence of NHPI is of obvious synthetic value. The products are intermediates of the Peterson alkenation.

Hydroacylation.[3] The Stetter-type reaction performed by a free radical pathway is realized when aldehydes and electron-deficient alkenes (acrylonitrile, acrylic esters...) are treated with Bz_2O_2 and NHPI.

[1]Hirabayashi, T., Sakaguchi, S., Ishii, Y. *ACIEE* **43**, 1120 (2004).
[2]Tayama, O., Iwahama, T., Sakaguchi, S., Ishii, Y. *EJOC* 2286 (2003).
[3]Tsujimoto, S., Sakaguchi, S., Ishii, Y. *TL* **44**, 5601 (2003).

Hydroxy(tosyloxy)iodobenzene. **14**, 179–180; **16**, 179; **17**, 150; **18**, 187; **19**, 170; **20**, 193; **21**, 223; **22**, 235

Oxidation.[1] The title reagent oxidizes benzylic alcohols under microwave irradiation.

Cyclization.[2] 2,1-Benzothiazine 2,2-dioxides are formed when 2-phenylethanesulfonamide derivatives are treated with the title reagent.

[1]Lee, J.C., Lee, J.Y., Lee, S.J. *TL* **45**, 4939 (2004).
[2]Misu, Y., Togo, H. *OBC* **1**, 1342 (2003).

Hypofluorous acid- acetonitrile.

Epoxidation. Efficient and general method for epoxidization of highly substituted enones[1] uses the HOF-MeCN reagent which is prepared from fluorine, water and MeCN. The usefulness of this reagent is demonstrated by the preparation of unusual epoxides, e.g., tropone give the trisepoxide in 82% yield.[2]

N-Oxidation. Amines[3] and azides[4] are converted into nitroalkanes. The rather rare 1,2-dinitroalkanes are readily made by this method.

[1]Rozen, S., Golan, E. *EJOC* 1915 (2003).
[2]Golan, E., Hagooly, A., Rozen, S. *TL* **45**, 3397 (2004).
[3]Golan, E., Rozen, S. *JOC* **68**, 9170 (2003).
[4]Rozen, S., Carmeli, M. *JACS* **125**, 8118 (2003).

Hypophosphorous acid - iodine. **21**, 223; **22**, 235

Dealkylation. The C-N bond of several types of amines are cleaved in hot HOAc.[1] Besides dealkylation of *N*-benzylamines, nucleosides are also degraded.

N−Bn $\xrightarrow[\text{HOAc } 110°]{H_3PO_2\ I_2}$ N−H

68%

[1]Meng, G., He, Y.-P., Chen, F.-E. *SC* **33**, 2593 (2003).

I

Indium. 14, 81; **16**, 181–182; **18**, 189; **19**, 171–173; **20**, 194–197; **21**, 224–227; **22**, 236–238

Reductions. Reduction of hydroxylamines to amines can be carried out with stoichiometric or catalytic amounts of indium, the latter version involves Zn or Al.[1] For deoxygenation of amine oxides[2] the In-$TiCl_4$ combination in THF is effective.

Allylation and propargylation. Vinyl epoxide gives 1,6-heptadien-4-ol in the In-induced allylation because of a precedent isomerization of the substrate.[3] Mucohalic acids react at the γ-position.[4]

Double allylation of dialdehydes with two different allyl bromides in one pot has been realized.[5] The stoppage at the monoallylation state is of synthetic value (Cf. difficulty for using Grignard reagents).

Vinylogous formamides such as 4-formylpyrazoles and 5-formyluracils deliver dienes[6] on dehydration subsequent to allylation, in the presence of $BF_3 \cdot OEt_2$.

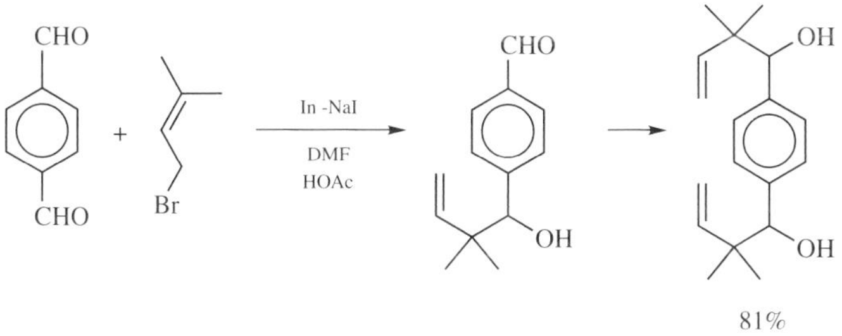

CHO CHO + Br —In -NaI / DMF HOAc→ CHO OH → OH OH

81%

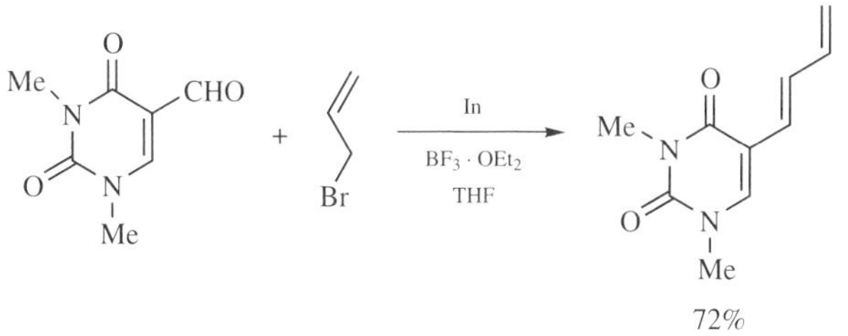

Me N O CHO O N Me + Br —In / $BF_3 \cdot OEt_2$ THF→ Me N O O N Me

72%

Using indium as catalyst the Barbier allylation of aldehydes with Mn-Me_3SiCl is achieved.[7] 2-Substituted crotyl bromides [both (Z)- and (*E*)-isomers] and aldehydes react in the presence of indium and $La(OTf)_3$ to afford predominantly *syn*-isomers.[8]

Fiesers' Reagents for Organic Synthesis, Volume 23. Edited by Tse-Lok Ho

Allylation of chiral hydrazones shows high diastereoselectivity.[9] (Note that indium acts as an initiator for alkyl radical addition to hydrazones in aqueous media.[10]) Activation by ClCOOPh enables quinoline and isoquinoline susceptible to attack by allylindium reagents.[11]

3-Ethenylidenechroman-4-ols are readily prepared by an intramolecular allenylation,[12] while the coupling method that gives rise to 2-substituted allylmetal species expands the scope of the reaction.[13]

Br CHO O In HOAc, DMF O OH 92%

PhCHO + =•= + PhI $Pd(OAc)_2$ In, 2-Fu_3P piperidine/DMF OH Ph Ph 83%

From a propargyl bromide the indium-mediated reaction with aldehydes can give rise to either allenyl carbinols or homopropargylic alcohols, the difference being that in the latter case InF_3 is present.[14] Interestingly, alkynyl ketones are obtained from reaction of aldehydes with 1-iodoalkynes, due to hydride transfer from the indium propargyloxides to aldehydes.[15] Very high diastereoselection (formation of *anti*-diols) is observed in the indium-mediated reaction of 1,4-dibromo-2-butyne with aldehydes in the presence of ZnF_2.[16]

A formal 1,4-addition of allyl (propargyl, allenyl) group to enones involves formation of siloxyallyl sulfonium salts and subsequent reaction with allylindiums.[17]

For sythesis of homoallylic amines by two consecutive reactions involving allylic and alkenylindium species the first for addition to imines and then Pd-catalyzed coupling.[18]

Ph NR + R′ Br I In $(Ph_3P)_4Pd$ LiCl, R″X Ph RNH R′ R″

Amides.[19] Mediated by indium the reaction of RCOCl with amines is facile. The method is suitable for peptide synthesis to avoid racemization.

(E)-3-Alkylidene oxindoles.[20] Reductive cyclization of *N*-alkynoyl 2-iodoarylamines is achieved by indium and Py-HBr_3.

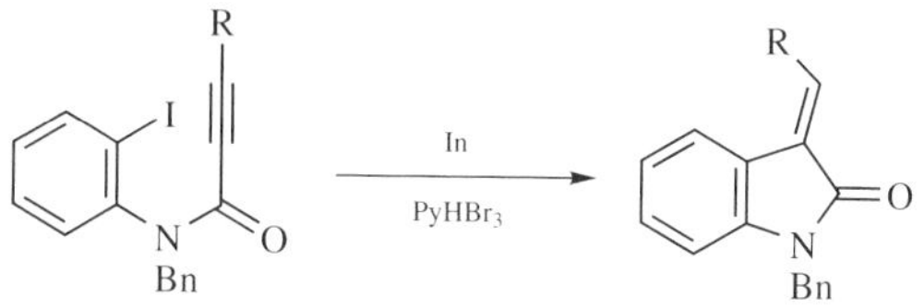

[1]Cicchi, S., Bonanni, M., Cardona, F., Revuelta, J., Goti, A. *OL* **5**, 1773 (2003).
[2]Yoo, B.W., Choi, K.-H., Choi, K.I., Kim, J.-H. *SC* **33**, 4185 (2003).
[3]Oh, B.K., Cha, J.H., Cho, Y.S., Choi, K.I., Koh, H.Y., Chang, M.H., Pae, A.N. *TL* **44**, 2911 (2003).
[4]Zhang, J., Blazecka, P.G., Berven, H., Belmont, D. *TL* **44**, 5579 (2003).
[5]Babu, S.A., Yasuda, M., Shibata, I., Baba, A. *SL* 1223 (2004).
[6]Kumar, V., Chimni, S.S., Kumar, S. *TL* **45**, 3409 (2004).
[7]Auge, J., Lubin-Germain, N., Marque, S., Seghrouchni, L. *JOMC* **679**, 79 (2003).
[8]Loh, T.-P., Yin, Z., Song, H.-Y., Tan, K.-L. *TL* **44**, 911 (2003).
[9]Cook, G.R., Maity, B.C., Kargbo, R. *OL* **6**, 1741 (2004).
[10]Miyabe, H., Ueda, M., Nishimura, A., Naito, T. *T* **60**, 4227 (2004).
[11]Lee, S.H., Park, Y.S., Nam, M.H., Yoon, C.M. *OBC* **2**, 2170 (2004).
[12]Kang, H.-Y., Kim, Y.-T., Yu, Y.-K., Cha, J.H., Cho, Y.S., Koh, H.Y. *SL* 45 (2004).
[13]Cleghorn, L.A.T., Cooper, I.R., Grigg, R., MacLachlan, W.S., Sridharan, V. *TL* **44**, 7969 (2003).
[14]Lin, M.-J., Loh, T.-P. *JACS* **125**, 13042 (2003).
[15]Auge, J., Lubin-Germain, N., Seghrouchni, L. *TL* **44**, 819 (2003).
[16]Miao, W., Lu, W., Chan, T.H. *JACS* **125**, 2412 (2003).
[17]Lee, K., Kim, H., Miura, T., Kiyota, K., Kusama, H., Kim, S., Iwasawa, N. *JACS* **125**, 9682 (2003).
[18]Hirashita, T., Hayashi, Y., Mitsui, K., Araki, S. *JOC* **68**, 1309 (2003).
[19]Cho, D.H., Jang, D.O. *TL* **45**, 2285 (2004).
[20]Yanada, R., Obika, S., Oyama, M., Takemoto, Y. *OL* **6**, 2825 (2004).

Indium - Indium(III) chloride. 21, 227; 22, 239

Allylation. Enones undergo 1,2-addition in DMF with allyl iodide when mediated by In/$InCl_3$.[1] Dihydropyran and dihydrofuran afford ω-alkenediols when they are mixed with In/$InCl_3$ and allyl bromide in water.[2]

Allyl alcohol and a number of derivatives (OAc, OCOOMe) and analogues (SO_2Ph) can be used as allyl source when $(Ph_3P)_4Pd$ is present.[3]

Elimination. Halohydrins are defunctionalized to give alkenes on treatment with In/$InCl_3$ and Ph_3P.[4]

[1]Kim, H.Y., Choi, K.I., Pae, A.N., Koh, H.Y., Choi, J.H., Cho, Y.S. *SC* **33**, 1899 (2003).
[2]Juan, S., Hua, Z.-H., Qi, S., Ji, S.-J., Loh, T.-P. *SL* 829 (2004).
[3]Jang, T.-S., Keum, G., Kang, S.B., Chung, B.Y., Kim, Y. *S* 775 (2003).
[4]Cho, S., Kang, S., Keum, G., Kang, S.B., Han, S.-Y., Kim, Y. *JOC* **68**, 180 (2003).

Indium(I) bromide. 21, 228; **22**, 239

Reductive coupling. α,α-Dichloro ketones undergo coupling that results in 1,4-diketones. In a very few cases $RCOCH_2CH_2COR'$ were made.[1]

[1]Peppe, C., Pavao das Chagas, R. *SL* 1187 (2004).

Indium(III) bromide. 22, 239

Aldehyde-acylal interconversion. Both acetylation[1] of RCHO and hydrolysis[2] of the derivatives are catalyzed by $InBr_3$. The former reaction is carried out in neat and the latter, in water.

Epoxide opening. The reaction of epoxides with $ArNH_2$ is promoted by $InBr_3$.[3]

Styrene dimerization. The catalyzed reaction is temperature dependent.[4]

Ph + $InBr_3$ → (CH_2Cl_2, 0°) Ph Ph 90%

Ph + $InBr_3$ → (CH_2Cl_2, r.t.) Ph 88%

Ph Ph + $InBr_3$ → (CH_2Cl_2, r.t.) Ph Ph Ph 72%

Flavones. 2′-Hydroxychalcones and flavanones are rapidly oxidized to flavones with $InBr_3$-SiO_2 in the neat.[5]

Alkynes. Different kinds of alkynes have been prepared or modified by means of $InBr_3$–catalyzed processes. Negishi-type coupling using alkynylindium species[6] (instead of alkynylzincs) is feasible.

Homopropargylic and propargylic alcohols are formed, using propargylic bromides and alkynes in combination with $InBr_3$ as promoter.[7] But Et_3N is required in the latter reaction that is also used to prepare propargylic amines from *N,O*-acetals.[8]

Reductive aldol reaction between enones and aldehydes is catalyzed by $InBr_3$. Et_3SiH provides hydride (perhaps to form InH_3) and $[Et_3Si]^+$ to silylate the products.[9]

[1]Yin, L., Zhang, Z.-H., Wang, Y.-M., Pang, M.-L. *SL* 1727 (2004).
[2]Zhang, Z.-H., Yin, L., Li, Y., Wang, Y.-M. *TL* **46**, 889 (2005).
[3]Rodriguez, J.R., Navarro, A. *TL* **45**, 7495 (2004).
[4]Peppe, C., Lang, E.S., de Andrade, F.M., de Castro, L.B. *SL* 1723 (2004).

[5]Ahmed, N., Ali, H., van Lier, J.E. *TL* **46**, 253 (2005).
[6]Sakai, N., Annaka, K., Konakahara, T. *OL* **6**, 1527 (2004).
[7]Lin, M.-J., Loh, T.-P. *JACS* **125**, 13042 (2003).
[8]Sakai, N., Hirasawa, M., Konakahara, T. *TL* **44**, 4171 (2003).
[9]Shibata, I., Kato, H., Ishida, T., Yasuda, M., Baba, A. *ACIEE* **43**, 711 (2004).

Indium(I) chloride.

Barbier reaction. Organozinc reagents derived from ordinary alkyl halides are apparently involved in a reaction with aldehydes in water which is promoted by InCl and CuI.

[1]Keh, C.C.K., Wei, C., Li, C.-J. *JACS* **125**, 4062 (2003).

Indium(III) chloride. 19, 173–174; 20, 197–198; 21, 228–230; 22, 240–242

Acylation. Acylation of ROH, ArOH, RSH, RR'NH is rapid using $InCl_3$ as catalyst.

Sulfonylation. Diaryl sulfones are prepared in TFA by sulfonylation of arenes at moderate temperatures in the presence of $InCl_3$ –TfOH.[2]

Substitutions. Direct displacement of a hydroxyl group of stabilized secondary alcohols (e.g., benzylic) and tertiary alcohols by an allyl residue from allyltrimethyl-silane is accomplished using $InCl_3$ as catalyst.[3] Alkynyl and allenyl group can also be introduced by using the proper silanes. An alternative method[4] employs benzil as a cocatalyst and $HSiMc_2Cl$ as the chlorine source.

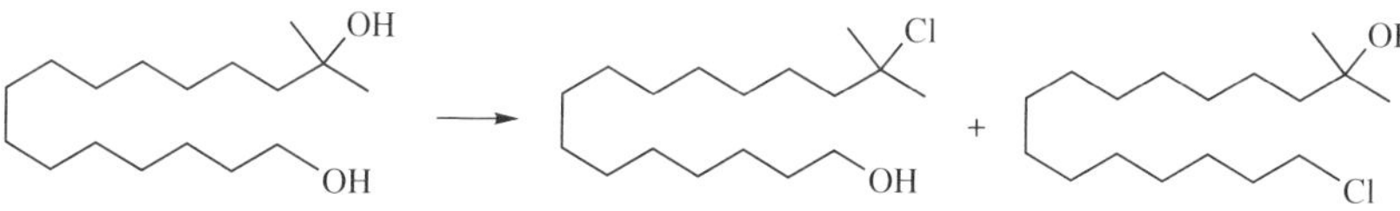

reagents:	(Cl / OH product)	(OH / Cl product)
$HSiMe_2Cl$, $InCl_3$ benzil	56%	–
Ph_3P CCl_4	–	80%

Aldol reaction. Partial hydrolysis and aldol reaction take place in sequence for silyl enol ethers of aryl methyl ketones (but no others) when they are exposed to $InCl_3$ and just enough water absorbed by the catalyst from moist air.[5]

Cross aldolization catalyzed by $InCl_3$ shows high degree of *syn*- diasteroselectivity.[6]

Combination of unsaturated ketones with ArCHO by a coupling reaction is thought to involve formation of oxycyclopropyl radicals from the former components and addition to

ArCHO is followed by ring cleavage.[7]

PhCHO + [enone] $\xrightarrow[\text{Me}_3\text{SiCl, THF}]{\text{InCl}_3\text{, Al}}$ Ph[...]

56%

Heterocycles. The catalyzed reaction of conjugated oximes with β-dicarbonyl compounds leads to substituted pyridines.[8] *trans*-2,6-Disubstituted tetrahydropyridines are assembled from silylated homoallylic amines and aldehydes by an aza-silyl-Prins reaction.[9] The excellent diastereoselectivity of the Prins cyclization is exploitable for a formal synthesis of ratjadone.[10]

$SiMe_3$ + $OHC–C_5H_{11}$ $\xrightarrow[\text{MeCN}]{\text{InCl}_3}$ C_5H_{11}

NH, Bn; N, Bn

70%

C_8H_{17}, N, O, HO, OTBS + CHO $\xrightarrow[\text{CH}_2\text{Cl}_2\text{, NH}_4\text{Cl aq.}]{\text{InCl}_3}$ O, H, H, O, TBSO

An allylchlorosilane provides the 3-carbon unit of *cis*-2,6-disubstituted 4-chlorotetrahydropyrans in the condensation with aldehydes.[11]

α-Diazo carbonyl compounds and alkynes participate in the 1,3-dipolar cycloaddition, giving pyrazole derivatives.[12]

Coupling reactions. A method for synthesizing 2-aryldihydropyrans involves conversion of dihydropyran-2-yllithium into the indium counterpart and a subsequent Pd-catalyzed coupling with ArX.[13]

An intramolecular reaction between an allylstannane moiety and an alkyne unit gives rise to bicyclic dienes.[14]

MeOOC, MeOOC, $CHCH_2SnBu_3$ $\xrightarrow[\text{MeCN}]{\text{InCl}_3}$ MeOOC, MeOOC

96%

1,2-Alkadien-4-ols are converted into dichloroindium 1,3-alkadien-2-oxides which can be used to condense with carbonyl compounds.[15]

HO Bu + PhCHO $\xrightarrow[CH_2Cl_2]{InCl_3}$ Bu O OH Ph

73%

Ph Ph OH + $SiMe_2Cl$ $\xrightarrow{InCl_3/CH_2Cl_2}$ Ph Ph

80%

[1]Chakraborti, A.K., Gulhane, R. *TL* **44**, 6749 (2003).
[2]Garzya, V., Forbes, I.T., Lauru, S., Maragni, P. *TL* **45**, 1499 (2004).
[3]Yasuda, M., Saito, T., Ueba, M., Baba, A. *ACIEE* **43**, 1414 (2004).
[4]Yasuda, M., Yamasaki, S., Onishi, Y., Baba, A. *JACS* **126**, 7186 (2004).
[5]Chancharunee, S., Perlmutter, P., Statton, M. *TL* **44**, 5683 (2003).
[6]Munoz-Muniz, O., Quintanar-Audelo, M., Juaristi, E. *JOC* **68**, 1622 (2003).
[7]Ohe, T., Ohse, T., Mori, K., Ohtaka, S., Uemura, S. *BCSJ* **76**, 1823 (2003).
[8]Saikia, P., Prajapati, D., Sandhu, J.S. *TL* **44**, 8725 (2003).
[9]Dobbs, A.P., Guesne, S.J.J., Hursthouse, M.B., Coles, S.J. *SL* 1740 (2003).
[10]Cossey, K., Funk, R. *JACS* **126**, 12216 (2004).
[11]Chan, K-P., Loh, T.-P. *TL* **45**, 8387 (2004).
[12]Jiang, N., Li, C.-J. *CC* 394 (2004).
[13]Lehmann, U., Awasthi, S., Minehan, T. *OL* **5**, 2405 (2003).
[14]Miura, K., Fujisawa, N., Hosomi, A. *JOC* **69**, 2427 (2004).
[15]Yu, C.-M., Kim, Y.-M., Kim, J.-M. *SL* 1518 (2003).

Indium(I) iodide. 21, 231; **22**, 243

Allylation. In the crotylation and cinnamylation of aldehydes, diastereomeric product distributions are opposite from the InI - $(Ph_3P)_4Pd$ and $InCl_3$ - Al -$(Ph_3P)_4Pd$ combinations.The reason is that with the latter $AlCl_3$ is formed and it prevents internal chelation.

OH CHO + Ph OAc $\xrightarrow[DMF]{(Ph_3P)_4Pd}$ OH Ph OH + OH Ph OH

InI	100%	96	4
$InCl_3$ Al	92%	2	98

Addition of allyl and benzyl groups to conjugated nitriles proceeds in the 1,2- fashion.[2]

Organochalcogenides. Reduction of PhXXPh (X = S, Se, Te) in the presence of electrophiles gives substitution or addition products: e.g., with RX and RCOCl,[3] enones,[4] 2-alkyn-1-ols.[5]

[1]Hirashita, T., Kamei, T., Satake, M., Horie, T., Shimizu, H., Araki, S. *OBC* **1**, 3799 (2003).
[2]Ranu, B.C., Mandal, T. *TL* **45**, 6875 (2004).
[3]Ranu, B.C., Mandal, T. *JOC* **69**, 5793 (2004).
[4]Ranu, B.C., Mandal, T. *SL* 1239 (2004).
[5]do Rego Barros, O.S., Lang, E.S., Peppe, C., Zeni, G. *SL* 1725 (2003).

Indium(III) triflate. 21, 231; **22**, 243–244

C-Glycosylation. Glycal derivatives (e.g., acetates) enter into S_N2' displacement by the In-catalyzed reaction with allylsilanes.[1]

Friedel-Crafts alkylation. Methyl trifluoropyruvate reacts with arenes in water in the presence of $In(OTf)_3$.[2] The title reagent is also an effective co-catalyst for the Pd-catalyzed dimerization (self-alkylation) of styrenes.[3]

Solvent-free conditions for acylation to prepare benzophenones include microwave irradiation and $In(OTf)_3$ catalysis.[4]

Alkenylation. Under the influence of $In(OTf)_3$ β-keto esters react with 1-alkynes regioselectively (bond formation with C-2 of the alkynes, except for silylalkynes).[5]

O
COOEt
In(OTf)3
Si - Ph
O
COOEt
94%
O
COOEt
Si−Ph
80%

[1]Ghosh, R., Chakraborty, A., Maiti, D.K. *SC* **33**, 1623 (2003).
[2]Ding, R., Zhang, H.B., Chen, Y.J., Liu, L., Wang, D., Li, C.-J. *SL* 555 (2004).
[3]Tsichimoto, T., Kamiyama, S., Negoro, R., Shirakawa, E., Kawakami, Y. *CC* 852 (2003).
[4]Koshima, H., Kubota, M. *SC* **33**, 3983 (2003).
[5]Nakamura, M., Endo, K., Nakamura, E. *JACS* **125**, 13002 (2003).

Iodine. **13**, 148–149; **14**, 181–182; **15**, 172–173; **16**, 182; **18**, 189–191; **19**, 174–175; **20**, 199–200; **21**, 231–232; **22**, 245–248

Protection and deprotection of functional groups. Iodine can mediate acetylation[1] and transesterification of β-keto esters.[2] Methods for deprotection of aryl acetates,[3] prenyl carbamates,[4] and acetals[5] involve iodine. More valuable are the reports of selective removal of MOM esters and an ortho MOM group from the corresponding 2-acylaryl ethers (without affecting a MOM ether at another position)[6] and cleavage of *N*-Boc substituents.[7]

Iodination. Iodine in combination with Et_3N is useful for iodination of enaminones.[8] For synthesis of certain iodinated naphthoquinones the morpholine-iodine complex finds use.[9]

To promote aromatic iodination, butyltriphenylphosphonium peroxodisulfate shows activity.[10] Decarboxylative iodination occurs when α-dithiomethylene-β-keto acids are treated with iodine.[11]

Cyclization. Several types of benzannulated heterocycles result when proper *o*-alkynylaryl derivatives are treated with iodine. Thus, 3-iodoindoles[12,13] and 4-iodoisochromenes[14] are prepared.

ω-Hydroxy carboxylic acids are activated for lactonization by reaction with di-2-thienyl carbonate and the cyclization step is accomplished on heating with iodine in toluene.[15] Hydroxychalcones cyclize to flavones on heating with iodine in triethyleneglycol.[16]

Functionalization of allylic and homoallylic alcohols by way of iodosulfation is preceded by treatment with Py-SO_3.[17]

Alkylation. Iodine catalyzes addition of indole to enones[18] and condensation with aldehydes[19] at room temperature. Protected homoallylic amines are formed when aldehydes, allylsilanes and *N*-protected amines (e.g., *N*-Boc derivatives) are mixed at room temperature with iodine.[20]

Addition. Formation of iodohydrins from allenes which bear a chalcogen group occurs at the double bond away from that substituent, and the products contain *cis*-related chalcogen and iodine atom.[21] When MeCN is added to the reaction milieu, acetamido derivatives are obtained.[22]

Oxidation. Conversion of alcohols to 2,2,2-trifluoroethyl esters[23] by iodine and K_2CO_3 in CF_3CH_2OH and of aldehydes to α-hydroxy acetals[24] by the I_2-KOH system in ROH are useful transformations.

Aldoses are oxidized by iodine in aq. NH_3. Only the aldehyde group is transformed into a primary amide.[25] The shock-sensitive nitrogen triiodide may be formed in situ therefore excessive reagent should be avoided.

A preparation of 1,2-diamine derivatives is based on oxidative coupling of organocuprates by iodine.[26] Advantage is taken in the access of such organocuprates from in situ *N*-protected amines via α-lithiation and metal exchange.

Oxidative dimerization of ester enolates is a known reaction, but the application to a ring closure is noteworthy.[27]

E E E E

LDA - I_2

THF -78°

E E E E

E=COOMe 58%

Aldol and Diels-Alder reactions. The use of iodine in the Mukaiyama version of aldol reaction[28] and the Diels-Alder reaction involving acetals of conjugated carbonyl compounds[29] has been reported.

Mannosylation.[30] β-Mannosides are formed with high stereoselectivity when mannosyl sulfoxides are treated with iodine and ROH in CH_2Cl_2.

[1]Phukan, P. *TL* **45**, 4785 (2004).
[2]Chavan, S.P., Kale, R.R., Shivasankar, K., Chandake, S.I., Benjamin, S.B. *S* 2695 (2003).
[3]Das, B., Banerjee, J., Ramu, R., Pal, R., Ravindranath, N., Ramwsh, C. *TL* **44**, 5465 (2003).
[4]Vatele, J.-M. *TL* **44**, 9127 (2003).
[5]Sun, J., Dong, Y., Cao, L., Wang, X., Wang, S., Hu, Y. *JOC* **69**, 8932 (2004).
[6]Keith, J.M. *TL* **45**, 2739 (2004).
[7]Nadia, K., Malika, B., Nawel, K., MedYazid, B., Zine, R., Aouf, N.-E. *JHC* **41**, 57 (2004).
[8]Kim, J.M., Na, J.E., Kim, J.N. *TL* **44**, 6317 (2003).
[9]Perez, A.L., Lamoureux, G., Herrera, A. *SC* **34**, 3389 (2004).
[10]Tajik, H., Esmaeili, A.A., Mohammadpoor-Baltork, I., Eshadi, A., Tajmehri, H. *SC* **33**, 1319 (2003).
[11]Zhao, Y.-L., Liu, Q., Sun, R., Zhang, Q., Xu, X.-X. *SC* **34**, 463 (2004).
[12]Amjad, M., Knight, D.W. *TL* **45**, 539 (2004).
[13]Yue, D., Larock, R.C. *OL* **6**, 1037 (2004).
[14]Yue, D., Della Ca, N., Larock, R.C. *OL* **6**, 1581 (2004).
[15]Oohashi, Y., Fukumoto, K., Mukaiyama, T. *CL* **34**, 72 (2005).
[16]Miyake, H., Takizawa, E., Sasaki, M. *BCSJ* **76**, 835 (2003).
[17]Inoue, M., Motomatsu, S., Nakada, M. *SC* **33**, 2857 (2003).
[18]Wang, S.-Y., Ji, S.-J., Loh, T.-P. *SL* 2377 (2003).
[19]Ji, S.-J., Wang, S.-Y., Zhang, Y., Loh, T.-P. *T* **60**, 2051 (2004).
[20]Phukan, P. *JOC* **69**, 4005 (2004).
[21]Ma, S., Hao, X., Meng, X., Huang, X. *JOC* **69**, 5720 (2004).
[22]Ma, S., Hao, X., Huang, X. *CC* 1082 (2003).
[23]Mori, N., Togo, H. *SL* 880 (2004).
[24]Zacuto, M.J., Cai, D. *TL* **46**, 447 (2005).
[25]Chen, M.-Y., Hsu, J.-L., Shie, J.-J., Fang, J.-M. *JCCS(T)* **50**, 129 (2003).
[26]Dieter, R.K., Li, S., Chen, N. *JOC* **69**, 2867 (2004).
[27]Ayats, C., Camps, P., Duque, M.D., Font-Bardia, M., Munoz, M.R., Solans, X., Vazquez, S. *JOC* **68**, 8715 (2003).
[28]Phukan, P. *SC* **34**, 1045 (2004).
[29]Chavan, S.P., Sharma, P., Krishna, G.R., Thakkar, M. *TL* **44**, 3001 (2003).
[30]Marsh, S.J., Kartha, K.P.R., Field, R.A. *SL* 1376 (2003).

Iodine(I) azide. 22, 248

Azyl azides.[1] By a free radical process aldehydes are converted into $RCON_3$ by IN_3 in MeCN. At higher reaction temperatures carbamoyl azides are formed.

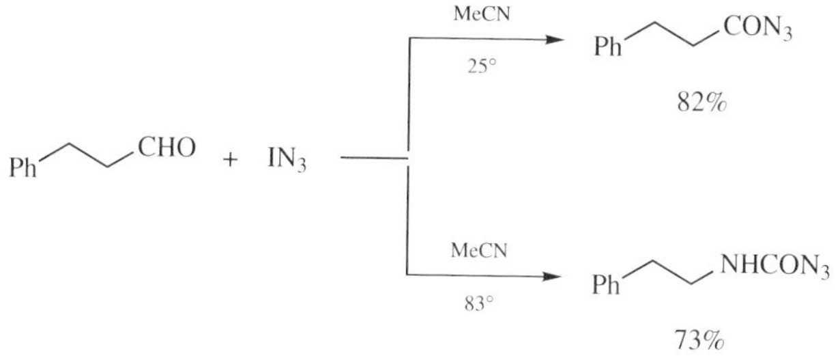

Ph–CH2CH2–CHO + IN_3 → (MeCN, 25°) Ph–CH2CH2–CON_3 82%; (MeCN, 83°) Ph–CH2CH2–$NHCON_3$ 73%

[1]Marinescu, L., Thinggaard, J., Thomsen, I.B., Bols, M. *JOC* **68**, 9453 (2003).

Iodine(I) chloride.

Iododesilylation.[1] In the context of a dendrimer synthesis, silylarenes are converted by ICl into the corresponding iodoarenes at 0°. The solvent system contains CH_2Cl_2 and Et_2O in a 10:1 ratio, with the ether to suppress side reactions.

Cyclization.[2] A convenient synthesis of phenanthrenes and related compounds from 2-alkynylbiaryls involves reaction with ICl.

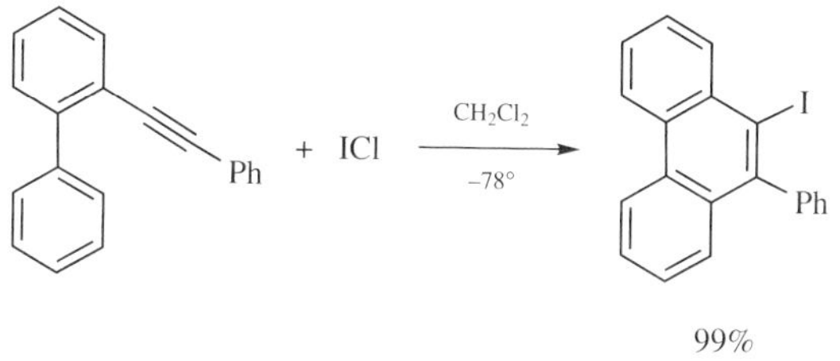

2-(phenylethynyl)biphenyl + ICl → (CH_2Cl_2, –78°) 9-iodo-10-phenylphenanthrene 99%

[1]Bo, Z., Qiu, J., Li, J., Schluter, A.D. *OL* **6**, 667 (2004).
[2]Yao, T., Campo, M.A., Larock, R.C. *OL* **6**, 2677 (2004).

Iodine pentafluoride. 22, 249

Fluorination. Fluorination of alkyl aryl sulfides with IF_5 involves arylthio group migration.[1]

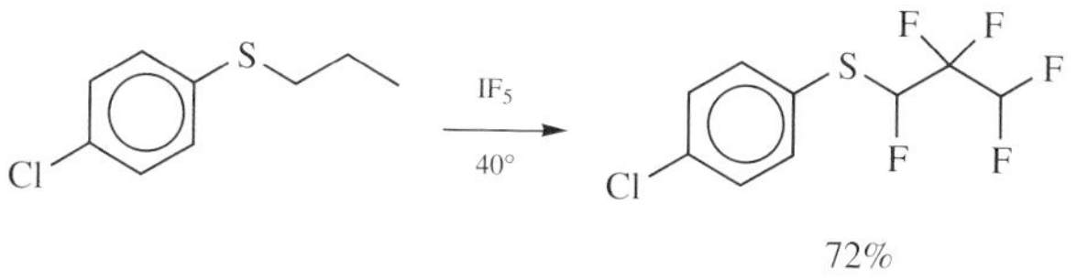

4-ClC6H4–S–CH2CH2CH3 → (IF_5, 40°) 4-ClC6H4–S–CHF–CF2–CHF2 72%

[1]Ayuba, S., Fukuhara, T., Hara, S. *OL* **5**, 2873 (2003).

B-Iodo-9-borabicyclo[3.3.1]nonane.

Demethylation.[1] Cleavage of ArOMe by the title reagent is compatible with compounds containing a double bond, while there is danger of bromoboration in using BBr_3. Another advantage is the ease in workup as addition of ethanolamine precipitates 9-BBN.

[1]Fürstner, A., Castanet, A.-S., Lehmann, C.W. *JOC* **68**, 1521 (2003).

***N*-Iodosaccharin.**

Iodination. Rapid iodination of enol acetates and β-diketones is accomplished using *N*-iodosaccharin as reagent.

[1]Dolenc, D. *SC* **33**, 2917 (2003).

***N*-Iodosuccinimide, NIS.** **16**, 185–186; **18**, 193–194; **19**, 177–178, **21**, 234; **22**, 249

Iodination and iodoetherification. On microwave irradiation and in the presence of TsOH, carbonyl compounds undergo α-iodination with NIS.[1]

Excellent stereocontrol for the double iodoetherification of σ-symmetric diene acetals enables a short synthesis of rubrenolide.[2]

Ph Ph Ph Ph H O O O O O O I I Me_3SiO Me_3SiO NIS H_2O MeCN

[1]Lee, J.C., Bae, Y.H. *SL* 507 (2003).
[2]Fujioka, H., Ohba, Y., Hirose, H., Murai, K., Kita, Y. *ACIEE* **44**, 734 (2005).

Iodosylbenzene. **13,** 151; **16**, 186; **18**, 194; **19**, 178; **20**, 201; **21**, 235; **22**, 249–250

Oxidations. In combination with SALEN-metal complexes the use of PhI = O to oxidize various compounds have been studied. Alcohols[1,2] and silyl ethers[3] give carbonyl products. Symmetrical 1,3- and 1,4-disilyl ethers afford optically active siloxy ketones.[3]

Sulfide to sulfoxide conversion proceeds under mild conditions [with (salen)-Cr complex as oxygen transfer agent].[4] Moderate asymmetric induction is observed in the presence of a chiral (salen)-Fe(III) complex.[5] Another oxidation system contains a triazacorrole-Mn(III) complex.[6]

Other additives exert different effects in the oxidation with PhIO. Primary alcohols form methyl esters in a system containing KBr and 0.5N HCl.[7] α-Tosyloxylation following oxidation of alcohols occurs when TsOH is present.[8]

N,S-Acetals are generated from sulfides when the oxidation products are further manipulated. Thus, oxidation of sulfolanes with pyrimidine derivatives (e.g., thymine) present leads to thiaribose analogues.[9]

[1]Kim, S.S., Kim, D.W. *SL* 1391 (2003).
[2]Kim, S.S., Borisova, G. *SC* **33**, 3961 (2003).
[3]Murahashi, S.-I., Noji, S., Hirabayashi, T., Komiya, N. *SL* 1739 (2004).
[4]Kim, S.S., Rajagopal, G. *S* 2461 (2003).
[5]Bryliakov, K.P., Talsi, E.P. *ACIEE* **43**, 5228 (2004).
[6]Wang, S.H., Mandimutsira, B.S., Todd, R., Ramdhanie, B., Fox, J.P., Goldberg, D.P. *JACS* **126**, 18 (2004).
[7]Tohma, H., Maegawa, T., Kita, Y. *SL* 723 (2003).
[8]Ueno, M., Nobaba, T., Togo, H. *JOC* **68**, 6424 (2003).
[9]Nishizono, N., Baba, R., Nakamura, C., Oda, K., Machida, M. *OBC* **1**, 3692 (2003).

***p*-Iodotoluene difluoride.** **19**, 178; **21**, 235–236; **22**, 250

Fluorination. On treatment with *p*-Tol-IF_2 and Et_3N-HF, cyclic ethers bearing an iodomethyl side chain at C-2 undergo ring expansion simultaneously with introduction of fluorine to afford fluorinated ethers.[1]

[1]Inagaki, T., Nakamura, Y., Sawaguchi, M., Yoneda, N., Ayuba, S., Hara, S. *TL* **44**, 4117 (2003).

***o*-Iodoxybenzamides.**

Oxidation.[1] The title reagents are prepared from the corresponding *o*-iodobenzamides and dimethyldioxirane. It has similar reactivity as IBX.

[1]Zhdankin, V.V., Koposov, A.Y., Netzel, B.C., Yashin, N.V., Rampel, B.P., Ferguson, M.J., Tykwinski, R.R. *ACIEE* **42**, 2194 (2003).

***o*-Iodoxybenzenesulfonamides.**

Oxidations.[1] Preliminary investigations on this class of compounds show similar oxidizing properties to IBX.

[1]Koposov, A.Y., Litvinov, D.N., Zhdankin, V.V. *TL* **45**, 2719 (2004).

***o*-Iodoxybenzoic acid, IBX.** **19**, 179; **21**, 236; **22**, 251–252

Oxidations. There are several modifications of the IBX oxidation of alcohols: addition of β-cyclodextrin,[1] in ionic liquid,[2,3] and solvent-free conditions.[4] The β-cyclodextrin protocol (to trap and align both reactants in it cavity) has also been applied to conversion of epoxides and *N*-tosylaziridines into α-ketols and α-*N*-tosylamino ketones, respectively.[5]

α-Acyloxylation of ketones (at CH_2 or CH_3) occurs on heating with IBX.[6]

KI/MeCN

60%

Transpositional oxidation of tertiary allylic alcohols to give enones[7] by IBX in DMSO is fast. The facile oxidation of amines to imines[8] and regeneration of carbonyl compounds from dithioacetals[8,9] are further uses of IBX. Note that dehydrogenation of amines is faster than oxidation of alcohols. Efficient oxidation of sulfides to sulfoxides[10] is expected.

Aromatization of dihydroquinones (e.g., Diels-Alder adducts) can be carried out with IBX in DMSO.[11]

Acylation of polymeric amines with 2-iodobenzoic acid followed by oxidation with Bu_4NSO_5H generates an IBX analogue with excellent oxidative activity.[12]

[1]Surendra, K., Krishnaveni, N.S., Reddy, M.A., Nageswa, Y.V.D., Rao, K.R. *JOC* **68**, 2058 (2003).
[2]Liu, Z., Chen, Z.-C., Zheng, Q.-G. *OL* **5**, 3321 (2003).
[3]Karthikeyan, G., Perumal, P.T. *SL* 2249 (2003).
[4]Moorthy, J.N., Singhal, N., Venkatakrishnan, P. *TL* **45**, 5419 (2004).
[5]Surendra, K., Krishnaveni, N.S., Reddy, M.A., Nageswa, Y.V.D., Rao, K.R. *JOC* **68**, 9119 (2003).
[6]Pan, Z., Liu, X., Liang, Y. *TL* **45**, 4101 (2004).
[7]Shibuya, M., Ito, S., Takahashi, M., Iwabuchi, Y. *OL* **6**, 4303 (2004).
[8]Nicolaou, K.C., Mathison, C.J.N., Montagnon, T. *ACIEE* **42**, 4077 (2003);.*JACS* **126**, 5192 (2004).
[9]Krishnaveni, N.S., Surendra, K., Nageswa, Y.V.D., Rao, K.R. *S* 2295 (2003).
[10]Shukla, V.G., Salgaonkar, P.D., Akamanchi, K.G. *JOC* **68**, 5422 (2003).
[11]Kotha, S., Mandal, B.K. *SL* 2043 (2004).
[12]Chung, W.-J., Kim, D.-K., Lee, Y.-S. *TL* **44**, 9251 (2003).

Ionic liquids. 22, 88–91

Novel species. Besides the most popular 1-butyl-3-methylimidazolium salts, several analogues in which the butyl group is replaced by other residues make their appearance during the 2003–2004 period. A polymer-supported salt is shown to facilitate nucleophilic substitution reactions, including synthesis of alkyl fluorides.[1] 2-(*N*-Butyl-*N*-methyl)imino-1,3-dimethylimidazolidine hexafluorophosphate is an ionic liquid that finds use in oxidation of alcohols by aq. NaOCl.[2]

Functionalized ionic liquids also serve as reagents. Thus, the triflate of 3-(3′-methylimidazol-1′-yl)propanesulfonic acid is active in catalyzing aromatic nitration.[3] An analogous sulfonyl chloride promotes Beckmann rearrangement.[4] [Acetalization employing 1-chlorosulfinyl-3-methylimidazolium chloride is also reported.[5]]

2-Methoxy-4-[(3′- methylimidazol-1′-yl)butoxy]benzaldehyde chloride can be used to form *N*-tosylated secondary amines via imine formation, reduction and tosylation, with final release from the ionic liquid molecule by treatment with aq. HPF_6.[6]

Also of obvious applications are the ionic liquids bearing a terminal phosphonyl group[7] and that carries a 2-(4-diacetoxyiodophenyl)ethyl substituent.[8]

A breeder-type preparation of ionic liquids involves warming 1-methylimidazole, alkyl halides and KPF_6 using the same ionic liquid as solvent.[9]

Selected applications. Immobilization of $Pd(OAc)_2$ in ionic liquid on silica can be used for the Heck reaction.[10] *N*-Pentylpyridinium tribromide is a useful brominating agent for unsaturated hydrocarbons (including arenes) and carbonyl compounds.[11]

[1]Kim, D.W., Chi, D.Y. *ACIEE* **43**, 483 (2004).
[2]Xie, H., Zhang, S., Duan, H. *TL* **45**, 2013 (2004).
[3]Qiao, K., Yokoyama, C. *CL* **33**, 808 (2004).
[4]Gui, J., Deng, Y., Hu, Z., Sun, Z. *TL* **45**, 2681 (2004).
[5]Li, D., Shi, F., Peng, J., Guo, S., Deng, Y. *JOC* **69**, 3582 (2004).
[6]de Kort, M., Tuin, A.W., Kuiper, S., Overkleeft, H.S., van der Marel, G.A., Buijsman, R.C. *TL* **45**, 2171 (2004).
[7]Mu, Z., Zhou, F., Zhang, S., Liang, Y., Liu, W. *HCA* **87**, 2549 (2004).
[8]Qian, W., Jin, E., Bao, W., Zhang, Y. *ACIEE* **44**, 952 (2005).
[9]Xu, D.-Q., Liu, B.-Y., Luo, S.-P., Xu, Z.-Y., Shen, Y.-C. *S* 2626 (2003).
[10]Hagiwara, H., Sugawara, Y., Isobe, K., Hoshi, T., Suzuki, T. *OL* **6**, 2325 (2004).
[11]Salazar, J., Dorta, R. *SL* 1318 (2004).

Iridium.

Reduction.[1] Hydrogenation of enones to allylic alcohols is accomplished in the presence of an Ir catalyst, which is prepared by calcinations of $Ir(acac)_3$ –impregnated zeolite.

[1]De Bruyn, M., Coman, S., Bota, R., Parvulescu, V.I., De Vos, D.E., Jacobs, P.A. *ACIEE* **42**, 5333 (2003).

Iridium(III) chloride.

Rearrangements. Epoxides rearrange to carbonyl compounds on heating with hydrated $IrCl_3$ in THF (scope seems limited to styrene oxides).[1] Allyl aryl ethers afford 2-substituted dihydrobenzofurans on treatment with $IrCl_3$ and AgOTf, as a result of a catalyzed Claisen rearrangement followed by cyclization.[2]

$IrCl_3$ - AgOTf

DCE, 60°

[1]Karame, I., Tommasino, M.L., Lemaire, M. *TL* **44**, 7687 (2003).
[2]Grant, V.H., Liu, B. *TL* **46**, 1237 (2005).

Iridium complexes. 21, 237–239; **22**, 253

Redox reactions. Different stereochemical consequences emerge in the hydrogenation of alkylideneprolinols on using either Raney nickel or a cationic iridium complex.[1] Apparently the oxygen atom exerts a directing effect on the iridium complex. Hydrogen-transfer reduction of aldehydes is observed in the presence of a carbene complex of iridium.[2]

For Oppenauer-type oxidation, Cp*Ir-complexes are useful catalysts. The other ligands can be 2,2-diphenyl-2-aminoethanol[3] or imidazol-2-ylidene and acetonitrile.[4]

Dehydrogenation of tertiary amines to give enamines is mediated by **1** with *t*-BuCH $=$ CH_2 present as hydrogen acceptor.[5] The promoting ability of ancillary group on nitrogen follows the trend: *i*-Pr $\gg$ Et $>$ Me.

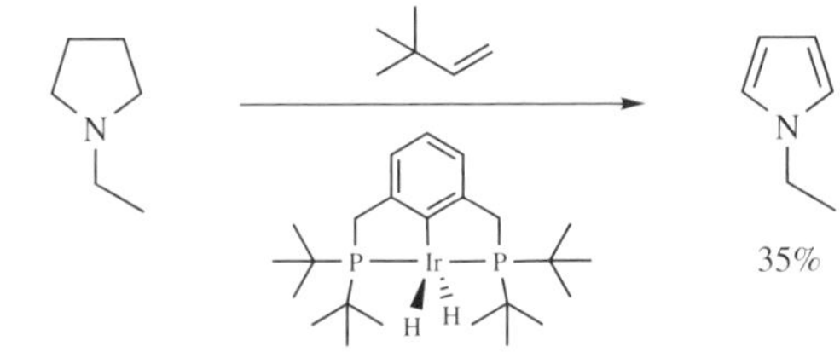

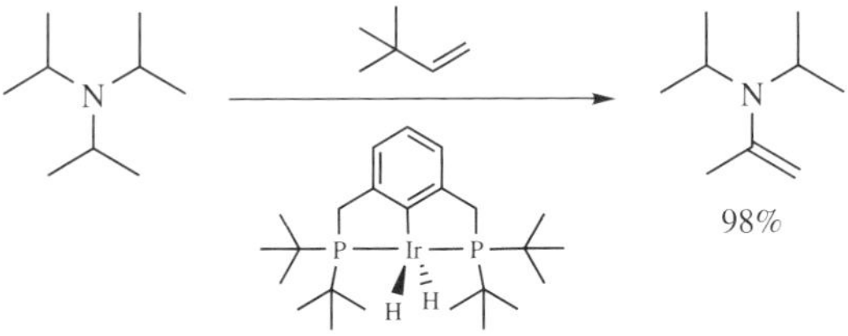

35%

98%

Rearrangement. A tris(cyclohexylphosphine)iridium complex catalyzes double bond migration of diallyl ethers thereby setting up for Claisen rearrangement at moderate temperatures.[6] $NaBPh_4$ is usually added to suppress side reactions.

Glutarimides.[7] A facile synthesis of glutarimides from alkyl cyanides and acrylonitrile involves heating the components with $(i\text{-}Pr_3P)_2IrH_5$.

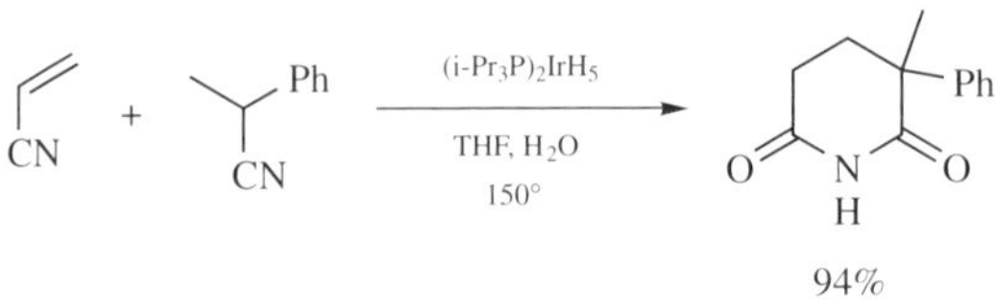

$(i\text{-}Pr_3P)_2IrH_5$, THF, H_2O, 150°

94%

[1]Del Valle, J.R., Goodman, M. *JOC* **68**, 3923 (2003).
[2]Miecznikowski, J.R., Crabtree, R.H. *OM* **23**, 629 (2004).
[3]Suzuki, T., Morita, K., Tsuchida, M., Hiroi, K. *JOC* **68**, 1601 (2003).
[4]Hanasaka, F., Fujita, K., Yamaguchi, R. *OM* **23**, 1490 (2004).
[5]Zhang, X., Fried, A., Knapp, S., Goldman, A.S. *CC* 2060 (2003).
[6]Nelson, S.G., Bungard, C.J., Wang, K. *JACS* **125**, 13000 (2003).
[7]Takaya, H., Yoshida, K., Isozaki, K., Terai, H., Murahashi, S.-I. *ACIEE* **42**, 3302 (2003).

Iron. **19**, 179–180; **20**, 203–204; **21**, 239; **22**, 253

Reduction. The advantage in the conversion $ArNO_2$ to $ArNH_2$ using nanoparticles of Fe in water[1] is questionable because its reaction temperature is high (210°).

The 1,2,4-oxadiazole ring is cleaved to amidines[2] by iron in aqueous HOAc-MeOH.

Allylation. In water the allylation of ArCHO by allyl bromides occurs in the presence of Fe powder and NaF.[3]

[1]Wang, L., Li, P., Wu, Z., Yan, J., Wang, M., Ding, Y. *S* 2001 (2003).
[2]Sendzik, M., Hui, H.C. *TL* **44**, 8697 (2003).
[3]Chan, T.K., Lau, C.P., Chan, T.H. *TL* **45**, 4189 (2004).

Iron(III) chloride. **13**, 133–134; **14**, 164–165; **15**, 158–159; **16**, 167–169, 190–191; **17**, 138–139; **18**, 197; **19**, 180–181; **20**, 204–205; **21**, 240–241; **22**, 254–254

Reduction. Reduction of $ArNO_2$ to $ArNH_2$ has been conducted with $FeCl_3$-In in aq. MeOH.[1]

Substitutions. S_N2' substitution is effected on Baylis-Hillman adducts[2] or their acetates[3] to give allylic chlorides by $FeCl_3$ in CH_2Cl_2 at room temperature. $InCl_3$ is also effective.[2]

1,6-Anhydroglucopyranose derivatives are formed from glycosides bearing a benzyl or MOM group at C-6 on treatment with $FeCl_3$.[4]

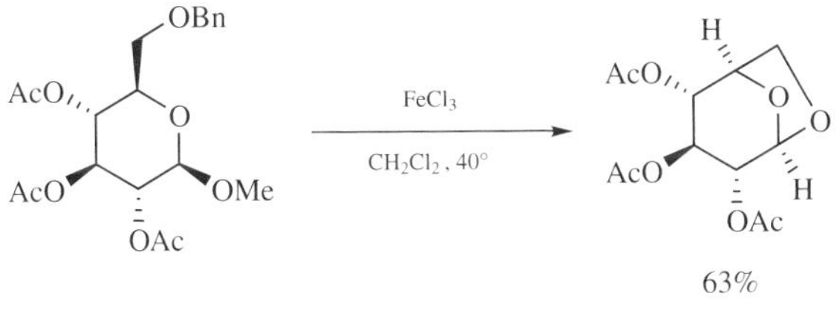

OBn
AcO
O
AcO
OMe
OAc
$FeCl_3$
CH_2Cl_2, 40°
H
AcO
O
O
AcO
H
OAc
63%

Homoallylic ethers. Modification of the allylation method by pretreating the carbonyl substrates with an alkyl (e.g., benzyl) silyl ether together $FeCl_3$ prior to addition of the allylsilane reagent leads to ether products.[5] [Alkyl ethers of cyanohydrins are also prepared analogously.[6]]

Ene reaction using acetals as partners is subject to catalysis by $FeCl_3$.[7]

$FeCl_3$ / CH_2Cl_2, 25°

92%

Michael reaction. Addition of carbamates to enones is accomplished in the presence of $FeCl_3$ and Me_3SiCl.[8]

Condensations. Aldol reactions in aqueous medium (surfactant added) involving silyl enol ethers as donors are promoted by $FeCl_3$ which is a water-compatible Lewis acid.[9]

[1]Yoo, B.W., Choi, J.W., Hwang, S.K., Kim, D.Y., Baek, H.S., Choi, K.Y., Kim, J.H. *SC* **33**, 2985 (2003).
[2]Das, B., Banerjee, J., Ravindranath, N., Venkataiah, B. *TL* **45**, 2425 (2004).
[3]Krishna, P.R., Kannan, V., Sharma, G.V.M. *SC* **34**, 55 (2004).
[4]Miranda, P.O., Brouard, I., Padron, J.I., Bermejo, J. *TL* **44**, 3931 (2003).
[5]Watahiki, T., Akabane, Y., Mori, S., Oriyama, T. *OL* **5**, 3045 (2003).
[6]Iwanami, K., Oriyama, T. *CL* **33**, 1324 (2004).
[7]Ladepeche, A., Tam, E., Ancel, J.-E., Ghosez, L. *S* 1375 (2004).
[8]Xu, L.-W., Xia, C.-G., Hu, X.-X. *CC* 2570 (2003).
[9]Aoyama, N., Manabe, K., Kobayashi, S. *CL* **33**, 312 (2004).

Iron(III) nitrate.

Oxidation. Solvent-free conditions using $Fe(NO_3)_3 \cdot 9H_2O$ and tungstophosphoric acid for alcohol oxidation are very convenient. Secondary alcohols are oxidized faster (but slower than benzylic alcohols).[1] Benzoins are oxidized to benzils with $Fe(NO_3)_3 \cdot 9H_2O$ on silica support.[2]

Aromatic substitutions. Mononitration of phenols is achieved with $Fe(NO_3)_3 \cdot 9H_2O$ in an ionic liquid,[3] whereas halogenation of arenes can use $Fe(NO_3)_3 \cdot 1.5N_2O_4$ on charcoal as catalyst (also as oxidant in case NaBr or NaI is employed as halogen source).[4]

[1]Firouzabadi, H., Iranpoor, N., Amani, K. *S* 408 (2003).
[2]Paul, A.M., Khandekar, A.C., Shenoy, M.A. *SC* **33**, 2581 (2003).
[3]Rajagopal, R., Srinivasan, K.V. *SC* **33**, 961 (2003).
[4]Firouzabadi, H., Iranpoor, N., Shiri, M. *TL* **44**, 8781 (2003).

Iron pentacarbonyl. **13**, 152; **18**, 196; **20**, 206–207; **21**, 242–243; **22**, 256–257

Carbonylation. Alkynes incorporate 2 molecules of CO on reaction with $Fe(CO)_5$ and Me_3NO to form cyclobutenediones[1] in moderate to good yields. Anhydrides are produced when Me_3NO is in excess.

1,2-Disubstituted cyclopropanes in which one substituent is an alkenyl group undergo photoinduced [5 + 1]cycloaddition. 2,5-Disubstituted cyclohexenones are obtained as major products on treatment of the reaction mixture with DBU.[2]

hv, $Fe(CO)_5$; DBU

OBn → OBn

>95% ee

[1]Periasamy, M., Mukkanti, A., Raj, D.S. *OM* **23**, 5323 (2004).
[2]Taber, D.F., Joshi, P.V., Kanai, K. *JOC* **69**, 2268 (2004).

Iron-porphyrin complexes.

Olefination. Carbonyl compounds are homologated to furnish conjugated esters when they react with diazoacetic esters in the presence of iron-porphyrin complexes and Ph_3P.[1,2] Benzoic acid is an important additive. The reaction is also suitable for olefination of trifluoromethyl ketones.[3]

Aziridines. Using bromamine-T [TsN(Br)Na] as nitrene source the cycloaddition to alkenes is effected by a Fe(III)-porphyrin complex.[4]

[1]Cheng, G., Mirafzal, G.A., Woo, I.K. *OM* **22**, 1468 (2003).
[2]Chen, Y., Huang, L., Zhang, X.P. *OL* **5**, 2493 (2003).
[3]Chen, Y., Huang, L., Zhang, X.P. *JOC* **68**, 5925 (2003).
[4]Vyas, R., Gao, G.-Y., Harden, J.D., Zhang, X.P. *OL* **6**, 1907 (2004).

Isocyanuric chloride. 21, 243; 22, 258

Oxidations. Oxidation of primary alcohols to methyl esters and diols to lactones (γ- and δ-) by isocyanuric chloride alone in MeOH has been achieved.[1] Heteroaromatic compounds such as pyridine yield *N*-oxides on reaction with the title reagent.[7] Pyrazolines undergo dehydrogenation (aromatization).[3]

Alkyl nitrates.[4] Chlorotriphenylphosphonium salt derived from Ph_3P and isocyanuric chloride activates alcohols in the preparation of nitrates (using $AgNO_3$). Similarly, nitrites and thiocyanates are obtained.

α,α-Dichloroketones.[5] Alkynes are transformed by isocyanuric chloride into $RCOCCl_2R'$ or the dimethyl acetals in one step. Water content in solvent determines the products.

[1]Hiegel, G.A., Gilley, C.B. *SC* **33**, 2003 (2003).
[2]Zhong, P., Guo, S., Song, C. *SC* **34**, 247 (2004).
[3]Zolfigol, M.A., Azarifar, D., Maleki, B. *TL* **45**, 2181 (2004).
[4]Hiegel, G.A., Nguyen, J., Zhou, Y. *SC* **34**, 2507 (2004).
[5]Hiegel, G.A., Bayne, C.D., Ridley, B. *SC* **33**, 1997 (2003).

L

Lanthanum. 22, 259

Alkyl phenyl selenides. Alcohols and their derivatives are transformed into phenyl selenides on heating with PhSeSePh and a mixture of La, I_2, Me_3SiCl and CuI in MeCN.[1]

[1]Nishino, T., Nishiyama, Y., Sonoda, N. *CL* **32**, 918 (2003).

Lanthanum(III) chloride.

2-Deoxy sugars.[1] A mixture of $LaCl_3{\cdot}7H_2O$, NaI and BnOH is effective for hydration of glycals.

Knoevenagel condensation.[2] The reaction is carried out with $LaCl_3$ in the neat at 80-85°.

[1]Rani, S., Agarwal, A., Vankar, Y.D. *TL* **44**, 5001 (2003).
[2]Narsaiah, A.V., Nagaiah, K. *SC* **33**, 3825 (2003).

Lanthanum(III) cyanide.

Acyloins.[1] A method for synthesis of unsymmetrical acyloins is based on the $La(CN)_3$-catalyzed reaction between RCHO and $R'COSiMe_3$.

PhCOSiEt3 + OHC-(2-furyl) → [La(CN)3, THF 25°] → PhCOCH(OSiEt3)(2-furyl)

88%

[1]Bausch, C.C., Johnson, J.S. *JOC* **69**, 4283 (2004).

Lanthanum(III) trifluoromethanesulfonate. 20, 209; **21**, 245; **22**, 259-260

PMB ethers.[1] Alcohols are derivatized by *p*-methoxybenzyl trichloroacetimidate with $La(OTf)_3$ as catalyst at room temperature within 5 minutes. Of similar effectiveness are $Sc(OTf)_3$ and $Yb(OTf)_3$.

[1]Rai, A.N., Basu, A. *TL* **44**, 2267 (2003).

Fiesers' Reagents for Organic Synthesis, Volume 23. Edited by Tse-Lok Ho

Lanthanum(III) hexamethyldisilazide. 22, 260

Hydrosilylation.[1] Alkenes and dienes undergo hydrosilylation in the Markovnikov sense with the La catalyst.

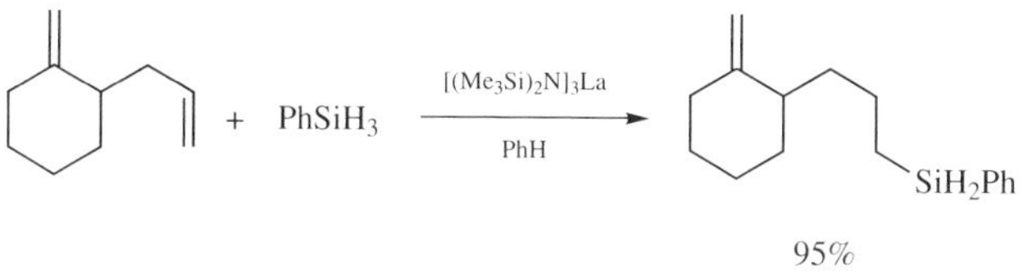

$[(Me_3Si)_2N]_3La$, PhH; + $PhSiH_3$ → SiH_2Ph product, 95%

[1]Horino, Y., Livinghouse, T. *OM* **23**, 12 (2004).

Lead.

Nitroarene reduction. Azoarenes are formed when $ArNO_2$ are treated with Pb and $HCOONEt_3H$.[1] *o*-Nitrobenzyl ketones give *N*-hydroxyindoles.[2]

[1]Srinivasa, G.R., Abiraj, K., Gowda, D.C. *TL* **44**, 5835 (2003).
[2]Wong, A., Kuethe, J.T., Davies, I.W. *JOC* **68**, 9865 (2003).

Lead(IV) acetate. 13, 155–156; **14**, 188; **16**, 193–194; **18**, 201–202; **19**, 184–185; **20**, 209–210; **21**, 245–246; **22**, 260–261

Oxidation. Dicarboxylic acids are degraded on mechanical mixing with $Pb(OAc)_4$ $-NH_4Cl$. Butyrolactone is formed from glutaric acid but octanedioic acid gives 1,6-dichlorohexane.[1]

Ring cleavage of 1-methylcycloalkanols in the presence of pyridines and quinolines leads to oxoalkylated products.[2]

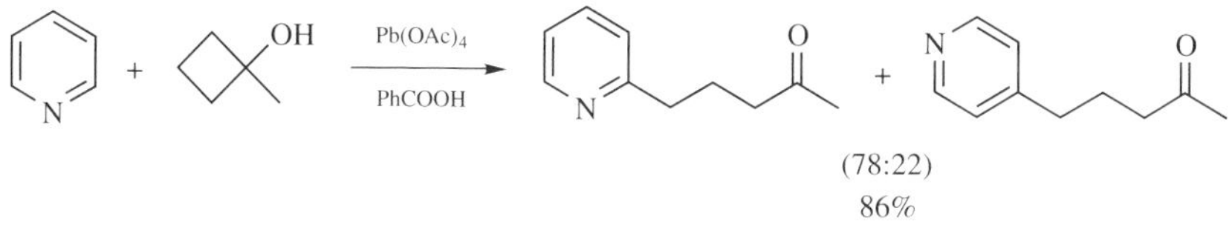

N + OH; $Pb(OAc)_4$, PhCOOH → N, O + N, O

(78:22)
86%

Decarboxylation. A facile route to 3.4-cyclobutanopyrroles involves cycloaddition of aziridines to cyclobutene-1,2-dicarboxylic esters, which is followed by saponification and treatment with $Pb(OAc)_4$.[3]

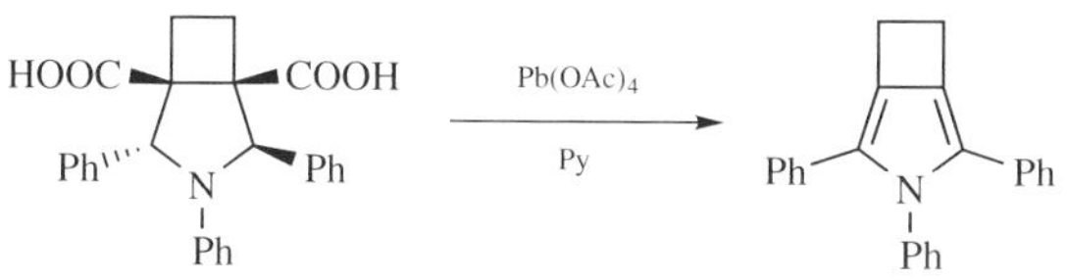

HOOC, COOH, Ph, Ph, N, Ph; $Pb(OAc)_4$, Py → Ph, Ph, N, Ph

[1]Nikishin, G.I., Sokova, L.L., Makhaev, V.D., Kapustina, N.I. *MC* 264 (2003).
[2]Nikishin, G.I., Sokova, L.L., Kapustina, N.I. *RCB* **51**, 1812 (2002).
[3]Matsumoto, K., Goto, S., Hayashi, N., Iida, H., Uchida, T., Kakehi, A. *EJOCC* 4667 (2004).

Lithium. 13, 157–158; **15**, 184; **18**, 205–206; **19**, 190–191; **20**, 212; **21**, 247; **22**, 262

Lithiation.[1] Lithium powder with *N,N*-dimethyl-1-naphthylamine works better than LDBB in lithiating certain alkenyl chlorides.

[1]Giannini, A., Coquerel, Y., Greene, A.E., Depres, J.-P. *TL* **45**, 6749 (2004).

Lithium - ethylenediamine.

Ether cleavage.[1] Aryl methyl ethers and benzyl ethers are readily cleaved. Since the method involves electron transfer to the aromatic ring highly hindered ethers (e.g., 2,6-di-*t*-butylphenyl methyl ether) are cleaved in excellent yields. Use of the 2,6-$Me_2C_6H_3CH_2$ group to protect alcohols is suggested.

1,3-Alkadiynes.[2] The triple bonds of internal conjugated diynes isomerize to the terminal position by treatment with the title reagent. Immediate application to coupling reaction with ArX is feasible.

[1]Shindo, T., Fukuyama, Y., Sugai, T. *S* 692 (2004).
[2]Balova, I.A., Sorokoumov, V.N., Morozkina, S.N., Vinogradova, O.V., Knight, D.W., Vasilevsky, S.F. *EJOC* 882 (2005).

Lithium - liquid ammonia. 13, 158; **17**, 161; **18**, 206; **20**, 213; **21**, 247; **22**, 262

Reductive cyclization.[1] δ,ε-Unsaturated esters cyclize, presumably via addition of acyl radical across the double bond.

COOMe
Li, NH_3(l)
OH
96%

[1]Srikrishna, A., Ramasastry, S.S.V. *TL* **45**, 379 (2004).

Lithium acetate.

Aldol and Mannich reactions. LiOAc in DMF containing a small amount of water is effective to promote reaction of ketene silyl acetals with aldehydes[1] and *N*-tosylimines.[2] β-Lactams are produced in the reaction with *N*-arylimines.[3]

[1]Nakagawa, T., Fujisawa, H., Mukaiyama, T. *CL* **32**, 696 (2003).
[2]Takahashi, E., Fujisawa, H., Mukaiyama, T. *CL* **33**, 936 (2004).
[3]Takahashi, E., Fujisawa, H., Yanai, T., Mukaiyama, T. *CL* **34**, 216 (2005).

Lithium aluminum hydride. **14**, 190–191; **18**, 207; **19**, 191; **20**, 213–214; **21**, 247–248; **22**, 263–264

Isoxazoline cleavage.[1,2] This heterocycle suffer reductive cleavage to afford 1,3-amino alcohols which can be used to synthesize other important compounds, e.g., β-amino acids.

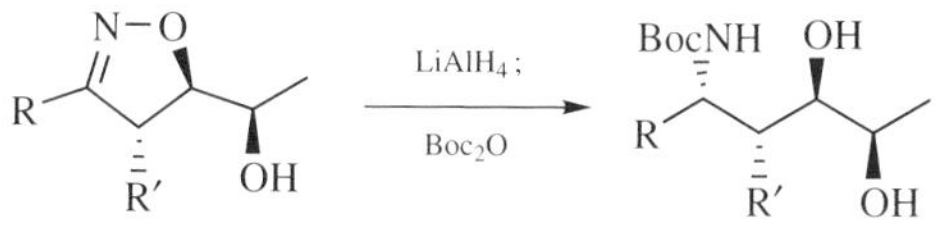

Debromination.[3] (*E*)-Styryl bromides are readily obtained on reduction of the dibromides with a reagent made of $LiAlH_4$ and EtOAc (1:1).

Conjugate dienes.[4] Cyclopropenyl carbinols fragment on heating with $LiAlH_4$ in benzene, the first stage of which is similar to hydroalumination of propargylic alcohols. Deuterated products can be obtained using $LiAlD_4$.

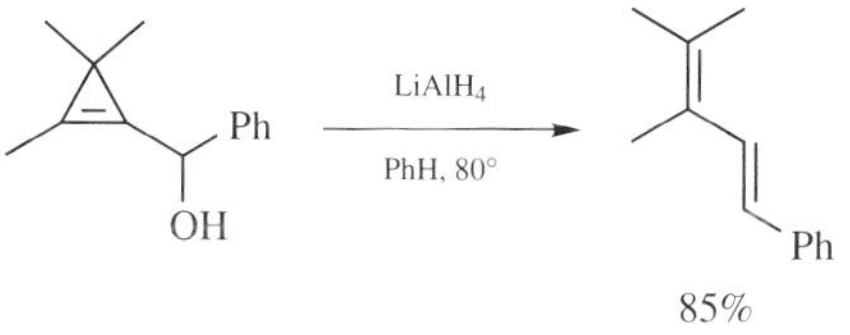

Selenocyclobutenes.[5] Formation of the small ring compounds from (propargylseleno)alkynes on *Se*-alkylation and reaction with $LiAlH_4$ involves allenyl selenoketene intermediate.

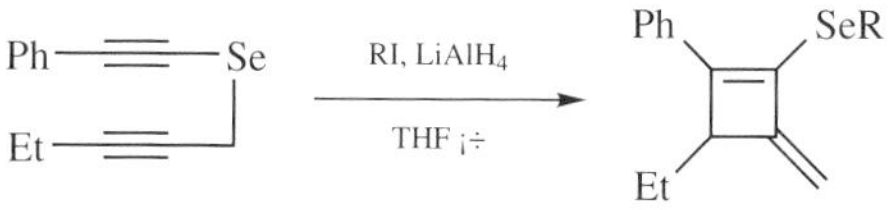

[1]Bickley, J.F., Roberts, S.M., Runhui, Y., Skidmore, J., Smith, C.B. *T* **59**, 5731 (2003).
[2]Minter, A.R., Fuller, A.A., Mapp, A.K. *JACS* **125**, 6846 (2003).
[3]Horibe, H., Kondo, K., Okuno, H., Aoyama, T. *S* 986 (2004).
[4]Zohar, E., Ram, M., Marek, I. *SL* 1288 (2004).
[5]Koketsu, M., Kanoh, M., Yamamura, Y., Ishihara, H. *TL* **46**, 1479 (2005).

Lithium aluminum hydride – niobium(IV) chloride.

Hydrodefluorination.[1] ArF are reduced by this reagent couple.

[1]Fuchibe, K., Akiyama, T. *SL* 1282 (2004).

Lithium benzyloxide.

Michael and Mannich reactions. Acting as a Lewis base LiOBn is a good activator for silyl enol ethers and silyl ketene acetals to react with enones[1] and *N*-tosylimines,[2] respectively. However, the emerging product with a negative-charged heteroatom is also a catalyst, which renders the reaction autocatalytic.[2]

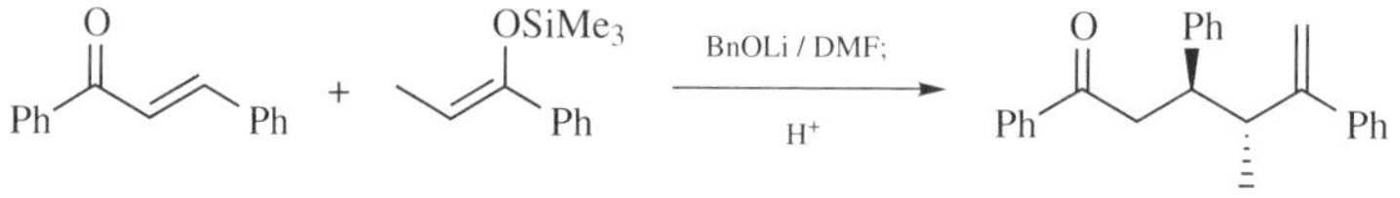

93% (*anti/syn* = 92 : 8)

[1]Mukaiyama, T., Tozawa, T., Fujisawa, H. *CL* **33**, 1410 (2004).
[2]Takahashi, E., Fujisawa, H., Mukaiyama, T. *CL* **33**, 1426 (2004).

Lithium borohydride.

Hydroboration. The presence of catalytic amounts of borane or Me_3SiCl (or $TiCl_4$) transforms $LiBH_4$ into a hydroborating agent. The key step is its exchange reaction between the product (alkylborane) to liberate more borane. An unusual reactivity pattern with respect to the alkenes is that tetramethylethene > 1-methylcyclohexene > cyclohexene.

[1]Villiers, C., Ephritikhine, M. *TL* **44**, 8077 (2003).

Lithium chloride.

Allylation. The CuCl-catalyzed reaction of zirconacyclopentadienes with allyl chloride gives different products in the absence or presence of LiCl.[1]

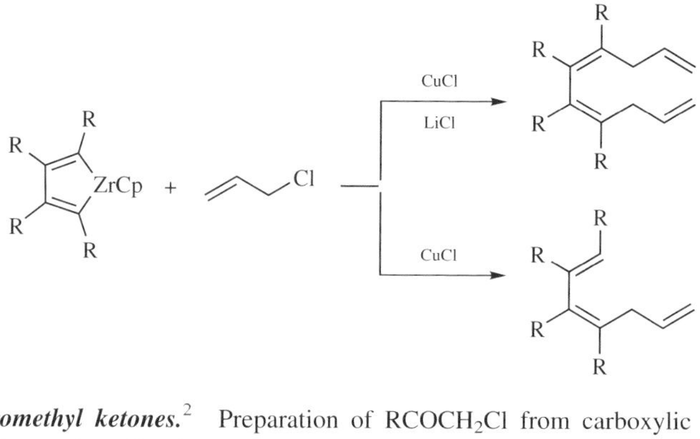

1-Chloromethyl ketones.[2] Preparation of $RCOCH_2Cl$ from carboxylic acids via acyl chlorides and α-diazoketones is not suitable for large scale operation. An alternative is by reaction of esters with dimethylsulfoxonium ylide and further reaction with LiCl in MeOH.

[1]Leng, L., Xi, C., Chen, C., Lai, C. *TL* **45**, 595 (2004).
[2]Schwinden, M.D., Radesca, L., Patel, B., Kronenthal, D., Huang, M.-H., Nugent, W.A., Wang, D. *JOC* **69**, 1629 (2004).

Lithium *N,N*-dialkylaminoborohydrides. 21, 249; **22**, 265–266

2-Dialkylaminopyridines. Displacement of a 2-fluoro group from the heterocyclic compounds by R_2N residue is readily accomplished using $LiBH_3NR_2$.[1]

[1]Thomas, S., Roberts, S., Pasumansky, L, Gamsey, S., Singaram, B. *OL* **5**, 3867 (2003).

Lithium 4,4'-di-*t*-butylbiphenylide. 13, 162–163; **16**, 195–196; **17**, 164; **18**, 210–211; **19**, 192–193; **20**, 216–217; **21**, 249–250; **22**, 266–267

Silanes. Benzylic and allylic thiols and sulfides are cleaved by Li-DTTB. In situ trapping with Me_3SiCl furnishes $RSiMe_3$.[1,2]

Reduction. In combination with $CuCl_2{\cdot}2H_2O$, carbonyl compounds and imines are reduced by Li-DTTB.[3] In the presence of bis(methoxyethyl)amine the reduction of pyrrole-2-carboxylic esters leads to pyrrolines[4] which can be alkylated at C-2.

Synthesis of spiroannulated tetrahydropyrans by decyanative S_N2 and S_N2' pathways is initiated by Li-DTTB.[5]

R O CN OMe
LiDBB
THF –40°; MeOH
R O
89%

Hydrodefluorination of RF is accomplished with Li-DTTB, using a substoichiometric amount of 1,2-bis(trimethylsilyl)benzene to activate the C-F bond.[6]

[1]Yus, M., Martinez, P., Guijarro, D. *SC* **33**, 2365 (2003).
[2]Streiff, S., Ribero, N., Desaubry, L. *JOC* **69**, 7592 (2004).
[3]Alonso, F., Vitale, C., Radivoy, G., Yus, M. *S* 443 (2003).
[4]Donohoe, T.J., House, D., Ace, K.W. *OBC* **1**, 3749 (2003).
[5]Rychnovsky, S.D., Takaoka, L.R. *ACIEE* **42**, 818 (2003).
[6]Guijarro, D., Martinez, P., Yus, M. *T* **59**, 1237 (2003).

Lithium diisopropylamide, LDA. 13, 163–164; **15**, 188–189; **16**, 196–197; **17**, 165–167; **18**, 212–214; **19**, 193–197; **20**, 218–220; **21**, 250–251; **22**, 267–268

Demesylation. Various aryl mesylates are cleaved by LDA in THF at 0°, therefore the use of mesyl group to protect phenols may be considered.[1]

Elimination. 2-Halo-1,1,1-trifluoroethanes (halogen = Br, F) are converted into $CF_2 = C(X)ZnCl$ via dehydrofluorination, lithiation and Li/Zn exchange, on treatment with LDA and $ZnCl_2$. The organozinc species are reactive in Negishi coupling.[2,3]

α-Cyanohydrin *O*-carbonates derived from ArCHO undergo transformation to give α-keto esters.[4]

Condensations. For a synthesis of alkenesulfonamides MsNHBoc is doubly deprotonated with LDA and reacted with $Ph_2P(O)Cl$ to give $Ph_2P(O)CH_2SO_2NHBoc$ which are then employed in a Horner reaction.[5] Silyldihalomethanes form carbanions that on reaction with nitriles furnish α-keto acylsilanes ($RCOCOSiR'_3$).[6]

Cross-Claisen condensation involving amino esters as acceptors and *t*-butyl acetate as donor is successful.[7] After deprotonation, *N*-benzylidene-α-amino esters react with *N*-(*p*-toluenesulfinyl)imines to give imidazolidines.[8]

Tol–S(O)–CH=CH–Ph + PhCH=N–CH(R)–COOMe —(LDA, THF)→ imidazolidine [Tol–S(O)–N; Ph; NH; Ph; R; COOMe]

A relay deprotonation by LDA at the benzylic position which is also attached to the nitrogen atom of properly substituted acrylamides initiates cyclization.[9] Intramolecular alkylation of *N*-mesyl-*N*-2-haloethylamines (via dianions) results in five-membered sultams.[10]

SPh; Ph; N; Me; O —(LDA, THF, –20°)→ Ph; N; Me; O

80%

The discovery of highly enhanced reactivity of lithium enolates of esters complexed to LDA toward aldimines is of synthetic significance.[11]

Cyclization. After deprotonation 2-(1-alkynyl)benzaldimines and 2-aza-2,4-heptadienyl-6-ynes undergo cyclization to furnish products possessing a new aminobenzene ring.[12]

Bu; N; Ph —(LDA)→ [Bu; Ph; CN] → Bu; Ph; NHR

Cannizzaro and Haller-Bauer reactions.[13] LDA catalyzes the reaction of aldehydes with lithium amides to provide amides together with the reduced alcohols. Benzyl ketones are cleaved to afford hydrocinnamides.

[1]Ritter, T., Stanek, K., Larrosa, I., Carreira, E.M. *OL* **6**, 1513 (2004).
[2]Anilkumar, R., Burton, D.J. *JFC* **125**, 561 (2004).
[3]Raghavanpillai, A., Burton, D.J. *JOC* **69**, 7083 (2004).
[4]Thasana, N., Prachyawarakorn, V., Tontoolarung, S., Ruchirawat, S. *TL* **44**, 1019 (2003).
[5]Reuter, D.C., McIntosh, J.E., Guinn, A.C., Madera, A.M. *S* 2321 (2003).
[6]Yagi, K., Tsunitani, T., Takami, K., Shinokubo, H., Oshima, K. *JACS* **126**, 8618 (2004).
[7]Katayama, S., Kojima, M., Suzuki, T., Izawa, K., Honda, Y. *TL* **44**, 3163 (2003).
[8]Viso, A., de la Pradilla, R.F., Garcia, A., Guerrero-Strachan, C., Alonso, M., Tortosa, M., Flores, A., Martinez-Ripoll, M., Fonseca, I., Andre, I., Rodriguez, A. *CEJ* **9**, 2867 (2003).
[9]Asaoka, M., Naitoh, R., Nakamura, Y., Katano, E., Nakamura, Y., Okada, E. *H* **63**, 1009 (2004).
[10]Lee, J., Zhong, Y.-L., Reamer, R.A., Askin, D. *OL* **5**, 4175 (2003).
[11]Hata, S., Iguchi, M., Iwasawa, T., Yamada, K., Tomioka, K. *OL* **6**, 1721 (2004).
[12]Sagar, P., Frohlich, R., Wurthwein, E.-U. *ACIEE* **43**, 5694 (2004).
[13]Ishihara, K., Yano, T. *OL* **6**, 1983 (2004).

Lithium hexafluorophosphate.

THP ethers. Tertiary alcohols are etherified by dihydropyran at room temperature or below using this catalyst.

[1]Hamada, N., Sato, T. *SL* 1802 (2004).

Lithium hexamethyldisilazide, LHMDS. 13, 165; **14**, 194; **18**, 215–216; **19**, 197–198; **20**, 221–222; **21**, 251–252; **22**, 268–269

Elimination. Dehydroiodination leading to semicyclic enol carbonates is of interest because the products are latent enolate ions.[1]

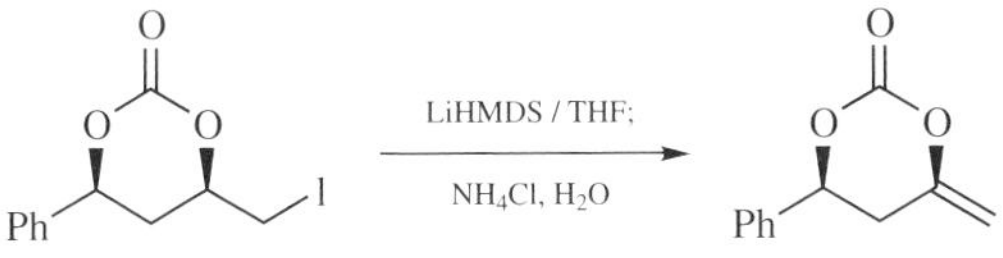

LiHMDS / THF; NH_4Cl, H_2O

91%

N,N-Bisbenzyloxycarbonyl amino acids.[2] The fully protected amino acid derivatives are formed by using the LHMDS-HMPA system as base at the second-staged reaction.

Aldol reaction.[3] *N*-(4-Methoxybenzyl) β-lactams are hydroxyalkylated using LHMDS as base. Choice of the *N*-protecting group is important to ensure complete enolization.

Cyclization. Intramolecular condensations of various types can be effected by LHMDS. Pairings of sulfoxide with lactam,[4] sulfoximine with conjugated ester,[5] and lactone with 2-aminoaryl ketone units[6] are exemplary, but the most significant result is the formation of β-lactam spiroannulated at C-3 of an indole nucleus.[7]

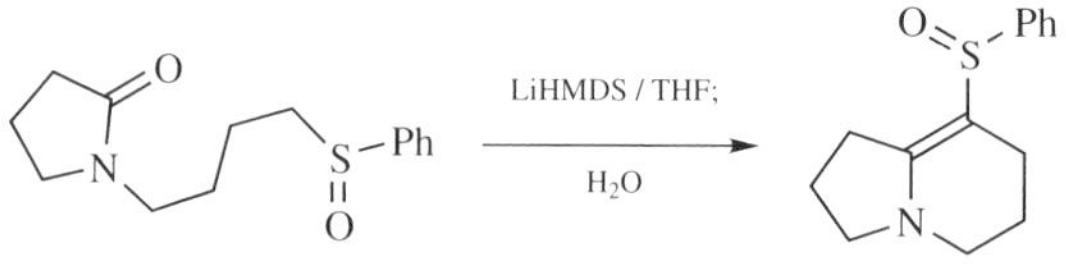

LiHMDS / THF; H_2O

85-90%

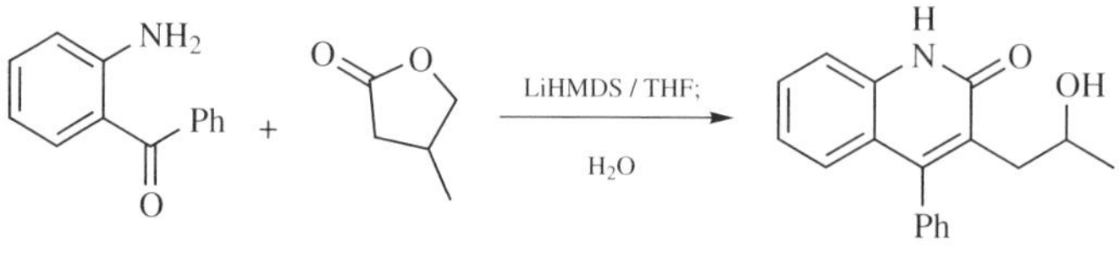

LiHMDS / THF; H_2O

82%

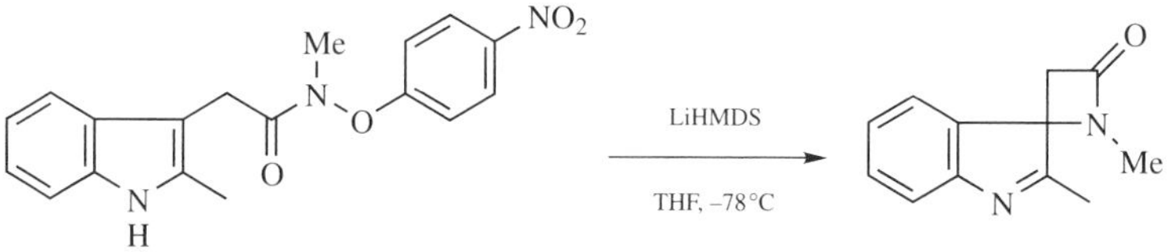

LiHMDS THF, –78°C

50%

With respect to regioselection in the intramolecular alkylation with an terminal allyl bromide unit a cyclooctene is formed readily from the (*Z*)-isomer. Steric constraint is responsible for the (*E*)-isomer to pursue an S_N2' reaction pathway.[8]

[1]Maddess, M.L., Lautens, M. *S* 1399 (2004).
[2]Hernandez, J.N., Martin, V.S. *JOC* **69**, 3590 (2004).
[3]Williams, D.R., Donnell, A.F., Kammler, D.C. *H* **62**, 167 (2004).
[4]Reutrakul, V., Pohmakotr, M., Numechai, P., Prateeptongkum, S., Tuchinda, P. *OBC* **1**, 3495 (2003).
[5]Harmata, M., Hong, X. *JACS* **125**, 5754 (2003).
[6]Wang, J., Discordia, R.P., Crispino, G.A., Li, J., Grosso, J.A., Polniaszek, R., Truc, V.C. *TL* **44**, 4271 (2003).
[7]Nishikawa, T., Kajii, S., Isobe, M. *SL* 2025 (2004).
[8]Kim, H., Choi, W.J., Jung, J., Kim, S., Kim, D. *JACS* **125**, 10238 (2003).

Lithium hydroxide. 22, 269

Desilylation.[1] A convenient and selective cleavage of aryl silyl ethers is by using LiOH. Aliphatic OTBS groups are preserved.

Conjugated nitriles.[2] The HWE reaction of carbonyl compounds with $(EtO)_2P(O)CH_2CN$ can be performed with the mild base LiOH.

[1]Ankala, S.V., Fenteany, G. *SL* 825 (2003).
[2]Lattanzi, A., Orelli, L.R., Barone, P., Massa, A., Iannece, P., Scettri, A. *TL* **44**, 1333 (2003).

Lithium iodide.

2-Oxazolidinones. Insertion of CO_2 into the less substituted C-N bond of an aziridine is mediated by LiI.[1,2] Adding HMPA improves the regioselectivity. (Other alkali halides and Bu_4NBr are useful catalysts and thiazolidine-2-thiones are similarly prepared using CS_2 instead of CO_2)[2].

1,2-Diamines. Aziridines react with iminium salts (e.g., $[Me_2N=CH_2]X$) in the presence of LiI to afford 1,2-diamines.[3]

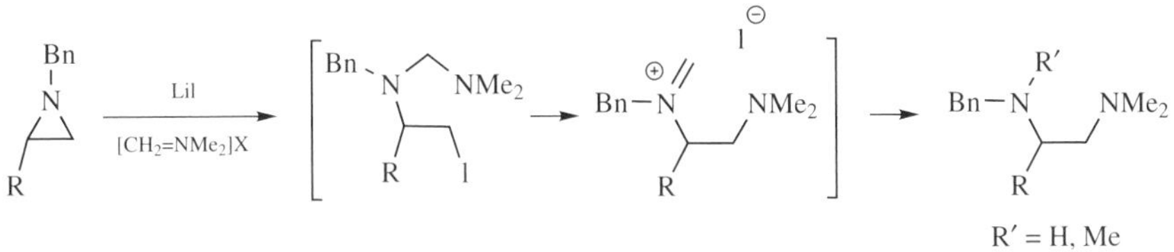

R′ = H, Me

[1]Hancock, M.T., Pinhas, A.R. *TL* **44**, 5457 (2003).
[2]Sudo, A., Morioka, Y., Koizumi, E., Sanda, F., Endo, T. *TL* **44**, 7889 (2003).
[3]Hancock, M.T., Pinhas, A.R. *TL* **44**, 7125 (2003).

Lithium naphthalenide, LN. 15, 190–191; **18**, 217–218; **19**, 199–200; **20**, 224–225; **21**, 252–254; **22**, 269–270

Detritylation. Both trityl ethers[1] and tritylamines[2] are cleaved by LN. Therefore, the method is very useful for situation that must avoid acid.

Reductive decomplexation. π-Allyltricarbonyliron complexes are converted into alcohols.[3]

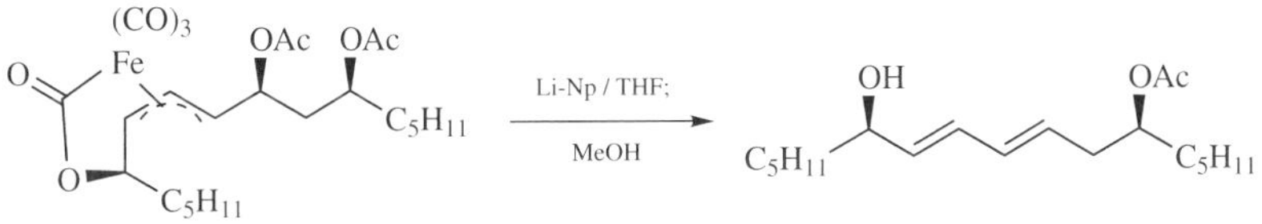

α-Cyano carbanions. 2,2-Disubstituted malononitriles undergo decyanation with LN at -40° and the carbanions can be alkylated.[4]

[1]Yus, M., Behloul, C., Guijarro, D. *S* 2179 (2003).
[2]Behloul, C., Guijarro, D., Yus, M. *S* 1274 (2004).
[3]Hollowood, C.J., Ley, S.V. *OBC* **1**, 3197 (2003).
[4]Tsai, T.-Y., Shia, K.-S., Liu, H.-J. *SL* 97 (2003).

Lithium perchlorate. 18, 218-219; **19**, 200-201; **20**, 224-225; **22**, 270-271

Silylation. The Lewis acidity of $LiClO_4$ is well recognized and its catalytic use in silylation of alcohols (including α,α-dimethylpropynol) with $(Me_3Si)_2NH$[1] and the formation of cyanohydrin silyl ethers (reagent: Me_3SiCN)[2] is perhaps unexceptional, but convenient operation that involves only filtration and evaporation after addition of a solvent (hexane or CH_2Cl_2) is meritorius. α-Siloxyphosphonates are obtained directly from the reaction of epoxides with Me_3SiCl and $(RO)_3P$,[3] owing to a prior isomerization to aldehydes.

$LiClO_4$ is said to accelerate the Michael addition of amines to electron-poor alkenes.[4]

Diaryl sulfoxides.[5] Arenes react with $SOCl_2$ in the presence of $LiClO_4$ (or $NaClO_4$) at room temperature to provide symmetrical diaryl sulfoxides.

Coupling reaction. The role played by $LiClO_4$ in electrolytic allylation is beyond being the electrolyte.[6]

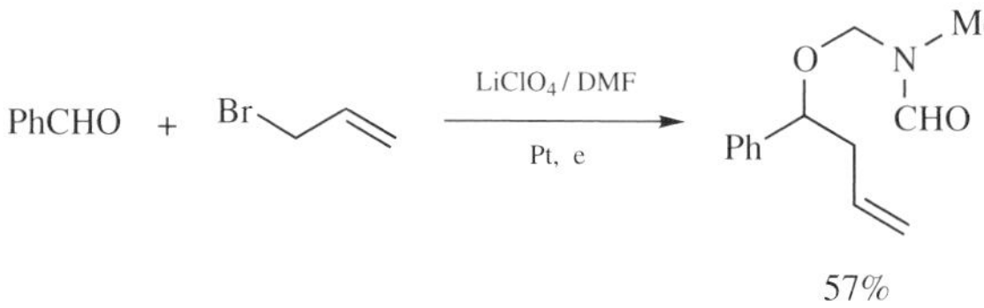

PhCHO + Br(allyl) → ($LiClO_4$ / DMF; Pt, e) → Ph-CH(OCH₂N(Me)CHO)CH₂CH=CH₂

57%

[1]Azizi, N., Saidi, M.R. *OM* **23**, 1457 (2004).
[2]Azizi, N., Saidi, M.R. *JOMC* **688**, 283 (2003).
[3]Azizi, N., Saidi, M.R. *TL* **44**, 7933 (2003).
[4]Azizi, N., Saidi, M.R. *T* **60**, 383 (2004).
[5]Bandgar, B.P., Makone, S.S. *SC* **34**, 743 (2004).
[6]Hilt, G. *ACIEE* **42**, 1720 (2003).

Lithium 2,2,6,6-tetramethylpiperidide, LTMP. 13, 167; **14**, 194–195; **17**, 171–172; **18**, 220–221; **19**, 202; **20**, 226–227; **21**, 256; **22**, 271

Directed lithiation. 3-Halobenzoic acids are lithiated with LTMP without forming a benzyne.[1] Chiral telluronium ylides are generated either by deprotonation with LTMP or LDA-LiBr. Different diastereoselectivities are observed in the subsequent cyclopropanation from the two.[2]

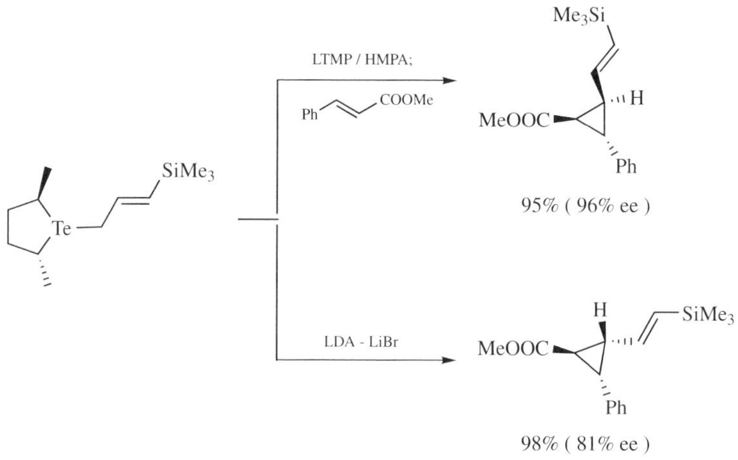

LTMP / HMPA; Ph/\/COOMe

95% (96% ee)

LDA - LiBr

98% (81% ee)

Reaction with epoxides. Enamines are generated from LTMP and 1,2-epoxyalkanes.[3] When the molecule contains a double bond separated from the terminal epoxide by 3 or 4 bonds, lithiation and cyclization ensue.[4] A further reaction mode is alkene formation.[5]

LTMP, *t*-BuOMe — 80%

$C_{10}H_{21}$-epoxide + Li-CH=CH-C_6H_{13} → (LTMP, hexane 0°) $C_{10}H_{21}$-CH=CH-CH=CH-C_6H_{13} 80%

[1]Gohier, F., Mortier, J. *JOC* **68**, 2030 (2003).
[2]Liao, W.W., Li, K., Tang, Y. *JACS* **125**, 13030 (2003).
[3]Hodgson, D.M., Bray, C.D., Kindon, N.D. *JACS* **126**, 6870 (2004).
[4]Hodgson, D.M., Chung, Y.K., Paris, J.-M. *JACS* **126**, 8664 (2004).
[5]Hodgson, D.M., Fleming, M., Stanway, S. *JACS* **126**, 12250 (2004).

Lithium triethylborohydride. 22, 272-273

Deacylation.[1] Amides and some tertiary carbamates are cleaved by $LiBHEt_3$ in THF between 0° and room temperature, the carbonyl function is reduced to a hydroxyl group.

Reduction. Formation of 2,3-dihydro-4-pyridones is observed when nitriles are treated with the title reagent and then Danishefsky's diene.[2] The imine products (perhaps still complexed to a metal) are capable of partake cycloaddition.

γ-Amino alcohols. Reduction of β-hydroxy sulfinimines with $LiBHEt_3$ gives derivatives of *anti*-γ-amino alcohols, whereas a similar reduction with catecholborane leads to predominantly *syn*-isomers.[3]

[1]Tanaka, H., Ogasawara, K. *TL* **43**, 4417 (2002).
[2]Kawecki, R. *S* 828 (2001).
[3]Kochi, T., Tang, T.P., Ellman, J.A. *JACS* **124**, 6508 (2002).

M

Magnesium. **13**, 170; **15**, 194; **16**, 198–199; **18**, 224–225; **19**, 205; **20**, 229–230; **21**, 258–259; **22**, 274–276

Reduction. Peroxides and ozonides are cleaved by Mg in MeOH,[1] while oximes are reduced to amines with Mg-$HCOONH_4$.[2]

Reductive coupling of two different carbonyl compounds to produce unsymmetrical pinacols is promoted by Mg-Me_3SiCl in DMF.[3] An intramolecular version to prepare 1,2-cyclobutanediol derivatives in good yields is also reported.

Anion radicals are generated from stilbenes that can add to carbonyl compounds.[4]

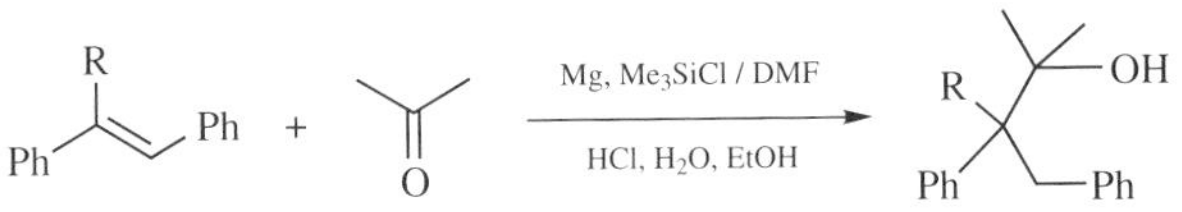

Defunctionalization of β-acetoxy sulfones by amalgamated Mg is as effective as Na-Hg, as shown by the preparation of alkylidenecyclopropanes.[5]

Coupling reactions. Unsymmetrical diaryl chalcogenides are formed when ArXXAr (X = S, Se) and Ar'I are treated with Mg and a copper catalyst.[6] An expeditious access to (trifluoromethyl)trimethylsilane is by the reaction of $PhSO_2CF_3$ and Me_3SiCl with Mg in DMF (the difluoro analogue is similarly obtained).[7]

Defluorinative *C,N*-disilylation of trifluoroacetimidoyl chlorides results when they are exposed to Mg and Me_3SiCl.[8]

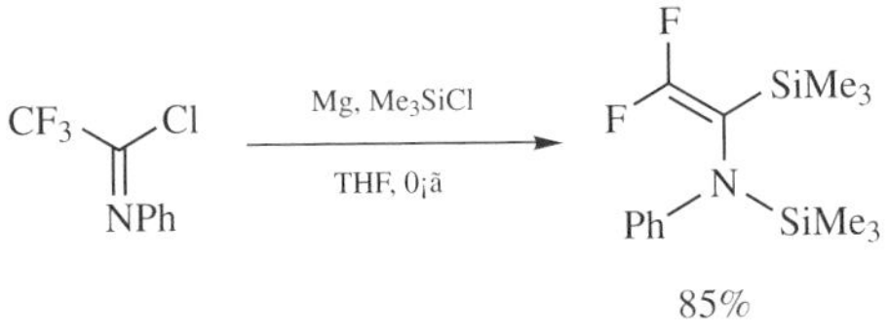

[1]Dai, P., Dussault, P.H., Trullinger, T.K. *JOC* **69**, 2851 (2004).
[2]Abiraj, K., Gowda, D.C. *SC* **34**, 599 (2004).
[3]Maekawa, H., Yamamoto, Y., Shimada, H., Yonemura, K., Nishiguchi, I. *TL* **45**, 3869 (2004).
[4]Yamamoto, Y., Kawano, S., Maekawa, H., Nishiguchi, I. *SL* 30 (2004).
[5]Benard, A.M., Frongia, A., Piras, P.P., Secci, F. *SL* 1064 (2004).
[6]Taniguchi, N., Onami, T. *JOC* **69**, 915 (2004).
[7]Prakash, G.K.S., Hu, J., Olah, G.A. *JOC* **68**, 4457 (2003).
[8]Kobayashi, T., Nakagawa, T., Amii, H., Uneyama, K. *OL* **5**, 4297 (2003).

Fiesers' Reagents for Organic Synthesis, Volume 23. Edited by Tse-Lok Ho

Magnesium bromide. 15, 194–196; **16**, 199; **17**, 174; **18**, 226–227; **19**, 206–207; **20**, 230–232; **21**, 260; **22**, 276–278

Hydrobromination. A simple way to α-bromo enamides (high *E:Z* ratios) is by mixing ynamides with $MgBr_2$ in wet CH_2Cl_2 at about room temperature.[1]

Bromine atom is incorporated into the Baylis-Hillman adducts when the reactants are aldehydes, ynones, and $MgBr_2$.[2]

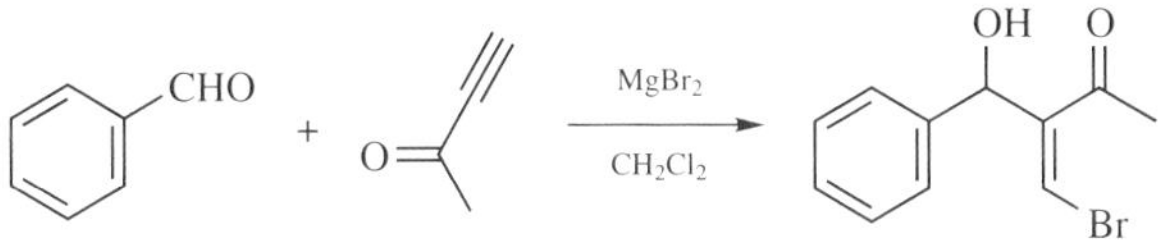

82% (*Z/E* 87:13)

Nitrones. Hydroxylamines condense with unsaturated aldehydes, the nitrone products may undergo 1,3-dipolar cycloaddition.[3]

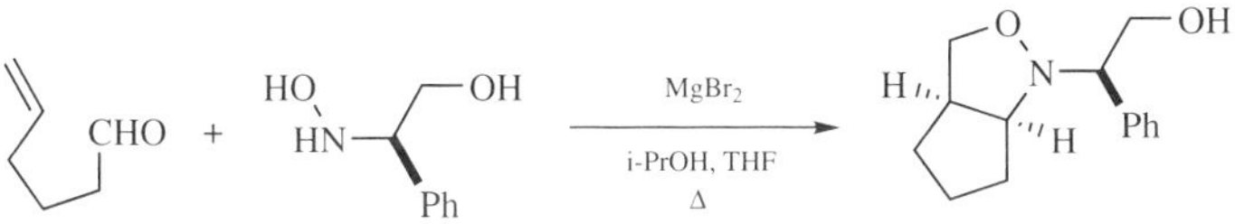

95%

Reduction.[4] $MgBr_2 \cdot OEt_2$ plays an important role in the reduction of 2-substituted phenyl acrylates to the substituted phenyl propanoates by Bu_3SnH in CH_2Cl_2. While substrates with an α-substituent are smoothly reduced, α-substituted acrylates are less well-suited for the reduction.

Diels-Alder reaction.[5] Chelation by $MgBr_2$ to bring diene and dienophile pairs together facilitates and renders stereo- and regiocontrol in the Diels-Alder reaction. The reaction of 2-cyclopentenone, a poor dienophile, to form an adduct is an example.

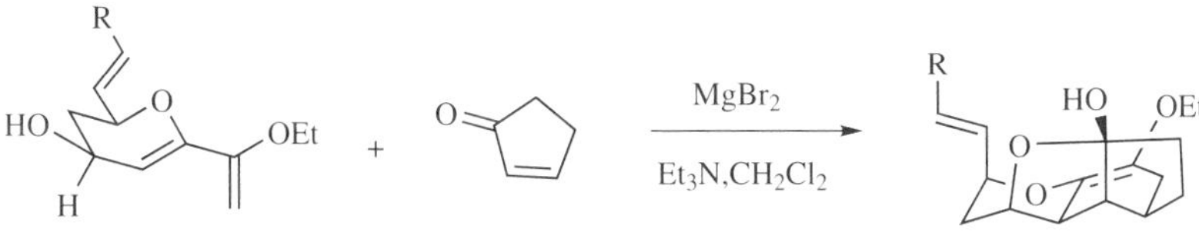

23%

[1]Mulder, J.A., Kurtz, K.C.M., Hsung, R.P., Coverdale, H., Frederick, M.O., Shen, L., Zificsak, C.A. *OL* **5**, 1547 (2003).
[2]Wei, H.-X., Jasoni, R.L., Hu, J., Li, G., Pare, P.W. *T* **60**, 10233 (2004).
[3]Hanselmann, R., Zhou, J., Ma, P., Confalone, P.N. *JOC* **68**, 8739 (2003).
[4]Hirasawa, S., Nagano, H., Kameda, Y. *TL* **45**, 2207 (2004).
[5]Barriault, L., Ang, P.J.A., Lavigne, R.M.A. *OL* **6**, 1317 (2004).

Magnesium chloride.

Michael reaction.[1] Dimethylsilyl enolates act as Michael donors in reaction with enones when promoted by $MgCl_2$ in DMF. Addition of chloride ion to silicon is the key. While $CaCl_2$ has approximately the same effect, $BaCl_2$, LiCl and Bu_4NCl are useless.

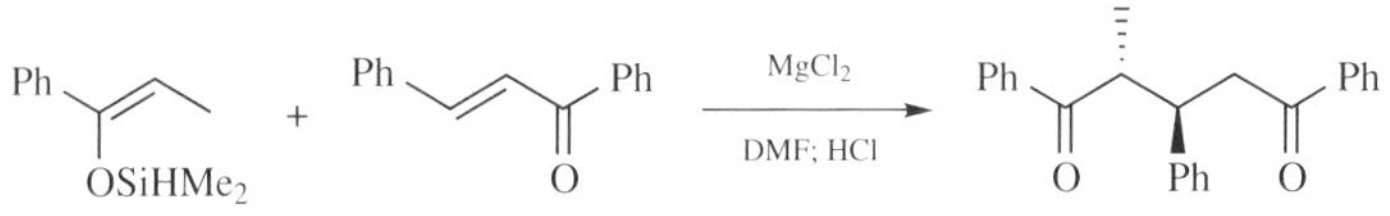

Ph OSiHMe$_2$ + Ph Ph O —$MgCl_2$ / DMF; HCl→ Ph Ph O Ph O

100% (*anti/syn* 98:2)

[1]Miura, K., Nakagawa, T., Hosomi, A. *SL* 2068 (2003).

Magnesium iodide. 20, 232; **21**, 261; **22**, 278

Transformations of cyclopropanes. 2-Methylenecyclopropanecarboxamides undergo isomerization to provide conjugated γ-lactams[1] on heating with MgI_2 in THF. In the presence of dipolarophiles, cycloaddition takes place.[2]

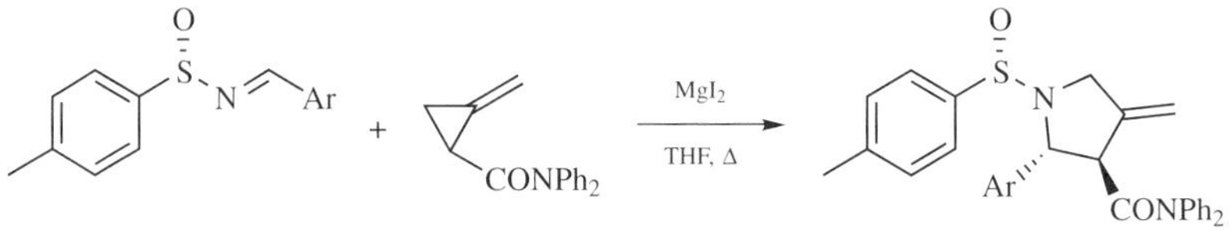

O S N Ar + CONPh$_2$ —MgI_2 / THF, Δ→ O S N Ar CONPh$_2$

Pyrrolidines are generated from imines and polarized cyclopropanes.[3] Heavily functionalized cyclopropanes also react with nitrones in a MgI_2-catalyzed process, tetrahydro-1,2-oxazines are formed.[4]

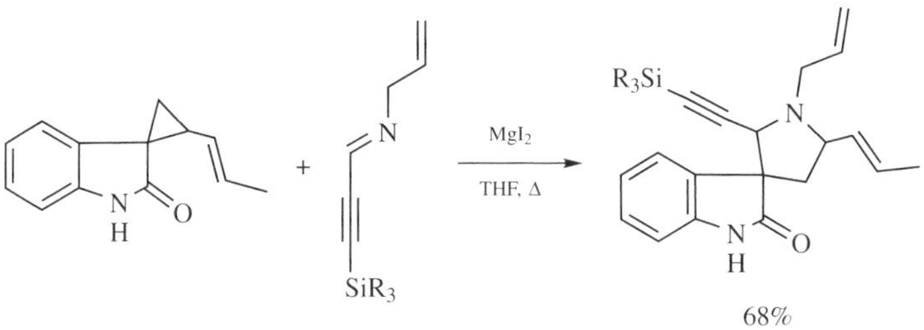

N H O + N SiR$_3$ —MgI_2 / THF, Δ→ R$_3$Si N N H O

68%

Aldol reactions. Direct aldol reaction proceeds in reasonably good yields between ketones and aldehydes by admixing with MgI_2-$(i\text{-}Pr)_2NEt$ in CH_2Cl_2.[5] However, the scope of the reaction has not been fully researched. Also the synthesis of β-iodinated Baylis-Hillman adducts from menthyl propynoate and aldehydes proceeds with only moderate asymmetric induction.[6]

[1]Lautens, M., Han, W., Liu, J.H.-C. *JACS* **125**, 4028 (2003).
[2]Scott, M.E., Han, W., Lautens, M. *OL* **6**, 3309 (2004).
[3]Meyers, C., Carreira, E. *ACIEE* **42**, 694 (2003).
[4]Ganton, M.D., Kerr, M.A. *JOC* **69**, 8554 (2004).
[5]Wei, H.-X., Li, K., Zhang, Q., Jasoni, R.L., Hu, J., Pare, P.W. *HCA* **87**, 2354 (2004).
[6]Wei, H.-X., Chen, D., Xu, X., Li, G., Pare, P.W. *TA* **14**, 971 (2003).

Magnesium monoperoxyphthalate, MMPP. 14, 197; **16**, 199–200; **18**, 228; **19**, 207–208; **22**, 279

Oxidation. MMPP oxidizes glycosyl sulfides to sulfoxides, short reaction time is achieved with microwave irradiation.[1]

It also shows great value in the selective cleavage of the N-N bond in hydrazides such as that links a β-lactam unit.[2]

OMe; O; N; N; BnO; + ; O; O; O; Mg; O; O; MeOH; O; NH; BnO; 88%; + ; OMe; N; O

[1]Chen, M.-Y., Patkar, L.N., Lin, C.-C *JOC* **69**, 2884 (2004).
[2]Fernandez, R., Ferrete, A., Llera, J.M., Magriz, A., Martin-Zamora, E., Diez, E., Lassaletta, J.M. *CEJ* **10**, 737 (2004).

Magnesium oxide. 22, 279

N-Sulfonylation. The previous reported acylation procedure for amines has been applied to sulfonylation in aqueous organic solvent systems in the presence of MgO.[1] Of course hydroxyl groups remain free.

[1]Kang, H.H., Rho, H.-S., Kim, D.-H., Oh, S.-G. *TL* **44**, 7225 (2003).

Magnesium perchlorate. 18, 228; **19**, 208; **22**, 279–280

Acylation. Alcohols, thiols, and anilines are successfully acylated in neat state using $Mg(ClO_4)_2$ as catalyst.[1,2] Acids and aliphatic RN=C=O combine (and decarboxylate in situ) to provide amides[3] by warming with $Mg(ClO_4)_2$. Calcium perchlorate also shows considerable catalytic activity.

t-Butyl esters are prepared by a $Mg(ClO_4)_2$-catalyzed reaction between carboxylic acids and Boc_2O at room temperature.[4]

1,3-Dienes.[5] 1,2-Disubstituted cyclopropyl tosylates ionize and suffer ring opening on contact with $Mg(ClO_4)_2$. In the presence of a base 2-substituted 1,3-dienes are formed.

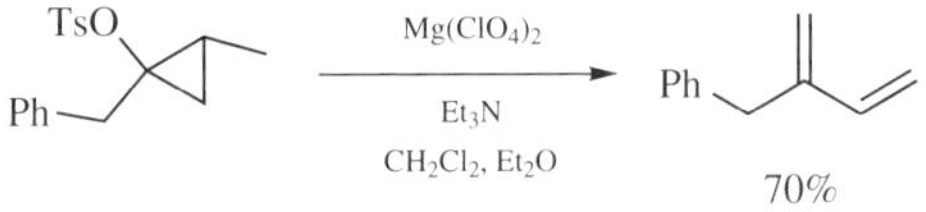

[1]Chakraborti, A.K., Sharma, L., Gulhane, R., Shivani. *T* **59**, 7661 (2003).
[2]Bartoli, G., Bosco, M., Dalpozzo, R., Marcantoni, E., Massaccesi, M., Rinaldi, S., Sambri, L. *SL* 39 (2003).
[3]Gurtler, C., Danielmeier, K. *TL* **45**, 2515 (2004).
[4]Goossen, L., Dohring, A. *ASC* **345**, 943 (2003).
[5]Kozyrkov, Y. Yu., Kulinkovich, O.G. *SL* 344 (2004).

Manganese. 20, 233–234; **21**, 261; **22**, 280

1,4-Diazacycloalkanes. Finely powdered Mn in CF_3COOH is useful for reductive cyclization of diimines derived from 1,2- and 1,3-diamines.[1]

Additive alkylation. Together with catalytic amounts of $PbCl_2$ and Me_3SiCl (both essential), manganese is employed to mediate radical addition to electron-deficient alkenes such as acrylonitrile and reduction to generate anionic species for addition to carbonyl compounds. A 3-component assembly process is realized.[2] A similar system promotes cyclopropanation with $ClCH_2I$.

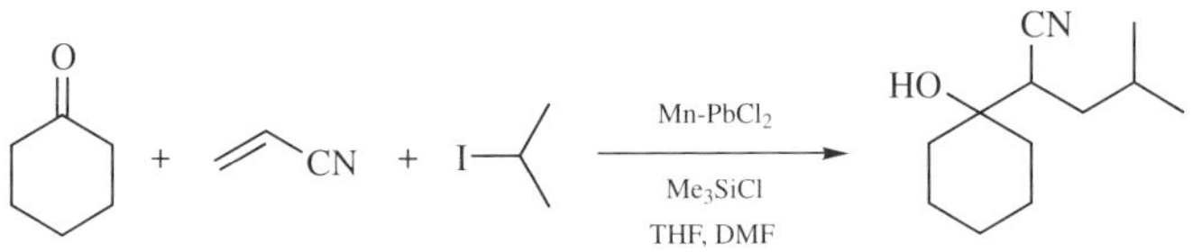

Pinacol coupling. Chemoselective coupling to form 3-methylene-1,2-cycloalkanediols is realized using a $CrCl_2$-catalyzed coupling by Mn – Me_3SiCl in DMF.[3] The double bond conjugated to the aldehyde survives such treatment.

Allylation. Asymmetric allylation and alkenylation use $CrCl_3$, Mn, Me_3SiCl, Et_3N and a salen-type chiral ligand.[4]

[1]Mercer, G.J., Sigman, M.S. *OL* **44**, 1591 (2003).
[2]Takai, K., Ueda, T., Ikeda, N., Ishiyama, T., Matsushita, H. *BCSJ* **76**, 347 (2003).
[3]Groth, U., Jung, M., Vogel, T. *SL* 1054 (2004).
[4]Berkessel, A., Menche, D., Sklorz, C.A., Schroder, M., Paterson, I. *ACIEE* **42**, 1032 (2003).

Manganese(III) acetate. 13, 171; **14**, 197–199; **16**, 200; **17**, 175–176; **18**, 229–230; **19**, 209–210; **20**, 234–235; **21**, 261–263; **22**, 280–281

Chlorobutyrolactones. Oxidation of monochloro- and dichloroacetic acids with $Mn(OAc)_3$· in the presence of alkenes leads to α-chlorinated γ-lactones in variable yields.[1]

Substitution of enones. α′-Acetoxylation of cycloalkenones with $Mn(OAc)_3$ is improved by using HOAc as a cosolvent (for benzene or MeCN).[2] Benzylation has also been realized.[3]

Biaryls. Homocoupling of $ArB(OH)_2$ is accomplished with $Mn(OAc)_3 \cdot 2H_2O$.[4]

[1]Snider, B.B., Che, Q. *H* **62**, 325 (2004).
[2]Demir, A.S., Reis, O., Igdir, A.C. *T* **60**, 3427 (2004).
[3]Tanyeli, C., Ozdemirhan, D. *TL* **44**, 731 (2003).
[4]Demir, A.S., Reis, O., Emrullahoglu, M. *JOC* **68**, 578 (2003).

Manganese dioxide. **14**, 200–201; **15**, 197–198; **18**, 230–231; **19**, 210; **20**, 237–238; **21**, 263; **22**, 282–283

Aromatic aldehydes.[1] $ArCH_2Br$ are oxidized in refluxing $CHCl_3$.

Bromination.[2] Monobromination of alkanes is achieved using MnO_2 as promoter.

Oxidation. Propargylic hydrazines on oxidation are transformed into allenyl azo compounds.[3] Exposure of the products to nucleophiles (e.g., MeOH) leads to pyrazole derivatives.

N, NH_2 — MnO_2, CH_2Cl_2, 0° — N=N, 92%

In the presence of noninterfering compounds that are reactive with the oxidation products tandem processes are sure to ensue. Besides Wittig reaction on a variety of phosphonium ylides, cyclopropanation of enals (generated from allyl alcohols) using stabilized sulfonium ylides is also viable.[4]

OH, Ph + S, COOEt — MnO_2, 4A-MS — Ph, O, COOEt, 100%

In the pyrolusite phase MnO_2 oxidizes only one alcohol function of pyridine-2,6-dimethanol. Bipyridinedimethanols are also oxidized in an analogous manner.[5]

[1]Goswami, S., Jana, S., Day, S., Adak, A.K. *CL* **34**, 194 (2005).
[2]Jiang, X., Shen, M., Tang, Y., Li, C. *TL* **46**, 487 (2005).
[3]Banert, K., Hagedorn, M., Schlott, J. *CL* **32**, 360 (2003).
[4]Oswald, M.F., Raw, S.A., Taylor, R.J.K. *OL* **6**, 3997 (2004).
[5]Ziessel, R., Nguyen, P., Douce, L., Cesario, M., Estournes, C. *OL* **6**, 2865 (2004).

Manganese(III) dipivaloylmethanate.

Hydrohydrazination.[1] Various alkenes and azodicarboxylic esters combine when treated with a hydrosilane and $Mn(dpm)_3$.

Boc–N=N–Boc + octahydronaphthalene → ($Mn(dpm)_3$, $PrSiH_3$ / i-PrOH, 0°) → Boc–N(NHBoc)-decalin, 74%

$Mn(dpm)_3$ =

[1]Waser, J., Carreira, E.M. *ACIEE* **43**, 4099 (2004).

Mercury(II) chloride.

Cyclization.[1] Thioureas in which one of the nitrogen atoms is bonded to an α-aminoacyl group cyclize to 2-iminoimidazolidin-4-ones by treatment with $HgCl_2$.

PhHN–C(=S)–NH–CH(CH$_2$CH(CH$_3$)$_2$)–C(=O)NH$_2$ → ($HgCl_2$, MeCN) → PhN=imidazolidinone, 99%

[1]Evindar, G., Batey, R.A. *OL* **5**, 1201 (2003).

Mercury(II) trifluoroacetate.

Overman rearrangement.[1] The trichloroacetimidate of 2,7-octadienol is induced to undergo an Overman rearrangement and mercurioamination by $Hg(OCOCF)_2$. The rearrangement does not occur in the presence of $Hg(OAc)_2$.

Hg(OCOCF$_3$)$_2$
THF
90%
Hg(OAc)$_2$

[1]Singh, O.M., Han, H. *OL* **6**, 3067 (2004).

Mercury(II) triflate.

Cyclization reactions. Transformation of many 1-alkyn-5-ones into 2-methylfurans[1] at room temperature requires only 5 mol% Hg(OTf)$_2$. 1,6-Enynes afford 2-hydroxyalkyl-1-methylenecyclopentanes and heterocyclic analogues in water-containing reaction media.[2] Five-membered ring formation seems most favorable as 1,7-enynes give mixtures and 1,8-enynes undergo hydration of the triple bond only.

Hg(OTf)$_2$
H$_2$O-MeCN-MeNO$_2$

X= O, NBs, CE$_2$ X=O 85%

A cyclization reaction study on polyenes reveals that the course is controlled by stability of the cationic intermediates.[3]

Hg(OTf)$_2$
TMU - MeNO$_2$;
NaCl, NaHCO$_3$
R = H 89%
R = Me 70%

4-Arylalkynes give dihydronaphthalenes on treatment with Hg(OTf)$_2$ – trimethylurea.[4]

[1]Imagawa, H., Kurisaki, T., Nishizawa, M. *OL* **6**, 3679 (2004).
[2]Nishizawa, M., Yadav, V.K., Skwarczynski, M., Takao, H., Imagawa, H., Sugihara, T. *OL* **5**, 1609 (2003).
[3]Takao, H., Wakabayashi, A., Takahashi, K., Imagawa, H., Sugihara, T., Nishizawa, M. *TL* **45**, 1079 (2004).
[4]Nishizawa, M., Takao, H., Yadav, V.K., Imagawa, H., Sugihara, T. *OL* **5**, 4563 (2003).

Methanesulfonic acid. 20, 240; **21**, 264–265; **22**, 284–285

Esterification. Deposition of MsOH on alumina is said to promote very selective formation of monoesters of diols at 80°.[1] More remarkably, ArCHO and 1,2-ethanediol are converted into $ArCOOCH_2CH_2OH$.[2]

Fries rearrangement.[3] A catalyst system consisting of MsOH and a metal triflate [$Y(OTf)_3$, $Sc(OTf)_3$] induces the Fries rearrangement at 100°.

[1]Sharghi, H., Sarvari, M.H. *T* **59**, 3627 (2003).
[2]Sharghi, H., Sarvari, M.H. *JOC* **68**, 4096 (2003).
[3]Mouhtady, O., Gaspard-Iloughmane, H., Roques, N., Le Roux, C. *TL* **44**, 6379 (2003).

***N*-Methanesulfonyl carboxamides.**

Acyl transfer.[1] The title compounds form salt with amines, which on thermolysis gives amides.

[1]Coniglio, S., Aramini, A., Cesta, M.C., Colagioia, S., Curti, R., D'Alessandro, F., D'Anniballe, G., D'Elia, V., Nano, G., Orlando, V., Allegretti, M. *TL* **45**, 5375 (2004).

1-Methoxycarbonylbenzoimidazole-3-oxide.

Methoxycarbonylation.[1] The title compound reacts with alcohols, usually at room temperature, to form mixed carbonates.

[1]Wuts, P.G.M., Ashford, S.W., Anderson, A.M., Atkins, J.R. *OL* **5**, 1483 (2003).

(Methoxycarbonylsulfamoyl)triethylammonium hydroxude.

Reaction with epoxides.[1] The Burgess reagent transforms epoxides into cyclic sulfamidates. Aryl-substituted epoxides give predominantly 7-membered ring products.

O + $Et_3\overset{+}{N}-SO_2\overset{-}{N}COOMe$ ⟶ OMe O N SO_2 O

[1]Rinner, U., Adams, D.R., dos Santos, M.L., Abboud, K.A., Hudlicky, T. *SL* 1247 (2003).

2-Methoxymethoxy-3-chloro-1-propene.

Methoxymethyl ethers.[1] The title reagent can be used for protection of alcohols as MOM ethers in an acid-catalyzed group exchange.

[1]Watanabe, Y., Ikemoto, T. *TL* **45**, 5795 (2004).

4-Methoxypyridine.

β-Amino acids.[1] A 3-step synthesis of the β-amino acids from the title reagent consists of *N*-benzoylation, Grignard reaction, and oxidative cleavage of the dihydropyridones with $NaIO_4$.

[1]Ege, M., Wanner, K.T. *OL* **6**, 3553 (2004).

Methylaluminum biphenyl-2,2′-bistriflamide.

Rearrangement.[1] Cyclization following a 1,2-rearrangement of γ-amino-α,β-unsaturated carbonyl compounds completes a novel synthesis of pyrroles.

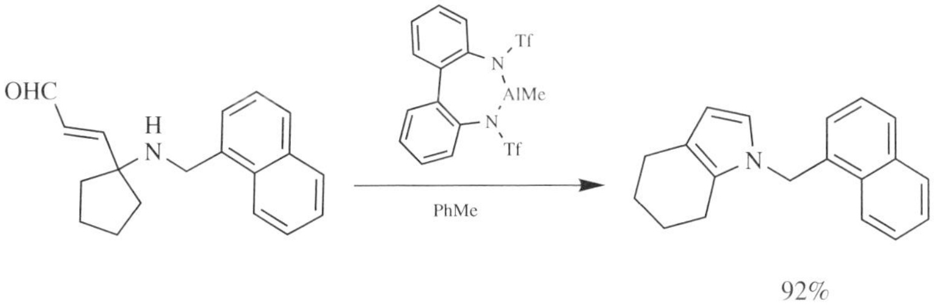

OHC H N Tf N AlMe N Tf PhMe N

92%

[1]Ooi, T., Ohmatsu, K., Ishii, H., Saito, A., Maruoka, K. *TL* **45**, 9315 (2004).

2-Methyl-6-nitrobenzoic anhydride. 22, 288

Esters and amides. This anhydride forms mixed anhydrides with carboxylic acids that are strong acylating agents for alcohols[1] and amines.[2]

[1]Shiina, I., Kubota, M., Oshiumi, H., Hashizume, M. *JOC* **69**, 1822 (2004).
[2]Shiina, I., Kawakita, Y. *T* **60**, 4729 (2004).

4-Methyl-1,2,4-triazoline-3,5-dione.

Indole protection.[1] The highly nucleophilic double bond between C-2 and C-3 of the indole nucleus is protected by forming adducts with the title compound. Regeneration of the indole system is by thermolysis.

[1]Baran, P.S., Guerrero, C.A., Corey, E.J. *OL* **5**, 1999 (2003).

Methyltrioxorhenium.

Alkylidenation. 3-Diazo-2-piperidinone reacts with aldehydes in the presence of $MeReO_3$ (with Ph_3P as additive) to afford the exocyclic conjugated lactams.

[1]Harrison, R., Mete, A., Wilson, L. *TL* **44**, 6621 (2003).

Methyltrioxorhenium - hydrogen peroxide. 19, 217; **20**, 248; **21**, 270; **22**, 288

Oxidations. 1,2-Diols are oxidized to the diketones in MeCN,[1] with higher reaction temperature and longer reaction time carboxylic acids are formed.[2]

There is also a report on Baeyer-Villiger oxidation by the title reagent in an ionic liquid.[3]

[1]Jain, S.L., Sharma, V.B., Sain, B. *TL* **45**, 1233 (2004).
[2]Jain, S.L., Sharma, V.B., Sain, B. *SC* **33**, 3875 (2003).
[3]Bernini, R., Coratti, A., Fabrizi, G., Goggiamani, A. *TL* **44**, 8991 (2003).

Methyltriphenylphosphinegold(I).

Addition to triple bond. Hydroamination of 1-alkynes with arylamines in the presence of $(Ph_3P)AuMe$ takes place to afford ketimines.

[1]Mizushima, E., Hayashi, T., Tanaka, M. *OLE* **5**, 3539 (2003).

Molecular sieves.

Benzhydryl ethers.[1] Acid-washed 4A-molecular sieves promote etherification of the primary hydroxyl group of sugars with $PhCH_2OH$ at room temperature. Tritylation is similarly accomplished.

Dipolar cycloaddition.[2] Hindered nitrile oxides and enolizable 1,3-cycloalkanediones give adducts. Molecular sieves shorten reaction times.

Cl, Cl, N, O, HO, O, +, i-PrOH / MS-4A, 50°, Cl, N, O, Cl, O, 80%

[1]Adinolfi, M., Barone, G., Iadonisi, A., Schiattarella, M. *TL* **44**, 3733 (2003).
[2]Matsuura, T., Bode, J.W., Hachisu, Y., Suzuki, K. *SL* 1746 (2003).

Molybdenum carbyne complexes.

Alkyne metathesis. The mixture of trinitridomolybdenum alkylidyne complex **1** and 2-trifluoromethylphenol is a highly active catalyst that converts ArCCR to ArCCCCAr. It is much better to use substrates with R = Et than R = Me because the 3-hexyne byproduct has a lower tendency to polymerize by the Mo catalyst, allowing its gainful turnover.

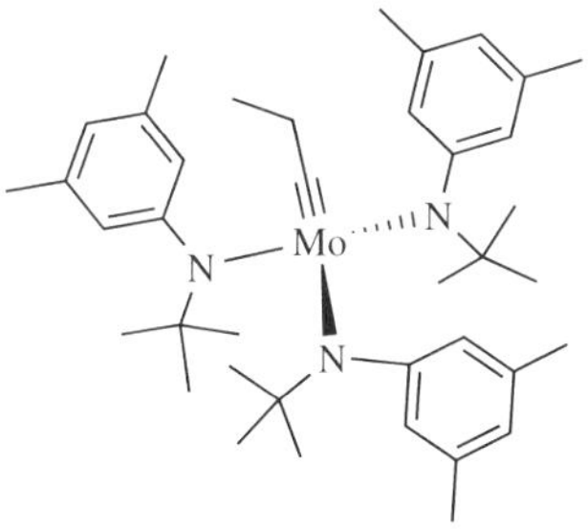

1

[1]Zhang, W., Kraft, S., Moore, J.S. *JACS* **126**, 329 (2004).

Molybdenum(V) chloride. 22, 291

Biaryls.[1] Alkylated anisoles are converted rapidly into biaryls using $MoCl_5$, forming 4,4′-dimethoxybiaryl derivatives. The title reagent is more effective than many others, the substrate may contain a *t*-butyl group.

Chlorination.[2] Aryloxyacetic esters undergo nuclear chlorination with $MoCl_5$. Because the activating effect of the sidechain on the reagent, an iodine atom present in the aromatic ring is retained. These substrates do not form biaryls.

[1]Mirk, D., Wibbeling, B., Frohlich, R., Waldvogel, S.R. *SL* 1970 (2004).
[2]Mirk, D., Kataeva, O., Frohlich, R., Waldvogel, S.R. *S* 2410 (2003).

Molybdenum hexacarbonyl. 13, 194–195; **15**, 225–226; **18**, 243–244; **19**, 221–222; **20**, 251–252; **21**, 273–274; **22**, 291–292

Deoxygenation.[1] α,β-Epoxy ketones and esters return to the conjugated carbonyl compounds on heating with $Mo(CO)_6$ in benzene.

Addition.[2] Typical of free-radical homologation process, polyhalo compounds (CCl_4, $Cl_3CCOOEt$) are split and add to alkenes in the presence of $Mo(CO)_6$.

Pauson-Khand reaction.[3] A bis(allene/yne) undergoes cycloaddition to give tetracyclic products. Large excess of $Mo(CO)_6$ is required to form two cyclopentenones.

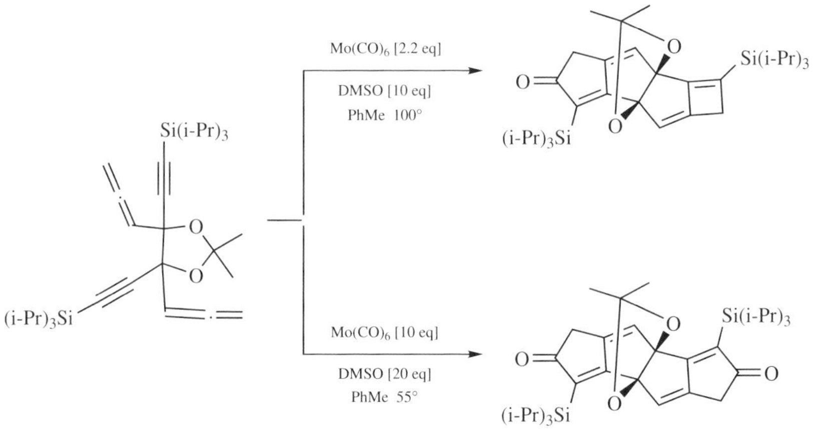

Si(i-Pr)$_3$

(i-Pr)$_3$Si

$Mo(CO)_6$ [2.2 eq]

DMSO [10 eq]

PhMe 100°

O

Si(i-Pr)$_3$

O

O

(i-Pr)$_3$Si

$Mo(CO)_6$ [10 eq]

DMSO [20 eq]

PhMe 55°

O

Si(i-Pr)$_3$

O

O

O

(i-Pr)$_3$Si

[1]Patra, A., Bandyopadhyay, M., Mal, D. *TL* **44**, 2355 (2003).
[2]Shvo, Y., Green, R. *JOMC* **675**, 77 (2003).
[3]Cao, H., Flippen-Anderson, J., Cook, J.M. *JACS* **125**, 3230 (2003).

N

Nafion-H. 14, 213; **18**, 246; **20**, 253–254; **21**, 275; **22**, 293

Addition and exchange reactions. Another version of Nafion-H (SAC-13) which is a silica nanocomposite, has much greater surface area and reactivity. It is useful for promoting addition of alcohols, thiols, and amines to enones.[1]

Mixed acetal formation from alcohols and $CH_2(OR)_2$ is simple to carry out with Nafion-H SAC-13.[2] Removal of TBS groups from nonaromatic silyl ethers is achieved using catalytic amount of Nafion-H and 1 equivalent of NaI.[3] Conversion of nonaromatic $ROSiMe_3$ to symmetrical ethers is usually complete within 1 h at room temperature by catalysis of Nafion-H.[4]

Glycosylation.[5] A simple operation involves activation of glycosyl donors as trichloroacetimidates in the presence of a polymer-supported amine and addition of acceptors together with Nafion-SAC.

[1]Wabnitz, T.C., Yu, J.-Q., Spencer, J.B. *SL* 1070 (2003).
[2]Ledneczki, I., Molnar, A. *SC* **34**, 3683 (2004).
[3]Rani, S., Babu, J.L., Vankar, Y.D. *SC* **33**, 4043 (2003).
[4]Zolfigol, M.A., Mohammadpoor-Baltork, I., Habibi, D., Mirjalili, B.F., Bamoniri, A. *TL* **44**, 8165 (2003).
[5]Oikawa, M., Tanaka, T., Fukuda, N., Kusumoto, S. *TL* **45**, 4039 (2004).

Neodymium(II) iodide.

Reductive alkylation.[1] NdI_2 is directly prepared from the elements by heating at 600°. It is stable in the solid state at room temperature, but once in contact with solvent it must be used immediately. The reduction power of NdI_2 is somewhere in between SmI_2-HMPA and alkali metal in liquid ammonia.

Ketones couple with RX to form tertiary alcohols on treatment with NdI_2.

[1]Evans, W.J., Workman, P.S., Allen, N.T. *OL* **5**, 2041 (2003).

Nickel. 12, 355; **13**, 197; **14**, 213; **18**, 246; **19**, 224; **20**, 253–254; **21**, 276; **22**, 294

Sonagashira coupling. Ultrafine nickel powder couples ArX (X = Br, I) and alkenyl iodides with 1-alkynes under heterogeneous conditions. Besides CuI and Ph_3P, the best results are obtained when the reaction is carried out in *i*-PrOH and with KOH as additive.[1]

[1]Wang, L., Li, P., Zhang, Y. *CC* 514 (2004).

Fiesers' Reagents for Organic Synthesis, Volume 23. Edited by Tse-Lok Ho

Nickel, Raney. 13, 265–266; **14**, 270; **15**, 278; **17**, 296; **18**, 246; **20**, 254; **21**, 276–277; **22**, 294–295

Reductions. With $HCOON_2H_5$ as hydrogen source azoarenes are reduced in the presence of Ra-Ni to either diarylhydrazines or anilines.[1] Reaction temperature and time determine which type of products. Transfer hydrogenation converts nitriles to amines very readily with Ra-Ni in 2-propanol containing 2% KOH. Amines are liberated from the *N*-alkyl-2-propylideneimines by hydrolysis.[2]

In acidic media ArBr are reduced to ArH and the C-X bond of $ArCH_2X$ (X = OH, OR, NHR) is cleaved by Ra-Ni at room temperature.[3] Carbonyl compounds are deoxygenated on refluxing with Raney alloy in water.[4]

Reductive alkylation of isatin with alcohols extends the previously reported alkylation of oxindoles.[5] In situ generation of oxindoles is implicated.

Diaryl ethers.[6] Unactivated ArI and phenols are coupled by heating with the Ra-Ni alloy, CuI and K_2CO_3. By this method even 2,4-di-*t*-butylphenol forms ether with PhI in 74% yield.

[1]Prasad, H.S., Gowda, S., Gowda, D.C. *SC* **34**, 1 (2004).
[2]Mebane, R.C., Jensen, D.R., Rickerd, K.R., Gross, B.H. *SC* **33**, 3373 (2003).
[3]Okimoto, M., Takahashi, Y., Nagata, Y., Satoh, M., Sueda, S., Yamashima, T. *BCSJ* **77**, 1405 (2004).
[4]Ishimoto, K., Mitoma, Y., Nagashima, S., Tashiro, H., Prakash, G.K.S., Olah, G.A., Yashiro, M. *CC* 514 (2003).
[5]Volk, B., Simig, G. *EJOC* 3991 (2003).
[6]Xu, L.-W., Xia, C.-G., Li, J.-W., Hu, X.-X. *SL* 2071 (2003).

Nickel-carbon. 21, 277; **22**, 295

Coupling reactions. Kumada coupling and *N*-arylation using Ni-C appear to be heterogeneous reactions, but a close study[1] has indicated that the nickel is leached into solution to perform its tasks but at the end readsorbed onto the carbon support.

[1]Lipshutz, B.H., Tasler, S., Chrisman, W., Spliethoff, B., Tesche, B. *JOC* **68**, 1177 (2003).

Nickel(II) acetylacetonate. 17, 201; **18**, 247–248; **19**, 225–226; **20**, 255–256; **21**, 277–280; **22**, 296–298

Coupling. The $Ni(acac)_2$–Et_2Zn couple reductively activates conjugated dienes so that aldimines are attacked that results in homoallylic amines.[1]

Me_3Si + H_2N–C_6H_4–OMe + OHC–furyl → ($Ni(acac)_2$, Et_2Zn THF)

81% (*syn:anti* >30:1)

[1]Kimura, M., Miyachi, A., Kojima, K., Tanaka, S., Tamaru, Y. *JACS* **126**, 14360 (2004).

Nickel bromide – amine complexes. 21, 280–281; **22**, 298–299

Arylations. Catalyzed by (bpy)$NiBr_2$ aryl halides are converted to arylzinc species which can be used to synthesize functionalized triarylphosphines[1] and aryl sulfides.[2]

Coupling reactions. Organosilicon compounds and unactivated secondary alkyl bromides are readily coupled in the presence of (diglyme)$NiBr_2$ and bathophenanthroline, CsF in DMSO.[3]

[1]Le Gall, E., Troupel, M., Nedelec, J.-Y. *T* **59**, 7497 (2003).
[2]Taniguchi, N. *JOC* **69**, 6904 (2004).
[3]Powell, D.A., Fu, G.C. *JACS* **126**, 7788 (2004).

Nickel bromide – DPPE - zinc. 21, 281; **22**, 299–300

Phthalides. Organozinc species derived from 2-haloaroic esters react aldehydes to form phthalides.[1] This Barbier-styled synthesis is simple because functional group compatibility is not an issue.

Cyclotrimerization. Allenes co-trimerize with arynes that are generated in situ. The method is suitable for synthesis of 9-methylene-9,10-dihydrophenanthrenes.[2]

SiMe$_3$ OTf + R —(dppe)$NiBr_2$, Zn - CsF, MeCN 80°→ R

Reductive coupling.[3] 2-Alkynoic esters form 3-zincio-2-alkenoic esters which behave as S_N2' nucleophiles toward 7-oxabicyclo[2.2.1]heptadienes.

[1]Rayabarapu, D.K., Chang, H.-T., Cheng, C.-H. *CEJ* **10**, 2991 (2004).
[2]Hsieh, J.-C., Rayabarapu, D.K., Cheng, C.-H. *CC* 532 (2004).
[3]Rayabarapu, D.K., Cheng, C.-H. *CEJ* **9**, 3164 (2003).

Nickel chloride. 22, 300

α-Ketol rearrangement. Catalyzed by $NiCl_2$ with a chiral ligand the rearrangement of certain α-ketols is efficient but unsatisfactory in asymmetric induction.[1]

Coupling reactions. With Mg as reducing agent, pinacol formation is a $NiCl_2$-mediated process.[2] Compared with Rieke nickel, this combination is much more practical.

Stille-type cross-coupling occurs between $ArSnCl_3$ and unactivated secondary alkyl halides by treatment with (bpy)$NiCl_2$ at moderate temperatures.[3]

Ni(O) reagents. Reduction of $NiCl_2$ (and other salts) by alkoxide-activated NaH generates Ni(0) species which mediates cyclization of (*o*-haloaryl)alkylamines in the presence of a hindered imidazolylidene.[4] This system promotes transfer hydrogenation of imines (a secondary sodium alkoxide as hydride source).[5]

[1]Brunner, H., Kagan, H.B., Kreutzer, G. *TA* **14**, 2177 (2003).
[2]Shi, L., Fan, C.-A., Tu, Y.-Q., Wang, M., Zhang, F.-M. *T* **60**, 2851 (2004).
[3]Powell, D.A., Maki, T., Fu, G.C. *JACS* **127**, 510 (2005).
[4]Omar-Amrani, R., Thomas, A., Brenner, E., Schneider, R., Fort, Y. *OL* **5**, 2311 (2003).
[5]Kuhl, S., Schneider, R., Fort, Y. *OM* **22**, 4184 (2003).

Nickel chloride – phosphine complexes. 14, 125; **15**, 122; **16**, 124; **18**, 250; **19**, 227–228; **20**, 258–259; **21**, 281–282; **22**, 300–301

Coupling reactions. The nickel-mediated coupling of arylboronic acids with various substituted arenes (ArX, ArOMs. . ..)[1] and that of aryl halides and 1-alkynes (also presence of CuI)[2] are alternative modes to the Pd-catalyzed Suzuki coupling and Sonagshira coupling, respectively.

A method for enyne synthesis involves coupling of alkenyltellurides with alkynyllithiums.[3] Alkynylzinc bromides react with ArCN to form alkynylarenes[4] in which process the cyano group is detached.

A 3-component coupling delivers 1,4-dienes stereoselectively.[5] The organic residues of ArI and alkenylzirconocene are added to the unsubstituted double bond of an allene. $Ni(cod)_2$ is totally inactive for the purpose.

R + ArI + Cl(Cp)$_2$Zr R′ —($(Ph_3P)_2NiCl_2$ - Zn, THF 50°)→ Ar, R, R′

E/Z ~ 99:1

Alkylzinc bromides prepared in situ (iodine-catalyzed) can be used to couple with aryl halides.[6]

Reformatsky reaction.[7] Nickel complexes exhibit as high a catalytic activity as $(Ph_3P)_3RhCl$ in the reaction promoted by Et_2Zn. β-Amino carbonyl compounds are obtained with imines generated in situ.

[1]Percec, V., Golding, G.M., Smidrkal, J., Weichold, O. *JOC* **69**, 3447 (2004).
[2]Beletskaya, I.P., Latyshev, G.V., Tsvetkov, A.V., Lukashev, N.V. *TL* **44**, 5011 (2003).
[3]Raminelli, C., Gargalaka, J., Silveira, C.C., Comasseto, J.V. *TL* **45**, 4927 (2004).
[4]Penney, J.M., Miller, J.A. *TL* **45**, 4989 (2004).
[5]Wu, M.-S., Rayabarapu, D.K., Cheng, C.-H. *JACS* **125**, 12426 (2003).
[6]Huo, S. *OL* **5**, 423 (2003).
[7]Adrian Jr, J.C., Snapper, M.L. *JOC* **68**, 2143 (2003).

Nickel chloride – zinc. 22, 301–302

Coupling reactions. Insertion of CO_2 into epoxides forms cyclic carbonates. A catalytic system composing $(Ph_3P)_2NiCl_2$, Ph_3P, Zn and Bu_4NBr is effective to bring about the transformation.[1]

Alkenylzirconium reagents submit the alkenyl residues to benzannulated 7-oxabicyclo[2.2.1]heptadienes while causing C-O bond scission.[2]

[1]Li, F., Xia, C., Xu, L., Sun, W., Chen, G. *CC* 2042 (2003).
[2]Wu, M.-S., Rayabarapu, D.K., Cheng, C.-H. *JOC* **69**, 8407 (2004).

Nickel cyanide.

Aryl cyanides.[1] The preparation from ArX may use $Ni(CN)_2$ in NMP with microwave irradiation instead of CuCN.

[1]Arvela, R.K., Leadbeater, N.E. *JOC* **68**, 9122 (2003).

Nickel peroxide.

Heteroaryl amides.[1] Substitution of heteroaryl halides by dialkylaminoacetonitrile anion followed by oxidation with NiO_2 gives rise to amides.

[1]Zhang, Z., Yin, Z., Kadow, J.F., Meanwell, N.A., Wang, T. *JOC* **69**, 1360 (2004).

Niobium(V) chloride.

Amidation. $NbCl_5$ promotes amide formation from acids and amines at moderate temperatures (40°–50°).[1]

Ether cleavage.[2] A hindered alkoxy group attached to an aromatic ring is cleaved selectively. ArOEt is more reactive than ArOMe.

R, OMe, OMe —($NbCl_5$, DCE)→ R, OH, OMe

R= Me, COOMe

ω-Allyllactams. Activation of ω-methoxylactams by $NbCl_5$ enables allyl transfer from allylsilanes.[3]

Pinacol coupling. The $NbCl_5$ -Zn system favors formation of the *dl*-1,2-diols from carbonyl compounds.[4]

Indenes. Reductive cyclization involving arylalkynes and aliphatic ketones is observed when they are heated with the thermally stable reagent derived from $NbCl_3$(dme) in 1,2-dichloroethane.[5] Interestingly, reaction in THF gives dienes.

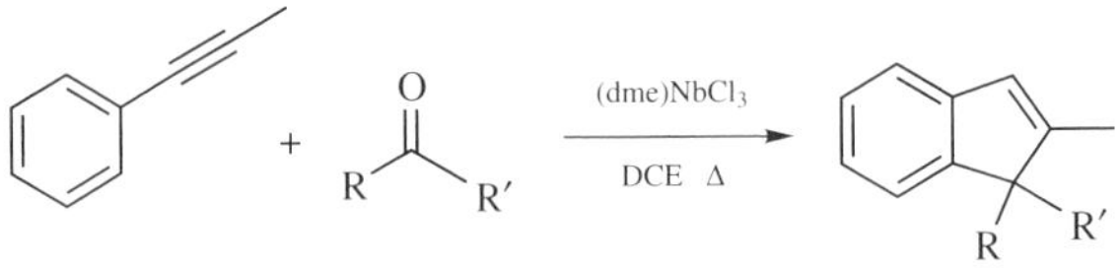

[1]Nery, M.S., Ribeiro, R.P., Lopes, C.C., Lopes, R.S.C. *S* 272 (2003).
[2]Arai, S., Sudo, Y., Nishida, A. *SL* 1104 (2004).
[3]Kleber, C., Andrade, Z., Matos, R.A.F. *SL* 1189 (2003).
[4]Arai, S., Sudo, Y., Nishida, A. *CPB* **52**, 287 (2004).
[5]Obora, Y., Kimura, M., Tokunaga, M., Tsuji, Y. *CC* 901 (2005).

Nitric acid. 18, 251–252; **19**, 228; **20**, 259; **21**, 281; **22**, 302

Oxidation. 2,2-Dichloroalkanals are oxidized to the corresponding 2,2-dichloroalkanoic acids by HNO_3 in dichloromethane.[1] Benzyl alcohols and their ethers are oxidized to carbonyl compounds.[2]

Nitration. A vanadyl phosphomolybdate catalyzes the nitration of alkanes with HNO_3 in HOAc.[3] For example, 1-nitroadamantane is obtained in 54% yield together with small amount of the 1,3-dinitro derivative. Pyridines undergo nitration in TFAA (product isolation after treatment with $Na_2S_2O_5$).[4] One or more halogen atoms of polyhalophenols is replaced by nitro group(s) on contact with HNO_3.[5] A new catalyst for arene nitration is lanthanum(III) *p*-nitrobenzenesulfonate.[6]

[1]Bellesia, F., De Buyck, L., Ghelfi, F., Pagnoni, U.M., Strazzolini, P. *SC* **34**, 1473 (2004).
[2]Strazzolini, P., Runcio, A. *EJOC* 526 (2003).
[3]Yamaguchi, K., Shinachi, S., Mizuno, N. *CC* 424 (2004).
[4]Katritzky, A.R., Scriven, E.F.V., Majumder, S., Akhmedova, R.G., Vakulenko, A.V., Akhmedov, N.G., Murugan, R., Abboud, K.A. *OBC* **3**, 538 (2005).
[5]Adimurthy, S., Vaghela, S.S., Vyas, P.V., Bhatt, A.K., Ramachandraiah, G., Bedekar, A.V. *TL* **44**, 6393 (2003).
[6]Parac-Vogt, T., Pachini, S., Nockemann, P., van Hecke, K., van Meervelt, L., Binnemans, K. *EJOC* 4560 (2004).

1-(*o*-Nitrophenyl)-2,2,2-trifluoroethanol.

Hydroxyl protection.[1] Ethers are derived from alcohols and the title reagent by a Mitsunobu reaction (Bu_3P - *N,N,N'N'*-tetramethylazodicarboxamide) but yields are not satisfactory. Therefore the synthetic utility of the protecting group must await further improvement of the etherification. Regeneration of the alcohols is performed photochemically.

[1]Specht, A., Goeldner, M. *ACIEE* **43**, 2008 (2004).

O

Organoaluminum reagents. 21, 287; **22**, 305, 462, 463–464

Homologation. Addition of Et_2AlCN to tosylhydrazones of aldehydes at room temperature furnishes the α-tosylhydrazino nitrile adducts. However, brief heating the mixture in THF also removes the TsN = NH unit.[1]

Transacylation. Me_3Al enables the reaction of carbamates with amines to produce substituted ureas, by transforming the amines to the very reactive Me_2Al-NR_2.[2] A double acylation of primary amines with (*Z*)-3-acylamino-2-alkenoic esters affords pyrimidinone derivatives.[3]

NHAc, COOMe + $PhNH_2$ → (Me_3Al, CH_2Cl_2, r.t.) N, NPh, O — 84%

Reactions of epoxides. Selenomethyl epoxides are attacked by organoaluminums via episelenonium intermediates. The reaction is regioselective and stereoselective.[4] Alkynylalanates open epoxides (with $BF_3 \cdot OEt_2$ as catalyst) at the more hindered site.[5]

Reduction accompanies rearrangement of a 1-benzyloxy-2,3-epoxide on treatment with Et_3Al to establish a quaternary center.[6]

BnO, O → (Et_3Al, THF) OBn, OH — 83%

Ethylation. Ethylation of imines with Et_3Al requires catalysts such as $Eu(dpm)_3$.[7]

Functionalization. The ate complex derived from LTMP and triisobutylaluminum metallates the *ortho* position of an n-donor substituent (OR, Cl, etc.) in aromatic compounds, enabling selective functionalization at that site.[8]

Deallylation. An allyl group at a fully substituted activated carbon atom is removable by a Ni-catalyzed process, using Et_3Al as reagent.[9]

Fiesers' Reagents for Organic Synthesis, Volume 23. Edited by Tse-Lok Ho

Et$_3$Al, (Ph$_3$P)$_2$NiBr$_2$

94%

Et$_3$Al, (Ph$_3$P)$_2$NiBr$_2$

Coupling reactions. Alkynyldimethylaluminum reagents are readily obtained from 1-alkynes and Me_3Al in the presence of Et_3N, and their use in a Pd-catalyzed coupling with ArX is reported.[10]

[1]Marczak, S., Wicha, J. *S* 1049 (2003).
[2]Lee, S.-H., Matsushita, H., Clapham, B., Janda, K.D. *T* **60**, 3439 (2004).
[3]Jeong, J.U., Chen, X., Rahman, A., Yamashita, D.S., Luengo, J.I. *OL* **6**, 1013 (2004).
[4]Sasaki, M., Hatta, M., Tanino, K., Miyashita, M. *TL* **45**, 1911 (2004).
[5]Zhao, H., Pagenkopf, B.L. *CC* 2592 (2003).
[6]Li, D.R., Xia, W.J., Tu, Y.Q., Zhang, F.M., Shi, L. *CC* 798 (2003).
[7]Tsvelikovsky, D., Gelman, D., Molander, G.A., Blum, J. *OL* **6**, 1995 (2004).
[8]Uchiyama, M., Naka, H., Matsumoto, Y., Ohwada, T. *JACS* **126**, 10526 (2004).
[9]Necas, D., Tursky, M., Kotora, M. *JACS* **126**, 10222 (2004).
[10]Wang, B., Bonin, M., Micouin, L. *OL* **6**, 3481 (2004).

Organocerium reagents. **13**, 206; **14**, 217–218; **15**, 221; **16**, 232; **17**, 205–207; **18**, 256; **19**, 231; **20**, 263–264; **22**, 305

Addition to 1,3-dichloroacetone.[1] Improved preparation of tertiary 1,3-dichloro-2-propanols is to employ combinations of RLi and $CeCl_3$. The alcohols are versatile precursors of epichlorohydrin and glycidol homologues.

[1]Chen, S.-T., Fang, J.-M. *JCCS(T)* **50**, 927 (2003).

Organocopper reagents. **13**, 207–209; **14**, 218–219; **15**, 221–227; **16**, 232–238; **17**, 207–218; **18**, 257–262; **19**, 232–235; **20**, 264–267; **21**, 287–290; **22**, 305–309

Reagents. Dilithium arylcuprates, obtained from ArTeBu by group excahnge with $Me_2Cu(CN)Li_2$, can be used to effect conjugate aryl transfer to enones.[1] The same technique is applicable to acquiring dilithium diallylcyanocuprate which is reactive toward enol triflates to produce 1,4-dienes.[2]

To access polyfunctional magnesium arylcuprates from ArI, exchange with **1** which is prepared from 1,5-dibromopentane (+Mg; CuCN·2LiCl) is recommended.[3]

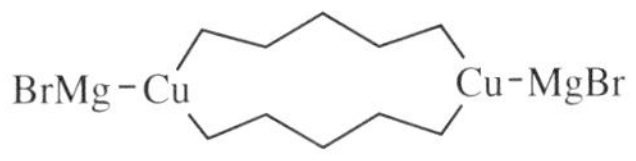

BrMg−Cu Cu−MgBr

1

2,3-Diiodoindole derivatives undergo cupration at C-2 selectively. After coupling (e.g., allylation) the iodine at C-3 becomes active toward cupration.[4]

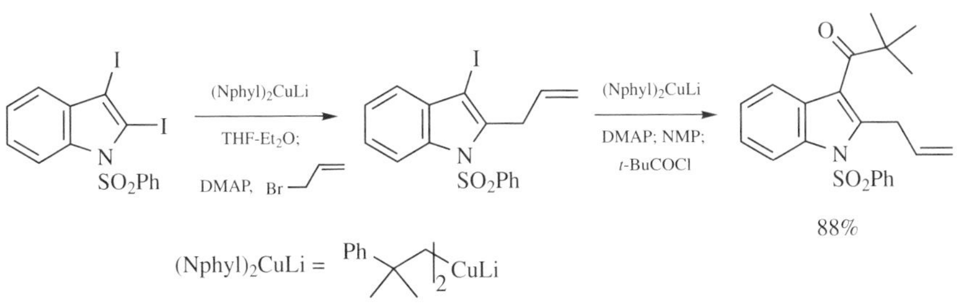

$(Nphyl)_2CuLi$; THF-Et_2O; DMAP, Br⁀allyl → ; $(Nphyl)_2CuLi$; DMAP; NMP; *t*-BuCOCl → 88%

$(Nphyl)_2CuLi$ =

Tin cuprates are readily prepared, e.g., from organolithiums. Their use is extensive in alkene synthesis and functionalization. Treatment of 4,5-dihydrofuran-2-yllithium with $(Bu_3Sn)_2Cu(CN)Li_2$ followed by quenching with NIS leads to (*Z*)-4-iodo-4-tributylstannyl-3-buten-1-ol.[5]

Additions. Stannylcopper reagents add to a triple bond of conjugate diynes[6] and thereby developing new synthetic possibilities. A new access to trifluoromethylalkenes is by way of regioselective and (*Z*)-selective carbocupration to 1,1,1-trifluoro-2-alkynes.[7] Regioselective elaboration of (*E*)-alken-1-yl carbamates from the alkynyl derivatives[8] may indicate chelating effect in operation.

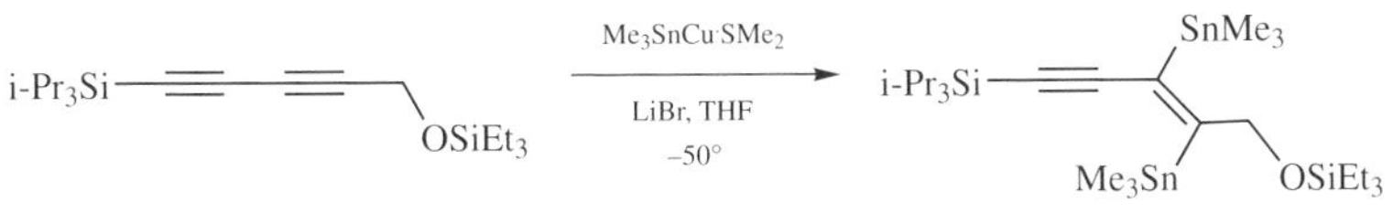

i-Pr_3Si–C≡C–C≡C–CH_2OSiEt_3 —($Me_3SnCu·SMe_2$; LiBr, THF, −50°)→ product with $SnMe_3$, Me_3Sn, $OSiEt_3$

Stannylcupration of allene by $Bu_3SnCu(CN)Li$ followed by reaction with proper electrophiles that replace the copper residue, allylic tin compounds are generated.[9] Higher order silylcuprates appear to react similarly, but equilibration of the adducts can give rise to alkenylsilanes or allylic silanes.[10]

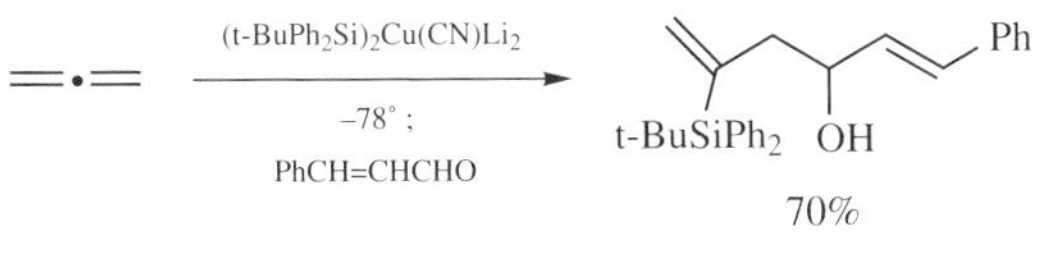

allene —($(t\text{-}BuPh_2Si)_2Cu(CN)Li_2$; −78°; PhCH=CHCHO)→ t-$BuSiPh_2$, OH, Ph product 70%

The steric course for conjugate addition of mono-organocopper reagents to chiral imides is remarkably dependent on the solvent and the presence of Me_3SiI.[11] In ether the substrates assume a *syn-s-cis* conformation during the addition, whereas in THF and presence of Me_3SiI the *anti-s-cis* conformation is preferred.

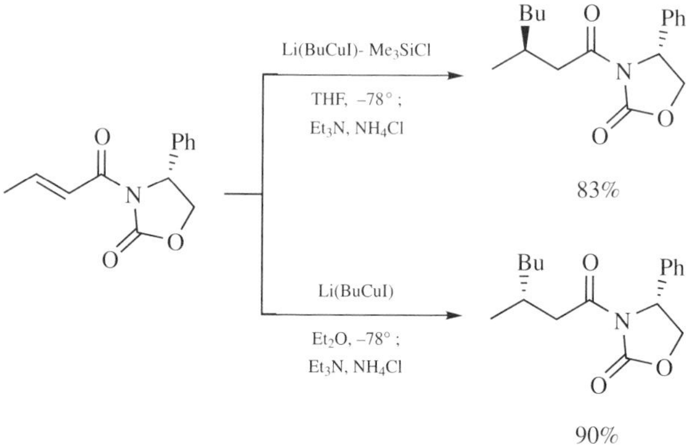

Li(BuCuI)- Me_3SiCl; THF, −78°; Et_3N, NH_4Cl → 83%

Li(BuCuI); Et_2O, −78°; Et_3N, NH_4Cl → 90%

The synthesis of 1-alkyl-3-hydroxy-2-naphthoic esters from 3-(2-alkoxycarbonylmethylphenyl)propynoic esters[12] in a tandem cuprate addition and Dieckmann cyclization is easily conceived but its realization is still valued.

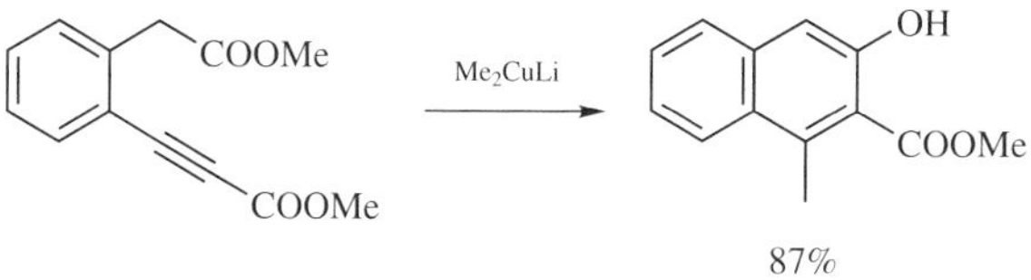

Me_2CuLi → 87%

Conjugate methylation of α,β-unsaturated esters is facilitated by chelation with a proximal triple bond.[13] The reaction of β,β-disubstituted α,β-unsaturated esters is best carried out with Me_2CuLi, Me_3SiCl in CH_2Cl_2.[14] While addition-elimination sequence on β-hetero enones has been known, perhaps the one-pot transformation of α-halo-β-diketones [with $(MeO)_3P$ and then cuprate reagents][15] still should be mentioned.

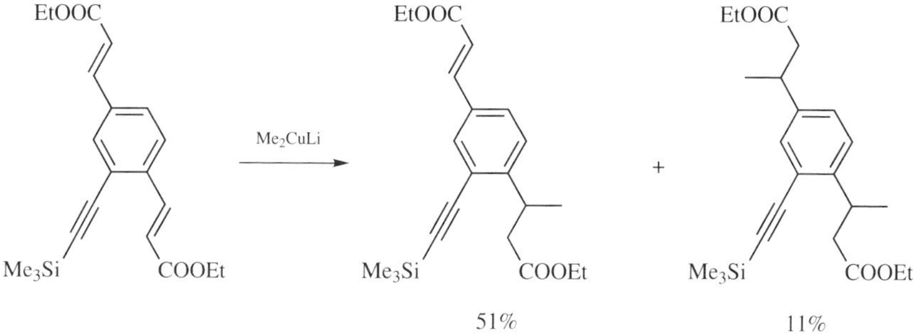

Me_2CuLi → 51% + 11%

A surrogate for "softened" enolate of acetic acid is the cyanocuprate reagent derived lithiated 2,4,4-trimethyloxazoline. It is a conjugate addend for enones with unhindered β-position.[16]

Amides. Carbamoyl chlorides are converted into tertiary amides on reaction with organocuprates.[17]

[1]Castelani, P., Luque, S., Comasseto, J.V. *TL* **45**, 4473 (2004).
[2]Castelani, P., Comasseto, J.V. *OM* **22**, 2108 (2003).
[3]Yang, X., Knochel, P. *SL* 81 (2004).
[4]Yang, X., Althammer, A., Knochel, P. *OL* **6**, 1665 (2004).
[5]Le Menez, P., Brion, J.-D., Betzer, J.-F., Pancrazi, A., Ardisson, J. *SL* 955 (2003).
[6]Simpkins, S.M.E., Kariuki, B.M., Arico, C.S., Cox, L.R. *OL* **5**, 3971 (2003).
[7]Konno, T., Daitoh, T., Noiri, A., Chae, J., Ishihara, T., Yamanaka, H. *OL* **6**, 933 (2004).
[8]Chechik-Lankin, H., Marek, I. *OL* **5**, 5087 (2003).
[9]Barbero, A., Pulido, F.J. *TL* **45**, 3765 (2004).
[10]Cuadrado, P., Gonzalez-Nogal, A.M., Sanchez, A., Sarmentero, M.A. *T* **59**, 5855 (2003).
[11]Dambacher, J., Anness, R., Pollock, P., Bergdahl, M. *T* **60**, 2097 (2004).
[12]Martinez, A.D., Deville, J.P., Stevens, J.L., Behar, V. *JOC* **69**, 991 (2004).
[13]Asao, N., Lee, S., Yamamoto, Y. *TL* **44**, 1803 (2003).
[14]Asao, N., Lee, S., Yamamoto, Y. *TL* **44**, 4265 (2003).
[15]Onishi, J.Y., Takuwa, T., Mukaiyama, T. *CL* **32**, 994 (2003).
[16]dos Santos, A.A., de Oliveira, A.R.M., Simonelli, F., de Marques, F.A., Clososki, G.C., Zarbin, P.H.G. *SL* 975 (2003).
[17]Lemoucheux, L., Seitz, T., Rouden, J., Lasne, M. *OL* **6**, 3703 (2004).

Organolithium reagents. 20, 268–269; **21**, 290–293; **22**, 309–311

Ketone synthesis. A practical route to certain α-amino ketones is via addition of ArLi to *N*-Boc-α-amino acids.[1] Enamino ketones are obtained from reaction of the corresponding esters with RLi in toluene.[2] [Solvent of low polarity such as toluene often favors reaction rates, also in addition to hindered and/or enolizable ketones.[3]]

Reaction with amides. *N*-Acylmorpholines are good precursors of acylsilanes (by reaction with R_3SiLi).[4] 2-Oxazolidinones, the cyclic carbamates, give amides from a low-temperature reaction.[5] *N*-Formylcarbazole furnishes products which expose the vulnerable aldehyde function only after reaction with RLi is complete, further manipulations aiming at formyl transformation can be carried out.[6]

N
CHO
RLi ;
$(EtO)_2P(O)CH_2COOEt$
R
COOEt

Substitutions. The benzenesulfonyl group of *N*-(α-sulfonylalkyl)carbamates is readily replaced. Specifically, the method is adaptable to a synthesis of β-amino alcohols.[7] 1,2-Disubstituted cyclobutenes are prepared from the benzenesulfonyl-cyclobutenes.[8]

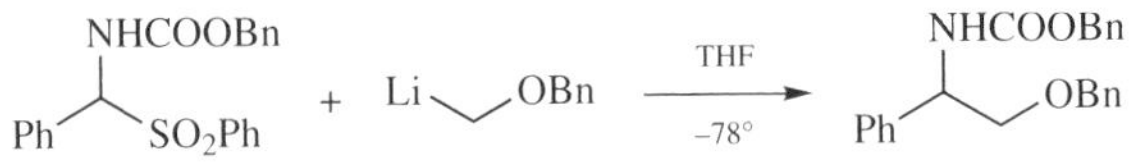

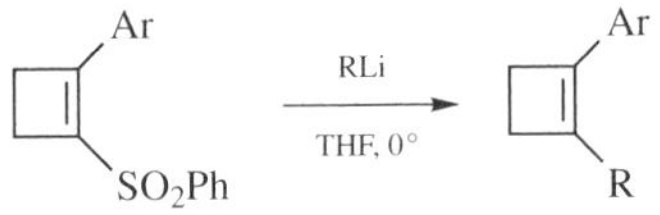

1-Boryl-1-silyl-1-alkenes are alternative precursors of acylsilanes. Their synthesis from the corresponding 1-boryl-1-bromoalkenes is straightforward. Attack by R_3SiLi at the boron atom initiates a 1,2-rearrangement and expulsion of the bromine atom.[9]

Active LiCN for nucleophilic opening of epoxides is generated from acetone cyanohydrin and RLi (MeLi is used as demonstration).[10] Elimination accompanies substitution in the reaction of β-alkoxyaziridines with organolithiums.[11]

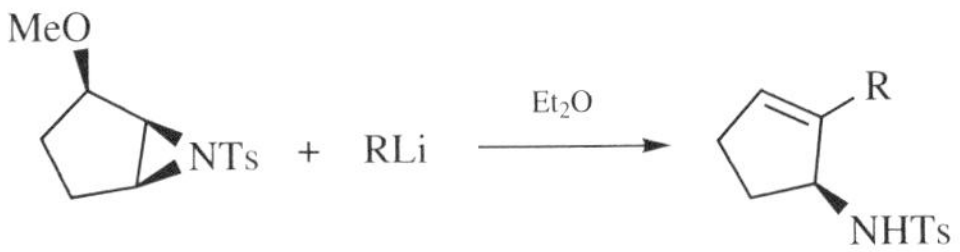

2-(2-Haloaryl)-4,4-dimethyloxazolines couple with ArLi without catalyst.[12] By allylic substitution of 1-trifluoromethyl-1-silylethenes with RLi functionalized 1,1-difluoro-1-alkenes are synthesized.[13] The silyl group can be used in coupling reactions, or removed on exposure to Bu_4NF. Perhaps by an addition-elimination sequence the amino group of β-(*N,N*-diemthylamino)-α,β-unsaturated ketones is replaced by a silyl residue on reaction with organosilyllithium reagents.[14]

Cyclization. *N*-Boc-2-vinylanilines are susceptible to attack by RLi. Addition of DMF to the resulting benzylic anions followed by acid workup affords 3-alkylindole derivatives.[15]

2,3-Disubstituted silacyclopentanes are formed on reaction of diphenylsilanes containing a vinyl and a homoallyl/propargyl groups on the silicon atom. The latter must also possess an anion-stabilizing residue at the far end.[16]

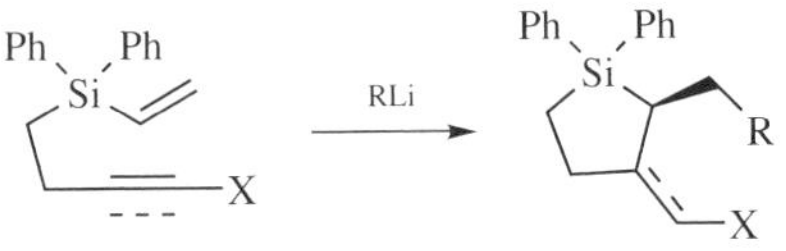

X = Ph, $SiMe_3$

2,5-Dihydropyrroles are formed in one step from the addition of lithiated alkoxyallenes to aldimines. A synthesis of (-)-detoxinine is based on this process.[17]

Li, OBn, + BnN, H, O, Ph, O, THF → OBn, H, O, Ph, N, Bn, H, O → OH, H, OH, N, H, COOH

53%

(–)-detoxinine

Rearrangement. Alkyllithiums induce conversion of alkynyloxiranes to allenyl ketones via dilithio ynenolates.[18]

R, O, R′ + BuLi → (H_3O^+) → O, R, C, R′

44–90%

[1]Florjancie, A.S., Sheppard, G.S. *S* 1653 (2003).
[2]Cimarelli, C., Palmieri, G., Volpini, E. *TL* **45**, 6629 (2004).
[3]Lecomte, V., Stephan, L., Le Bideau, F., Jaouen, G. *T* **59**, 2169 (2003).
[4]Clark, C.T., Milgram, B.C., Scheidt, K.A. *OL* **6**, 3977 (2004).
[5]Jones, S., Norton, H.C. *SL* 338 (2004).
[6]Dixon, D.J., Lucas, A.C. *SL* 1092 (2004).
[7]Jung, D.Y., Ko, C.H., Kim, Y.H. *SL* 1315 (2004).
[8]Knapp, K.M., Goldfuss, B., Knochel, P. *CEJ* **9**, 5259 (2003).
[9]Bhatt, N.G., Tamm, A., Gorena, A. *SL* 297 (2004).
[10]Ciaccio, J.A., Smrtka, M., Maio, W.A., Rucando, D. *TL* **45**, 7201 (2004).
[11]Rosser, C.M., Coote, S.C., Kirby, J.P., O'Brien, P., Caine, D. *OL* **6**, 4817 (2004).
[12]Astley, D., Saygi, H., Gezer, S., Astley, S.T. *TL* **45**, 7315 (2004).
[13]Ichikawa, J., Ishibashi, Y., Fukui, H. *TL* **44**, 707 (2003).
[14]Fleming, I., Marangon, E., Roni, C., Russell, M.G., Chamudis, S.T. *CC* 200 (2003).
[15]Coleman, C.M., O'Shea, D.F. *JACS* **125**, 4054 (2003).
[16]Wei, X., Taylor, R.J.K. *TL* **44**, 7143 (2003).
[17]Flogel, O., Gjislaine, M., Amombo, O., Reissig, H.-U., Zahn, G., Brudgam, I., Hartl, H. *CEJ* **9**, 1405 (2003).
[18]Denichoux, A., Ferreira, F., Chemla, F. *OL* **6**, 3509 (2004).

Organomagnesium reagents.

Addition reactions.[1] Lithium triorganomagnesiates exhibit higher reactivities toward carbonyl compounds than organolithiums and Grignard reagents. Due to their lower basicities the substrates are less prone to enolization in their presence. Mixed organomagnesiates show chemoselectivity such that the R group of RMe_2MgLi is preferentially transferred.

[1]Hatano, M., Matsumura, T., Ishihara, K. *OL* **7**, 573 (2005).

Organomanganese reagents. 22, 311

Coupling reactions.[1] *o*-Acylated chloroarenes undergo Cu-catalyzed nuclear substitution with RMnCl. From *o*-chloroaroyl chlorides stepwise reactions, first on the acyl group, can be performed. The different reactivities permit the use of two different organomanganese reagents in the transformation.

[1]Cahiez, G., Luart, D., Lecomte, F. *OL* **6**, 4395 (2004).

Organotitanium reagents. 21, 295; **22**, 312

Reaction with aldehydes. Pentadienyllithium species are converted into the corresponding titanium reagents on addition of (i-PrO)$_3$TiCl. Regioselective reaction with aldehydes compounds gives bishomoallylic alcohols (reaction at the central carbon atom).[1] Functionalized organotitanium reagents show chemoselectivity such that aldehydes are attacked cleanly in the presence of ketones.[2]

Coupling reaction.[3] Reagents such as (i-PrO)$_3$TiR are useful for cross-coupling to convert ArX (X = OTf, halide) to ArR which is catalyzed by Pd complexes.

[1]Zellner, A., Schlosser, M. *SL* 1016 (2001).
[2]Pastor, I.M., Yus, M. *T* **57**, 2365 (2001).
[3]Han, J.W., Tokunaga, N., Hauashi, T. *SL* 871 (2002).

Organozinc reagents. 13, 220–222; **14**, 233–235; **15**, 238–240; **16**, 246–248; **17**, 228–234; **18**, 264–265; **19**, 240–241; **20**, 270–275; **21**, 295–299; **22**, 312–315

Preparation. Using I/Zn exchange method to prepare organozinc reagents addition of nucleophilic catalysts such as Li(acac) and Cs(acac) increases the rates.[1]

Diorganozinc reagents in which the zinc atom is separated from an imino group by four CC bonds are obtained from imines after lithiation, metal exchange and alkylation (with RLi).[2] Such reagents are ready for Negishi coupling and probably other reactions.

NAr
LDA
$ZnCl_2$; BuLi
ArN ZnBu
$H_2C{=}CH_2$
hexanes 50°
ArN Zn Bu

Radical reactions. α-Radical of THF is generated in the presence of Me_2Zn and air. Subsequent reaction with aldehydes occurs at the β-position due to intramolecular hydrogen abstraction by the 2-peroxy radical to generate a radical at C-3.[3] More remarkable is an aminoalkylation of alkanes.[4]

THF + ArCHO → (Me_2Zn, O_2; THF, heat) → Ar, OH, O, OH

cyclopentane + Ph–CH=NTs → (Me_2Zn, O_2; BF_3Et_2O, r.t.) → NHTs, Ph

68%

Substitutions. Lithium organozincates open alkenyl epoxides at the allylic position by an S_N2 pathway.[5] Successive treatments of glycals with dimethyldioxirane and R_2Zn give *C*-glycosides,[6] while 2,3-dehydro analogues are prepared by a direct S_N2' substitution (catalyst: $BF_3 \cdot OEt_2$).[7]

Either S_N2 or S_N2' substitution can be controlled. (*Z*)-Allylic pentafluorobenzoates favor the S_N2' pathway.[8]

Cu(neophyl)Li, COOEt

OAc, I; THF, Et_2O → H, I, COOEt

77% (98% ee)

$ZnBr_2$; Bu, $OCOC_6F_6$ → Bu, H, COOEt

85% (95% ee)

α-Chloro ketones are transformed into α-alkylketones on reaction with Grignard reagents and $ZnCl_2$ (forming $RZnCl \cdot MgCl_2$ in situ) with $Cu(acac)_2$ as catalyst.[9] Amination of R_2Zn with $R'_2N\text{-}OBz$ is also catalyzed by Cu(II) salts.[10]

Addition to C=X compounds. In the presence of $CuCN \cdot 2LiCl$, organozinc reagents react with acid chlorides to form ketones.[1,11] Another useful catalyst is $CoBr_2$ (for converting ArZnX to ArCOR[12] and ArZnBr are prepared from ArBr, Zn, $ZnBr_2$, $CoBr_2$ and CF_3COOH in MeCN[13]).

Alkynylation mediated by organozinc and chiral ligands shows only moderate success (in terms of ee).[14,15]

Metallation of 1-boryl-1-bromoalkenes with R_2Zn leads to valuable synthetic intermediates. For example, further reaction with RCHO gives allylic alcohols.[16] The hydroxyalkylation also causes deboronation.

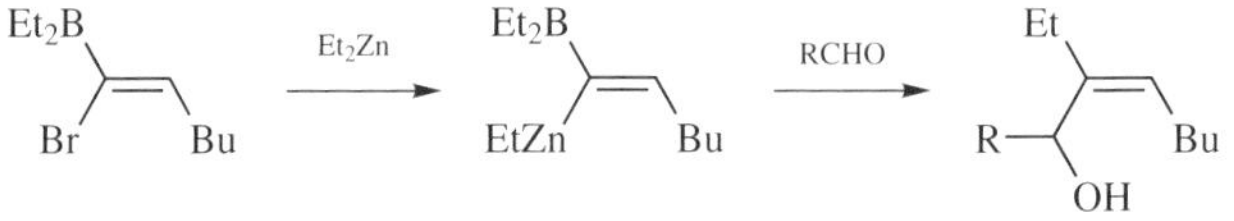

Alkenylboranes are transformed into zinciacyclopentenes which on reaction with carbonyl compounds furnish *cis*-3-hexene-1,6-diols.[17]

Alkenylzinc species are involved in the synthesis of propargylic amines starting from hydrozirconation of 1-alkynes and treatment with Me_2Zn prior to reaction with aldimines.[18] Formation of allylic and propargylic hydroxylamines from nitrones employing alkenylboronates[19] and 1-alkynes,[20] respectively, is subject to promotion by R_2Zn.

Organozinc reagents derived from 2-bromoallyltrimethylsilane and analogues (via lithiation with *t*-BuLi and Li/Zn exchange) are allylzinc species.[21] There must be a rapid transposition of [Me_3Si] and [ZnCl] groups.

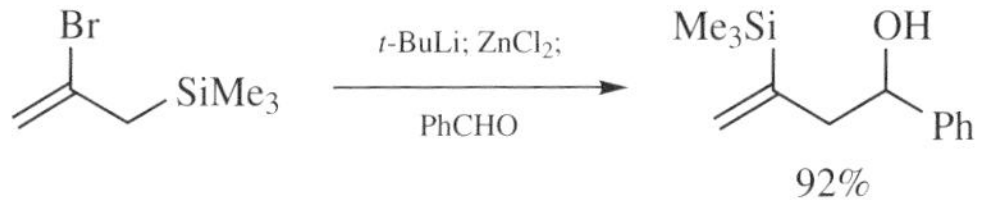

Addition to CC multiple bonds. Alkynyl sulfoxides undergo Cu-catalyzed conjugate addition with R_2Zn, placing the entrant group *trans* to the sulfur atom.[22] Arylzinc species derived from $ArB(OH)_2$ and Et_2Zn are useful for conjugate addition to enones.[23]

Unactivated alkynes are adequate acceptors for zincioenamine of β-aminocrotonamides to form 2-alkylideneacetoacetamides.[24] A regioselective hydrosilylation of 1-alkynes to provide 2-triorganosilyl-1-alkenes employs a zinc complex of biphenyl-2,2′-dioxide.[25]

Addition of *N*-zincio-alkenylhydrazines to alkenylboronates is stereoselective. The adducts can be captured by electrophiles (allyl halides, enones) to assemble highly functionalized molecules.[26]

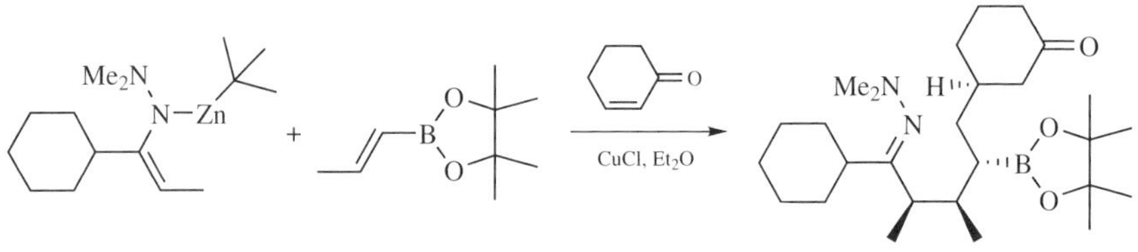

Cyclopropanes. Cyclopropanes are formed from organozinc carbenoids prepared from *N*-diethoxymethyl amides.[27] The reaction of ArCN with Et_2Zn using $(i\text{-PrO})_3TiMe$ as catalyst and some additives (LiI, *i*-PrOLi) leads to cyclopropylamines.[28]

$$\text{N-(diethoxymethyl)pyrrolidin-2-one} + \text{CH}_2\text{=CHCH}_2\text{Ph} \xrightarrow[\text{ZnCl}_2\text{ - Me}_3\text{SiCl},\ \text{Et}_2\text{O}\ \Delta]{\text{Zn(Hg)}} \text{N-(2-benzylcyclopropyl)pyrrolidin-2-one}$$

[1]Kneisel, F.F., Dochnahl, M., Knochel, P. *ACIEE* **43**, 1017 (2004).
[2]Nakamura, M., Hatakeyama, T., Nakamura, E. *JACS* **126**, 11820 (2004).
[3]Yamamoto, Y., Yamada, K., Tomioka, K. *TL* **45**, 795 (2004).
[4]Yamada, K., Yamamoto, Y., Maekawa, M., Chen, J., Tomioka, K. *TL* **45**, 6595 (2004).
[5]Equey, O., Vrancken, E., Alexakis, A. *EJOC* 2151 (2004).
[6]Xue, S., Han, K.-Z., He, L., Guo, Q.-X. *SL* 870 (2003).
[7]Xue, S., He, L., Han, K.-Z., Zheng, X.-Q., Guo, Q.-X. *CR* **340**, 303 (2005).
[8]Calaza, M.I., Yang, X., Soorukram, D., Knochel, P. *OL* **6**, 529 (2004).
[9]Malosh, C.F., Ready, J.M. *JACS* **126**, 10240 (2004).
[10]Berman, J.M., Johnson, J.S. *JACS* **126**, 5680 (2004).
[11]Yus, M., Gomis, J. *T* **59**, 4967 (2003).
[12]Fillon, H., Gosmini, C., Perichon, J. *T* **59**, 8199 (2003).
[13]Fillon, H., Gosmini, C., Perichon, J. *JACS* **125**, 3867 (2003).
[14]Cozzi, P.G. *ACIEE* **42**, 2895 (2003).
[15]Liu, L., Kang, Y., Wang, R., Zhou, Y., Chen, C., Ni, M., Gong, M. *TA* **15**, 3757 (2004).
[16]Chen, Y.K., Walsh, P.J. *JACS* **126**, 3702 (2004).
[17]Garcia, C., Libra,E.R., Carroll, P.J, Walsh, P.J. *JACS* **125**, 3210 (2003).
[18]Wipf, P., Kendall, C., Stephenson, C.R. *JACS* **125**, 761 (2003).
[19]Pandya, S.U., Pinet, S., Chavant, P.Y., Vallee, Y. *EJOC* 3621 (2003).
[20]Patel, S.K., Py, S., Pandya, S.U., Chavant, P.Y., Vallee, Y. *TA* **14**, 525 (2003).
[21]Viseux, E.M.E., Parsons, P.J., Pavey, J.B.J. *SL* 861 (2003).
[22]Maezaki, N., Sawamoto, H., Suzuki, T., Yoshigami, R., Tanaka, T. *JOC* **69**, 8387 (2004).
[23]Dong, L., Xu, Y.-J., Gong, L.-Z., Mi, A.-Q., Jiang, Y.-Z. *S* 1057 (2004).
[24]Nakamura, M., Fujimoto, T., Endo, K., Nakamura, E. *OL* **6**, 4837 (2004).
[25]Nakamura, S., Uchiyama, M., Ohwada, T. *JACS* **126**, 11146 (2004).
[26]Nakamura, S., Hatakeyama, T., Hara, K., Fukudome, H., Nakamura, E. *JACS* **126**, 14344 (2004).
[27]Begis, G., Cladingboel, D., Motherwell, W.B. *CC* 2656 (2003).
[28]Wiedemann, S., Frank, D., Winsel, H., de Meijere, A. *OL* **5**, 753 (2003).

Osmium.

Dihydroxylations and oxidation. Alkenes undergo dihydroxylation or cleavage (different conditions) by nanosized Os network and Oxone. The highly active and reusable Os material is prepared from decomposition of $Os_3(CO)_{12}$ in the presence of a mesoporous silicate.

[1]Lee, K., Kim, Y.-H., Han, S.B., Kang, H., Park, S., Seo, W.S., Park, J.-T., Kim, B., Chang, S. *JACS* **125**, 6844 (2003).

Osmium tetroxide. **13**, 222–225; **14**, 233–239; **15**, 240–241; **16**, 249–253; **17**, 236–240; **18**, 265–267; **19**, 241–242; **20**, 275–276; **21**, 301–303; **22**, 315–316

Dihydroxylations. Poly(ethylene glycol) is said to be a recyclable medium for dihydroxylation with OsO_4-NMO.[1] For improved asymmetric version the chiral ligands are quaternized with BnBr, and either PEG or an ionic liquid is added.[2] There is another report[3] heralding the benefit of ionic liquids in the same context.

Chiral α-ketols are prepared from allenes[4] and alkenyl sulfones provide α-hydroxy aldehydes[5] using the asymmetric dihydroxylation system. 1,5-Dienes furnish cis-2,5-bis(hydroxymethyl)tetrahydrofurans.[6]

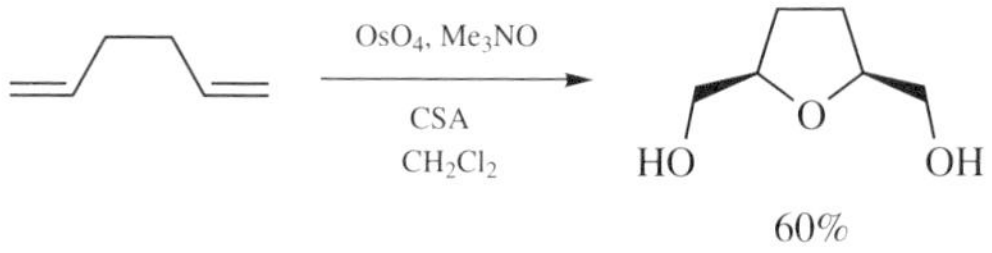

Modifications. In the catalytic system use of NaOCl as the re-oxidant makes it more practical.[7] Immobilization of OsO_4 onto Amberlite resins via osmylation of the residual vinyl groups gives rise to safe and recyclable catalysts[8] whose efficiency in catalyzing asymmetric dihydroxylation is not diminished. Microencapsulation of OsO_4 in polyurea is another possibility.[9]

The cyclic osmate ester derived from a perfluoroalkyl-extended tetramethylethene is an excellent and reusable catalyst because the derived trioxoosmium species (from oxidation with NMO) does not hydrolyze in situ, but the osmate esters after cycloaddition to substrate alkenes readily decompose to regenerate the catalyst and release the products. Furthermore, facile separation of the fluorous catalyst is another distinct advantage.[10]

[1]Chandrasekhar, S., Narsihmulu, C., Sultana, S.S., Reddy, N.R. *CC* 1716 (2003).
[2]Jiang, R., Kuang, Y., Sun, X., Zhang, S. *TA* **15**, 743 (2004).
[3]Branco, L.C., Afonso, C.A.M. *JOC* **69**, 4381 (2004).
[4]Fleming, S.A., Carroll, S.M., Hirschi, J., Liu, R., Pace, J.L., Redd, J.T. *TL* **45**, 3341 (2004).
[5]Evans, P., Leffray, M. *T* **59**, 7973 (2003).
[6]Donohoe, T.J., Butterworth, S. *ACIEE* **42**, 948 (2003).
[7]Mehltretter, G.M., Bhor, S., Klawonn, M., Dobler, C., Sundermeier, U., Eckert, M., Militzer, H.-C., Beller, M. *S* 295 (2003).
[8]Jo, C.H., Han, S.-H., Yang, J.W., Roh, E.J., Shin, U.-S., Song, C.E. *CC* 1312 (2003).
[9]Ley, S.V., Ramarao, C., Lee, A.-L., Ostergaard, N., Smith, S.C., Shirley, I.M. *OL* **5**, 185 (2003).
[10]Huang, Y., Meng, W.-D., Qing, F.-L. *TL* **45**, 1965 (2004).

Osmium tetroxide – sodium periodate.

Degradation. Efficiency of this well-known combination for cleaving double bonds is dramatically improved by adding 2,6-lutidine. As a comparison, yields of aldehydes from 5 terminal alkenes increase from 21–44% to 71–99%.[1] Lutidine suppresses formation of α-hydroxy ketone side products, increases rates of the desired reaction as to dramatically enhancing the yields. It also neutralizes acids therefore labile protecting groups are not harmed.

AcO◀(cyclohexene)◀OTBS —(OsO_4, $NaIO_4$; 2, 6-lutidine, dioxane, H_2O)→ OHC–CH(OTBS)–CH₂CH₂–CH(OAc)–CHO 77%

Aldehydes possessing a quaternary center near the formyl group and oxygen function are degraded by one carbon unit.

1-(COOMe)-1-(CH_2CHO)cyclohexane —(OsO_4, $NaIO_4$; NMO, acetone, H_2O)→ 1-(COOMe)-1-(CHO)cyclohexane 61%

[1]Yu, W., Mei, Y., Kang, Y., Hua, Z., Jin, Z. *OL* **6**, 3217 (2004).
[2]Belotti, D., Andreatta, G., Pradaux, F., BouzBouz, S., Cossy, J. *TL* **44**, 3613 (2003).

Osmium trichloride.

Hydroxylation reagent.[1] An Os/Cu-Al-hydrotalcite prepared from adding aq. $OsCl_3$, $CuCl_2$, $AlCl_3$ to Na_2CO_3 in 1M NaOH is active in catalyzing hydroxylation (together with NMO) of alkenes.

[1]Friedrich, H.B., Govender, M., Makhoba, X., Ngcobo, T.D., Onani, M.O. *CC* 2922 (2003).

Oxabis(triphenylphosphonium) triflate.

Sulfonamides and sulfonic esters.[1] Sulfonic acid salts are activated such that formation of their amides and esters is accomplished by addition of amines and alcohols, respectively.

[1]Caddick, S., Wilden, J.D., Judd, D.B. *JACS* **126**, 1024 (2004).

Oxalic acid.

Amides.[1] Heating ketones with $NH_2OH{\cdot}HCl$ and anhydrous oxalic acid without solvent gives amides.

[1]Cahndrasekhar, S., Gopalaiah, K. *TL* **44**, 7437 (2003).

Oxalyl bromide.

Bicyclic amidines.[1] An efficient synthesis of bicyclic amidines from *N*-ω-azidoalkyllactams is by reaction with oxalyl bromide.

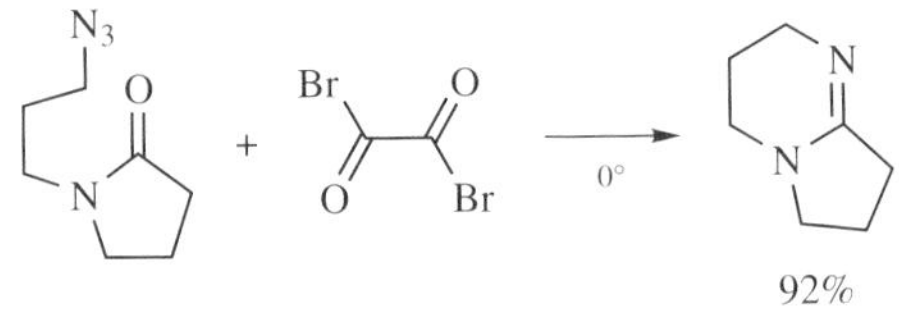

92%

[1]Kumagai, N., Matsunaga, S., Shibasaki, M. *ACIEE* **43**, 478 (2004).

Oxalyl chloride. 17, 241–242; **18**, 267–268; **19**, 243; **20**, 277; **21**, 304; **22**, 316–317

Aryl isocyanates.[1] When an aryl amine and oxalyl chloride are treated with HCl and then heated in dichlorobenzene at 130°, ArN = C = O is formed (9 examples, 71–98%).

β-Chloro cinnamic acids.[2] 3-Arylpropynoic acids undergo hydrochlorination in DMF at 0° with oxalyl chloride. While the reaction mixtures are quenched with water to provide chlorocinnamic acids, also a direct synthesis of esters and amides is easily achieved.

Sulfides.[3] On following the course of Swern oxidation and modifying the reactants (sulfoxide and alcohol) it is easy to put the focus on deoxygenation of sulfoxides that contain an α-hydrogen atom. The sulfonium ylides can be reduced with isopropanol.

Imidazoles.[4] Under the Swern oxidation conditions 2-arylmethylimidazolines (prepared from 2-aryl-1,1-dibromoethenes and ethylenediamine) undergo dehydrogenation.

[1]Oh, L.M., Spoors, P.G., Goodman, R.M. *TL* **45**, 4769 (2004).
[2]Urdaneta, N.A., Salazar, J., Herrera, J.C., Lopez, S.E. *SC* **34**, 657 (2004).
[3]Bhatia, G.S., Graczyk, P.P. *TL* **45**, 5193 (2004).
[4]Huh, D.H., Ryu, H., Kim, Y.G. *T* **60**, 9857 (2004).

Oxygen. 18, 268–269; **19**, 243–244; **20**, 277–279; **21**, 305–308; **22**, 317–321

Epoxidations. Several more oxidizing systems based on Ru complexes, *i*-PrCHO and molecular oxygen are found effective for epoxidation of cycloalkenes.[1-3] A tetraarylporphyrin-$RuCl_2$ complex and 2,6-dichloropyridine *N*-oxide catalyze epoxidation of styrenes with air. Because the metal ion is a Lewis acid the isolated products are arylacetaldehydes.[4]

Peroxides and hydroperoxides. A superior catalyst for aerobic epoxidation and hydroperoxysilylation of alkenes (to form R_3SiOOR') is $Co(dpm)_2$.[5] Propargylic hydroperoxides arise from the corresponding ethers via allenylzinc intermediates; oxygenation in the presence of $ZnCl_2$ and Me_3SiCl accomplishes the transformation.[6] Photooxygenation of 1,1-diarylethenes with TiO_2 catalyst leads to 3,3,6,6-tetraaryl-1,2-dioxanes.[7]

Oxidation of heterofunctionalities. Disulfide formation can be catalyzed by CsF/celite[8] or $VOCl_3$.[9] Co-salen complexes are useful for oxidation of tertiar amines to *N*-oxides,[10] whereas SnI_4 promotes oxygenation of tertiary arylphosphines.[11]

Oxidations. Numerous variants of catalytic systems for aerobic oxidation of alcohols continue to appear, some are listed as follows. Ru/Al_2O_3 in $PhCF_3$,[12] V_2O_5-K_2CO_3,[13] $VO(acac)_2$/polyaniline,[14] Mo/polyaniline[15] and poly(ethylene glycol)-stabilized Pd nanoparticles (in supercritical CO_2)[16] represent supported catalysts. TEMPO-mediated reactions are those involving (bipy)$CuBr_2$,[17] $NaNO_2$-Br_2,[18] or CAN.[19] Applications of *N*-hydroxyphthalimide are seen in the more recent development: with CAN,[20] and $VO(acac)_2$,[21] and Co(II) carboxylates that also oxidize silyl ethers to carbonyl compounds[22] and acetals (to esters).[23] Cu complexes coupled with azodicarboxylic ester under basic conditions are useful for oxidation of various primary alcohols.[24] Salicyl alcohols undergo photo-oxidation in the presence of a Ru(salen) complex in which the metal is also bonded to NO and Cl ligands.[25] A closely related system (a complex with an apical hydroxo ligand) is found to oxidize primary alcohols selectively in the presence of activated secondary alcohols.[26] However, vanadate apatite-calcium phosphate is a system suitable for oxidation of propargylic alcohols.[27]

For conversion of aldehydes to acids a pentanuclear yttrium hydroxo cluster is found effective.[28] Asymmetric induction by α-amino acids is observed (with moderate ee) in photohydroxylation at the α-position of aldehydes.[29]

Under basic conditions, mandelic acids[30] and α-oximino ketones[31] suffer oxidative cleavage by a Co(II)-catalyzed process and action of photogenerated singlet oxygen, respectively. β-Dicarbonyl compounds are hydroxylated at the α-position in a reaction catalyzed by $CeCl_3.7H_2O$.[32]

Benzylic oxidation has also enjoyed substantial attention. Activated carbon promotes oxidation of diarylmethanes including fluorene, xanthine, and thioxanthine to the corresponding ketones.[33] Routes for converting $ArCH_3$ to aroic acids based on $Co(OAc)_2$ with ancillary additives,[34,35] and via photoinitiated bromination[36] have been established.

Aerobic oxidation catalyzed by $PdCl_2$ in the presence of (-)-sparteine is enantioselective and useful for securing certain chiral alcohols, the solvent of choice is *t*-BuOH because yields are better.[37]

Dehydrogenation of primary amines to nitriles is successful employing the above-mentioned Ru/Al_2O_3 in $PhCF_3$ system[38] and $CuCl_2$.[39]

Coupling reactions. $Pd(OAc)_2$ performs oxidative coupling of alkynylsilanes[40] and $ArB(OH)_2$ with alkenes.[41] The former reaction requires a phosphine ligand.

[1]Qi, J.Y., Aiu, Q.Q., Lam, K.H., Yip, C.W., Zhou, Z.Y., Chan, A.S.C. *CC* 1058 (2003).
[2]Ragagnin, G., Knochel, P. *SL* 951 (2004).
[3]Srikanth, A., Nagendrapper, G., Chandrasekaran, S. *T* **59**, 7761 (2004).
[4]Chen, J., Che, C.-M. *ACIEE* **43**, 4950 (2004).
[5]O'Neill, P.M., Hindley, S., Pugh, M.D., Davies, J., Bray, P.G., Park, B.K., Kapu, D.S., Ward, S.A., Stocks, P.A. *TL* **44**, 8135 (2003).
[6]Harada, T., Kutsuwa, E. *JOC* **68**, 5716 (2003).
[7]Maeda, H., Miyamoto, H., Mizuno, K. *CL* **33**, 462 (2004).
[8]Shah, S.T.A., Khan, K.M., Fecker, M., Voelter, W. *TL* **44**, 6789 (2003).
[9]Kirihara, M., Okubo, K., Uchiyama, T., Kato, Y., Ochiai, Y., Matsushita, S., Hatano, A., Kanamori, K. *CPB* **52**, 625 (2004).

[10]Jain, S.L., Sain, B. *ACIEE* **42**, 1265 (2003).
[11]Levason, W., Patel, R., Reid, G. *JOMC* **688**, 280 (2003).
[12]Yamaguchi, K., Mizuno, N. *CEJ* **9**, 4353 (2003).
[13]Velusamy, S., Punniyamurthy, T. *OL* **6**, 217 (2004).
[14]Reddy, S.R., Das, S., Punniyamurthy, T. *TL* **45**, 3561 (2004).
[15]Velusamy, S., Ahamed, M., Punniyamurthy, T. *OL* **6**, 4821 (2004).
[16]Hou, Z., Theyssen, N., Brinkmann, A., Leitner, W. *ACIEE* **44**, 1346 (2005).
[17]Gamez, P., Arends, I.W.C.E., Reedijk, J., Sheldon, R.A. *CC* 2414 (2003).
[18]Liu, R., Liang, X., Dong, C., Hu, X. *JACS* **126**, 4112 (2004).
[19]Kim, S.S., Jung, H.C. *S* 2135 (2003).
[20]Kim, S.S., Rajagopal, G. *SC* **34**, 2237 (2004).
[21]Figiel, P.J., Sobczak, J.M., Ziolkowski, J.J. *CC* 244 (2004).
[22]Karimi, B., Rajabi, J. *OL* **6**, 2841 (2004).
[23]Karimi, B., Rajabi, J. *S* 2373 (2003).
[24]Marko, I.E., Gautier, A., Dumeunier, R., Doda, K., Philippart, F., Brown, S.M., Urch, C.J. *ACIEE* **43**, 1588 (2004).
[25]Tashiro, A., Mitsuishi, A., Irie, R., Katsuki, T. *SL* 1868 (2003).
[26]Egami, H., Shimizu, H., Katsuki, T. *TL* **46**, 783 (2005).
[27]Maeda, Y., Washitake, Y., Nishimura, T., Iwai, K., Yamauchi, T., Uemura, S. *T* **60**, 9031 (2004).
[28]Roesky, P.W., Canseco-Melchor, G., Zulys, A. *CC* 738 (2004).
[29]Cordova, A., Sunden, H., Engqvist, M., Ibrahem, I., Casas, J. *JACS* **126**, 8914 (2004).
[30]Favier, I., Dunach, E., Hebrault, D., Desmurs, J.-R. *NJC* **28**, 62 (2004).
[31]Ocal, N., Yano, L.M., Erden, I. *TL* **44**, 6947 (2003).
[32]Christoffers, J., Werner, T., Unger, S., Frey, W. *EJOC* 425 (2003).
[33]Kawabata, H., Hayashi, M. *TL* **45**, 5457 (2004).
[34]Yang, F., Sun, J., Zheng, R., Qui, W., Tang, J., He, M. *T* **60**, 1225 (2004).
[35]Hirai, N., Sawatari, N., Nakamura, S., Sakaguchi, S., Ishii, Y. *JOC* **68**, 6587 (2003).
[36]Itoh, A., Kodama, T., Hashimoto, S., Masaki, Y. *S* 2289 (2003).
[37]Mandal, S.K., Jensen, D.R., Pugsley, J.S., Sigman, M.S. *JOC* **68**, 4600 (2003).
[38]Yamaguchi, K., Mizuno, N. *ACIEE* **42**, 1480 (2003).
[39]Maeda, Y., Nishimura, T., Uemura, S. *BCSJ* **76**, 2399 (2003).
[40]Yoshida, H., Yamaryo, Y., Ohshita, J., Kunai, A. *CC* 1510 (2003).
[41]Jung, Y.C., Mishra, R.K., Yoon, C.H., Jung, K.W. *OL* **5**, 2231 (2003).

P

Palladacycles. 21, 310–312; **22**, 322–323

Alkynone synthesis. For acylation of 1-alkynes with RCOCl, a palladacycle derived from benzhydrylhydroxylamine finds service as catalyst.[1]

H-Phosphonic acids. A water-tolerant and recyclable Pd complex of bisphosphine-substituted, polymer-linked dibenzo-1,4-oxazine promotes addition of H_3PO_2 to alkenes.[2]

Coupling reactions. Synthesis of allylstannanes[3] and alkynylstannanes[4] from the activated chlorides and $Bu_3SnSnBu_3$ is accomplished in the presence of a pincer complex derived from $PdBr_2$ and 1,3-bis(dimethylaminomethyl)benzene. Another pincer complex used for Heck reaction contains N,P-ligands: 2,2′-bis(diphenylphosphino) diphenylamine.[5] Various kinds of couplings can be mediated by Pd complexes based on di-2-pyridylmethylamines.[6]

Palladacycles having additional *N*-heterocyclic carbene ligand as exemplified by **1** are active coupling catalysts.[7] Double arylation of dimethylaminoethoxyl vinyl ether is under chelation control while catalyzed by an *o*-diarylphosphinylbenzylpalladium acetate dimer.[8]

Palladacycles containing a sulfiliminoaryl moiety,[9] a 2-(2'-dimethylaminobiphenyl) group,[10] and those constructed from 1,2-bis(2'-pyrazineethynyl)benzene[11] and ferrocene-based phosphinimine-phosphine ligand **2**,[12] cyclobutene-3,4-diphosphorane **3**,[13] and the ferrocene derivative **4**.[14]

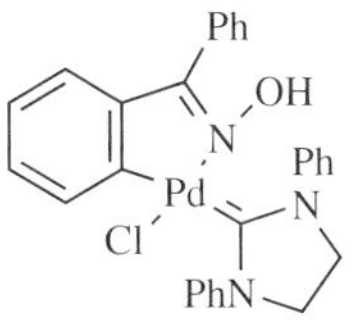

1 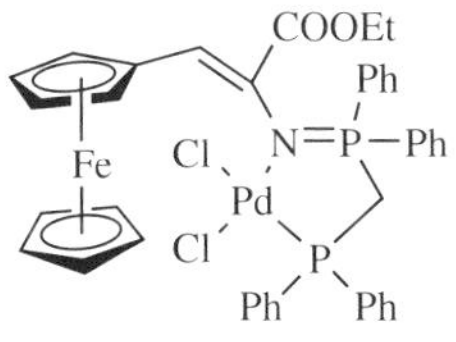 **2** 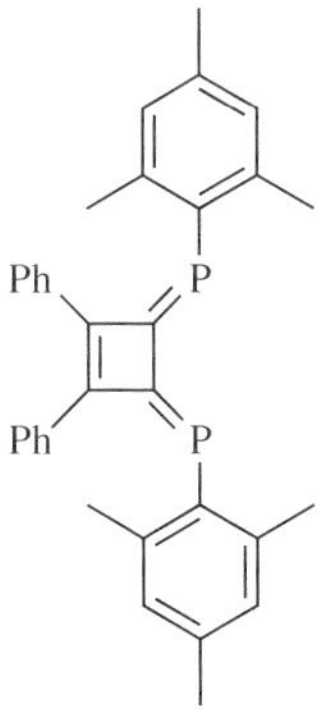 **3** 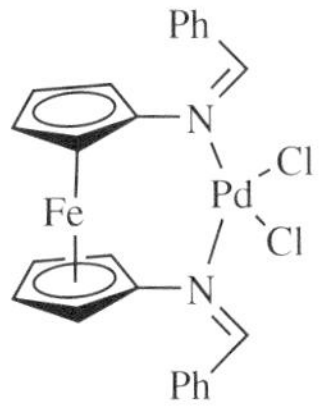 **4**

Fiesers' Reagents for Organic Synthesis, Volume 23. Edited by Tse-Lok Ho

Alkynes, acyl chlorides and aldimines are brought together to form substituted pyrroles in one step. It is suitable for synthesis of a large number of analogues. A palladacycle directs the efficient construction.[15]

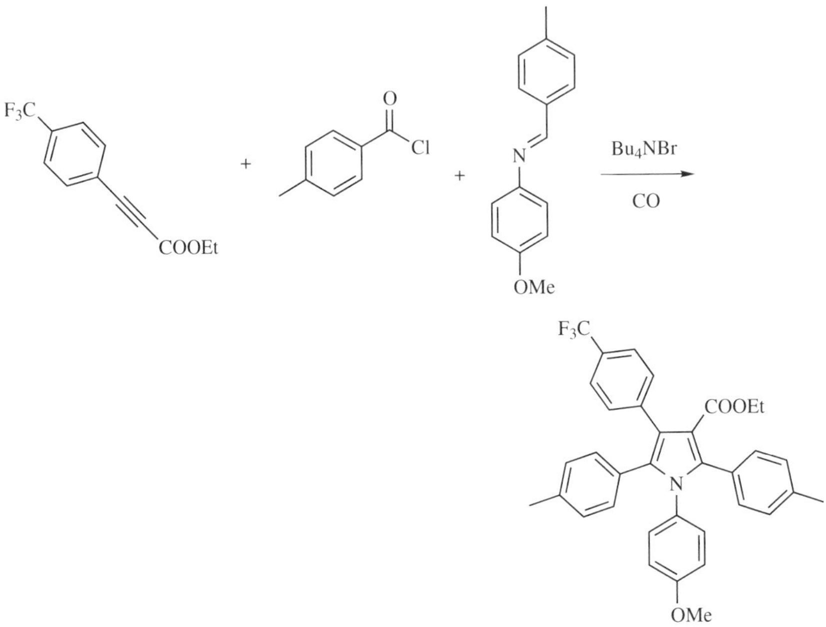

***N*-Arylation.** Many improvements on catalysts have made this reaction very efficient. 2-(2′-Di-*t*-butylphosphinylbiphenyl)palladium acetate is stable to air and heat, and it promotes arylation of amines by ArCl.[16] Room temperature, solvent-free *N*-arylation is made possible using a π-allylpalladium complex to a diphosphinidenecyclobutene ligand.[17]

Allylation. Palladacycles derived from 1,8-bis(diphenylphosphino)anthracene with Pd atom at C-9 are useful for allylation reactions, after exchanging the remaining ligand on Pd by an allyl group from allylstannanes.[18]

[1]Alonso, D.A., Najera, C., Pacheco, M.C. *JOC* **69**, 1615 (2004).
[2]Deprele, S., Montchamp, J. *OL* **6**, 3805 (2004).
[3]Wallner, O.A., Szabo, K.J. *OL* **6**, 1829 (2004).
[4]Kjellgren, J., Sunden, H., Szabo, K.J. *JACS* **126**, 474 (2004).
[5]Huang, M.-H., Liang, L.-C. *OM* **23**, 2813 (2004).
[6]Najera, C., Gil-Molto, J., Karlström, S., Falvello, L.R. *OL* **5**, 1451 (2003).
[7]Iyer, S., Jayanthi, A. *SL* 1125 (2003).
[8]Svennebring, A., Nilsson, P., Larhed, M. *JOC* **69**, 3345 (2004).
[9]Thakur, V.V., Kumar, N.S.C.R., Sudalai, A. *TL* **45**, 2915 (2004).

[10]Navarro, O., Kelly, III, R.A., Nolan, S.P. *JACS* **125**, 16194 (2003).
[11]Schultheiss, N., Barnes, C.L., Bosch, E. *SC* **34**, 1499 (2004).
[12]Arques, A., Aunon, D., Molina, P. *TL* **45**, 4337 (2004).
[13]Gajare, A.S., Jensen, R.S., Toyota, K., Yoshifuji, M., Ozawa, F. *SL* 144 (2005).
[14]Weng, Z., Koh, L.L., Hor, T.S.A. *JOMC* **689**, 18 (2004).
[15]Dhawan, R., Arndtsen, B.A. *JACS* **126**, 468 (2004).
[16]Zim, D., Buchwald, S.L. *OL* **5**, 2413 (2003).
[17]Gajare, A.S., Toyota, K., Yoshifuji, M., Ozawa, F. *JOC* **69**, 6504 (2004).
[18]Solin, N., Kjellgren, J., Szabo, K.J. *ACIEE* **42**, 3656 (2003).

Palladium$_o$

Substitution. Excellent results for enantioselective allylic substitution catalyzed by Pd nanoparticles (prepared by decomposition of $(dba)_3Pd_2$ by H_2 at room temperature in the presence of a chiral xylofuranoside diphosphite) are obtained.[1]

Highly active hydrogenation catalysts can be obtained by incarceration in polymers. One such catalyst prepared from $(Ph_3P)_4Pd$ and epoxide-containing copolymer shows good activity in hydrogenation and allylation.[2]

Coupling reactions. Palladium black in water with a PTC efficiently catalyzes the Suzuki coupling.[3] For Sonagashira coupling a combination of Pd powder with KF, CuI and Ph_3P are warmed with the substrates in aq. THF.[4]

Other types of Pd nanoparticles include those formed on treating $Pd(OAc)_2$ with Bu_4NOAc and the Pd-dodecanethiolate, the former are suitable for introducing another aryl group at the β-position (*E*-oriented) of cinnamic esters in ionic liquids,[5] and the latter, for Suzuki coupling at ambient temperature.[6]

[1]Jansat, S., Gomez, M., Phillippot, K., Müller, G., Guiu, E., Claver, C., Castillon, S., Chaudret, B. *JACS* **126**, 1592 (2004).
[2]Akiyama, R., Kobayashi, S. *JACS* **125**, 3412 (2003).
[3]Kuznetsov, A.G., Korolev, D.N., Bumagin, N.A. *RCB* **52**, 1882 (2003).
[4]Wang, L., Li, P. *SC* **33**, 3679 (2003).
[5]Calo, V., Nacci, A., Monopoli, A., Laera, S., Cioffi, N. *JOC* **68**, 2929 (2003).
[6]Lu, F., Ruiz, J., Astruc, D. *TL* **45**, 9443 (2004).

Palladium/carbon. **13**, 230–232; **15**, 245; **18**, 273; **19**, 247; **20**, 280–281; **21**, 312–314; **22**, 323–326

Hydrogenation. Catalytic hydrogenation of γ-hydroxy-α-methylenealkanoic esters by Pd/C in the presence of $MgBr_2$ gives predominantly the *syn* isomers,[1] the very high diastereoselectivity is due to chelation-control.

Hydrogenation of benzyl α-cyano-*o*-nitroarylacetates furnishes indoles directly.[2] This method has some merit because the substrates are readily prepared from *o*-nitroaryl halides.

Aromatic aldehydes are reduced in the transfer-hydrogenation system of Pd/C-HCOOK, but ketones are inert.[3]

Deprotection. PEG is a reusable medium for hydrogenolysis of *O*-benzyl group.[4] Selective cleavage of a simple benzyl group from phenolic ethers in preference to a *p*-methoxybenzyl residue is achievable using pyridine as catalyst poison.[5]

BnO ... OMe → Pd/C, Py, H_2, MeOH → HO ... OMe

98%

Allyl aryl ethers afford phenols under the catalytic hydrogenation conditions (in the presence of KOH in MeOH via a single electron transfer process), nitro group and benzylic allyl ethers are preserved.[6] Desilylation by Pd/C-catalyzed hydrogenation is performed in MeOH, but EtOAc nullifies the catalytic ability.[7] Accordingly, for cleavage of a benzyl ether while retaining a siloxy group that solvent should be chosen. With strictly neutral Pd/C desilylation involves hydrogen,[8] as acid released from certain commercial catalysts is responsible for solvolytic desilylation and the report that no hydrogen is required is erroneous.

In hydrogenolysis of ArX either hydrazine[9] or HCOONa[10] can be used as hydrogen source. In the latter report activated haloarenes are saturated (e.g., chlorophenol gives cyclohexanol) when the reaction is performed at 100°.

Low-valent sulfur compounds are poisonous to noble metal-catalysts, but there is a report of converting thio esters to aldehydes with Pd/C-Et_3SiH.[11]

Carbonylation and decarbonylation. Pd/C is an efficient catalyst for transforming ArI into ArCOOR under CO in the presence of Et_3N.[12] Under hydrothermal conditions Pd-catalyzed decarbonylation/decarboxylation occurs.[13] In the presence of base RCH_2CHO undergo aldol reaction, dehydration, decarbonylation and hydrogenation to afford $R(CH_2)_3R$.

HO ... COOH → Pd/C, H_2O, 250¡æ, 4-5MPa → HO ...

89%

HO ... CHO → Pd/C, H_2O, 250¡æ, 4-5MPa → HO ...

86%

Coupling reactions. The capability of Pd/C to catalyze useful couplings is recognized. With the proper additive(s) such as phosphines and copper salts, its application to Suzuki[14] and Sonagashira couplings[15-18] to produce valuable synthetic intermediates has appeared. Reductive Ullmann coupling[19] of ArI with zinc in water is catalyzed by Pd/C and under CO_2.

Thioesters react with diorganozinc reagents to give ketones in the presence of Pd/C, Zn, and Br_2.[20]

[1]Nagano, H., Yokota, M., Iwazaki, Y. *TL* **45**, 3035 (2004).
[2]Walkington, A., Gray, M., Hossner, F., Kitteringham, J., Voyle, M. *SC* **33**, 2229 (2003).
[3]Baidossi, M., Joshi, A.V., Mukhopadhyay, S., Sasson, Y. *SC* **34**, 643 (2004).
[4]Chandrasekhar, S., Shyamsunder, T., Chandrasekhar, G., Narsihmulu, C. *SL* 522 (2004).
[5]Sajiki, H., Hirota, K. *CPB* **51**, 320 (2003).
[6]Ishizaki, M., Yamada, M., Watanabe, S., Hoshino, O., Nishitani, K., Hayashida, M., Tanaka, A., Hara, H. *T* **60**, 7973 (2004).
[7]Sajiki, H., Ikawa, T., Hattori, K., Hirota, K. *CC* 654 (2003).
[8]Sajiki, H., Ikawa, T., Hirota, K. *TL* **44**, 7407 (2003).
[9]Cellier, P.P., Spindler, J.-F., Taillefer, M., Cristau, H.-J. *TL* **44**, 7191 (2003).
[10]Arcadi, A., Cerichelli, G., Chiarini, M., Vico, R., Zorzan, D. *EJOC* 3404 (2004).
[11]Miyazaki, T., Han-ya, Y., Tokuyama, H., Fukuyama, T. *SL* 477 (2004).
[12]Ramesh, C., Nakamura, R., Kubota, Y., Miwa, M., Sugi, Y. *S* 501 (2003).
[13]Matsubara, S., Yakota, Y., Oshima, K. *OL* **6**, 2071 (2004).
[14]Tagata, T., Nishida, M. *JOC* **68**, 9412 (2003).
[15]Fairlamb, I.J.S., Lu, F.J., Schmidt, J.P. *S* 2564 (2004).
[16]Novak, Z., Szabo, A., Repasi, J., Kotschy, A. *JOC* **68**, 3327 (2003).
[17]Pal, M., Subramanian, V., Yeleswarapu, K.R. *TL* **44**, 8221 (2003).
[18]Pal, M., Subramanian, V., Batchu, V.R., Dager, I. *SL* 1965 (2004).
[19]Li, J.-H., Xie, Y.-X., Yin, D.-L. *JOC* **68**, 9867 (2003).
[20]Mori, Y., Seki, M. *TL* **45**, 7343 (2004).

Palladium/inorganic support.

Hydrogenation/dehydrogenation. Semihydrogenation of alkynes catalyzed by $Pd/CaCO_3$ can be carried out in PEG.[1]

While transfer hydrogenation is readily performed with a Pd deposited onto mesoporous silicate molecular sieves,[2] Pd bound to calcium hydroxyapatite is good for dehydrogenation of indolines[3] and alcohols.[4]

Coupling reactions. Oxidative carbonylation of amines is a convenient way to symmetrical ureas (without using phosgene), and it can be achieved by Pd catalysis (zirconia sulfate supported) in the presence of oxygen.[5]

Suzuki coupling occurs on Pd-doped KF/Al_2O_3. Reaction conditions involve microwave irradiation without solvent.[6]

Solid catalysts by incorporation of Pd into the cages of zeolites perform excellently in coupling reactions. Thus, Heck reaction using such catalysts involves extraction of Pd into solution to activate aryl halides,[7–9] including ArCl.[8,9] Suzuki reaction also successfully employs such a catalyst.[10] The Sonagashira coupling catalyzed by copper-free Pd-zeolites generally proceeds well,[11] and a direct indole synthesis involving

2-iodoanilides[12] is achievable.

I
NHAc
+
Ph
Pd-zeolite
LiCl, Cs_2CO_3
DMF
Ph
N
H
82%

Another version of the supported catalyst is that prepared from in situ reduction of a Pd colloid layer inside the channels of mesoporous silica. Its utility is proven in the Heck reaction.[13]

Allylation. Homoallylic alcohols are formed by combining epoxides and allyl bromide in the presence of mesoporous silica-supported Pd(0) catalyst and indium(I) chloride. Epoxide rearrangement precedes allylation in the process.[14]

[1]Chandrasekhar, S., Narsihmulu, C., Chandrasekhar, G., Shyamsunder, T. *TL* **45**, 2421 (2004).
[2]Selvam, P., Sonavane, S.U., Mohapatra, S.K., Jayaram, R.V. *TL* **45**, 3071 (2004).
[3]Hara, T., Mori, K., Mizugaki, T., Ebitani, K., Kaneda, K. *TL* **44**, 6207 (2003).
[4]Mori, K., Hara, T., Mizugaki, T., Ebitani, K., Kaneda, K. *JACS* **126**, 10657 (2004).
[5]Shi, F., Dong, Y., SiMa, T., Yang, H. *TL* **42**, 2161 (2001).
[6]Kabalka, G.W., Wang, L., Pagni, R.M., Hair, C.M., Namboodiri, V. *S* 217 (2003).
[7]Waghmode, S.B., Wagholikar, S.G., Sivasanker, S. *BCSJ* **76**, 19897 (2003).
[8]Prockl, S.S., Kleist, W., Gruber, M.A., Kohler, K. *ACIEE* **43**, 1881 (2004).
[9]Srivastava, R., Venkatathri, N., Srinivas, D., Ratnasamy, P. *TL* **44**, 3649 (2003).
[10]Artok, L., Bulut, H. *TL* **45**, 3881 (2004).
[11]Djakovitch, L., Rollet, P. *TL* **45**, 1367 (2004).
[12]Hong, K.B., Lee, C.W., Yum, E.K. *TL* **45**, 693 (2004).
[13]Li, L., Shi, J.-I., Yan, J.-n. *CC* 1990 (2004).
[14]Jiang, N., Hu, Q., Reid, C.S., Lu, Y., Li, C.-J. *CC* 2318 (2003).

Palladium/polymer support.

Reductions. Palladium black deposited on polypropylene sheet is a highly selective catalyst for hydrogenation of alkenes.[1] Nanosized Pd catalyst supported on amphiphilic polymer is useful for hydrogenation and hydrodehalogenation.[2] A polymer-incarcerated Pd prepared from $(Ph_3P)_4Pd$ and a copolymer of styrene, and monomers of an epoxide and tetraethylene glycol is recoverable, reusable, highly active and sulfur-tolerant. Its utility in catalytic hydrogenation has been demonstrated.[3] With a recyclable polyurea-encapsulated $Pd(OAc)_2$ catalyst and HCOOH transfer hydrogenation of aryl ketones is achieved.[4]

Coupling reactions. The above catalyst is highly active for promoting Suzuki coupling in the presence of a triarylphosphine.[5] A different version of the Suzuki coupling catalyst is a poly(*N,N*-dialkylcarbodiimide)-stabilized Pd nanoparticle composite.[6] In pharmaceutical applications minimization of Pd contamination is of paramount importance, and a simple, recyclable, polymer-supported phosphine chelated Pd catalyst for Suzuki coupling leaves <1.1 ppm of Pd residue in the products can be achieved.[7]

A heterogeneous catalyst assembled from $(NH_4)_2PdCl_4$ and non-crosslinked amphiphilic polymer ligand is highly active in promoting the Heck reaction.[8]

[1]Maki, S., Okawa, M., Makii, T., Hirano, T., Niwa, H. *TL* **44**, 3717 (2003).
[2]Nakao, R., Rhee, H., Uozumi, Y. *OL* **7**, 163 (2005).
[3]Okamoto, K., Akiyama, R., Kobayashi, S. *JOC* **69**, 2871 (2004).
[4]Yu, J.-Q., Wu, H.-C., Ramarao, C., Spencer, J.B., Ley, S.V. *CC* 678 (2003).
[5]Okamoto, K., Akiyama, R., Kobayashi, S. *OL* **6**, 1987 (2004).
[6]Liu, Y., Khemtong, C., Hu, J. *CC* 398 (2004).
[7]Shieh, W.-C., Shekhar, R., Blacklock, T., Tedesco, A. *SC* **32**, 1059 (2002).
[8]Yamada, Y.M.A., Takeda, K., Takahashi, H., Ikegami, S. *T* **60**, 4097 (2004).

Palladium(II) acetate. **13**, 232–233; **14**, 248; **15**, 245–247; **16**, 259–263; **17**, 255–259; **18**, 274–277; **19**, 248–251; **20**, 281–283; **21**, 314–320; **22**, 326–330

Coupling reactions. Biphenylcarboxylic esters have been prepared from a Suzuki coupling of iodobenzoic esters which contain an ionic liquid tag (as part of the alkoxy group).[1] Alcoholysis of the products removes the *N*-(hydroxyalkyl)imidazolium residue. Polymeric support loaded with imidazolium end-groups has been used to ligate $Pd(OAc)_2$ and the resulting catalyst tests successfully in the Suzuki coupling.[2] Sterically hindered aryl chlorides undergo Suzuki coupling in the presence of the ligand **1**.[3] New ligands to supplant phosphines for Pd include the bishydrazone **2**[4] and dicyclohexylamine.[5] A valuable observation with $Pd(OAc)_2$-$(Cy_2NH)_2$ is that selective coupling of electron-deficient ArBr proceeds at room temperature and electron-rich ArBr at elevted temperature. Grinding facilitates aryl coupling reactions.[6]

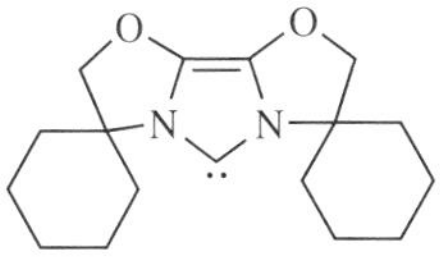

1  **2**

An improved preparation of arylboronates involves cross coupling of ArX with (bispinacolato)diboron which is catalyzed by $Pd(OAc)_2$. Direct addition of other ArX and $(Ph_3P)_4Pd$ completes the synthesis of biaryls.[7]

Aryltrifluoroborate salts find more frequent applications as coupling partners in $Pd(OAc)_2$-catalyzed reactions, e.g., with ArBr,[8] and with acetylated Baylis-Hillman adducts.[9]

New ligands for Heck reaction are thiourea **3**,[10] 1-(2-iodophenyl)-1*H*-tetrazole (**4**),[11] and carbene from **5**.[12] For ligand-free Heck reaction catalyzed by $Pd(OAc)_2$ either amberlite IRA-400 (basic) resin[13] or K_3PO_4[14] is present.

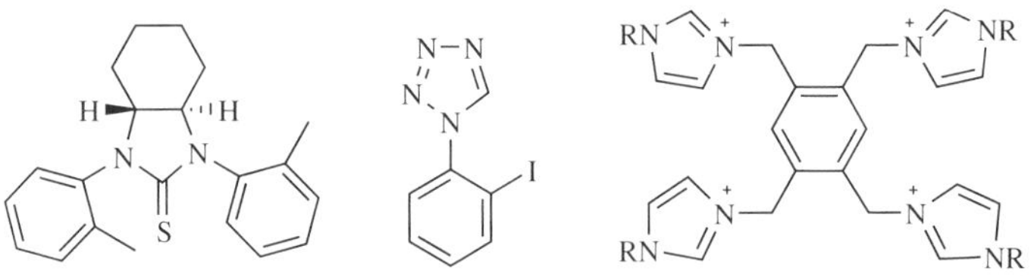

3 4 5

R = 2,6-diisopropylphenyl

New application of the Heck reaction include arylation of 1-(3-hydroxy)propoxy-1,3-dienes that are available from treatment of 2-alkenyl-1,3-dioxanes with BuLi - *t*-BuOK at low temperature to provide acetals of 4-aryl-2-butenals,[15] and indole synthesis from *o*-iodoarylamines and allyl acetate.[16]

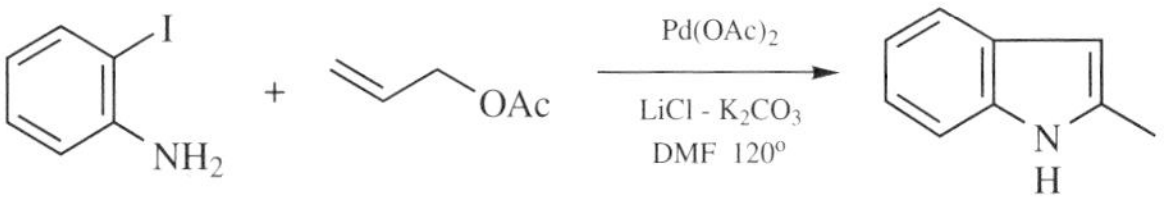

$Pd(OAc)_2$; $LiCl - K_2CO_3$, DMF 120°

47%

Intramolecular coupling via 1,4-palladium migration forms substituted fluorenes, dibenzofurans and related polycyclic skeletons.[17] Norbornene-mediated sequential coupling reactions[18,19] build up complex structures with little effort.

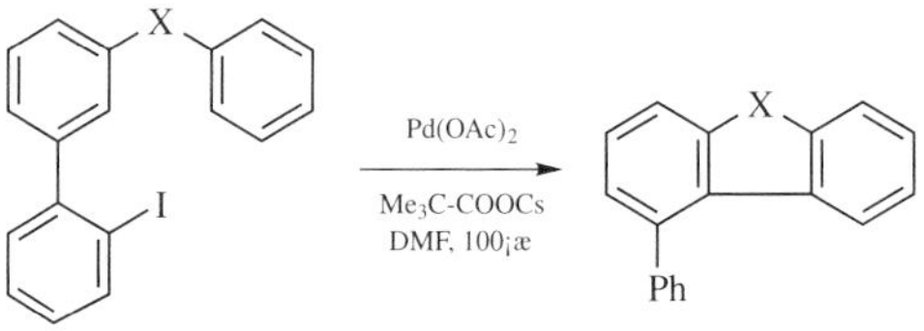

$Pd(OAc)_2$; $Me_3C\text{-}COOCs$, DMF, 100¡æ

X = O, CH_2

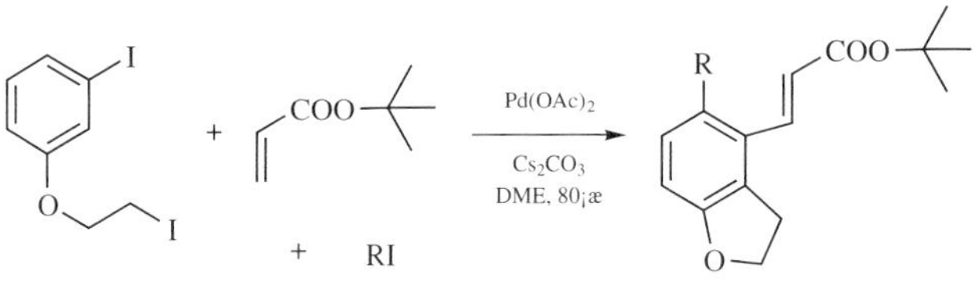

$Pd(OAc)_2$; Cs_2CO_3, DME, 80¡æ

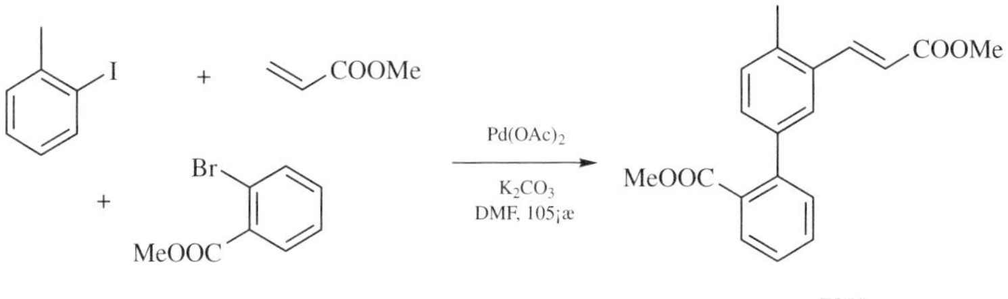

+ RI

$Pd(OAc)_2$; K_2CO_3, DMF, 105¡æ

73%

2-Nitrobiaryls form carbazoles when exposed to $Pd(OAc)_2$ in DMF at 140° under CO which acts as a reductant for the nitro group.[20] The only byproduct of the reaction is CO_2.

Addition reactions. Hydroselenation of alkynes in the Markovnikov sense is catalyzed by $Pd(OAc)_2$ – Py.[21] In forming cyclic carbonates from 1-trifluoromethyl-1-alkyn-4-ols, Na_2CO_3 can be source of CO_2.[22]

Reaction of allylic alcohols with TsN = C = O followed by treatment with $Pd(OAc)_2$–LiBr–$CuCl_2$ delivers 4-chloromethyloxazolidin-2-ones.[23] Catalytic chloroamination of alkynes by *N,N*-dichlorosulfonamides also results in allylic chlorination.[24]

Ph—≡— + $PhSO_2NCl_2$ —[$Pd(OAc)_2$ / MeCN, 80¡æ]→ Ph(Cl)C=C(CH_2Cl)($NHSO_2Ph$)

61%

Transfer hydrogenation of imines in situ is easily performed in the presence of $Pd(OAc)_2$ – HCOOK.[25] Reducing system using ion exchange resin loaded with formate anion to provide hydride is useful for saturating conjugated double bonds.[26]

Cyclization reactions. Coumarin synthesis from phenols and 2-alkynoic acids[27] or esters[28] involves a net C-H insertion and is consistent with the principle of atom economy. Addition of benzyl group to a double bond of diallylmalonic esters followed by coupling reaction furnishes cyclopentene derivatives.[29] Diallyl amines similarly afford pyrrolines.[30]

TsN(allyl)(butenyl) + Cl–CH_2Ph —[$Pd(OAc)_2$ / Bu_3N / DMF ¦¤]→ pyrroline product

76%

In the presence of $Pd(OAc)_2$ α-cyano-β-3-butenylcycloalkanones readily cyclize to afford methylencyclopentano-annulated products.[31] Under oxidative conditions benzofurans are obtained from aryl allyl ethers[32] and ring closure to the indole nucleus involving an unsaturated side chain is also achieved.[33,34] Optimal conditions specify benzoquinone as oxidant and ethyl nicotinate as ligand, the latter compound makes the Pd species more electrophilic (reactive).

Pd(OAc)$_2$, ethyl nicotinate, O_2, HOAc, EtOAc — 73%

Pd(OAc)$_2$, ethyl nicotinate, O_2, HOAc, EtOAc — 74%

Cyclopentadienone acetals. [3 +2]Cycloaddition between cyclopropenone acetals and alkynes takes place at room temperature when catalyzed by Pd(OAc)$_2$ in THF.[35]

COOMe, Pd(OAc)$_2$, THF, R, COOMe, COOMe

Oxidation. 1-Alkenes are converted into linear allylic acetates with Pd(OAc)$_2$ – benzoquinone/HOAc-DMSO in the air.[36] The role of DMSO is activation of the catalyst for C-H bond oxidation, regiocontrol and suppressing Wacker oxidation.

Fused δ-lactones and pyrans are readily formed from cyclic dienes containing oxygenated 3-carbon chains under oxidative conditions.[37]

COOH, Pd(OAc)$_2$ - MnO_2, benzoquinone, LiCl, LiOAc, HOAc, Me_2CO, AcO, OAc — 75 : 25, 90%

Ketone synthesis. Carbopalladation of nitriles by arenes after C-H bond activation by Pd(OAc)$_2$ is the key step of an aryl ketone synthesis.[38] Three-component coupling involving chloropyridines, CO and ArB(OH)$_2$ gives access to aroylpyridines.[39]

Racemization. Racemization of undesired enantiomers after optical resolution is important industrially, methods for which are therefore in demand. Allenes rapidly racemize on exposure to Pd(OAc)$_2$ and LiBr in MeCN.[40]

[1]Miao, W., Chan, T.K. *OL* **5**, 5003 (2003).
[2]Byun, J.-W., Lee, Y.-S. *TL* **45**, 1837 (2004).
[3]Altenhoff, G., Goddard, R., Lehmann, C.W., Glorius, F. *ACIEE* **42**, 3690 (2003).
[4]Mino, T., Shirae, Y., Sakamoto, M., Fujita, T. *SL* 882 (2003).
[5]Tao, B., Boykin, D.W. *TL* **44**, 7993 (2003); *JOC* **69**, 4330 (2004).
[6]Klingensmith, L.M., Leadbeater, N.E. *TL* **44**, 765 (2003).
[7]Zhu, L., Duquette, J., Zhang, M. *JOC* **68**, 3729 (2003).
[8]Molander, G.A., Biolatto, B. *JOC* **68**, 4302 (2003).
[9]Kabalka, G.W., Venkataiah, B., Dong, G. *OL* **5**, 3803 (2003).
[10]Dai, M., Liang, B., Wang, C., Chen, J., Yang, Z. *OL* **6**, 221 (2004).
[11]Gupta, K., Song, C.H., Oh, C.H. *TL* **45**, 4113 (2004).
[12]Liu, J., Zhao, Y., Zhou, Y., Li, L., Zhang, T.Y., Zhang, H. *OBC* **1**, 3227 (2003).
[13]Solabannavar, S.B., Helavi, V.B., Desai, U.V., Mane, R.B. *SC* **33**, 361 (2003).
[14]Yao, Q., Kinney, E.P., Yang, Z. *JOC* **68**, 7528 (2003).
[15]Deagostino, A., Prandi, C., Venturello, P. *OL* **5**, 3815 (2003).
[16]Hong, C.S., Seo, J.Y., Yum, E.K., Sung, N.-D. *H* **63**, 631 (2004).
[17]Huang, Q., Campo, M.A., Tai, T., Tian, Q., Larock, R.C. *JOC* **69**, 8251 (2004).
[18]Pache, S., Lautens, M. *OL* **5**, 4827 (2003).
[19]Faccini, F., Motti, E., Catellani, M. *JACS* **126**, 78 (2004).
[20]Smitrovich, J.H., Davies, I.W. *OL* **6**, 533 (2004).
[21]Kamiya, I., Nishinaka, E., Ogawa, A. *JOC* **70**, 696 (2005).
[22]Jiang, Z.-X., Qing, F.-L. *JFC* **123**, 57 (2003).
[23]Lei, A., Lu, X., Liu, G. *TL* **45**, 1785 (2004).
[24]Karur, S., Kotti, S.R.S.S., Xu, X., Cannon, J.F., Headley, A., Li, G. *JACS* **125**, 13340 (2003).
[25]Basu, B., Jha, S., Bhuiyan, M.M.H., Das, P. *SL* 555 (2003).
[26]Basu, B., Bhuiyan, M.M.H., Das, P., Hossain, I. *TL* **44**, 8931 (2003).
[27]Kotani, M., Yamamoto, K., Oyamada, J., Fujiwara, Y., Kitamura, T. *S* 1466 (2003).
[28]Trost, B.M., Toste, F.D., Greenman, K. *JACS* **125**, 4518 (2003).
[29]Hu, Y., Zhou, J., Lian, H., Zhu, C., Pan, Y. *S* 1177 (2003).
[30]Hu, Y., Zhou, J., Long, X., Han, J., Zhu, C., Pan, Y. *TL* **44**, 5009 (2003).
[31]Kung, L.-R., Tu, C.-H., Shia, K.-S., Liu, H.-J. *CC* 2490 (2003).
[32]Zhang, H., Ferreira, E., Stoltz, B. *ACIEE* **43**, 6144 (2004).
[33]Ferreira, E., Stoltz, B. *JACS* **125**, 9578 (2003).
[34]Baran, P.S., Guerrero, C.A., Corey, E.J. *JACS* **125**, 5628 (2003).
[35]Isobe, H., Sato, S., Tanaka, T., Tokuyama, H., Nakamura, E. *OL* **6**, 3569 (2004).
[36]Chen, M.S., White, M.C. *JACS* **126**, 1346 (2004).
[37]Verboom, R.C., Persson, B.A., Backvall, J.-E. *JOC* **69**, 3102 (2004).
[38]Zhou, C., Larock, R.C. *JACS* **126**, 2302 (2004).
[39]Maerten, E., Hassouna, F., Couve-Bonnaire, S., Mortreux, A., Capentier, J.-F., Castanet, Y. *SL* 1874 (2003).
[40]Horvath, A., Backvall, J.-E. *CC* 964 (2004).

Palladium(II) acetate – oxygen.

Coupling reactions. Molecular oxygen is used as oxidant for Pd(0) [initial reagent: $Pd(OAc)_2$] in the Heck reaction involving arylboronic acids in the synthesis of cinnamic esters,[1] styryl sulfones,[2] and styrylphosphonates.[3] Addition and coupling reactions operate in tandem in the formation of stilbene derivatives.

$B(OH)_2$ + EtOOC—≡—$SiMe_3$ → ($Pd(OAc)_2$; O_2, 4A-MS/DMSO, 50¡æ) EtOOC, $SiMe_3$ 72%

With PhCOOH as additive, Heck reaction conditions that are halide-free and without solvent are established. The only by-product is water.[5] Oxidative coupling of arenes (instead of ArX) with electron-deficient alkenes is achieved in the presence of $Pd(OAc)_2$, molybdovanadophosphoric acid in the air.[6]

Nazarov reaction. Different cyclopentenones are formed from a cross-conjugated dienone under different conditions of Pd-catalysis.[7]

$Pd(OAc)_2$; O_2, DMSO, 80¡æ → O, OEt, Ph 50%

O, OEt, Ph

$(MeCN)_2PdCl_2$; O, (H_2O), rt → OH, O, Ph 91%

Oxidation. An imidazole-carbene complex of $Pd(OAc)_2$ is capable of catalyzing aerobic oxidation of alcohols.[8]

[1]Andappan, M.M.S., Nilsson, P., Larhed, M. *CC* 218 (2004).
[2]Kabalka, G.W., Guchhait, S.K. *TL* **45**, 4021 (2004).
[3]Kabalka, G.W., Guchhait, S.K., Naravane, A. *TL* **45**, 4685 (2004).
[4]Zhou, C., Larock, R.C. *OL* **7**, 259 (2005).
[5]Dams, M., De Vos, D.E., Celen, S., Jacobs, P.A. *ACIEE* **42**, 3512 (2003).
[6]Tani, M., Sakaguchi, S., Ishii, Y. *JOC* **69**, 1221 (2004).
[7]Bee, C., Leclerc, E., Tius, M.A. *OL* **5**, 4927 (2003).
[8]Jensen, D.R., Schultz, M.J., Mueller, J.A., Sigman, M.S. *ACIEE* **42**, 3810 (2003).

Palladium(II) acetate – phase-transfer catalyst. 20, 284–286; **21**, 320; **22**, 330–331

Coupling reactions. Phosphine-free Heck reaction for synthesis of 2-arylbenzo[*b*]thiophenes has been developed.[1] Slight variations of additives enable the production of either cinnamaldehydes[2] or dihydrocinnamic esters[3] from acrolein diethyl acetal.

Coupling occurs in trigonal carbon atom of diiodides and a preparation of cross-conjugated trienes is based on coupling-elimination tandem.[4]

Ph Ph I I + MeOOC

$Pd(OAc)_2$ / Bu_4NCl -$NaHCO_3$ DMF 100°

Ph Ph MeOOC

80%

Ligand-free Pd(II) species are found to promote Suzuki[5] and Sonagashira couplings[6] at room temperature when exposed to air. The latter coupling also does not require copper salt and amine additives. Three different steps are involved in the formation of 3,3-disubstituted 2,3-dihydrobenzofurans.[7] Conditions for Suzuki coupling in water using microwave and conventional heating (at 150°) have been scrutinized.[8]

OH I + Br COOMe + $PhB(OH)_2$

$Pd(OAc)_2$ / Bu_4NCl -$NaHCO_3$ DMF 80°

O COOMe Ph

Stille coupling of functionalized triorganotin halides also can avoid use of phosphine ligands.[9]

Bu_2Sn Br O H + Br

$Pd(OAc)_2$ / Bu_4NF / NMP

HO

54%

Arylphosphonic acids are reactive toward coupling with alkenes under oxidative conditions (Me_3NO as oxidant).[10]

Conjugate addition. Introduction of an aryl substituent into the β-position of an enone or nitroalkene by Pd catalysis in MeCN is aided by $SbCl_3$ and Bu_4NF.[11] The antimony salt coordinates to the carbonyl group in proximity of C-bound Pd species after the palladoarylation therefore conversion to Pd enolates is facilitated. While many Pd(II) salts can be employed in the reaction, inhibitory phosphine ligands must be avoided.

Cyclocarbonylation. *o*-Iodostyrenes give indanones on exposure to $Pd(OAc)_2$ - Py and Bu_4NCl in DMF under CO. Similarly, 2-cyclopentenones are prepared from dienyl derivatives.[12] Internal alkynes coupled with *o*-iodophenols and then carbonylated to afford coumarins.[13]

[1]Chabert, J.F.D., Joucla, L., David, E., Lemaire, M. *T* **60**, 3221 (2004).
[2]Battistuzzi, G., Cacchi, S., Fabrizi, G. *OL* **5**, 777 (2003).
[3]Battistuzzi, G., Cacchi, S., Fabrizi, G., Bernini, R. *SL* 1133 (2003).
[4]Shi, M., Shao, L.-X. *SL* 807 (2004).
[5]Deng, Y., Gong, L., Mi, A., Liu, H., Jiang, Y. *S* 337 (2003).
[6]Urgaonkar, S., Verkade, J.G. *JOC* **69**, 5752 (2004).
[7]Szlosek-Pinaud, M., Diaz, P., Martinez, J., Lamaty, F. *TL* **44**, 8657 (2003).
[8]Leadbeater, N.E., Marco, M. *JOC* **68**, 888 (2003).
[9]Thiele, C.M., Mitchell, T.N. *AOC* **18**, 83 (2004).
[10]Inoue, A., Shinokubo, H., Oshima, K. *JACS* **125**, 1484 (2003).
[11]Denmark, S.E., Amishiro, N. *JOC* **68**, 6997 (2003).
[12]Gagnier, S.V., Larock, R.C. *JACS* **125**, 4804 (2003).
[13]Kadnikov, D.V., Larock, R.C. *JOC* **68**, 9423 (2003).

Palladium(II) acetate – silver carbonate.

Coupling reactions. Multistep coupling-cyclization can be achieved using this reagent combination, enabling synthesis of tetrasubstituted naphthalenes[1] and dihydrophenanthrenes[2] in one operation.

I + EtOOC—≡—COOEt → ($Pd(OAc)_2$, Ag_2CO_3, DMF, 100¡æ) COOEt, COOEt, COOEt, COOEt

71%

Ph, SO_2Ph

PhI + SO_2Ph → ($Pd(OAc)_2$, Ag_2CO_3, DMF, 120¡æ)

69%

[1]Kawasaki, S., Satoh, T., Miura, M., Nomura, M. *JOC* **68**, 6836 (2003).
[2]Mauleon, P., Nunez, A.A., Alonso, I., Carretero, J.C. *CEJ* **9**, 1511 (2003).

Palladium(II) acetate – tertiary phosphine. **13**, 91, 233–234; **14**, 249, 250–253; **15**, 247–248; **16**, 264–268; **17**, 259–269; **18**, 277–281; **19**, 252–256; **20**, 286–289; **21**, 321–324; **22**, 331–337

Coupling reactions. Schlenk technique is the key to performing Suzuki coupling using 1,2-dibromoethane as source of bromoethene.[1] To prepare diarylethynes in situ dehalogenation of 1-bromo-2-chloroethane is involved.[2] Ionic liquids promote regioselective Heck reaction between aryl halides and electron-rich alkenes, thus α/β ratio in the arylation of butyl vinyl ether reaches a level of >99:1.[3]

Migration of Pd from an alkyl to aryl position after one Heck reaction enables a second coupling in *o*-iodoaryl 2-(α-styryl)ethyl ethers. Certain polycyclic compounds are generated in one operation.[4]

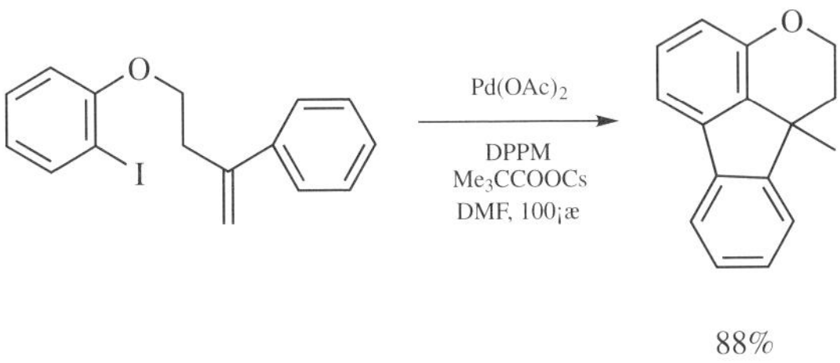

$Pd(OAc)_2$, DPPM, $Me_3CCOOCs$, DMF, 100¡æ

88%

Solid-phase synthesis of indole-3-carboxylic esters[5] from *o*-iodoarylamines via supported β-aminoacrylates is readily accomplished. Detachment of the products from polymer backbone is by methanolysis. Two aryl groups can be introduced into the thiazole ring (at C-2 and C-5) in one Heck reaction.[6] Potassium diaryldifluoroborates[7] and organobismuth compounds (e.g., **1**)[8] are a new class of coupling reaction partners.

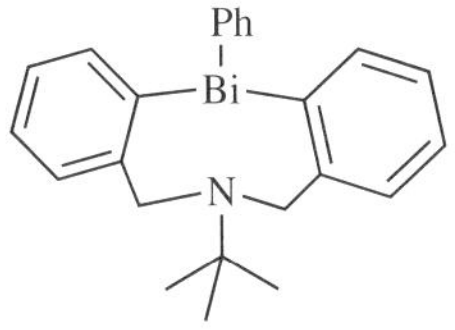

Ph Bi N

1

Arylation of pyrroles (at C-2) is achieved by submitting the sodium salts to $ZnCl_2$, $Pd(OAc)_2$ and a hindered phosphine ligand.[9] In introduction of an aryl group to C-2 of *N*-substituted indoles by reaction with ArX the choice of an added base has a critical influence (CsOAc is best).[10] A lactam-derived α-boryl enamide is useful for conversion into the α-aryl analogues.[11] α,α-Disubstituted arylmethanols lose the carbinol unit to enter coupling reaction.[12] Enantioselective Suzuki coupling at room temperature to create quaternary carbon centers is encouraging (ee 73–85% ee).[13]

Research and development in this area in recent years also emphasize efficiency, practicality and simplification of procedures. With poly(ethylene oxide) as cosolvent for Suzuki coupling the product is extracted into hexane and the polar phase containing the catalyst is reused.[14]

Several new ligands show distinct advantages for the Pd-catalyzed Suzuki couplings. Di-*t*-butyl-(*N*-phenylpyrrol-2-yl)phosphine is highly efficient.[15] Tris(4,6-dimethyl-3-sulfonatophenyl)phosphine trisodium salt enables the reaction in an aqueous medium.[16]

Hindered biarylphosphine-prototypes continue to be evaluated as effective ligands for Pd in Suzuki coupling. Ligand **1**[17] is representative, the ferrocenylphosphine **2**[18] is air-stable. Isopropylaminodiphenylphosphine proves its value also.[19]

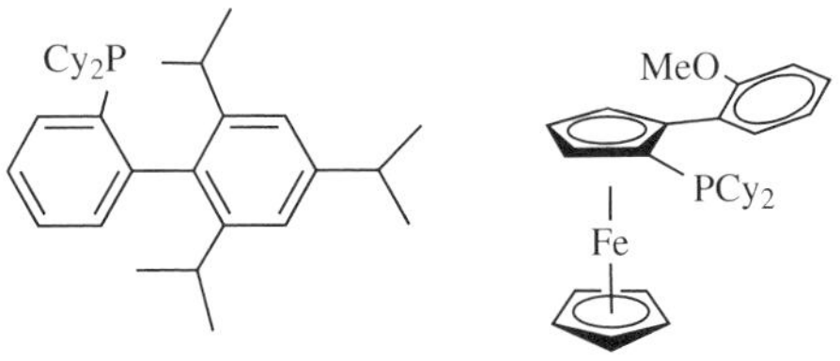

Cy_2P MeO PCy_2 Fe

1 **2**

Diarylmethane synthesis[20] is achieved by Pd-catalyzed coupling between $ArB(OH)_2$ and $ArCH_2X$ in the presence of Ph_3P. Heck-type benzylation of alkenes by benzyl trifluoroacetates has been realized.[21] A copper-free Sonagashira coupling system contains *N*-benzyl-*N,N*-bis(di-*t*-butylphosphinylmethyl)amine.[22]

Certain cyclobutanols undergo enantioselective cleavage and coupling (arylation, alkenylation, and allenylation).[23]

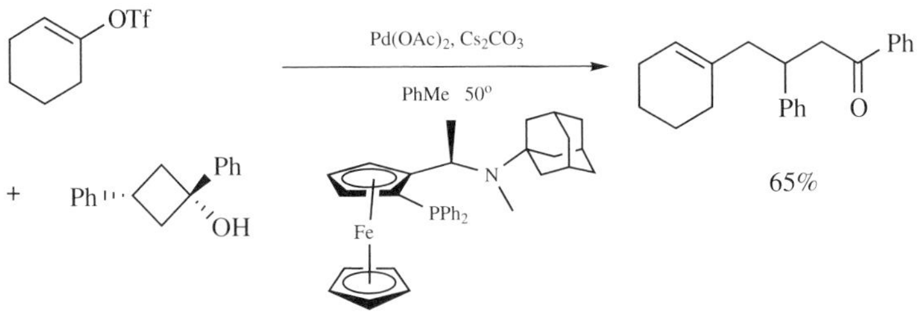

OTf + Ph Ph OH — $Pd(OAc)_2$, Cs_2CO_3 / PhMe 50° → Ph Ph O 65%

PPh₂ N Fe

Concerning *N*-arylation by the coupling technique the presence of **1** to assist $Pd(OAc)_2$, aryl sulfonates become viable substrates instead of halides.[24] Arylation of piperazines succeeds on employing a catalytic system that contains (2'-dimethylaminobiphenyl)dicyclohexylphosphine.[25] In supercritical CO_2 *N*-silylamines can be used as cross-coupling partners for electron-deficient ArX.[26]

By temperature-controlled microwave heating aminoheteroarenes (e.g., aminopyridines from chloropyridines) are rapidly accessed.[27] Usually, Pd-catalyzed *N*-arylation of sultams is more effective than the Cu-catalyzed reaction.[28]

An intramolecular coupling is involved in the synthesis of 1-aryl-1*H*-indazoles from 2-bromobenzaldehydes and arylhydrazines.[29]

Conjugated diynes are formed by homocoupling of lithium alkynyltriisoproxyborates (CuI present).[30]

Heterocycles. Hydroarylation of 4-hydroxy-2-alkynoic esters with $ArB(OH)_2$ gives either α-aryl or β-aryl butenolides, depending on the phosphine ligand (t-Bu_3P vs. DPPE).[31]

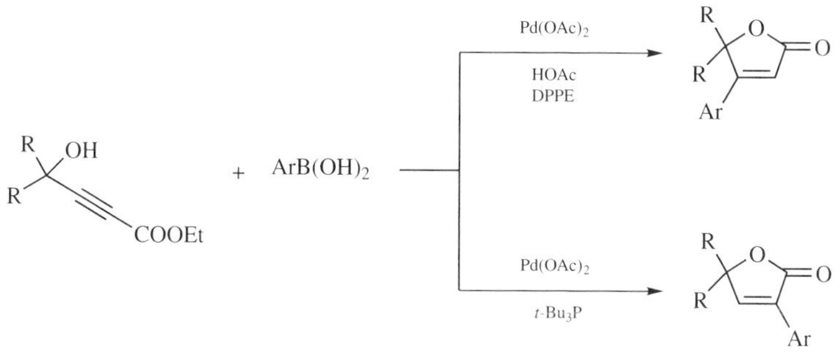

2-Haloarylamines and internal alkynes combine and cyclize to give 2,3-disubstituted indoles under the influence of $Pd(OAc)_2$. Both phosphine [such as 1,1'-bis(di-*t*-butylphosphino)ferrocene][32] and carbene ligands[33] are useful.

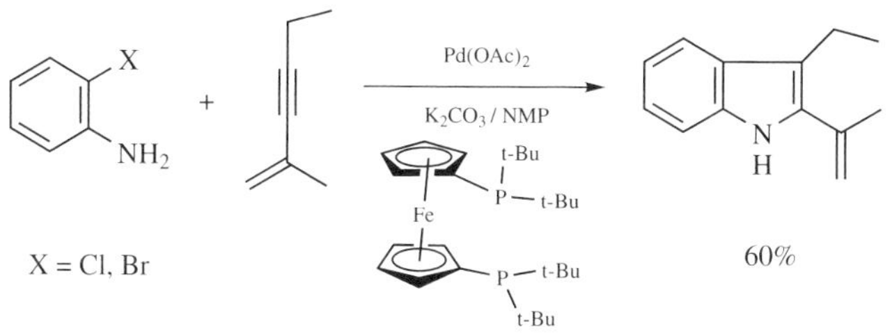

Substituted oxindoles are obtained from *N*-chloroacetyl arylamines, thus providing a milder method than the Friedel-Crafts alkylation.[34]

Tetrahydroisoquinolines are assembled by a coupling-Michael addition tandem.[35] More interestingly, in the synthesis of dibenzofurans and carbazoles from *o*-iodophenols and *o*-iodoarylamines, respectively, the coupling partner *o*-trimethylsilylphenyl triflate potentially generates benzyne in situ (in the presence of CsF).[36]

COOMe, I + H_2N Ph + C (allene) → $Pd(OAc)_2$; Ph_3P, K_2CO_3, PhMe 80° → COOMe, N, Ph

73%

Aryl cyanides and phosphines. Conversion of ArX to ArCN by Pd-catalysis requires lower temperature than the Cu-mediated process. With TMEDA as additive even aryl chlorides undergo substitution.[37,38] A new cyanation agent for this reaction is $K_4[Fe(CN)_6]$.[39]

It is much easier to prepare functionalized $ArPPh_2$ by exchanging one of the phenyl groups of Ph_3P.[40,41]

Substitution. Arylation of acetoacetic esters[42] and azlactones[43] is readily achieved by Pd-catalyzed reactions. Bulky phosphine ligands such as 2'-substituted 2-biphenyldi(*t*-butyl)phosphines and Bis(1-adamantyl)-*t*-butylphosphine are ligands of choice, although the cage phoshinamide (**1**) also is effective.[44] For diarylation of acetylarenes Ph_3P suffice.[45]

N, P, N, N, N

1

Dienyl alcohols serve as alkylating agents for bis(2,2,2-trifluoroethyl) malonates which are more acidic than diethyl malonate. No matter conjugated or interrupted pentadienols are used only linear products are formed.[46]

Catalyzed by $Pd(OAc)_2$, allylic ether synthesis from aliphatic alcohols and allyl methyl carbonate[47] as well as ferrier rearrangement[48] are quite efficient. In the latter reaction β-glycosides are highly favored of glycals possessing a β-acetate at the allylic position if 2- biphenyldi-*t*-butylphosphine is used as ligand.

On treatment with Et_2Zn and $Pd(OAc)_2$, 2-allyloxylated cyclic ethers are converted to 1-alkene-4,ω-diols.[49] Splitting of the substrates into π-allylpalladium species and hydroxyaldehydes (zinc salts) is followed by a recombination reaction.

O, O → $Pd(OAc)_2$, Bu_3P; Et_2Zn, PhMe r.t. → OH, OH

91%

Addition reactions. 2-Methyl-1-alken-3-ynes are products from addition of 1-alkynes to allene using the $Pd(OAc)_2$ – CuI catalytic system.[50] Similar reaction is observed with allenylphosphine oxides.[51]

Ph + =C=C(Ph)P(O)Ph$_2$ → (Pd(OAc)$_2$, Ar$_3$P; THF, r.t.) 77%

Ar_3P = (2,6-dimethoxyphenyl)$_3$P

Diaryl dichalcogenides are split and add to alkynes to generate *cis*-adducts.[52] Hexane favors hydrostannylation of highly hindered alkynes because in that solvent the competing reaction that generates $Bu_3SnSnBu_3$ and H_2 is suppressed.[53]

Ketone synthesis. Pyrid-2-yl esters react with $ArB(OH)_2$ to afford aryl ketones and with *B*-benzyl-9-BBN, benzyl ketones.[54] The nitrogen atom of the pyridine ring plays a crucial role in directing formation of acylpalladium intermediates.

Isomerization. A double bond of 2-vinylidenehydrofurans migrates into the ring to generate a 1,3-diene unit on exposure to $Pd(OAc)_2$ – DPPP.[55] The substrates are readily available by alkylation of cycloalkanone-α,α'-dicarboxylic esters with 1,4-dibromo-2-butyne.

[1]Lando, V.R., Monteiro, A.L. *OL* **5**, 2891 (2003).
[2]Abele, E., Abele, R., Arsenyan, P., Lukevics, E. *TL* **44**, 3911 (2003).
[3]Mo, J., Xu, L., Xiao, J. *JACS* **127**, 751 (2005).
[4]Huang, Q., Fazio, A., Dai, G., Campo, M.A., Larock, R.C. *JACS* **126**, 7460 (2004).
[5]Yamazaki, K., Nakamura, Y., Kondo, Y. *JOC* **68**, 6011 (2003).
[6]Yokooji, A., Okazawa, T., Satoh, T., Muira, M., Nomura, M. *T* **59**, 5685 (2003).
[7]Ito, T., Iwai, T., Mizumo, T., Ishino, Y. *SL* 1435 (2003).
[8]Yamazaki, O., Tanaka, T., Shimada, S., Suzuki, Y., Tanaka, M. *SL* 1921 (2004).
[9]Rieth, R.D., Mankad, N.P., Calimano, E., Sadighi, J.P. *OL* **6**, 3981 (2004).
[10]Lane, B.S., Sames, D. *OL* **6**, 2897 (2004).
[11]Ferrali, A., Guarna, A., Lo Galbo, F., Occhiato, E.G. *TL* **45**, 5271 (2004).
[12]Terao, Y., Wakui, H., Nomoto, M., Satoh, T., Miura, M., Nomura, M. *JOC* **68**, 5236 (2003).
[13]Willis, M.C., Powell, L.H.W., Claverie, C.K., Watson, S.J. *ACIEE* **43**, 1249 (2004).
[14]Nobre, S.M., Wolke, S.I., da Rosa, R.G., Monteiro, A.L. *TL* **45**, 6527 (2004).
[15]Zapf, A., Jackstell, R., Rataboul, F., Riermeier, T., Monsees, A., Fuhrmann, C., Shaikh, N., Dingerdissen, U., Beller, M. *CC* 38 (2004).
[16]Moore, L.R., Shaughnessy, K.H. *OL* **6**, 225 (2004).
[17]Nguyen, H.N., Huang, X., Buchwald, S.L. *JACS* **125**, 11818 (2003).
[18]Jensen, J.F., Johannsen, M. *OL* **5**, 3025 (2003).

[19]Cheng, J., Wang, F., Xu, J.-H., Pan, Y., Zhang, Z. *TL* **44**, 7095 (2003).
[20]Nobre, S.M., Monteiro, A.L. *TL* **45**, 8225 (2004).
[21]Narahashi, H., Yamamoto, A., Shimizu, I. *CL* **33**, 348 (2004).
[22]Mery, D., Heuze, K., Astruc, D. *CC* 1934 (2003).
[23]Matsumura, S., Maeda, Y., Nishimura, T., Uemura, S. *JACS* **125**, 8862 (2003).
[24]Huang, X., Anderson, K.W., Zim, D., Jiang, L., Klapars, A., Buchwald, S.L. *JACS* **125**, 6653 (2003).
[25]Michalik, D., Kumar, K., Zapf, A., Tillack, A., Arlt, M., Heinrich, T., Beller, M. *TL* **45**, 2057 (2004).
[26]Smith, C.J., Early, T.R., Holmes, A.B., Shute, R.E. *CC* 1976 (2004).
[27]Maes, B.U.W., Loones, K.T.J., Lemiere, G.L.F., Dommisse, R.A. *SL* 1822 (2003).
[28]Steinhuebel, D., Palucki, M., Askin, D., Dolling, U. *TL* **45**, 3305 (2004).
[29]Cho, C.S., Lim, D.K., Heo, N.H., Kim, T.-J., Shim, S.C. *CC* 104 (2004).
[30]Oh, C.H., Reddy, V.R. *TL* **45**, 5221 (2004).
[31]Oh, C.H., Park, S.J., Ryu, J.H., Gupta, A.K. *TL* **45**, 7039 (2004).
[32]Shen, M., Li, G., Lu, B.Z., Hossain, A., Roschangar, F., Farina, V., Senanayake, C.H. *OL* **6**, 4129 (2004).
[33]Ackermann, L., Kaspar, L., Gschrei, C. *CC* 2824 (2004).
[34]Hennessy, E.J., Buchwald, S.L. *JACS* **125**, 12084 (2003).
[35]Gai, X., Grigg, R., Koppen, I., Marchbank, J., Sridharan, V. *TL* **44**, 7445 (2003).
[36]Liu, Z., Larock, R.C. *OL* **6**, 3739 (2004).
[37]Sundermeier, M., Zapf, A., Beller, M. *ACIEE* **42**, 1661 (2003).
[38]Sundermeier, M., Zapf, A., Mutyala, S., Baumann, W., Sans, J., Weiss, S., Beller, M. *CEJ* **9**, 1828 (2003).
[39]Schareina, T., Zapf, A., Beller, M. *CC* 1388 (2004).
[40]Kwong, F.Y., Lai, C.W., Yu, M., Tian, Y., Chan, K.S. *T* **59**, 10295 (2003).
[41]Kwong, F.Y., Lai, C.W., Yu, M., Chan, K.S. *T* **60**, 5635 (2004).
[42]Zeevaart, J.G., Parkinson, C.J., de Koning, C.B. *TL* **45**, 4261 (2004).
[43]Liu, X., Hartwig, J.F. *OL* **5**, 1915 (2003).
[44]You, J., Verkade, J.G. *ACIEE* **42**, 5051 (2003).
[45]Churruca, F., SanMartin, R., Tellitu, I., Dominguez, E. *TL* **44**, 5925 (2003).
[46]Takacs, J.M., Jiang, X., Leonov, A.P. *TL* **44**, 7075 (2003).
[47]Haight, A.R., Stoner, E.J., Peterson, M.J., Grover, V.K. *JOC* **68**, 8092 (2003).
[48]Kim, H., Men, H., Lee, C. *JACS* **126**, 1336 (2004).
[49]Kimura, M., Shimizu, M., Shibata, K., Tazoe, M., Tamaru, Y. *ACIEE* **42**, 3392 (2003).
[50]Bruyere, D., Grigg, R., Hinsley, J., Hussain, R.K., Korn, S., De la Cierva, C.O., Sridharan, V., Wang, J. *TL* **44**, 8669 (2003).
[51]Rubin, M., Markov, J., Chuprakov, S., Wink, D.J., Gevorgyan, V. *JOC* **68**, 6251 (2003).
[52]Ananikov, V.P., Beletskaya, I.P. *RCB* **53**, 561 (2004).
[53]Semmelhack, M.F., Hooley, R.J. *TL* **44**, 5737 (2003).
[54]Tatamidani, H., Kakiuchi, F., Chatani, N. *OL* **6**, 3597 (2004).
[55]Wavrin, L., Nicolas, C., Viala, J., Rodriguez, J. *SL* 1820 (2004).

Palladium(II) acetate – tertiary phosphine – carbon monoxide. 20, 292; **21**, 327–328; **22**, 337–338

Esters and lactones. 3,4,5-Trichloropyridine-2,6-dicarboxylic esters are formed on subjecting pentachloropyridine to Pd-catalyzed carbonylation under atmospheric CO.[1] 1-Alkynes are converted into 2-alkynoic esters in a carbonylation using O_2 as oxidant.[2] When *N*-hydroxysuccinimide is present activated esters of aroic acids are generated from ArX (X = I, OTf).[3]

Synthesis of phthalides[4] from *o*-bromoarylmethanols can utilize $Mo(CO)_6$ and DMF as CO sources. Dimethyl carbonate is shown to be an appropriate solvent in lactone formation from unsaturated alcohols.[5]

Lactone synthesis from allenyl carbinols[6] is greatly influenced by the ligand in that γ-lactone or δ-lactone formation seems controllable.

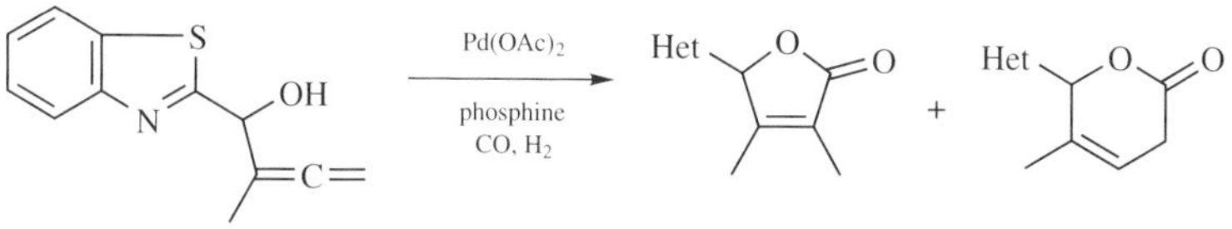

$Pd(OAc)_2$ / phosphine / CO, H_2

phosphine: BINAP 95% / —; DPPB — / 80%

Amides and lactams. α,β-Unsaturated carboxamides are synthesized from Pd-catalyzed assembly of amines, 1-alkynes and CO, with regiochemistry controlled by reaction conditions.[7] Formation of benzolactams from Pd-catalyzed carbonylation in the presence of $Cu(OAc)_2$ [but no phosphine] shows different regioselectivity with substrates bearing 3,4-dimethoxy and 3,4-methylenedioxy substituents.[8] The results are attributed to steric factors.

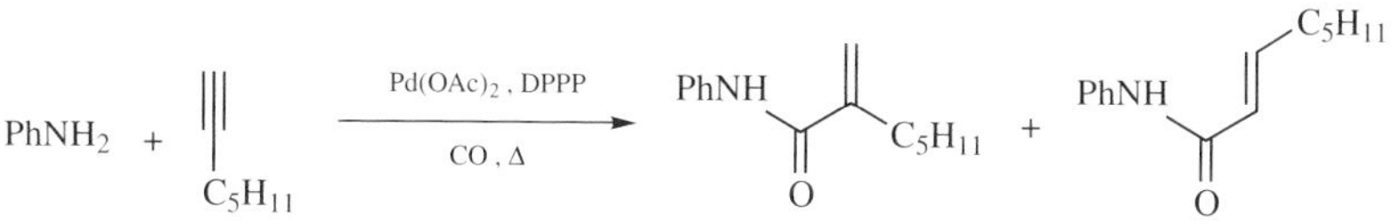

$PhNH_2$ + C_5H_{11}C≡CH —[$Pd(OAc)_2$, DPPP; CO, Δ]→ PhNHCO–C(=CH$_2$)–C_5H_{11} + PhNHCO–CH=CH–C_5H_{11}

From allyl bromides and imines the Pd-catalyzed carbonylation and cycloaddition leads to α-alkenyl β-lactams.[9]

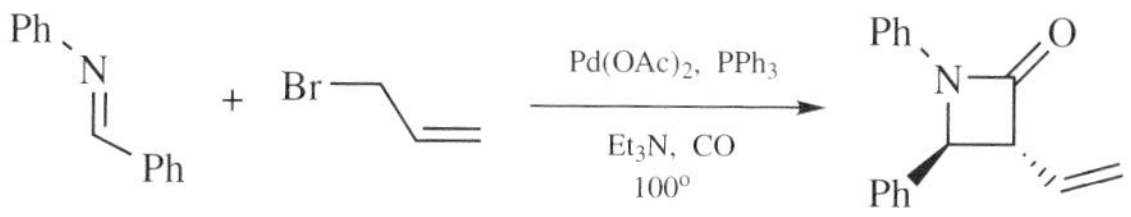

$Pd(OAc)_2$, PPh_3 / Et_3N, CO / 100°

84%

Cycloaddition. A convenient method for synthesis of bicycle[3.3.0]octadienones is by carbonylation of 1,6-diynes.[10]

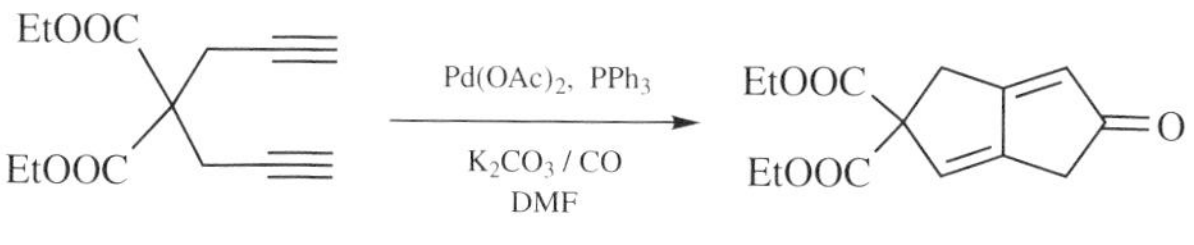

EtOOC, EtOOC —[$Pd(OAc)_2$, PPh_3; K_2CO_3 / CO, DMF]→ EtOOC, EtOOC

74%

[1]Hull, Jr, J.W., Wang, C. *H* **63**, 411 (2004).
[2]Izawa, Y., Shimizu, I., Yamamoto, A. *BCSJ* **77**, 2033 (2004).
[3]Lou, R., VanAlstine, M., Sun, X., Wentland, M.P. *TL* **44**, 2477 (2003).
[4]Wu, X., Mahalingam, A.K., Wan, Y., Alterman, M. *TL* **45**, 4635 (2004).
[5]Vasapollo, G., Mele, G., Maffei, A., Del Sole, R. *AOC* **17**, 835 (2003).
[6]Granito, C., Troisi, L., Ronzini, L. *H* **63**, 1027 (2004).
[7]El Ali, B., Tijani, J. *AOC* **17**, 921 (2003).
[8]Orito, K., Horibata, A., Nakamura, T., Ushito, H., Nagasaki, H., Yuguchi, M., Yamashita, S., Tokuda, M. *JACS* **126**, 14342 (2004).
[9]Troisi, L., De Vitis, L., Granito, C., Epifani, E. *EJOC* 1357 (2004).
[10]Grigg, R., Zhang, L., Collard, S., Keep, A. *CC* 1902 (2003).

Palladium(II) acetylacetonate.

Allylation.[1] The 1,4-benzooxazine skeleton is formed when 2-aminophenols and 2-butene-1,4-diols react under the influence of $Pd(acac)_2$.

HO NH$_2$ + OH OH → Pd(acac)$_2$, PPh$_3$ / (^{i}PrO)$_4$Ti, MS-4A / PhH Δ → O NH (59%) + O N ... N O (27%)

Coupling reactions. Alkenylzirconium reagents and RX are coupled thereby alkenes are produced. The coupling is compatible to synthesis of functionalized alkenes (e.g., the alkyl halides may contain double bond, several kinds of ethers and ester, amide group).[2]

A route to enyne esters is based on oxidative alkynylation of acrylic esters under oxygen.[3] 1-Alkynes can be generated from their acetone adducts in situ.

Ph—≡—C(Me)$_2$OH + COOEt → Pd(acac)$_2$, Py / MS-3A, O$_2$ / PhMe, 80°C → Ph—≡—CH=CH—COOEt 61%

[1]Yang, S.-C., Lai, H.-C., Tsai, Y.-C. *TL* **45**, 2693 (2004).
[2]Wiskur, S.L., Korte, A., Fu, G.C. *JACS* **126**, 82 (2004).
[3]Nishimura, T., Araki, H., Maeda, Y., Uemura, S. *OL* **5**, 2997 (2003).

Palladium(II) bis(trifluoroacetate).

Transvinylation. Vinyl ethers of primary, secondary, and tertiary alcohols can be prepared from commercially available butyl vinyl ether by an exchange reaction, using as catalyst $Pd(OCOCF_3)_2$ complexed with 4,7-diphenyl-1,10-phenanthroline.[1] A similar transfer to nitrogen nucleophiles forms enamides.[2]

Addition reactions. Alkylamines add to vinylarenes in the Markovnikov sense[3] when subjected to a catalytic system consisting of $Pd(OCOCF_3)_2$, DPPF, and CF_3SO_3H in dioxane.

Intramolecular haloamination[4] of unsaturated *N*-tosylamines is the result of Pd/Br exchange of the initial adducts by $CuBr_2$.

Coupling reactions. Oxidative carbocyclization[5] involving an allene and an alkene unit occurs on exposure to $Pd(OCOCF_3)_2$.

COOEt COOEt $Pd(OCOCF_3)_2$ benzoquinone THF Δ H H COOEt COOEt 94%

Heck-type arylation of 2-cycloalkenones (at C-3) by aroic acids is via decarboxylative palladation.[6] Phosphine-free conditions give the highest yield.

OMe COOH MeO + O $Pd(OCOCF_3)_2$, Ag_2CO_3 DMSO-DMF 80°C O OMe MeO 92%

Oxidative cyclization. 2-Alkenyltetrahydrofurans and γ-alkenyl-γ-lactones are readily assembled from unsaturated alcohols and carboxylic acids using $Pd(OCOCF_3)_2$ – O_2 as the oxidant system.[7]

CONHTs $Pd(OCOCF_3)_2$, Py, O_2 MS-3A, PhMe, 80°C O NH 88%

OH $Pd(OCOCF_3)_2$, Py, O_2 O 93%

Arene functionalization. Carboxylation of benzene derivatives by HCOOH is catalyzed by $Pd(OCOCF_3)_2$ in the presence of $K_2S_2O_8$ in CF_3COOH and CH_2Cl_2. Biphenyl is hydroxylated in one ring and carboxylated in the other. Small amount of water plays an important role in the hydroxylation step.[8]

[1]Bosch, M., Schlaf, M. *JOC* **68**, 5225 (2003).
[2]Brice, J.L., Meerdink, J.E., Stahl, S.S. *OL* **6**, 1845 (2004).
[3]Utsunomiya, M., Hartwig, J.F. *JACS* **125**, 14286 (2003).
[4]Manzoni, M.R., Zabawa, T.P., Kasi, D., Chemler, S.R. *OM* **23**, 5618 (2004).
[5]Franzen, J., Backvall, J.-E. *JACS* **125**, 6056 (2003).
[6]Tanaka, D., Myers, A.G. *OL* **6**, 433 (2004).
[7]Trend, R.M., Ramtohul, Y.K., Ferreira, E.M., Stoltz, B.M. *ACIEE* **42**, 2892 (2003).
[8]Shibahara, F., Kinoshita, S., Nozaki, K. *OL* **6**, 2437 (2004).

Palladium(II) bromide. 22, 338–339

Coupling reactions. Using the $PdBr_2$ – $(t\text{-}Bu)_2PMe$ catalytic system Hiyama coupling of arylsilanes and alkyl halides (Br, I) to give Ar-R proceeds at room temperature.[1] Salt-free Heck reaction uses enol esters of aroic acids[2] (instead of ArX) with promotion by $PdBr_2$ and a PTC.

[1]Lee, J.-Y., Fu, G.C. *JACS* **125**, 5616 (2003).
[2]Goossen, L.J., Paetzold, J. *ACIEE* **43**, 1095 (2004).

Palladium(II) chloride. 13, 234–235; **15**, 248–249; **16**, 268–269; **18**, 282; **19**, 257–258; **20**, 293–394; **21**, 329; **22**, 339–340

Coupling reactions. Oxazoline complexes of Pd for catalyzing various coupling reactions have been reported.[1,2] It is also found that $PdCl_2$ – pyridine and K_2CO_3 are sufficient to accomplish the Suzuki coupling.[3] Even simpler and milder conditions for coupling of $ArCH_2X$ with $ArB(OH)_2$ consist of $PdCl_2$ in aq. acetone.[4] The Pd-Chugaev carbene complex is a modular, air-stable catalyst.[5]

A 3-component coupling is very useful for the synthesis of highly substituted 1,3-dienes and 1,3,5-trienes.[6]

I + Ph—≡—Ph + $PhB(OH)_2$ → ($PdCl_2$, KF; DMF, H_2O, 100°C air) 87%

The $PdCl_2$–catalyzed Heck reaction involving Ar_3Sb also requires t-BuOOH.[7] 2,3-Alkadienoic acids and allylic halides form β-allylic butenolides.[8]

2,5-Dialkynylfurans are accessible by the $PdCl_2$-catalyzed coupling of 2,5-bis(butyl-telluro)furan. It is possible to introduce two different alkynyl groups.[9]

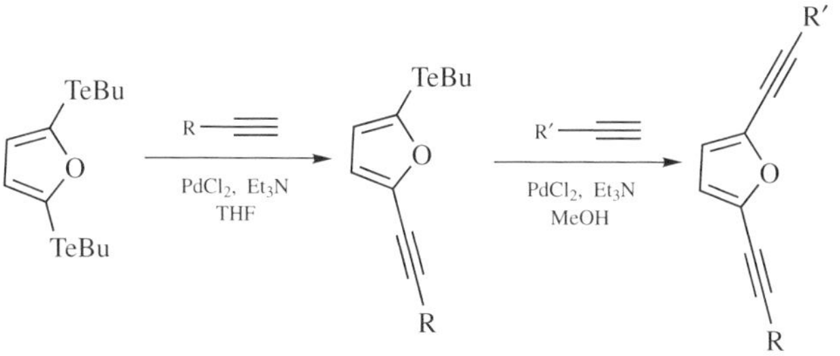

Hydroarylation. Addition of an aryl group to conjugated nitroalkenes is achieved in the Pd-catalyzed reaction with arylstannanes and $NaBAr_4$.[10] The Lewis acid $BiCl_3$ is added to improve yields in some cases.

[1]Mazet, C., Gade, L.H. *EJIC* 1161 (2003).
[2]Gossage, R.A., Jenkins, H.A., Yadav, P.N. *TL* **45**, 7689 (2004).
[3]Tao, X., Zhao, Y., Shen, D. *SL* 359 (2004).
[4]Bandgar, B.P., Bettigeri, S.V., Phopase, J. *TL* **45**, 6952 (2004).
[5]Moncada, A.I., Khan, M.A., Slaughter, L.M. *TL* **46**, 1399 (2005).
[6]Zhang, X., Larock, R.C. *OL* **5**, 2993 (2003).
[7]Moiseev, D.V., Gushchin, A.V., Morugova, V.A., Dodonov, V.A. *RCB* **52**, 2081 (2003).
[8]Ma, S., Yu, Z. *JOC* **68**, 6149 (2003).
[9]Zeni, G., Nogueira, C.W., Silva, D.O., Menezes, P.H., Braga, A.L., Stefani, H.A., Rocha, J.B.T. *TL* **45**, 1387 (2004).
[10]Ohe, T., Uemura, S. *BCSJ* **76**, 1423 (2003).

Palladium(II) chloride – copper salts. 22, 341

Esterification. The acetyl group of vinyl acetate is readily transferred to alcohols at room temperature (except tertiary alcohols and phenols) in the presence of $PdCl_2$-$CuCl_2$.[1] Amines are not affected.

Cyclotrimerization. Internal alkynes furnish substituted benzenes and the regioselective reaction is promoted by CO_2.[2,3]

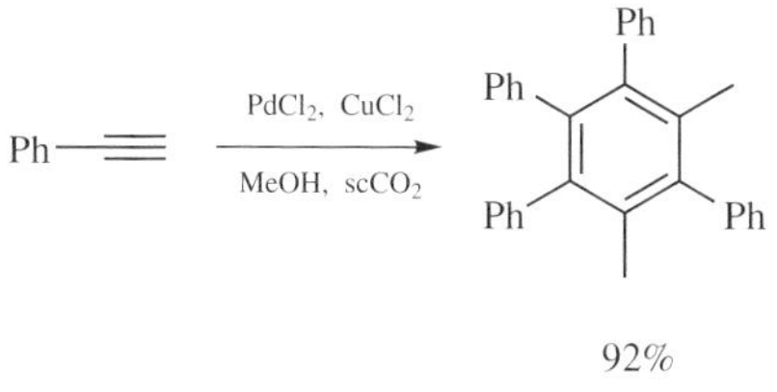

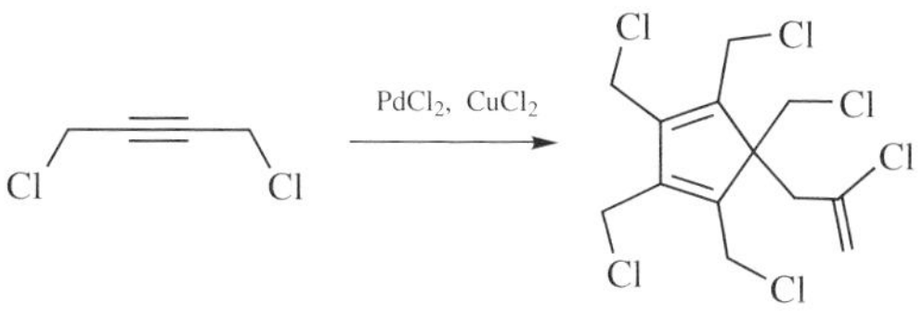

92%

Alkynyl ketones. A facile synthesis of alkynyl ketones from ArCOCl and 1-alkynes involves the $(Ph_3P)_2PdCl_2$ – CuI catalyst system and a mild base such as Et_3N.[4] If the reaction is carried out in water a surfactant and K_2CO_3 are added.[5] The method is suitable for synthesis ferrocenylethynyl ketones.[6]

Wacker oxidation. 1,5-Alkadienes are oxidized to aldehydes instead of methyl ketones under standard Wacker oxidation conditions, due to participation of the other double bond.[7]

$PdCl_2$, CuCl, O_2

DMF - H_2O

CHO

[1]Bosco, J.W.J., Saikia, A.K. *CC* 1116 (2004).
[2]Li, J.-H., Xie, Y.-X. *SC* **34**, 1737 (2004).
[3]Cheng, J.-S., Jiang, H.-F. *EJOC* 643 (2004).
[4]Cox, R.J., Ritson, D.J., Dane, T.A., Berge, J., Charmant, J.H.P., Kantacha, A. *CC* 1037 (2005).
[5]Chen, L., Li, C.-J. *OL* **6**, 3151 (2004).
[6]Yin, J., Wang, X., Liang, Y., Wu, X., Chen, B., Ma, Y. *S* 331 (2004).
[7]Ho, T.-L., Chang, M.H., Chen, C. *TL* **44**, 6955 (2003).

Palladium(II) chloride – copper(II) chloride – carbon monoxide. 20, 294; **21**, 330; **22**, 341–342

Carbonylation. Under CO and in the presence of $PdCl_2$ - $CuCl_2$, propargylic alcohols are converted into (*Z*)-α-chloroalkylidene-β-lactones.[1]

C_6H_{13} OH

$PdCl_2$, $CuCl_2$

CO / THF

C_6H_{13} Cl O O

86%

For the preparation of succinic esters from 1-alkenes using the catalyst system $PdCl_2$-CuCl, thiourea-based ligands find service.[2]

[1]Ma, S., Wu, B., Zhao, S. *OL* **5**, 4429 (2003).
[2]Dai, M., Wang, C., Dong, G., Xiang, J., Luo, T., Liang, B., Chen, J., Yang, Z. *EJOC* 4346 (2003).

Palladium(II) chloride – copper(I) iodide – (tertiary phosphine).

Coupling reactions. A common catalytic system for the Sonagashira coupling consists of $(Ph_3P)_2PdCl_2$ – CuI. An easily accessible resin-supported Pd-catalyst is useful,[1] and aqueous ammonia is found be be an activator.[2] In one method the additive is polymethylhydrosiloxane.[3]

3-Arylpropargylamines are prepared in a one-pot operation from amines, propargyl bromide and ArX.[4,5] 2-Substituted indoles are synthesized from 1-alkynes and 2-iodoarylamines by the Sonagashira coupling.that is promoted by Bu_4NF.[6]

Homocoupling of 1-alkynes is improved by avoiding stoichiometric additive.[7] [Note that phosphine-free condition employing Me_3NO as oxidant has been established also.[8]] A mixture of 1-alkynes and activated alkynes react in the Michael addition sense in water at room temperature (CuBr is the cocatalyst).[9]

Cross-coupling of alkenyl tellurides with 1-alkynes by $PdCl_2$ – CuI to give enynes[10,11] is chemoselective, regioselective, and stereoselective. Since the tellurides can be made from alkynes via hydroalumination and Al/Te exchange with RTeBr, two alkynes are employed to construct enynes. Negishi coupling of ArZnCl with alkenyl tellurides provides stilbenes.[12]

HO + C_5H_{11} BuTe SMe → ($PdCl_2$, CuI; Et_3N / MeOH) C_5H_{11} SMe HO

81%

Oxidative cyclization. In the presence of $PdCl_2$ – CuI, β′-hydroxy-α,β-unsaturated ketones afford 2,3-dihydro-4*H*-pyran-4-ones.[13]

Ketone synthesis. Thiol esters react with 1-alkynes to provide alkynones.[14]

[1]Gonthier, E., Breinbauer, R. *SL* 1049 (2003).
[2]Ahmed, M.S.M., Mori, A. *T* **60**, 9977 (2004).
[3]Gallagher, W.P., Maleczka, Jr, R.E. *SL* 537 (2003).
[4]Olivi, N., Spruyt, P., Peyrat, J.-F., Alami, M., Brion, J.-D. *TL* **45**, 2607 (2004).
[5]Russo, O., Alami, M., Brion, J.-D., Sicsic, S., Berque-Bestel, I. *TL* **45**, 7069 (2004).
[6]Suzuki, N., Yasaki, S., Yasuhara, A., Sakamoto, T. *CPB* **51**, 1170 (2003).
[7]Fairlamb, I.J.S., Bauerlein, P.S., Marrison, L.R., Dickinson, J.M. *CC* 632 (2003).
[8]Li, J.-H., Liang, Y., Zhang, X.-D. *T* **61**, 1903 (2005).
[9]Chen, L., Li, C.-J. *TL* **45**, 2771 (2004).
[10]Zeni, G., Nogueira, C.W., Pena, J.M., Pilissao, C., Menezes, P.H., Braga, A.L., Rocha, J.B.T. *SL* 579 (2003).
[11]Zeni, G., Alves, D., Pena, J.M., Braga, A.L., Stefani, H.A., Nogueira, C.W. *OBC* **2**, 803 (2004).
[12]Zeni, G., Alves, D., Braga, A.L., Stefani, H.A., Nogueira, C.W. *TL* **45**, 4823 (2004).
[13]Reiter, M., Ropp, S., Gouverneur, V. *OL* **6**, 91 (2004).
[14]Tokuyama, H., Miyazaki, T., Yokoshima, S., Fukuyama, T. *SL* 1512 (2003).

Palladium(II) chloride – di-*t*-butylphosphinous acid.

Coupling reactions. This catalytic system performs essentially the same tasks as $PdCl_2$–R_3P, e.g., in Stille coupling,[1] Sonagashira coupling[2] and that involving $ArSi(OMe)_3$.[3,4] It is stable in air and also easier to carry out the reactions in water.

[1]Wolf, C., Lerebours, R. *JOC* **68**, 7551 (2003).
[2]Wolf, C., Lerebours, R. *OBC* **2**, 2161 (2004).
[3]Wolf, C., Lerebours, R., Tanzini, E.H. *S* 2069 (2003).
[4]Wolf, C., Lerebours, R. *OL* **6**, 1147 (2004).

Palladium(II) chloride – tertiary phosphine. **19**, 261; **20**, 295–298; **21**, 332–334; **22**, 343–345

Coupling reactions. Promotion of Suzuki coupling involving perfluorooctanesulfonyloxyarenes by microwaves has been advocated.[1] (Reductive removal of the sulfonyloxy group needs only replacement of the coupling partners with HCOOH, thereby phenols can be deoxygenated).[2] Microwaves also enhance formation of electron-rich arylboronates from bis(pinaconato)diboron.[3] Also under microwave activation synthesis of biaryls from ArX and $NaBAr_4$ is accomplished.[4] Multifold Suzuki coupling (up to 8-fold) of polyhaloarenes can be very efficient.[5] Bulky ligands for $PdCl_2$ include (2,6-dimesitylphenyl)dimethyphosphine[6] and 1,1'-bis(di-*t*-butylphosphino)ferrocene.[7]

Copper-free Sonagashira coupling is achieved with an amine (piperidine, pyrrolidine, Et_3N) as additive.[8] With Ag_2O it allows coupling of $ArB(OH)_2$ with 1-alkynes (via silver acetylides) in air at room temperature.[9] Coupling reactions involving alkynylsilanols as surrogates for 1-alkynes are promoted by Me_3SiOK.[10] Silicone that contains Ph-Si units[11] and $[MeBF_3]K$[12] are suitable as coupling partners for ArX.

Sol-gel entrapped $(Ph_3P)_2PdCl_2$ catalyzes Heck reaction. Phenanthrenes can be synthesized by photocyclization without isolation of the stilbenes.[13]

One catalyst adequately serves in Sonagashira and Heck coupling reactions to assemble arylated fused furans,[14] and *N*-(1-alkynyl)-2-iodoarylamines cyclize on incorporation of amines to furnish 2-aminoindoles.[15]

Many 2,4-disubstituted oxazoles are obtained by applying selective and consecutive coupling reactions to 4-bromomethyl-2-chlorooxazole.[16]

Cinnamides and acetoacetamides are synthesized from chloroacetamides by the coupling reactions with styrenes and vinyl ethers, respectively, in the latter case the primary products are the enol ethers.[17] Note the different regiochemical selectivity from the radical type reaction involving the vinyl ethers.

Chelated diorganoaluminum 2,2-dimethylaminoethoxides cross-couple with haloarenes and tricarbonylchromium complexes of chloroarenes.[18] A Pd-Fe bimetallic catalyst is also effective to accomplish the reaction.[19]

Ar-Boronato groups are replaced by allyl residues of allyl acetates through coupling (additive: KF).[20] Alkynylstannanes and benzylic halides combine to give enynes, three CC bonds are formed.[21]

Ar⌒Br + $SnBu_3$–C≡C–R $\xrightarrow[\text{NMP, 50°C}]{(Ph_3P)_2PdCl_2}$ product (Ar, R, Ar, R)

A Pd-catalyzed three-component coupling to construct pyrrolidines starts from Michael addition of allylic or propargylic amines to conjugated carbonyl compounds, with cyclization-coupling (ArI) to conclude the synthesis.[22]

By stereoselective Negishi couplings 1,1-dibromalkenes serve as platforms for construction of carbon chains. The first reaction replaces the (*E*)-oriented bromine atom.[23]

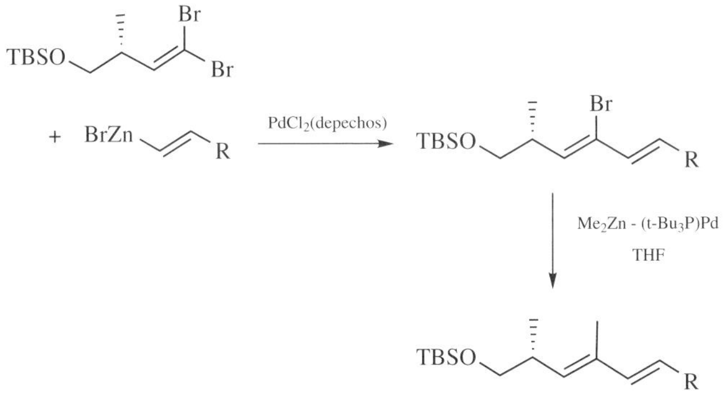

Geminal coupling involving *t*-butyl isocyanide, bromoalkenes or bromoarenes, and amines leads to amidines and imidates. Both open-chain and cyclic products are accessible by the method.[24,25]

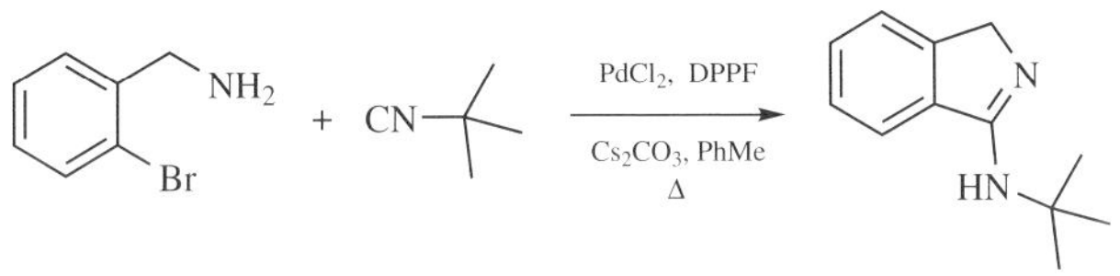

A convenient synthesis of aryldimethylsilanols is based on coupling of ArBr with *sym*-diethoxytetramethyldisilane.[26]

Oxidative decarboxylation. Alkenes result when the mixed anhydrides of pivalic and carboxylic acids are heated with phosphine-complexed $PdCl_2$.[27]

Substitution reactions. Secondary benzylic alcohols are converted into ethers, sulfides and amines in a Pd-catalyzed, Ag-assisted process.[28] 1,4-Heterobridged naphthalenes undergo ring opening by $ArB(OH)_2$ to afford *cis*-2-aryl-1-hetero-1,2-dihydronaphthalenes.[29] Arylation of active methylene compounds by Pd-catalyzed coupling[30] is most convenient.

Ketone synthesis. Because of the relative mild conditions and chemoselectivity the Pd-catalyzed reaction between $ArB(OH)_2$ and unsaturated acid chlorides is highly suitable for synthesis of aryl alkenyl ketones.[31] Friedel-Crafts acylation is clearly unsuitable to access such products.

Both aryl pyridyl ketones[32] and alkynyl aryl ketones[33] can be constructed using CO as the linchpin.

Hydrostannylation. Major products from *syn*-addition of Bu_3SnH to propargylic alcohols catalyzed by the Pd complex have the tin group near the hydroxyl group.[34] For diols flanking both sides of the triple bond the tin group attaches itself to the less hindered side preferentially.

$(Ph_3P)_2PdCl_2$, Bu_3SnH, THF

93%

Depropargylation.[35] *O/N*-Propargyl derivatives are cleaved in aqueous media using a catalyst system containing $(Ph_3P)_2PdCl_2$ and Et_3N.

[1]Zhang, W., Chen, C.H.-T., Lu, Y., Nagashima, T. *OL* **6**, 1473 (2004).
[2]Zhang, W., Nagashima, T., Lu, Y., Chen, C.H.-T. *TL* **45**, 4611 (2004).
[3]Appukkuttan, P., Van der Eycken, E., Dehaen, W. *SL* 1204 (2003).
[4]Wang, J.-X., Yang, Y., Wei, B. *SC* **34**, 2063 (2004).
[5]Stulgies, B., Prinz, P., Magull, J., Rauch, K., Meindl, K., Ruhl, S., de Meijere, A. *CEJ* **11**, 308 (2004).
[6]Smith, R.C., Woloszynek, R.A., Chen, W., Ren, T., Protasiewicz, J.D. *TL* **45**, 8327 (2004).
[7]Colacot, T., Shea, H. *OL* **6**, 3731 (2004).
[8]Leadbeater, N.E., Tominack, B.J. *TL* **44**, 8653 (2003).
[9]Zou, G., Zhu, J., Tang, J. *TL* **44**, 8709 (2003).
[10]Denmark, S.E., Tymonko, S.A. *JOC* **68**, 9151 (2003).
[11]Koike, T., Mori, A. *SL* 1850 (2003).
[12]Molander, G.A., Yun, C.-S., Ribagorda, M., Biolatto, B. *JOC* **68**, 5534 (2003).
[13]Hamza, K., Abu-Reziq, R., Avnir, D., Blum, J. *OL* **6**, 925 (2004).
[14]Bossharth, E., Desbordes, P., Monteiro, N., Balme, G. *OL* **5**, 2441 (2003).
[15]Witulski, B., Alayrac, C., Tevsadze-Saeftel, L. *ACIEE* **42**, 4257 (2003).
[16]Young, G.L., Smith, S.A., Taylor, R.J.K. *TL* **45**, 3797 (2004).
[17]Glorius, F. *TL* **44**, 5751 (2003).
[18]Schumann, H., Kaufmann, J., Schmalz, H.-G., Bottcher, A., Gotov, B. *SL* 1783 (2003).
[19]Shenglof, M., Gelman, D., Heymer, B., Schumann, H., Molander, G.A., Blum, B. *S* 302 (2003).
[20]Ortar, G. *TL* **44**, 4311 (2003).
[21]Pottier, L.R., Peyrat, J.-F., Alami, M., Brion, J.-D. *TL* **45**, 4035 (2004).
[22]Martinon, L., Azoulay, S., Monteiro, N., Kundig, P., Balme, G. *JOMC* **689**, 3831 (2004).
[23]Zeng, X., Qian, M., Hu, Q., Negishi, E. *ACIEE* **43**, 2259 (2004).
[24]Kishore, K., Tetela, R., Whitby, R.J., Light, M.E., Hurtshouse, M.B. *TL* **45**, 6991 (2004).
[25]Saluste, C.G., Crumpler, S., Furber, M., Whitby, R.J. *TL* **45**, 6995 (2004).
[26]Denmark, S.E., Kallemeyn, J.M. *OL* **5**, 3483 (2003).
[27]Goossen, L.J., Rodriguez, N. *CC* 724 (2004).
[28]Miller, K.J., Abu-Omar, M.M. *EJOC* 1294 (2003).
[29]Lautens, M., Dockendorff, C. *OL* **5**, 3695 (2003).
[30]Gao, C., Tao, X., Qian, Y., Huang, J. *SL* 1716 (2003).
[31]Urawa, Y., Nishiura, K., Souda, S., Ogura, K *S* 2882 (2003).
[32]Couve-Bonnaire, S., Carpentier, J.-F., Mortreux, A., Castanet, Y. *T* **59**, 2793 (2003).
[33]Ahmed, M.S.M., Mori, A. *OL* **5**, 3057 (2003).

[34]Marshall, J.A., Bourbeau, M.P. *TL* **44**, 1087 (2003).
[35]Pal, M., Parasuraman, K., Yeleswarapu, K.R. *OL* **5**, 349 (2003).

Palladium(II) hydroxide/carbon. 19, 262; **20**, 299–300; **21**, 335–336; **22**, 346–347

Hydrogenation. Saturation of the pyridine ring by catalytic hydrogenation using $Pd(OH)_2/C$ is subject to asymmetric 1,5-induction. A chiral auxiliary at C-2 is subsequently hydrogenolyzed.[1]

H_2, $Pd(OH)_2/C$, HOAC

93% (91% ee)

Conversion of *N*-benzylamines, *N*-benzhydrylamines and *N*-tritylamines to *N*-Boc-amines[2] is done in one step by admixture of the substrates with $Pd(OH)_2/C$, PMHS, and $(Boc)_2O$ in EtOH. Under essentially the same conditions Δ^2-isoxazolines are cleaved to give *N*-Boc-1,3-amino alcohols.[3]

Coupling reactions. The catalytic activity of $Pd(OH)_2/C$ is comparable to that of many other Pd(0) species in promoting cross-couplings (e.g., Suzuki and Sonagashira couplings). It is also useful for ketone synthesis from thio esters and alkyl iodides.[4]

$Pd(OH)_2/C$, THF, PhMe, DMF

84%

Oxidation. Enones and their acetals are oxidized to 1,4-enediones/monoacetals with $Pd(OH)_2/C$ and *t*-BuOOH in excellent yields.[5]

[1]Glorius, F., Spielkamp, N., Holle, S., Goddard, R., Lehrmann, C.W. *ACIEE* **43**, 2850 (2004).
[2]Chandrasekhar, S., Babu, B.N., Reddy, C.R. *TL* **44**, 2057 (2003).
[3]Chandrasekhar, S., Babu, B.N., Ahmed, M., Reddy, M.V., Srihari, P., Jagadeesh, B., Prabhakar, A. *SL* 1303 (2004).
[4]Mori, Y., Seki, M. *JOC* **68**, 1571 (2003).
[5]Yu, J.-Q., Corey, E.J. *JACS* **125**, 3232 (2003).

Palladium(II) iodide. 18, 283

Ureas. Symmetrical and unsymmetrical ureas are prepared from amines by catalysis of PdI_2–KI.[1] The carbonylation agent consists of CO, CO_2 and air. A similar procedure is used to form 2-oxazolidinones from 2-amino alcohols.[2]

[1]Gabriele, B., Mancuso, R., Salerno, G., Costa, M. *CC* 486 (2003).
[2]Gabriele, B., Mancuso, R., Salerno, G., Costa, M. *JOC* **68**, 601 (2003).

Pentacarbonylchlororhenium(I).

Cyclization.[1] $ReCl(CO)_5$ coordinates with and activates triple bonds toward electrophilic attack. Intramolecular reaction with electron-rich double bond leads to cyclic products. A reflex process gives bicyclic structures. The reaction is also photoassisted, and it likely involves displacement of one CO ligand from the Re center.

OTIPS

$ReCl(CO)_5$

hν PhMe

TIPSO H + TIPSO H

(7:3)

92%

[1]Kusama, H., Yamabe, H., Onisawa, Y., Hoshino, T., Iwasawa, N. *ACIEE* **44**, 468 (2005).

Periodic acid. 13, 238–239; **16**, 292; **18**, 285–286; **20**, 302; **22**, 349

Oxidation. Functional group oxidation by periodic acid is catalyzed by chromium salts: sulfides to sulfones by CrO_3[1] and alcohols to carbonyl compounds by $Cr(acac)_3$.[2] More remarkable is the oxidation at the benzylic position by a system containing $CrO_2(OAc)_2$ and H_5IO_6[3]

Dealkylation.[4] Using chromium(III) acetate hydroxide as catalyst *N*-alkylsulfonamides and *N,N*-dialkylsulfonamides are totally dealkylated by periodic acid.

[1]Xu, L., Cheng, J., Trudell, M.L. *JOC* **68**, 5388 (2003).
[2]Xu, L., Trudell, M.L. *TL* **44**, 2553 (2003).
[3]Lee, S., Fuchs, P.L. *JACS* **124**, 13978 (2002).
[4]Xu, L., Zhang, S., Trudell, M.L. *SL* 1901 (2004).

Phase-transfer catalysts. 13, 239–240; **15**, 252–253; **18**, 286–289; **19**, 264–267; **20**, 302–303; **21**, 338–341; **22**, 349–350

Acylation and deacylation. Amides are formed in micelle[1] and using cyclodextrins as inverse phase-transfer catalysts.[2]

In the presence of Bu_4NHSO_4, powdered NaOH is able to saponify esters.[3]

Alkylation. *t*-Butyl 2-phenyloxazoline-4-carboxylate suffers ring cleavage in the presence of *t*-BuOK. For alkylation phase-transfer conditions are viable.[4] α-Alkylserine derivatives of very high optical purity are produced using a chiral PTC (**1**) of the spirocyclic bis(binaphthylmethyl)ammonium type.[5]

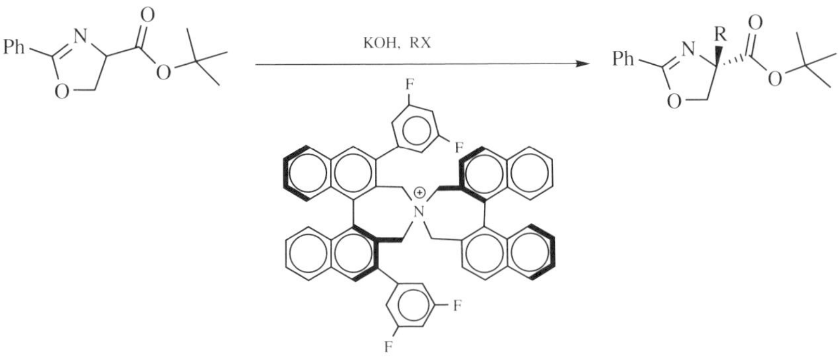

KOH, RX

(**1**)

By this technique enantioselective construction of quaternary centers by alkylation and Michael reaction of β-keto esters is accomplished.[6,7] The bis(binaphthylmethyl) ammonium bromide bears two aryl substituents at C-3 and C-3′ positions of *one* binaphthalene segment.

Enantioselective alkylation of glycine derivatives has the implication of access to natural and unnatural α-amino acids. The chiral PTC method has been systematically developed so that effect of substituents on the binaphthalene skeleton is liable to exploitation.[8] Symmetrical 4,4′,6,6′-tetraaryl analogues of **1** are found to be good catalysts[9] and the corresponding fluorous-tagged alkylsilyl derivative is recyclable.[10] Tartaric acid-derived bisquaternary ammonium salts have certain advantages.[11]

Protected glycinamides also undergo asymmetric alkylation using an unsymmetrical 3,3′-diaryl-substituted salt for catalysis.[12] The products are readily reduced to chiral 1,2-diamines.

Further work has uncovered the effectiveness of **2** for the same purpose of asymmetric alkylation,[13] while inferior results are obtained employing cinchona alkaloid-based PTC.[14,15]

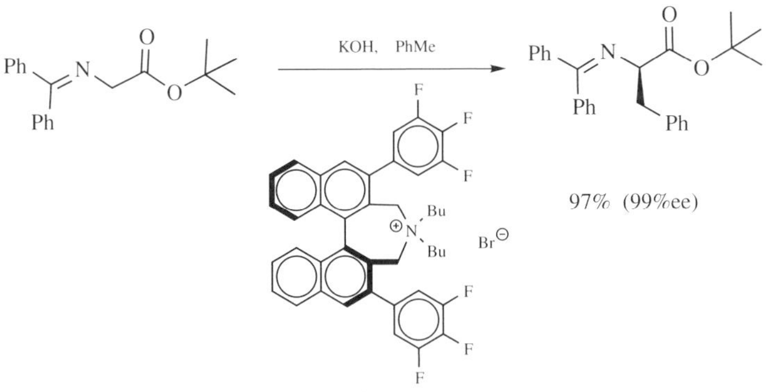

KOH, PhMe

97% (99%ee)

(**2**)

Preparation of racemic α-amino acid derivatives from a polymer-bound glycine ester involves attachment of reagents from the liquid phase.[16] The imino bonding to the polymer is easily broken by aq. acid after the alkylation step.

Etherification of phenols in water with K_2CO_3 as base and Bu_4NCl present is assisted by 2,6-lutidine.[17] In a study of alkylation and related processes,[18] 2-methyltetrahydrofuran emerges as a superior solvent to CH_2Cl_2. Borane complexes of secondary phosphines are alkylated by PTC.[19]

Related is α-benzenesulfenylation of arylacetic esters.[20]

Michael reaction. With nitroalkanes[21] or silyl nitronates[22] as donors and a chiral spirocyclic ammonium salt as PTC, excellent results are obtained. In the former case Cs_2CO_3 is the base, whereas in the latter a $[HF_2]^-$ counterion for the salt serves to release the nucleophile.

Condensation reactions. A PTC version of the Darzens reaction is achieved in MeCN.[23] A test of *N,N'*- bis(naphthylmethylquinuclidinium) bromide in promoting Darzens condensation can be called a modest success.[24] Equilibration of *t*-butyl glycidates under PTC conditions leads to the *cis* isomers.[25]

The HWE-reaction using methyl bis(2,2,2-trifluoroethoxy)phosphonylacetate shows very high (*Z*)-selectivity when using tris(methoxyethoxyethyl)amine to coordinate with the potassium ion of the base.[26] However, with (*Z*)-3-triorganostannylacrolein results are somewhat unexpected.[27]

Crossed benzoin condensation[28] from acylsilanes and aldehydes requires catalytic amount of KCN and 18-crown-6 only. The silyl group is transferred to the hydroxyl function of the product.

Addition and cycloaddition. PTC is applicable to diastereoselective allylation of carbonyl compounds with trifluoroborate salts.[29]

Trialkyl(3-sulfopropyl)ammonium betaines are useful for dichlorocarbene generation from $CHCl_3$ and cycloaddition to alkenes.[30]

Coupling reactions. Pd-catalyzed cross-coupling of ArX in water is viable by the addition of a surfactant.[31] Alternatively, biphasic Suzuki coupling using a PEG-oxylated benzylphosphine ligand has the advantage of easy recovery of the catalyst.[32]

Radical-mediated carbonylation of RI in aq. media (+ CTAB) leads to RCOOR.[33]

Oxidoreductions. Oxidiperoxovanadate ion converts benzylic halides into carbonyl compounds under PTC conditions. The corresponding benzylic alcohols are oxidized.[34]

Monoaza-15-crown-5 lariat ethers constructed from D-glucose and D-mannose have been used in asymmetric epoxidation of chalcones.[35]

Nitroalkanes are reduced to oximes on *O*-benzylation followed by elimination of PhCHO in situ under PTC conditions at room temperature.[36] This process can be used to prepare chiral oximes.

Substitutions. Preparation of β-fluoroalkyl amines[37] from aziridines is easily accomplished by treatment with $KF.2H_2O$ in the presence of Bu_4NHSO_4.

Generation of tosyl azide for azo transfer reactions is facilitated by the presence of a PTC.[38]

OTIPS

SPh

N
H

PhI(CN)OTf
lutidine
MeCN, –40°C

O

SPh

N

40%

[1]Naik, S., Bhattacharjya, G., Talukdar, B., Patel, B.K. *EJOC* 1254 (2004).

[2]Kunishima, M., Watanabe, Y., Terao, K., Tani, S. *EJOC* 4535 (2004).

[3]Crouch, R.D., Burger, J.S., Zietek, K.A., Cadwallader, A.B., Bedison, J.E., Smielewska, M..M. *SL* 991 (2003).

[4]Park, H., Lee, J., Kang, M.J., Lee, Y.-J., Jeong, B.-S., Lee, J.-H., Yoo, M.-S., Choi, S., Jew, S. *T* **60**, 4243 (2004).

[5]Jew, S., Lee, Y.-J., Lee, J., Kang, M.J., Jeong, B.-S., Lee, J.-H., Yoo, M.-S., Kim, M.-J., Choi, S., Ku, J.-M., Park, H. *ACIEE* **43**, 2382 (2004).

[6]Ooi, T., Miki, T., Taniguchi, M., Shiraishi, M., Takeuchi, M., Maruoka, K. *ACIEE* **42**, 3796 (2003).

[7]Ooi, T., Miki, T., Maruoka, K. *OL* **7**, 191 (2005).

[8]Hashimoto, T., Maruoka, K. *TL* **44**, 3313 (2003).

[9]Hashimoto, T., Tanaka, Y., Maruoka, K. *TA* **14**, 1599 (2003).

[10]Shirakawa, S., Tanaka, Y., Maruoka, K. *OL* **6**, 1429 (2004).

[11]Ohshima, T., Shibuguchi, T., Fukuta, Y., Shibasaki, M. *T* **60**, 7743 (2004).

[12]Ooi, T., Sakai, D., Takeuchi, M., Tayama, E., Maruoka, K. *ACIEE* **42**, 5868 (2003).

[13]Kitamura, M., Shirakawa, S., Maruoka, K. *ACIEE* **44**, 1549 (2005).

[14]Yu, H., Koshima, H. *TL* **44**, 9209 (2003).

[15]Kumar, S., Ramachandran, U. *TA* **14**, 2539 (2003).

[16]Park, H., Kim, M.-J., Park, M.-K., Jung, H.-J., Lee, J., Lee, Y.-J., Jeong, B.-S., Lee, J.-H., Yoo, M.-S., Jew, S. *TL* **46**, 93 (2005).

[17]Aki, S., Nishi, T., Minamikawa, J. *CL* **33**, 940 (2004).

[18]Ripin, D.H.B., Vetelino, M *SL* 2353 (2003).

[19]Lebel, H., Morin, S., Paquet, V. *OL* **5**, 2347 (2003).

[20]Marzorati, L., da Silva, M.A., Wladislaw, B., Di Vitta, C. *SC* **33**, 3491 (2003).

[21]Ooi, T., Fujioka, S., Maruoka, K. *JACS* **126**, 11790 (2004).

[22]Ooi, T., Doda, K., Maruoka, K. *JACS* **125**, 9022 (2003).

[23]Wang, Z.-T., Xu, L.-W., Xia, C.-G., Wang, H.-Q. *HCA* **87**, 1958 (2004).

[24]Arai, S., Tokumaru, K., Aoyama, T. *TL* **45**, 1845 (2004).

[25]Jonczyk, A., Zomerfeld, T. *TL* **44**, 2359 (2003).

[26]Touchard, F.T. *TL* **45**, 5519 (2004).

[27]Franci, X., Martina, S.L.X., McGrady, J.E., Webb, M.R., Donald, C., Taylor, R.J.K. *TL* **44**, 7735 (2003).

[28]Linghu, X., Johnson, J.S. *ACIEE* **42**, 2534 (2003).

[29]Thadani, A.N., Batey, R.A. *TL* **44**, 8051 (2003).

[30]Jayachandran, J.P., Wang, M.-L. *SC* **33**, 2463 (2003).

[31]Arcadi, A., Cerichelli, G., Chiarini, M., Correa, M., Zorzan, D. *EJOC* 4080 (2003).

[22]an der Heiden, M., Plenio, H. *EJOC* 1789 (2004).

[33]Sugiura, M., Hagio, H., Kobayashi, S. *CL* **32**, 898 (2003).

[34]Li, C., Zheng, P., Li, J., Zhang, H., Shao, Q., Ji, X., Zhang, J., Zhao, P., Xu, Y. *ACIEE* **42**, 5063 (2003).

[35]Bako, P., Bako, T., Meszaros, A., Keglevich, G., Szollosy, A., Bodor, S., Mako, A., Toke, L. *SL* 643 (2004).
[36]Czekelius, C., Carreira, E.M. *ACIEE* **44**, 612 (2005).
[37]Fan, R.-H., Zhou, Y.-G., Zhang, W.-X., Hou, X.-L., Dai, L.-X. *JOC* **69**, 335 (2004).
[38]Arai, S., Hasegawa, K., Nishida, A. *TL* **45**, 1023 (2004).

Phenyl cyanate.

Cyanation.[1] The title reagent is a electrophilic cyano donor, e.g., for synthesis of cyanoarenes from ArLi.

[1]Sato, N., Yue, Q. *T* **59**, 5831 (2003).

Phenyl(cyano)iodine(III) triflate.

Oxidative cyclization.[1] 2-Phenylthioindoles that contain a nucleophilic sidechain at C-3 (allyl silane, siloxyalkene,...) give spirocyclic products on exposure to PhI(CN)OTf. The transformation cannot be brought about with $PhI(OCOCX_3)_2$.

Alkynyliodonium triflates.[2] The unstable iodonium salts are generated by an exchange process from alkynylstannanes. They are precursors of alkenylcarbenes.

$SnBu_3$ Bu_3Sn N O $SnBu_3$ — PhI(CN)OTf, CH_2Cl_2 → $SnBu_3$ Bu_3Sn N O I OTf Ph — $ArSO_2Na$ → $SnBu_3$ $SnBu_3$ N O SO_2Ar

[1]Feldman, K.S., Vidulova, D.B. *TL* **45**, 5035 (2004).
[2]Feldman, K.S., Perkins, A.L., Masters, K.M. *JOC* **69**, 7928 (2004).

Phenyliodine(III) bis(trifluoroacetate). **13**, 241–242; **14**, 257; **15**, 257–258; **16**, 274–275; **18**, 289–290; **19**, 267–268; **20**, 305; **21**, 342–343; **22**, 351–352

Alkene functionalization. (Methylthio)acetonitrile is oxidized to the cyanomethylenesulfonium salt by $PhI(OCOCF_3)_2$. 1-Alkenes react with this salt and thereby extending the chain by a 2-carbon unit. Oxidative elimination of the thio group gives $\alpha,\beta;\gamma,\delta$-unsaturated nitriles.[1]

Cyclization. 3-Arylpropanols in which the aromatic ring contains activated (e.g., methoxy) groups are converted into chromans[2] directly by $PhI(OCOCF_3)_2$. A polymer-supported reagent is used for synthesis of dihydrobenzoxathiepins[3] from aryloxyethylthioacetic esters.

Intramolecular oxidative coupling of 1-benzyltetrahydroisoquinoline derivatives to afford morphinandienones[4] is efficiently accomplished using a combination of $PhI(OCOCF_3)_2$ and heteropoly acid. Spiroannulation by oxidation of an *N*-methoxy-3-arylpropanamide is the key step in a synthesis of dysibetaine.[5] Nitrenium ion is formed and captured.

Aromatic substitutions. Oxidation of iodine and PhSeSePh by $PhI(OCOCF_3)_2$ in MeCN in the presence of activated arenes gives ArI and ArSePh, respectively.[6]

[1]Huang, H.-Y., Wang, H.-M., Kang, I.-J., Chen, L.-C. *JCCS(T)* **50**, 1053 (2003).
[2]Hamamoto, H., Hata, K., Nambu, H., Shiozaki, Y., Tohma, H., Kita, Y. *TL* **45**, 2293 (2004).
[3]Huang, H.-S., Wang, H.-M., Hou, R.-S., Chen, L.-C. *JCCS(T)* **51**, 1025 (2004).
[4]Hamamoto, H., Shiozaki, Y., Nambu, H., Hata, K., Tohma, H., Kita, Y. *CEJ* **10**, 4977 (2004).
[5]Wardrop, D.J., Burge, M.S. *CC* 1230 (2004).
[6]Panunzi, B., Rotiroti, L., Tingoli, M. *TL* **44**, 8753 (2003).

Phenyliodine(III) diacetate. **13**, 242–243; **14**, 258–259; **15**, 258; **16**, 275–276; **17**, 280–281; **18**, 290–291; **19**, 268–270; **20**, 305–307; **21**, 343–344; **22**, 352–354

Oxidations. The $PhI(OAc)_2$ – TEMPO combination is useful for selective oxidation of primary alcohols, permitting the preparation of lactones from polyfunctional compounds.[1] A new TEMPO variant is a fluorous-tagged **1**.[2]

$[(PivO)_2Rh]_2$, $PhI(OAc)_2$

83%

1

Generation of electrophilic bromine species from LiBr for aromatic bromination[3] is conveniently achieved with $PhI(OAc)_2$.

Cycloaddition. Oxidation of aromatic ketones in RCN under acidic ($CFSO_3H$) conditions leads to 5-aryloxazoles,[4] as the cyano group of RCN is incorporated into the products.

Because of the enolizability, nitroacetic esters are readily converted into an iodonium ylide by $PhI(OAc)_2$ and in turn a carbenoid with $[Rh(OPiv)_2]_2$. Trapping by alkenes (e.g., styrene) affords 1-nitro-1-cyclopropanecarboxylic esters.[5]

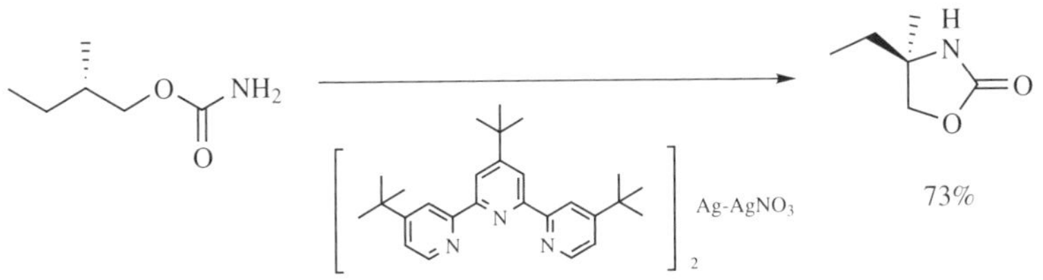

Nitrene derived from *N*-phthaloylhydrazine is trapped to form aziridines.[6] Using a chiral (bisoxazoline)-Cu complex to direct aziridination from nitrenes generated from arenesulfonamides with $PhI(OAc)_2$ optically active products (ee 45–75%) are obtained.[7]

[1]Hansen, T.M., Florence, G.J., Lugo-Mas, P., Chen, J., Abrams, J.N., Forsyth, C.J. *TL* **44**, 57 (2003).
[2]Pozzi, G., Cavazzini, M., Holczknecht, O., Quici, S., Shepperson, I. *TL* **45**, 4249 (2004).
[3]Braddock, D.C., Cansell, G., Hermitage, S.A. *SL* 461 (2004).
[4]Kang, I.-J., Wang, H.-M., Lin, M.-L., Chen, L.-C. *JCCS(T)* **49**, 1031 (2002).
[5]Wurz, R.P., Charette, A.B. *OL* **5**, 2327 (2003).
[6]Li, J., Liang, J.-L., Chan, P.W.H., Che, C.-M. *TL* **45**, 2685 (2004).
[7]Kwong, H.-L., Liu, D., Chan, K.-Y., Lee, C.-S., Huang, K.-H., Che, C.-M. *TL* **45**, 3965 (2004).

Phenyliodine(III) diacetate - iodine. 20, 307; **21**, 345–347; **22**, 354–355

Demethoxylation. Selective oxidation of a methoxy group via hydrogen transfer to a proximal hydroxyl radical should find use in carbohydrate chemistry. Generation of hypoiodites by the $PhI(OAc)_2 - I_2$ combination is convenient.[1]

Iodopyrazoles. Rapid iodination of pyrazoles is done under mild conditions.

[1]Boto, A., Hernandez, D., Hernandez, R., Suarez, E. *OL* **6**, 3785 (2004).
[2]Cheng, D.-P., Chen, Z.-C., Zheng, Q.-G. *SC* **33**, 2671 (2003).

Phenyliodine(III) diacetate – metal salts.

C-H bond functionalization. Nitrogen atom-directed remote oxidation of an unactivated carbon (both sp^3 and sp^2 hybridized) by the $PhI(OAc)_2 - Pd(OAc)_2$ combination is synthetically significant.[1,2] The $PhI(OAc)_2 - AgNO_3$ combination (with a terpyridine ligand) is able to convert an amide into a cyclic product.[3]

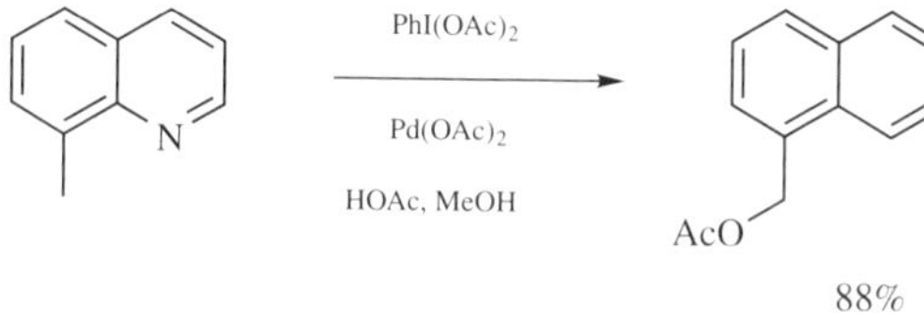

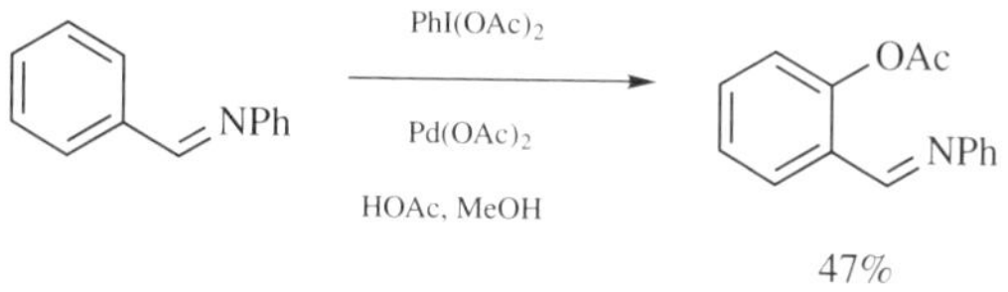

PhI(OAc)$_2$
Pd(OAc)$_2$
HOAc, MeOH
AcO
88%

NPh
PhI(OAc)$_2$
Pd(OAc)$_2$
HOAc, MeOH
OAc
NPh
47%

Aziridines.[4] In the presence of alkenes the oxidation of 2-pyridinesulfonamides by $PhI(OAc)_2$ and Cu(II) trifluoroacetylacetonate gives rise to aziridines. Removal of the *N*-substituent of the adduct is easily accomplished with Mg in MeOH. Regioselectivity in ring opening with nucleophiles such as Me_2CuLi is different from that of the benzenesulfonyl derivative.

[1]Desai, L.V., Hull, K.L., Sanford, M.S. *JACS* **126**, 9542 (2004).
[2]Dick, A.R., Hull, K.L., Sanford, M.S. *JACS* **126**, 2300 (2004).
[3]Cui, Y., He, C. *ACIEE* **43**, 4210 (2004).
[4]Han, H., Bae, I., Yoo, E.J., Lee, J., Do, Y., Chang, S. *OL* **6**, 4109 (2004).

Phenyliodine(III) dichloride. 22, 355

Iodoalkoxylation. Addition of [I/OR] to alkenes is achieved with $PhICl_2$ and iodine in ROH at room temperature.[1] Iodohydrins can also be obtained.

[1]Yasubov, M.S., Yasubova, R.J., Filimonov, V.D., Chi, K.-W. *SC* **34**, 443 (2004).

Phenyliodine(III) difluoride.

Esterification. Diazoalkanes generated in situ from *N*-(*t*-butyldimethyl-silyl)hydrazones on reaction with $PhIF_2$ are used to esterify carboxylic acids [with promotion by $Sc(OTf)_3$ and 2-chloropyridine].

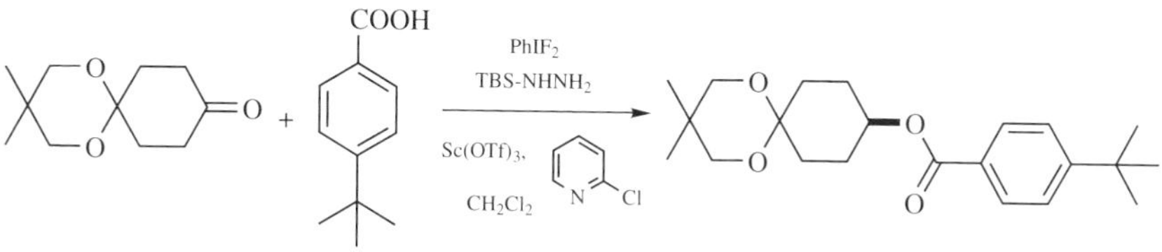

O
O
O +
COOH
$PhIF_2$
TBS-NHNH$_2$
$Sc(OTf)_3$,
CH_2Cl_2
N Cl
O
O
O
O

[1]Furrow, M.E., Myers, A.G. *JACS* **126**, 12222 (2004).

Phenyliodine(III) tosylimide.

Tosylimination. Rapid transfer of the [TsN] group from PhI = NTs to sulfoxides[1] and pyridines[2] is accomplished in the presence of $Cu(OTf)_2$.

For unknown reasons an *N,N*-ditosylamino group is introduced into furan, thiophen and *N*-tosylpyrrole in the reaction catalyzed by a porphyrin-Ru(III) complex.[3]

2-Acylpyrrolidines.[4] 2-Alkoxy-3,4-dihydro-2*H*-pyrans undergo aminative ring contraction by a Cu-catalyzed reaction with PhI = NTs. Diastereomeric products are produced in the reaction with TsN(Na)Cl and NBS.

[1]Leca, D., Song, K., Amatore, M., Fensterbank, L., Lacote, E., Malacria, M. *CEJ* **10**, 906 (2004).
[2]Jain, S.L., Sharma, V.B., Sain, B. *TL* **44**, 4385 (2003).
[3]He, L., Chan, P.W.H., Tsui, W.-M., Yu, W.-Y., Che, C.-M. *OL* **6**, 2405 (2004).
[4]Armstrong, A., Cumming, G.R., Pike, K. *CC* 812 (2004).

1-Phenyl-1*H*-tetrazol-5-yl alkyl sulfones.

Alkenylsilanes. The sulfones react with acyltrimethylsilanes after deprotonation to give alkenylsilanes. However, with acyltriphenylsilanes the Brook rearrangement pathway can take precedence to afford silyl enol ethers.

[1]Jankowski, P., Plesniak, K., Wicha, J. *OL* **5**, 2789 (2003).

N-(Phenylthio)-ε-caprolactam.

Glycosylation.[1] The title reagent together with Tf_2O are used to activate thioglycosides during glycosylation at room temperature. This couple is superior to PhSOTf which is unstable therefore requires in situ preparation from PhSCl and AgOTf.

[1]Duron, S.G., Polat, T., Wong, C.-H. *OL* **6**, 839 (2004).

Phenyl tributylstannyl selenide.

Phenylselenation.[1] The title reagent reacts with acyl chlorides [catalyzed by $(Ph_3P)_4Pd$] to afford acyl selenides and also with α-halo carbonyl compounds to deliver the corresponding phenylseleno derivatives.

[1]Nishiyama, Y., Kawamatsu, H., Funato, S., Tokunaga, K., Sonoda, N. *JOC* **68**, 3599 (2003).

9-Phenylxanthen-9-yl chloride.

Protection.[1] In oligonucleotide synthesis protection of the 5′-hydroxy group of monomers in the form of phenylxanthenyl ether (ROPx) or *p*-tolylxanthenyl ether (ROTx) is reemphasized. Such ethers offer advantage to 4,4′-dimethoxytrityl ethers for being highly crystalline. Cleavage of these protecting groups is by exposing the derivatives to $Cl_2CHCOOH$ in CH_2Cl_2.

[1]Reese, C.B., Yan, H. *TL* **45**, 2567 (2004).

Phosphoric acid.

N-C bond cleavage. *t*-Butoxycarbonyl derivatives of amines can be degraded by aq. H_3PO_4, interestingly without hydrolyzing TBS ethers, acetonides, and some other N-protecting groups (CBZ, Fmoc).[1] The cleavage of aminals by this method provides differentially substituted diamino esters.[2]

O, S Tol
N Ph
Ph
NH
MeOOC
H_3PO_4
H_2O, THF
NHSOTol
Ph
NH_2
76%

[1]Li, B., Bemish, R., Buzon, R.A., Chiu, C.K.-F., Colgan, S.T., Kissel, W., Le, T., Leeman, K.R., Newell, L., Roth, J. *TL* **44**, 8113 (2003).
[2]Viso, A., de la Pradilla, R.F., Lopez-Rodriguez, M.L., Garcia, A., Flores, A., Alonso, M. *JOC* **69**, 1542 (2004).

Phosphoryl chloride.

Condensation. $POCl_3$ on montmorillonite is used to convert acylals into dithioacetals.[1] Cyclodehydration of β-amino acids to give β-lactams[2] is also achieved using $POCl_3$ - Et_3N.

Dehydration Nitriles are directly obtained from *N*-(*t*-butyl)pyridinecarboxamides by heating with $POCl_3$ in toluene,[3] dehydration of formamides to isonitriles at $-20°$ does not affect other secondary amides.[4]

Sulfenylchlorination. *p*-Toluenesulfinyl amides are activated by $POCl_3$ to provide reagents for derivatizing alkenes.[5]

$POCl_3$, CH_2Cl_2

95%

[1]Jin, T.-S., Sun, G., Li, Y.-W., Li, T.-S. *SC* **34**, 4105 (2004).
[2]Sharma, S.D., Anand, R.D., Kaur, G. *SC* **34**, 1855 (2004).
[3]Markevitch, D.Y., Rapta, M., Hecker, S.J., Renau, T.E. *SC* **33**, 3285 (2003).
[4]Zhao, G., Bughin, C., Bienayme, H., Zhu, J. *SL* 1153 (2003).
[5]Krasnova, L.B., Yudin, A.K. *JOC* **69**, 2584 (2004).

***O*-Phthalimidomethyl trichloroacetimidate.**

Hydroxyl protection[1]. The reagent is available from phthalimidomethanol (base-catalyzed addition to Cl_3CCN) and it forms *N*-(alkoxymethyl)phthalimides (RO-Pim) in a Me_3SiOTf-catalyzed reaction with primary and secondary alcohols. Hydrazinolysis liberates alcohols (deprotection) from the ethers.

[1]Ali, I.A.I., Abdel-Rahman, A.A.-H., El Ashry, E.S.H., Schmidt, R.R. *S* 1065 (2004).

Platinum and complexes. 22, 358

Hydrogenation. The Pt/Al_2O_3 catalyst directs selective hydrogenation of the activated ketone adjacent to the trifluorofluoromethyl group of 1,1,1-trifluoro-2-4-alkanediones.[1] However, asymmetric induction by *O*-methylcinchonidine is moderate.

Silylcyanation.[2] An NCN-pincer complex of Pt prepared from 1,3-benzenedialdehyde and cyclohexylamine promotes addition of Me_3SiCN to aldehydes and imines.

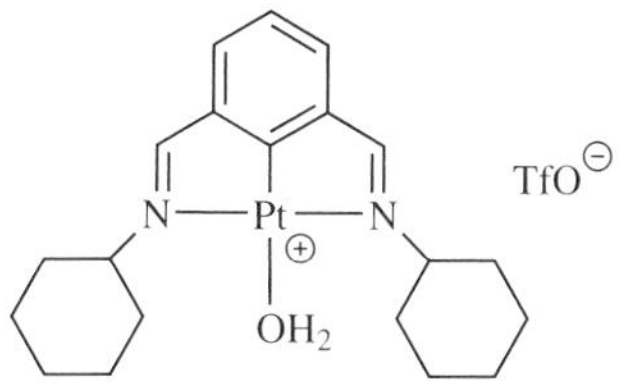

TfO⊖ N Pt N OH$_2$ ⊕

1

[1]Hess, R., Diezi, S., Mallat, T., Naiker, A. *TA* **15**, 251 (2004).
[2]Fossey, J.S., Richards, C.J. *TL* **44**, 8773 (2003).

Platinum(II) chloride. 19, 272; **21**, 352–353; **22**, 358–359

Addition to CC multiple bonds. Acetal formation from alkynes and alcohols is catalyzed by $PtCl_2$.[1] A six-membered ring is formed in an intramolecular alkylation of the indole nucleus[2] by an unsaturated side chain.

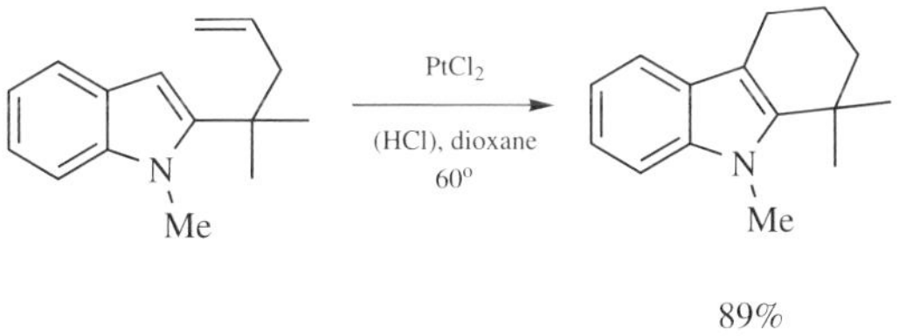

$PtCl_2$, (HCl), dioxane, 60°

89%

Cycloisomerization. Two properly separated CC multiple bonds are brought together in a ring-closing process by $PtCl_2$. Further transformations deliver some intriguing products.[3,4]

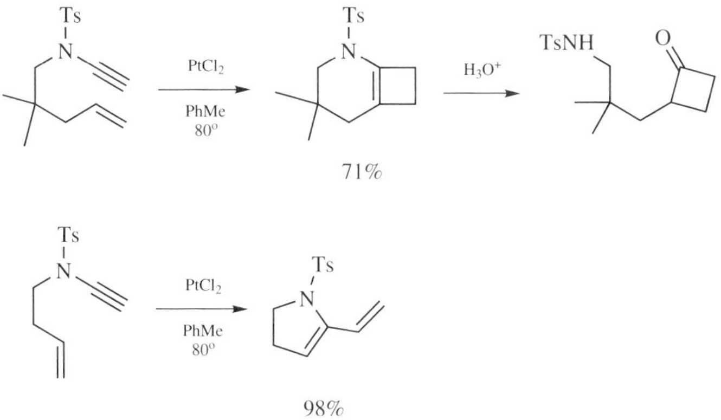

$PtCl_2$, PhMe, 80° — 71% — H_3O^+

$PtCl_2$, PhMe, 80° — 98%

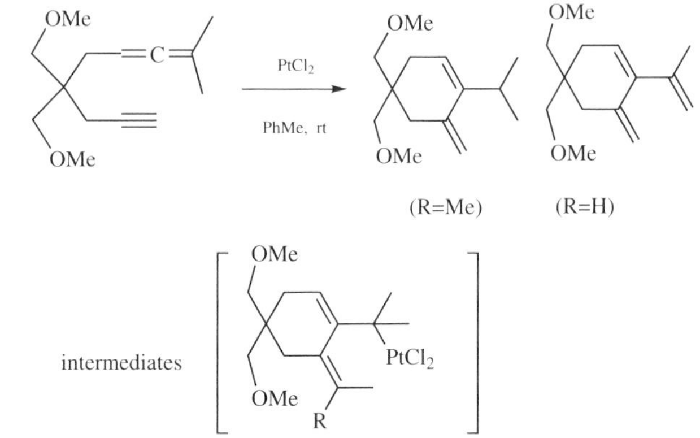

OMe OMe OMe
$PtCl_2$
PhMe, rt
OMe OMe OMe
(R=Me) (R=H)

OMe
intermediates
$PtCl_2$
OMe
R

Formation of bicycle[3.1.0]hexanes from enynes by this type of transformation[5,6] is synthetically significant. Nature of the propargylic oxygen functionality (alkoxy vs. acyloxy group) also determines the position of the ketone group that appears in the products.[7]

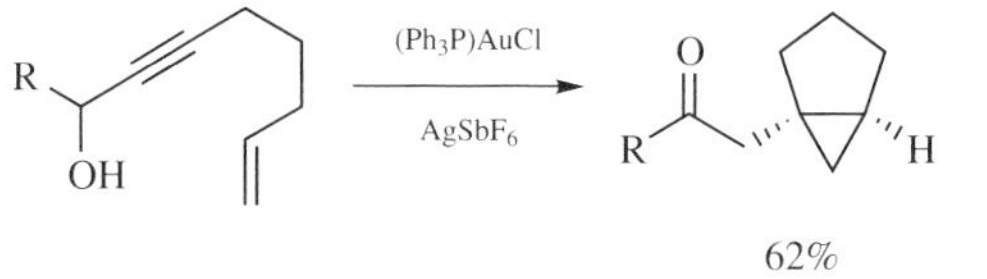

R
OH
$(Ph_3P)AuCl$
$AgSbF_6$
O
R
H
62%

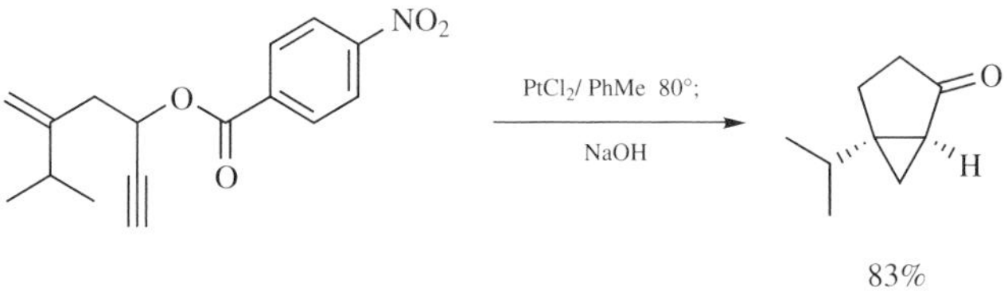

NO_2
O
O
$PtCl_2$/ PhMe 80°;
NaOH
O
H
83%

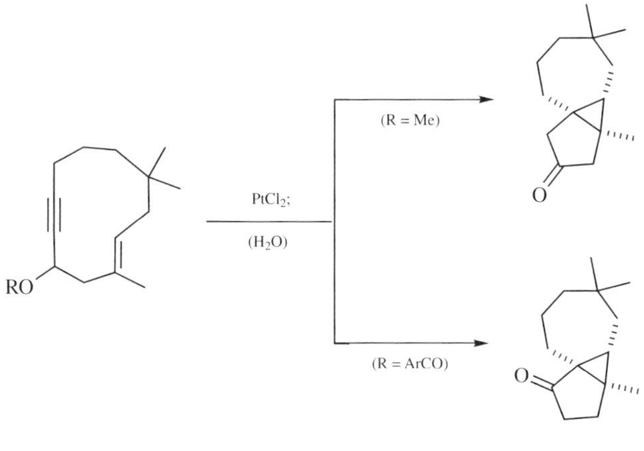

$PtCl_2$;
(H_2O)
(R = Me)
RO
O
(R = ArCO)
O

2-Alkynylbenzaldehyde acetals cyclize to 1,2-dialkoxy-1*H*-indenes on exposure to $PtCl_2$.[8] Under essentially the same conditions very extensive molecular reorganization of cyclic acetal is brought about.[7] Also notable is that dithioacetals give 1,3-dithio-1*H*-indenes although the catalyst is PdI_2.

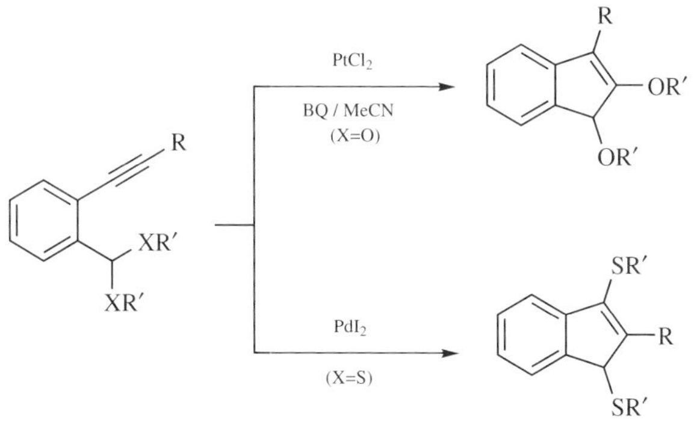

R
XR′
XR′
$PtCl_2$
BQ / MeCN
(X=O)
R
OR′
OR′
PdI_2
(X=S)
SR′
R
SR′

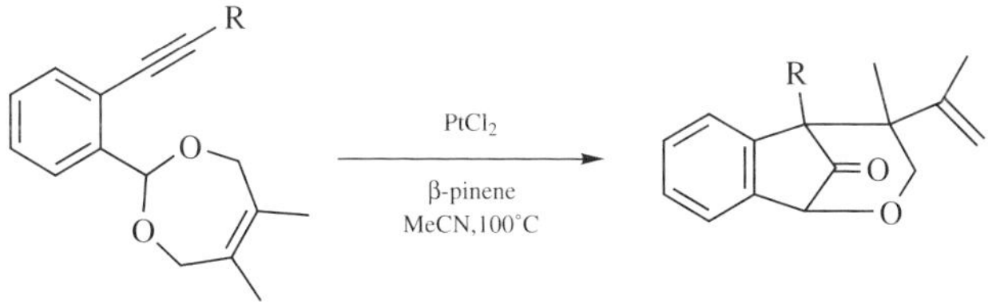

R
O
O
$PtCl_2$
β-pinene
MeCN,100°C
R
O
O

2-Alkynylbiaryls undergo cyclization to afford phenanthrenes on heating with $PtCl_2$ (or $GaCl_3$, $InCl_3$).[9] Cyclization toward indole and thiophene rings is also possible.

N-Acyl-α-amino acids. Carbonylation of enamides that are generated in situ from aldehydes and amides by CO in NMP is catalyzed by K_2PtCl_4 (with Ph_3P added as ligand).[10]

Cyclization. *N*-Acyl-2-alkynylarylamines undergo cyclization with attendant transfer of the acyl group to C-3 of the resulting indoles.[11]

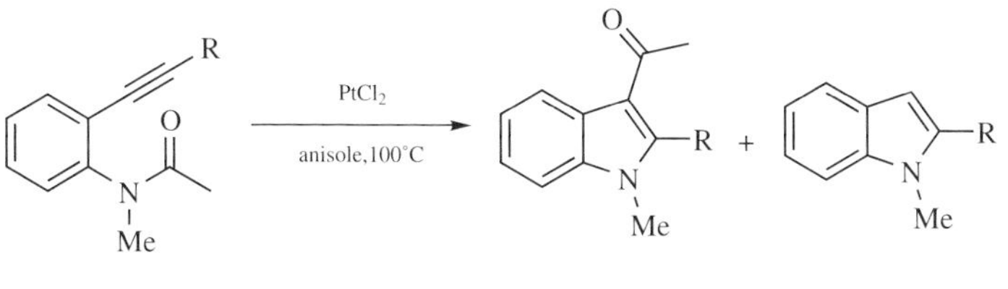

R
O
N
Me
$PtCl_2$
anisole,100°C
O
R
N
Me
+
R
N
Me

An intramolecular cyclopropanation[12] involves a propargyl unit that behaves as an alkenylcarbene.

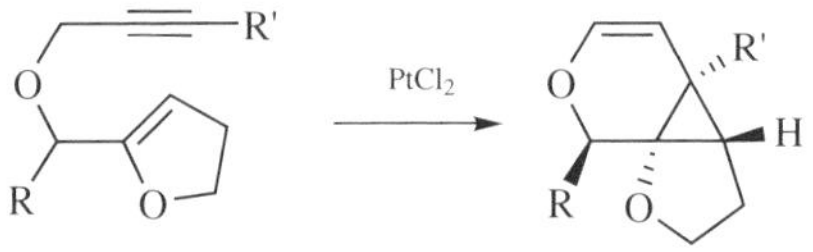

N-Protoected enamines add intramolecularly to alkynes (preferably electron-deficient) to form dienes, using either $PtCl_2$ or AgOTf as catalyst.[13]

[1]Hartman, J.W., Sperry, L. *TL* **45**, 3787 (2004).
[2]Liu, C., Han, X., Wang, X., Widenhoefer, R.A. *JACS* **126**, 3700 (2004).
[3]Marion, F., Coulomb, J., Courillon, C., Fensterbank, L., Malacria, M. *OL* **6**, 1509 (2004).
[4]Cadran, N., Cariou, K., Herve, G., Aubert, C., Fensterbank, L., Malacria, M. *JACS* **126**, 3408 (2004).
[5]Mamane, V., Gress, T., Krause, H., Furstner, A. *JACS* **126**, 8654 (2004).
[6]Harrak, Y., Blaszykowski, C., Bernard, M., Cariou, K., Mainetti, E., Mouries, V., Dhimane, A.-L., Fensterbank, L., Malacria, M. *JACS* **126**, 8656 (2004).
[7]Blaszykowski, C., Harrak, Y., Goncalves, M.-H., Cloarec, J.-M., Dhimane, A.-L., Fensterbank, L., Malacria, M. *OL* **6**, 3771 (2004).
[8]Nakamura, I., Bajracharya, G.B., Wu, H., Oishi, K., Mizushima, Y., Gridnev, I., Yamamoto, Y. *JACS* **126**, 15423 (2004).
[9]Mamane, V., Hannen, P., Furstner, A. *CEJ* **10**, 4556 (2004).
[10]Sagae, T., Sugiura, M., Hagio, H., Kobayashi, S. *CL* **32**, 160 (2003).
[11]Shimada, T., Nakamura, I., Yamamoto, Y. *JACS* **126**, 10546 (2004).
[12]Nevado, C., Ferrer, C., Echavarren, A.M. *OL* **6**, 3191 (2004).
[13]Harrison, T.J., Dake, G.R. *OL* **6**, 5023 (2004).

Platinum(IV) oxide. 21, 353–354; 22, 360

Hydrogenation. Acylated pyridines and quinolines undergo hydrogenation in the presence of PtO_2. Reaction conditions are crucial: with 1 eq. of HCl in EtOH at 55 bar hydrogen the heterocycle is saturated, but in CF_3COOH only the ketone group is reduced to the alcohol.[1]

[1]Solladie-Cavallo, A., Roje, M., Baram, A., Sunjic, V. *TL* **44**, 8501 (2003).

Poly(methylhydrosiloxane), PMHS. 20, 311; 21, 354–355; 22, 360

Reductive transformations. Reductive etherification of carbonyl compounds is accomplished by reaction with PMHS and catalytic $B(C_6F_5)_3$ while alkyl trimethylsilyl ethers furnish the alkyl residues.[4] Copper hydride is generated in the conjugate reduction of acyclic enones with PMHS and excellent asymmetric induction in the presence of a chiral ferrocenyldiphosphine is observed.[2]

[1]Chandrasekhar, S., Chandrashekar, G., Babu, B.N., Vijeender, V., Reddy, K.V. *TL* **45**, 5497 (2004).
[2]Lipshutz, B.H., Servesko, J.M. *ACIEE* **42**, 4789 (2003).

Potassium chloroplatinate.

N-Acyl α-amino acids. A catalyst system of K_2PtCl_4, Ph_3P, and HCl converts aldehydes to protected amino acids in the presence of CO and $RCONH_2$.

[1]Sagae, T., Sugiura, M., Hagio, H., Kobayashi, S. *CL* **32**, 160 (2003).

Potassium cyanide.

Unsymmetrical α-ketols. Cleavage of benzils with KCN in the presence of different aromatic aldehydes leads to unsymmetrical benzoins.[1] Alternatively, condensation of acylsilanes with aldehydes using the KCN/18-crown-6 catalytic system leads to silyl ethers of α-ketols.[2]

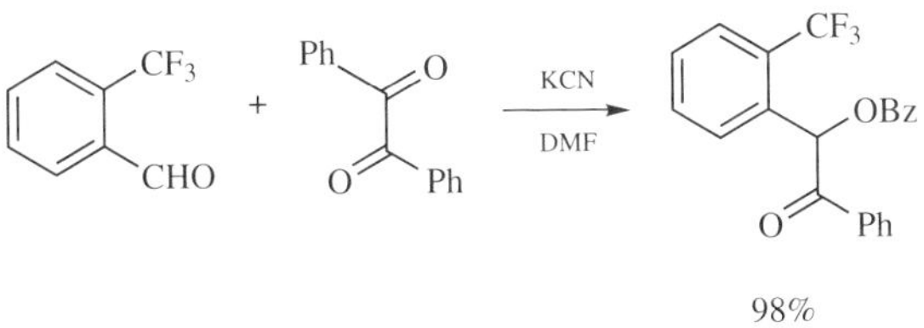

CF3, CHO, +, Ph, O, O, Ph, KCN, DMF, CF3, OBz, O, Ph

98%

3-Fluoroquinolines.[3] Aldimines derived from 2-amino-β,β-difluorostyrenes undergo cyclization on exposure to KCN. The ring closure is supposed to start from CN-addition to the imino group.

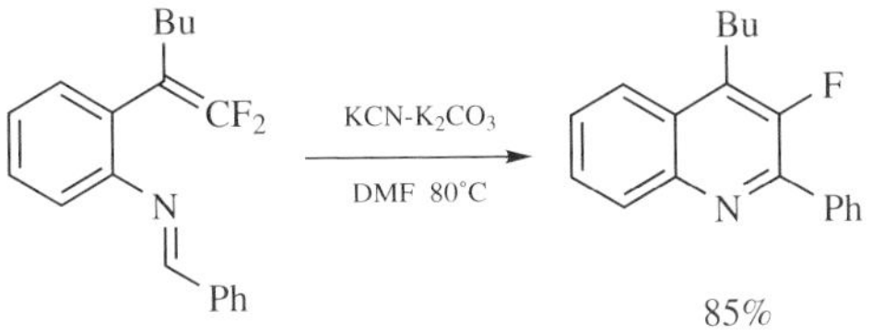

Bu, CF2, N, Ph, KCN-K2CO3, DMF 80°C, Bu, F, N, Ph

85%

[1]Demir, A.S., Reis, O. *T* **60**, 3803 (2004).
[2]Linghu, X., Johnson, J.S. *ACIEE* **42**, 2534 (2003).
[3]Mori, T., Ichikawa, J. *CL* **33**, 590 (2004).

Potassium fluoride. **13**, 256–257; **15**, 272; **18**, 297–298; **19**, 275–276; **20**, 313; **21**, 359; 22, 363

Esterification. Alcohols (but not phenols) can be acetylated on warming with HOAc and KF.[1] In ionic liquids KF promotes the conversion of carboxylic acids into benzyl esters with BnCl.[2]

Cyclopropanation. Stabilized arsonium ylides are readily generated by proton abstraction of the arsonium halides with KF and subsequent reaction with activated alkenes affords cyclopropanes.[3]

[1]Bosco, J.W.J., Raju, B.R., Saikia, A.K. *SC* **34**, 2849 (2004).
[2]Brinchi, L., Germani, R., Savelli, G. *TL* **44**, 6583 (2003).
[3]Ren, Z., Cao, W., Ding, W., Wang, Y., Wang, L. *SC* **34**, 3785 (2004).

Potassium fluoride / alumina. 20, 313; **21**, 359; **22**, 363

***N*-Formylation.** KF/Al_2O_3 promotes conversion of secondary amines into formamides in $CHCl_3$ by serving as base to generate dichlorocarbene.[1]

Benzylidenation.[2] Free α-positions of cycloalkanones are benzylidenated in the presence of KF/Al_2O_3. Ultrasound assists the reaction.

[1]Mihara, M., Ishino, Y., Minakata, S., Komatsu, M. *S* 2317 (2003).
[2]Li, J.-T., Yang, W.-Z., Chen, G.-F., Li, T.-S. *SC* **33**, 2619 (2003).

Potassium hydride. 21, 360–361; **22**, 364–365

Indoles.[1] 2-Alkynylarylamines cyclize to indoles on treatment with KH in NMP at room temperature.

[1]Koradin, C., Dohle, W., Rodriguez, A.L., Schmid, B., Knochel, P. *T* **59**, 1571 (2003).

Potassium monoperoxysulfate, Oxone®. 13, 259; **14**, 267; **15**, 274–275; **16**, 285; **18**, 300; **19**, 277; **20**, 313–315; **21**, 361–362; **22**, 365–357

Oxidation. Oxidation of alcohols with Oxone® in a phase-transfer system can be mediated by tris[(2-oxazolinyl)phenato]manganese(III).[1] Facile oxidation of aldehydes to acids is performed with Oxone in DMF and to esters in alcohols.[2]

Treatment with Oxone achieves oxidative desulfurization[3] of thio amides, thioureas, and thio esters. Conversion of *N*-Boc-indoles to the oxindoles via lithiation and borylation is completed by oxidation with Oxone.[4]

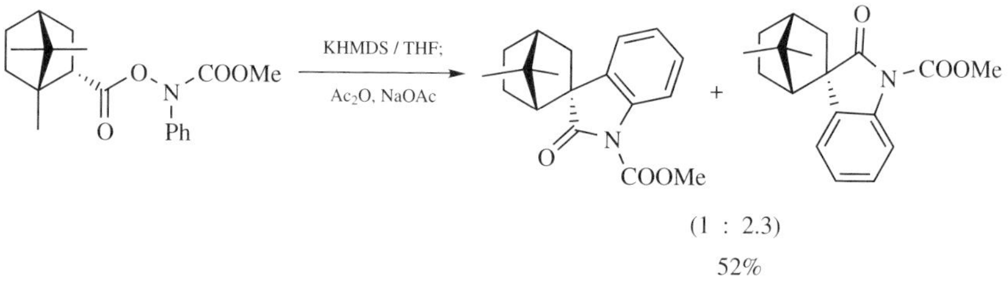

KHMDS / THF; Ac_2O, NaOAc

(1 : 2.3)
52%

Positive iodine. Iodide salts provide I^+ in the presence of Oxone. Thus, NH_4I and KI provide respective iodinating reagent for arenes[5] and iodolactonization/iodoetherification[6] under such conditions.

[1]Bagherzadeh, M. *TL* **44**, 8943 (2003).
[2]Travis, B.R., Sivakumar, M., Hollist, G.O., Borhan, B. *OL* **5**, 1031 (2003).
[3]Mohammadpoor-Baltork, I., Sadeghi, M.M., Esmayilpour, K. *SC* **33**, 953 (2003).
[4]Vazquez, E., Payack, J.F. *TL* **45**, 6549 (2004).
[5]Mohan, K.V.V.K., Narender, N., Kulkarni, S.J. *TL* **45**, 8015 (2004).
[6]Curini, M., Epifano, F., Marcotullio, M.C., Montanari, F. *SL* 368 (2004).

Potassium nitrite – sulfuryl chloride.

Sulfonyl chlorides. Triisopropylsilyl thioethers are oxidized with KNO_3 – SO_2Cl_2 in MeCN.

[1]Gareau, Y., Pellicelli, J., Laliberte, S., Gauvreau, D. *TL* **44**, 7821 (2003).

Potassium permanganate. **13**, 258–259; **14**, 267; **15**, 273–274; **18**, 301; **19**, 277–278; **20**, 315–316; **21**, 362–363; **22**, 367

Tetrahydropyrans.[1] Previously unobserved oxidative cyclization of 1,6-dienes with $KMnO_4$ is actually highly stereoselective.

KMnO4; HOAc, Me2CO; – 15°

[1]Cecil, A.R.L., Brown, R.C.D. *TL* **45**, 7269 (2004).

Potassium phthalimide.

Mannich-type reaction.[1] As Lewis bases alkali metal salts are effective in promoting the condensation of trimethylsilyl enolates with aldimines in DMF at room temperature. Potassium phthalimide or BzNHLi can be used.

[1]Fujisawa, H., Takahashi, E., Nakagawa, T., Mukaiyama, T. *CL* **32**, 1036 (2003).

(*S*)-Proline. **22**, 368

Aldol reaction. Recognition of the power of (*S*)-proline as an inexpensive catalyst for asymmetric synthesis has stimulated much interest. Aldol reaction is the most relevant in its applicability. An expedient synthesis of chiral *trans*-2-hydroxycycloalkanecarbaldehydes involves proline-catalyzed cyclization of dialdehydes.[1]

The aldol reaction between simple ketones and aldehydes has been investigated under different conditions: high pressure,[2] in aq. micelles,[3] and in PEG.[4] Cross-aldol reaction of two aldehydes (MeCHO as acceptor) has been carried out with great success (*anti*-aldol, ee >99%) in the presence of an ionic liquid.[5] Freezing water surrounding the reaction tube to induce high pressure has been applied to the asymmetric aldol reaction.[6]

(*S*)-proline

[bmin]PF6
DMF 4°
15-17 h

OH O H

73% (99% ee)

O H + O H

(*S*)-proline

[bmin]PF_6
DMF 4°
38 h

O OH OH

(49% ee)

Water enhances the reaction rates between ketones and aldehydes to enable the use of stoichiometric (1:1) amounts of the two reactants.[7] Using zinc prolinate the reaction proceeds in water alone.[8] Diacetone alcohol is useful as a surrogate for acetone in the aldol reaction with aldehydes because it forms *N*-isopropenylproline rapidly.[9]

N,N-Dibenzyl-α-amino aldehydes derived from natural amino acids condense with acetone to give *anti*-4-hydroxy-5-amino-2-alkanones.[10] Higher interest has been allotted to using 1-oxy-2-alkanones as donors in the cross-aldol reaction.[11,12] It has been shown that α-oxyacetaldehydes behave as acceptors for other aldehydes and the self-condensation of benzyloxyacetaldehyde (98% ee) constitutes the first step of an extremely efficient synthesis of a sugar.[12]

Dual catalyst control is observed in an asymmetric Baylis-Hillman reaction using (*S*)-proline and an octapeptide.[13]

Performing several different transformations in tandem with the aldol reaction much increases the value of the proline-catalyzed process. There is distinct possibility of combining Michael, Knoevenagel, Wittig, Diels-Alder, and Huisgen reactions.[14]

O PPh_3 + ArCHO + Ar'CHO + O O O O

(*S*)-proline

EtOH 65°C

Ar O O O O O Ar O

99% (dr>100:1)

O Ar → HOOC N O Ar   Ar O O O

For a demonstration of supported ionic liquid asymmetric catalysts with respect to their effectiveness and recoverability, proline in aldol reaction is used.[15]

Mannich reaction. Extension of the success of asymmetric aldol reaction to Mannich reaction is well anticipated. Aldimines are good electrophiles,[16] including those derived from polyfluoroaldehydes[17] and glyoxylic esters.[18] Naturally, the direct 3-component reaction works as well.[19–21]

Azodicarboxylic esters are found to react with aldehydes and subsequent aldol reaction with ketones completes a synthesis of hydroxyl ketones that also bear a hydrazine moiety.[22]

O + H(O) + BnOOC-N=N-COOBn —(S)-proline, MeCN→ O OH NHCOOBn N COOBn

85% (*anti/syn* = 54:46)
ee >99% (anti)
ee 34% (syn)

Aminoxylation. Nitrosobenzene reacts with aldehydes[23–25] and ketones[26,27] using the oxygen atom as electrophilic site. Asymmetric induction approaches 100% in most cases. Besides acquisition of chiral 1,2-amino alcohols by adding a reducing agent (e.g., $NaBH_4$) after completion of the condensation, allylation is also allowed.[28] Consecutive to enantioselective aminoxylation of aldehydes, Emmons-Wadsworth reaction can be achieved in the same flask.[29]

C_5H_{11} CHO + PhN=O —(S)-proline, CH_2=$CHCH_2Br$, In, NaI→ C_5H_{11} ONHPh OH ; C_5H_{11} ONHPh OH [4 : 1] (70%)

Other reactions. Perhaps the Michael reaction between ketones and nitroalkenes[30] and allylic benzyloxylation of alkenes[31] in the presence of (*S*)-proline are disappointing in terms of asymmetric induction, product yields are reasonable. Intramolecular aldehyde-enone addition gives much better results.[32] For the Knoevenagel reaction, proline is just a general acid-base catalyst.[33]

Chiral cycloalkanecarbaldehydes are formed in an intramolecular alkylation in the presence of (*S*)-α-methylproline.[34]

[1]Pidathala, C., Hoang, L., Vignola, N., List, B. *ACIEE* **42**, 2785 (2003).
[2]Sekiguchi, Y., Sasaoka, A., Shimomoto, A., Fujioka, S., Kotsuki, H. *SL* 1655 (2003).
[3]Peng, Y.-Y., Ding, Q.-P., Li, Z., Wang, P.G., Cheng, J.-P. *TL* **44**, 3871 (2003).
[4]Chandrasekhar, S., Narsihmulu, C., Reddy, N.R.R., Sultana, S.S.S. *TL* **45**, 4581 (2004).
[5]Cordova, A. *TL* **45**, 3949 (2004).
[6]Hayashi, Y., Tsuboi, W., Shoji, M., Suzuki, N. *TL* **45**, 4353 (2004).
[7]Nyberg, A.I., Usano, A., Pihko, P.M. *SL* 1891 (2004).
[8]Darbre, T., Machuqueiro, M. *CC* 1090 (2003).
[9]Chandrasekhar, S., Narsihmulu, C., Reddy, N.R., Sultana, S.S. *CC* 2450 (2004).
[10]Pan, Q., Zou, B., Wang, Y., Ma, D. *OL* **6**, 1009 (2004).
[11]Liu, H., Peng, L., Zhang, T., Li, Y. *NJC* **27**, 1159 (2003).
[12]Northrup, A.B., Mangion, I.K., Hettche, F., MacMillan, D.W.C. *ACIEE* **43**, 2152 (2004).
[13]Imbriglio, J.E., Vasbinder, M.M., Miller, S.J. *OL* **5**, 3741 (2003).
[14]Ramachary, D.B., Barbas III, C.F. *CEJ* **10**, 5323 (2004).
[15]Gruttadauria, M., Riela, S., Lo Meo, P., D'Anna, F., Noto, R. *TL* **45**, 6113 (2004).
[16]Cordova, A. *SL* 1651 (2003).
[17]Funabiki, K., Nagamori, M., Goushi, S., Matsui, M. *CC* 1928 (2004).
[18]Notz, W., Tanaka, F., Watanabe, S., Chowdari, N.S., Turner, J.M., Barbas III, C.F., Thayumanavan, R. *JOC* **68**, 9624 (2003).
[19]Hayashi, Y., Tsuboi, W., Ashimine, I., Urushima, T., Shoji, M., Sakai, K. *ACIEE* **42**, 3677 (2003).
[20]Cordova, A. *CEJ* **10**, 1987 (2004).
[21]Cordova, A. *ACIEE* **43**, 6528 (2004).
[22]Chowdari, N.S., Ramachary, D.B., Barbas III, C.F. *OL* **5**, 1685 (2003).
[23]Brown, S.P., Brochu, M.P., Sinz, C.J., MacMillan, D.W.C. *JACS* **125**, 10808 (2003).
[24]Zhong, G. *ACIEE* **42**, 4247 (2003).
[25]Hayashi, Y., Yamaguchi, J., Hibino, K., Shoji, M. *TL* **44**, 8293 (2003).
[26]Hayashi, Y., Yamaguchi, J., Sumiya, T., Hibino, K., Shoji, M. *JOC* **69**, 5966 (2004).
[27]Hayashi, Y., Boevig, A., Sunden, H., Cordova, A. *ACIEE* **43**, 1109 (2004).
[28]Zhong, G. *CC* 606 (2004).
[29]Zhong, G., Yu, Y. *OL* **6**, 1637 (2004).
[30]List, B., Pojarliev, P., Martin, H.J. *OL* **3**, 2423 (2001).
[31]Le Bras, J., Muzart, J. *TA* **14**, 1911 (2003).
[32]Fonseca, M.T.H., List, B. *ACIEE* **43**, 3958 (2004).
[33]Cardillo, G., Fabbroni, S., Gentilucci, L., Gianotti, M., Tolomelli, A. *SC* **33**, 1587 (2003).
[34]Vignola, N., List, B. *JACS* **126**, 450 (2004).

Pyridinium trihalides. 22, 369

Bromination. Bromohydrins are formed when alcohols are treated with $(t\text{-BuO})_3Al$ and $PyHBr_3$.[1] There is an Oppenauer oxidation prior to bromination and the bromoketones are reduced subsequently.

[1]Cami-Kobeci, G., Williams, J.M.J. *SL* 124 (2003).

Pyridonaphthyridine.

Acylation. This conformationally rigid analogue of DMAP is prepared from 1,6-naphthyridine in 4 steps and it catalyzes acylation when DMAP causes many side reactions.[1]

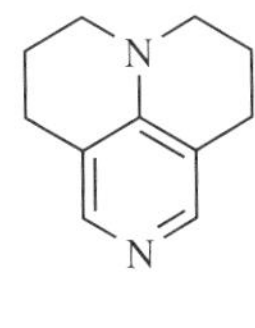

1

[1]Heinrich, M.R., Klisa, H.S., Mayr, H., Steglich, W., Zipse, H. *ACIEE* **42**, 4826 (2003).

Pyrrole.

Aldehyde protection.[1] Formation of *N*-hydroxyalkylpyrroles from aldehydes masks the formyl group and it is selective for ketoaldehydes. The aldehydes are regenerated by treatment with NaOMe.

[1]Dixon, D.J., Scott, M.S., Luckhurst, C.A. *SL* 2317 (2003).

R

Rhenium(V) oxychloride.

Propargylation. The dppm complex of $ReOCl_3$ is stable to air and moisture. It catalyzes propargylation of alcohols without affecting groups like Cl, OMe, COOR, and double bond.[1] Formation of 1,5-enynes from propargylic alcohols and allylsilanes is accomplished.[2] Efficient propargylation of arenes by this method shows excellent synthetic potential, and this reaction remediates the deficiency of the Ru-catalyzed process that anisole and 1,3-dimethoxybenzene fail.[3]

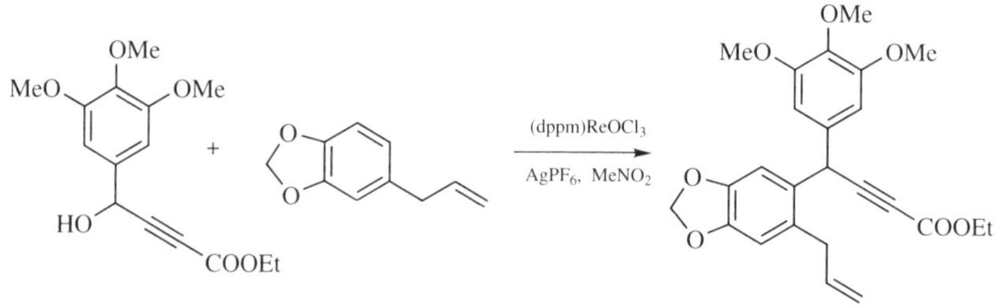

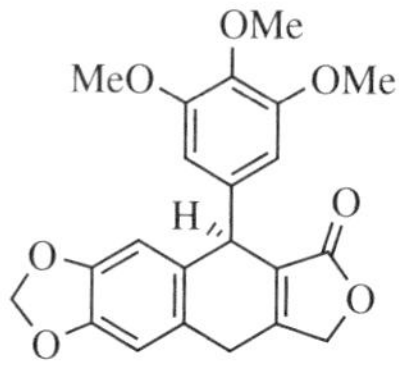

(dppm)$ReOCl_3$
$AgPF_6$, $MeNO_2$

β-apopicropodophyllin

2-Deoxy-α-glycosides.[4] A heterocomplex of $ReOCl_3$ promotes regioselective and stereoselective addition of ROH to glycals. Actually, other nucleophiles (RSH, RSO_2NH_2) also can serve as addends.

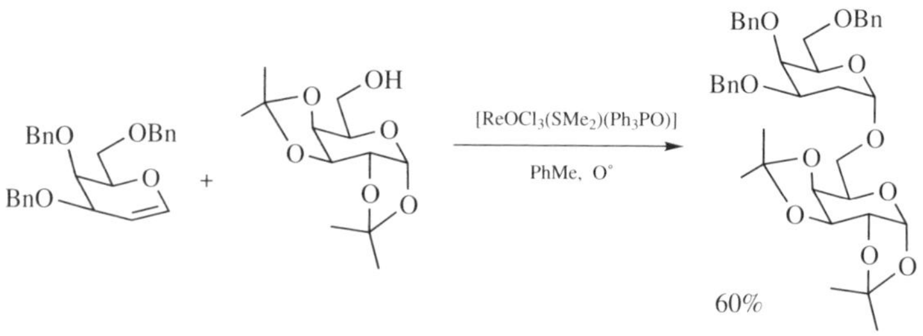

[$ReOCl_3(SMe_2)(Ph_3PO)$]
PhMe, 0°

60%

Fiesers' Reagents for Organic Synthesis, Volume 23. Edited by Tse-Lok Ho

[1]Sherry, B.D., Radosevich, A.T., Toste, F.D. *JACS* **125**, 6076 (2003).
[2]Luzung, M.R., Toste, F.DHH. *JACS* **125**, 15760 (2003).
[3]Kennedy-Smith, J.J., Young, L.A., Toste, F.D. *OL* **6**, 1325 (2004).
[4]Sherry, B.D., Loy, R.N., Toste, F.D. *JACS* **126**, 4510 (2004).

Rhodium carbonyl clusters. **13**, 288; **15**, 334; **18**, 305–306; **19**, 280–281; **20**, 317–318; **21**, 366; **22**, 371–372

Hydroalkoxycarbonylation.[1] The atom-economic combination of alkenes with CO and pyridine-2-methanol to generate esters is catalyzed by $Rh_4(CO)_{12}$. The nitrogen atom of the pyridine ring is critical for forming a chelated Rh-H species that adds to 1-alkenes.

[3 + 2 + 1]Cycloaddition.[2] Bicyclic cyclohexenones are formed when 1,3,8-alkatrienes, CO and $Rh_4(CO)_{12}$ are heated in toluene. Prolonged reaction causes deconjugated.

Pauson-Khand reaction. Nanosized Rh_2Co_2 prepared from $Rh_2Co_2(CO)_{12}$ has the unusual ability to promote cyclopentenone formation from enals which are poor substrates for the Pauson-Khand reaction.[3] The adsorded enals slowly generate CO and alkenes in situ. While styrene is a poor reactant, successful reaction of cinnamaldehyde is noted.

[1]Yokota, K., Tatamidani, H., Fukumoto, Y., Chatani, N. *OL* **5**, 4329 (2003).
[2]Lee, S.I., Park, J.H., Chung, Y.K., Lee, S.-G. *JACS* **126**, 2714 (2004).
[3]Park, K.H., Jung, I.G., Chung, Y.K. *OL* **6**, 1183 (2004).

Rhodium(II) carboxamidates. 22, 372

Hetero Diels-Alder reaction.[1] A new and effective chiral Rh catalysts to induce formation of dihydro-4-pyrones is **1**.

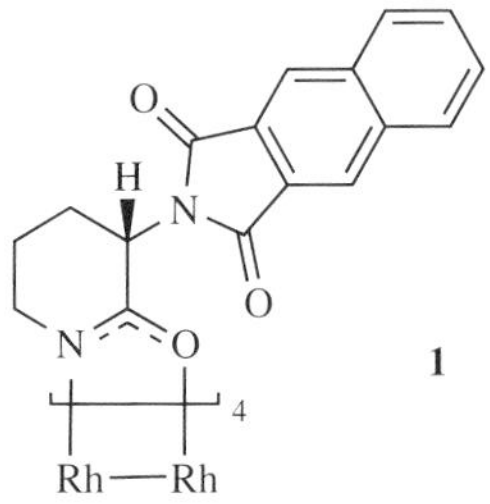

[1]Ananda, M., Washio, T., Shimada, N., Kitagaki, S., Nakajima, M., Shiro, M., Hashimoto, S. *ACIEE* **43**, 2665 (2004).

Rhodium(II) carboxylates. 13, 266; **15**, 278–286; **16**, 289–292; **17**, 298–302; **18**, 306–307; **19**, 281–285; **20**, 318–320; **21**, 367–369; **22**, 372–378

Carbenoid insertions. Intramolecular C–H insertion forms lactams, and in some cases β-lactams, from α-dialkoxyphosphoryl-α-diazo amides in good yields.[1]

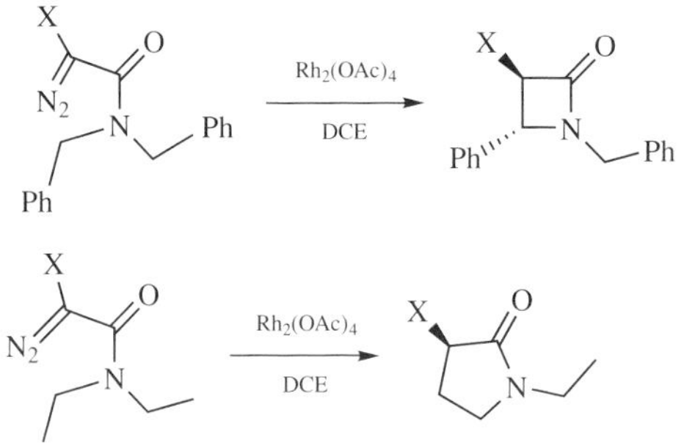

When sulfamide, sulfamate and urea groups are activated to form RN=IPh species by a Rh(II) carboxylate catalyst, 6-membered heterocycles are formed. There is an interesting dependence of regioselectivity on the carboxyl groups of the catalysts (e.g., AcO vs. Ph_3CCOO).[2] The designed catalyst $[Rh_2(esp)_2]$ constituting from benzene-1,3-bis(2,2-dimethylpropanoic acid) has superior catalytic activity.[3]

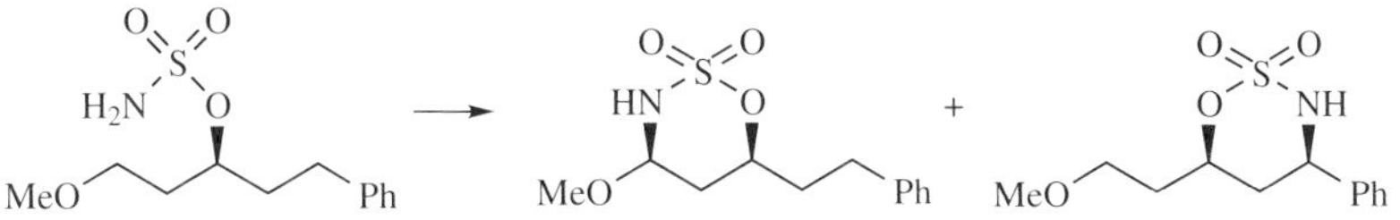

$Rh_2(OAc)_4$	1 :	1
$Rh_2[OCOCPh_3]$	15 :	1

Asymmetric insertion of a stabilized carbenoid to the α-position of an amine derivative[4] by catalysis of the *N*-(dodecylbenzenesulfonyl)prolinate complex $Rh_2(S\text{-DOSP})_4$ has the same result of a Mannich reaction, but it is of a different scope. Similarly, insertion into a C—H bond adjacent to a silyl ether leads to aldol-type products.[5] Furthermore, combined C—H insertion and Cope rearrangement have been observed.[6]

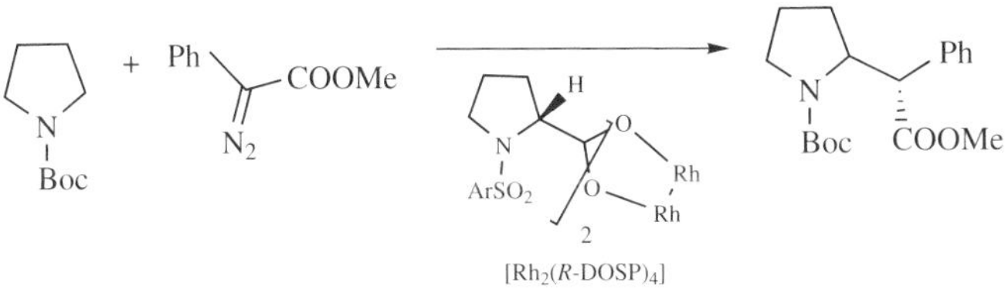

$[Rh_2(R\text{-DOSP})_4]$

$[Rh_2(S\text{-DOSP})_4]$

Insertion of conjugated carbenoids into the C—H bond of 1,2-dihydronaphthalenes is a mechanistically interesting process (involving two Cope rearrangments).[7]

95% (>98% de
99.5% ee)

When the C—X bond of halides (acyl halides, benzyl halides and CH_2X_2) is inserted by carbenoids derived from diazodicarbonyl compounds, α-halo enones are produced.[8]

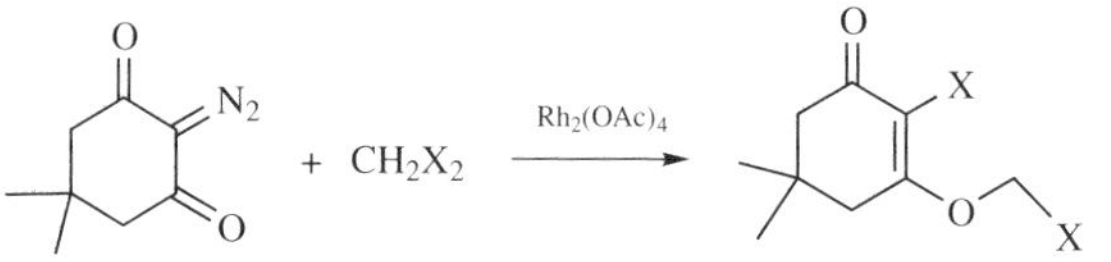

X=Cl 81%
X=Br 75%

Different types of products are obtained from reactions of α-diazocarbonyl compounds with AcSH, when promoted by $Rh_2(OAc)_4$ or $BF_3.OEt_2$. α-Acetyl thiocarbonyl compounds and α-acetoxyl carbonyl compounds are formed as exclusive or major products under respective conditions.[9]

The N–H bond of both amines[10] and amides[11] is insertable. Products from the latter process are easily transformed into oxazoles and thiazoles.

A new utility of the intramolecular insertion reaction leading to isomünchnone dipoles pertains to a synthesis of mappicine ketone.[12] Trapping of such a dipole generate an α-pyridone.

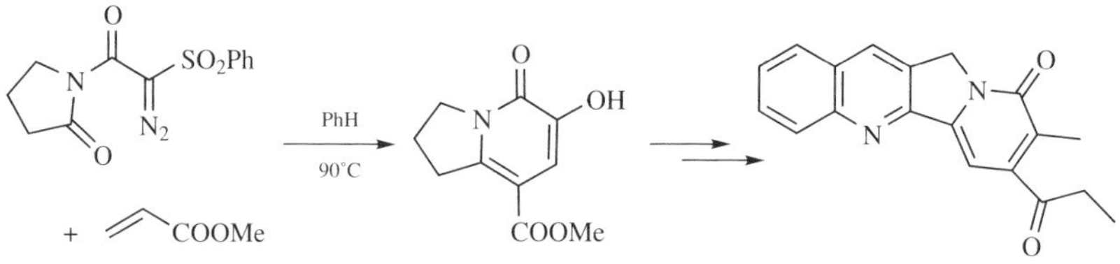

mappicine ketone

2-(*o*-Hydroxyarylation) of indoles is accomplished by the reaction with 2-diazo-3,5-cyclohexadienones. 2-Siloxyindoles afford 3-aryloxindoles.[13]

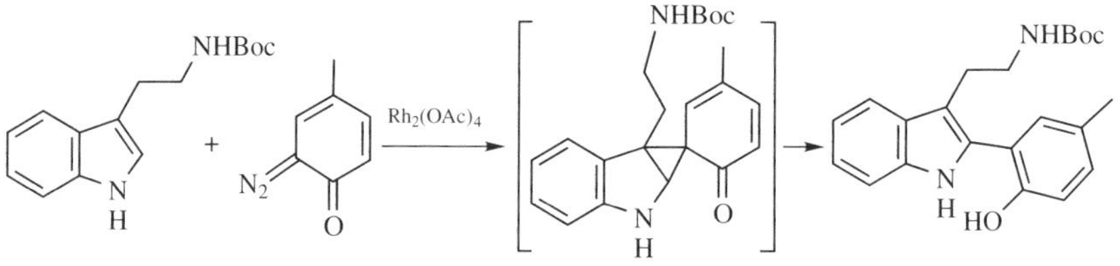

60-70%

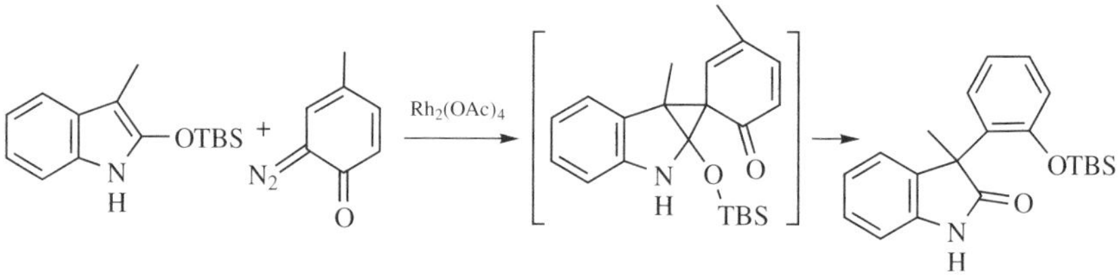

54%

Functionalization of heteroatoms. Sulfides and sulfoxides are readily converted into sulfilimine and sulfoximine derivatives, respectively, by reaction with amides (e.g., CF_3CONH_2, $NsNH_2$) in the presence of $PhI(OAc)_2$, MgO and $Rh_2(OAc)_4$.[14] The Rh-catalyzed reductive acylation of azodicarboxylic esters with aldehydes possibly involves a radical mechanism.[15]

Pyrrolines. The decomposition of α-diazo-β,γ-unsaturated esters and capture with imines afford pyrrolines. Catalysts govern the reaction pathways and hence products.[16] Rh-carbenoids react with imines to generate ylides while copper-iminium species are formed and attacked by the diazo compounds.

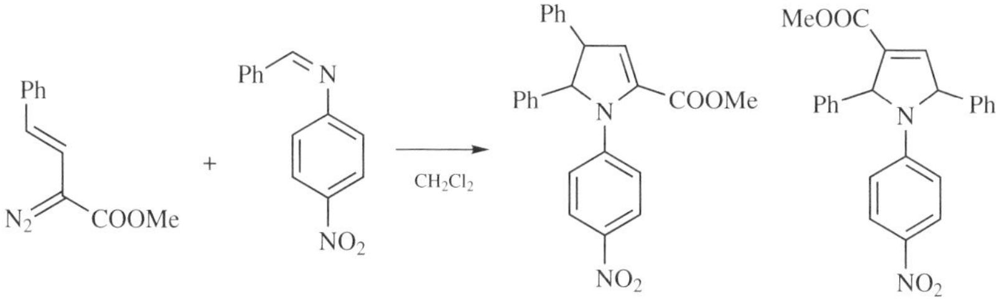

$Rh_2(OAc)_4$	66% (*trans/cis* 98:2)	--
$Cu(OTf)_2$	--	67% (*trans/cis* 91:9)

Cycloalkenes. Using $Rh_2(S\text{-DOSP})_4$ as catalyst enantioselective cycloaddition of Rh-carbenoids to alkynes is realized.[17] An intramolecular reaction involving a tricarbonyliron-complexed diene unit and acylcarbenoid forms a cyclopentenone.[18]

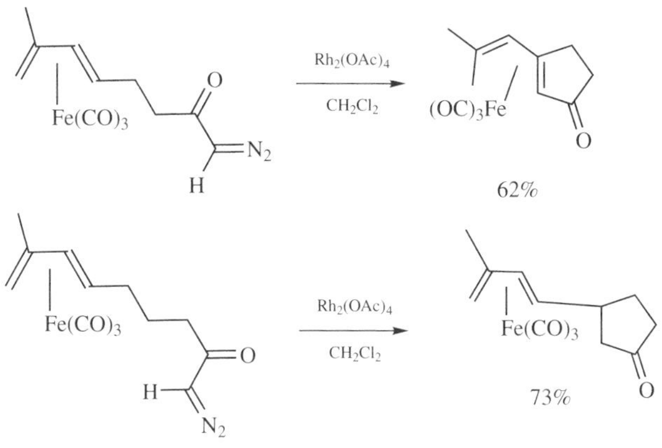

Alkenation. Through transformation of Rh-ylide to As-ylide intermediates, a Wittig-type reaction between $[C_6F_5CH{=}NNTs]Na$ and ArCHO is accomplished.[19] In other words, Ph_3As serves as a reagent and $Rh_2(OAc)_4$ as catalyst.

[1]Gois, P.M.P., Afonso, C.A.M. *EJOC* 3798 (2003).
[2]Fiori, K., Fleming, J.J., Du Bois, J. *ACIEE* **43**, 4349 (2004).
[3]Espino, C.G., Fiori, K., Kim, M., Du Bois, J. *JACS* **126**, 15378 (2004).
[4]Davies, H.M.L., Venkataramani, C., Hansen, T., Hopper, D.W. *JACS* **125**, 6462 (2003).
[5]Davies, H.M.L., Beckwith, R.E.J., Antoulinakis, E.G., Jin, Q. *JOC* **68**, 6126 (2003).
[6]Davies, H.M.L., Beckwith, R.E.J. *JOC* **69**, 9241 (2004).
[7]Davies, H.M.L., Jin, Q. *JACS* **126**, 10862 (2004).
[8]Lee, Y.R., Cho, B.S., Kwon, H.J. *T* **59**, 9333 (2003).
[9]Yao, W., Liao, M., Zhang, X., Xu, H., Wang, J. *EJOC* 1784 (2003).
[10]Muthusamy, S., Srinivasan, P. *TL* **46**, 1063 (2005).
[11]Davies, J.R., Kane, P.D., Moody, C.J. *T* **60**, 3967 (2004).
[12]Raolji, G.B., Garcon, S., Greene, A.E., Kanazawa, A. *ACIEE* **42**, 5059 (2003).
[13]Sawada, T., Fuerst, D.E., Wood, J.L. *TL* **44**, 4919 (2003).
[14]Okamura, H., Bolm, C. *OL* **6**, 1305 (2004).
[15]Lee, D., Otte, R.D. *JOC* **69**, 3569 (2004).
[16]Doyle, M.P., Yan, M., Hu, W., Gronenberg, L.S. *JACS* **125**, 4692 (2003).
[17]Davies, H.M.L., Lee, G.H. *OL* **6**, 1233 (2004).
[18]Franck-Neumann, M., Geoffroy, P., Gassmann, D., Winling, A. *TL* **45**, 5407 (2004).
[19]Zhu, S., Liao, Y., Zhu, S. *OL* **6**, 377 (2004).

Rhodium(III) chloride. 21, 370; **22**, 378

Coupling reactions. Heck-type reaction between $ArB(OH)_2$ and unsubstituted acrylic esters and acrylonitrile can be effected by $RhCl_3$ - Ph_3P.[1] This process is sensitive to steric effects and the optimal reactant ratio is 2:1 in favor of the alkenes. Formation of aryl arylethyl ketones[2] from $ArB(OH)_2$ and ArCH=CHCHO is also unusual.

Sulfonylation.[3] Low temperature (65°) reaction of methane with oleum (to give methanesulfonic acid) is initiated by H_2O_2 - urea and promoted by $RhCl_3$.

[1]Zou, G., Wang, Z., Zhu, J., Tang, J. *CC* 2438 (2003).
[2]Wang, Z., Zou, G., Tang, J. *CC* 1192 (2004).
[3]Mukhopadhyay, S., Bell, A.T. *ACIEE* **42**, 2990 (2003).

Ruthenium-carbene complexes. 18, 308; **19**, 285–289; **20**, 320–323; **21**, 370–376; **22**, 378–387

General aspects and new metathesis catalysts. A succinct and invaluable delineation of a model for cross metathesis is published.[1] The most widely used metathesis catalysts are the Grubbs I (**1**) and Grubbs II (**2**) complexes. Catalyst **3** has several modifications: an air-stable, inexpensive **4** is easily prepared from α-asarone;[2] fluorous-tailed[3] and dimeric complexes with PEG-connecter[4] are recyclable, **5** has an end group similar to a common ionic liquid,[5] and **6** incorporates a BINAMOL unit.[6] Introduction of an *o*-phenyl substituent to the isopropoxy group of **3** makes a robust catalyst (**3P**) that tolerates the presence of Lewis basic groups in the substrates.[7] An exceptionally efficient catalyst is **7** because it does not have a phosphine ligand to disengage before entering the catalytic cycle.[8]

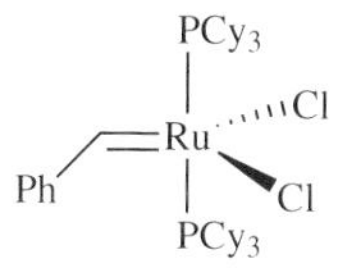

1 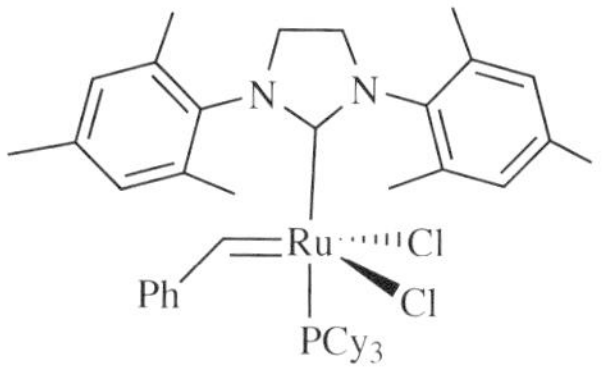 2

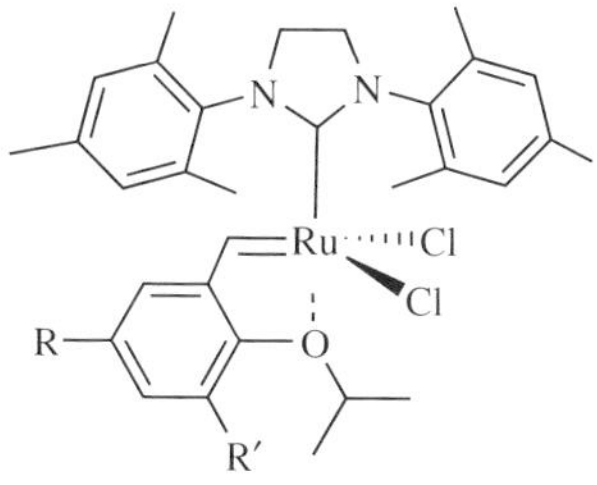

3 R = R′ = H

3N R = NO_2 , R′ = H

3P R = H , R′ = Ph

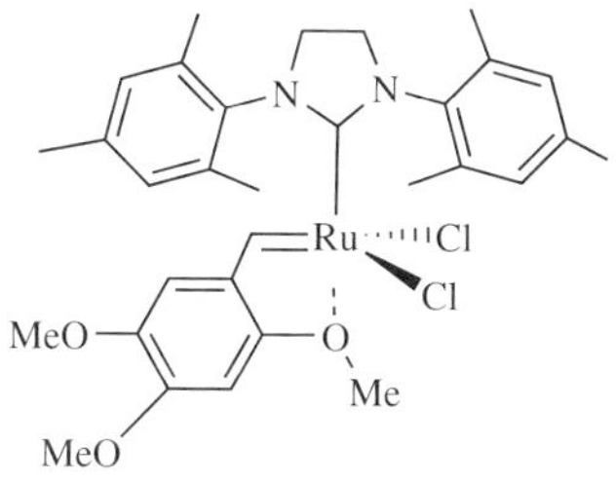

4 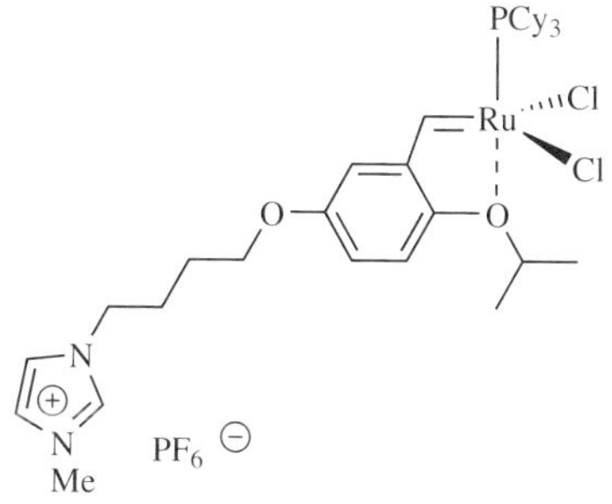 5

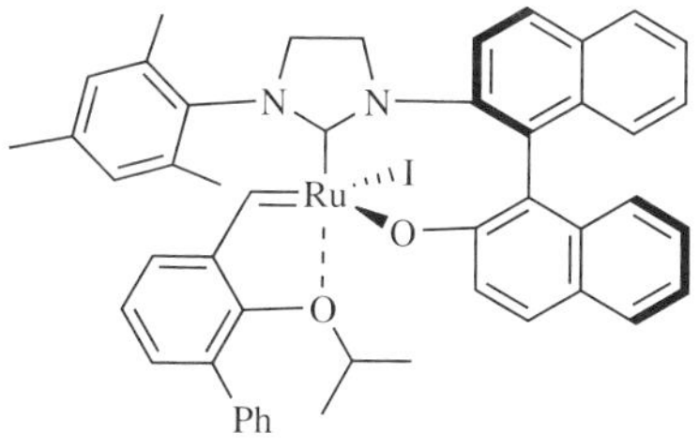

6 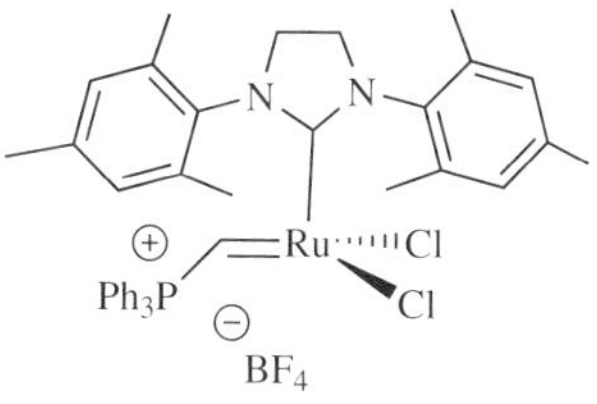 7

Grubbs I catalyst dispersed in paraffin can be kept active in the air for at least 22 months.[9] Some ring-closing metathesis (RCM) reactions assisted by microwaves are carried out neat[10] or in a solvent.[11]

Addition of polymer-bound phosphine to reaction media avoids contamination of colored impurities, and the products are free of Ru.[12]

Cross-metathesis reactions. Grubbs II precatalyst in the presence of CuCl is used to advantage to cross acrylonitrile with other alkenes, because the copper salt significantly enhances turnover and it also acts as phosphine scavenger.[13]

Cross-metathesis is most effective to access alkenes with special substituents: 1,1-dibromo-1,3-alkadienes,[14] 1-trichlorosilylalkenes,[15] alkenyl sulfoxides and sulfones,[16] phosphine oxides,[17,18] (most interstingly in cross-metathesis of trivinylphosphine oxide with three different 1-alkenes[19]) and functionalized alkenyl boronates.[20]

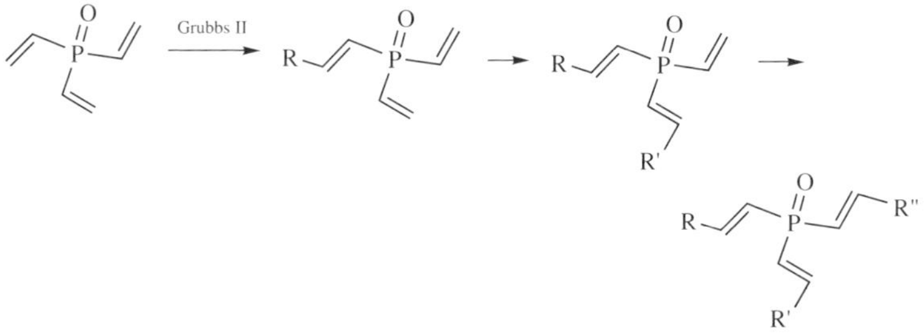

Grubbs II

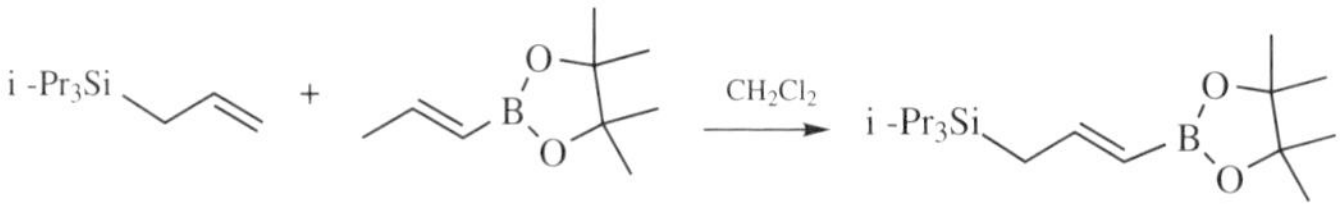

R R' R''

i -Pr$_3$Si + CH_2Cl_2 i -Pr$_3$Si

99% (*E/Z* 1 : 1)

1,3-Alkadienes participate in cross-metathesis with electron-deficient alkenes at the terminal double bond.[21] Conjugated enynes are a new type of substrates for metathesis.[22,23] Previous difficulties are due to strong binding of the triple bond to Ru center of the catalyst. Complex **8** is very reactive and it enables the desired reaction to proceed.[22]

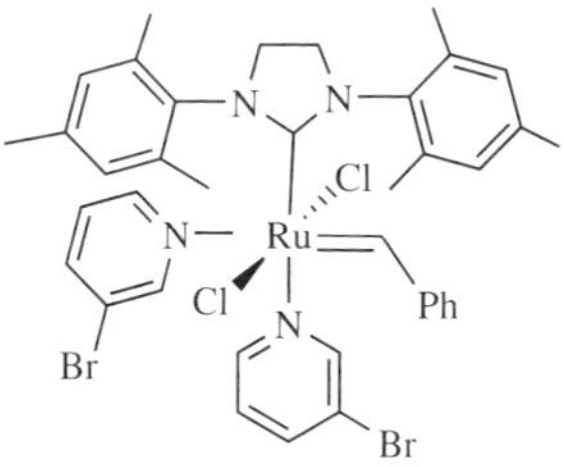

8

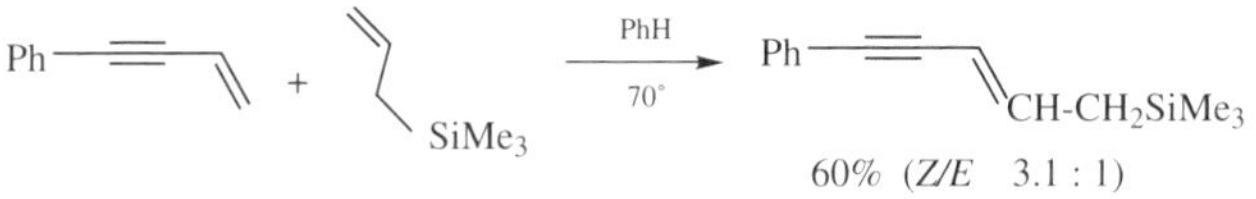

Ph + SiMe$_3$ PhH 70° Ph CH-CH$_2$SiMe$_3$

60% (*Z/E* 3.1 : 1)

AcO Ph + **2** AcO Ph H

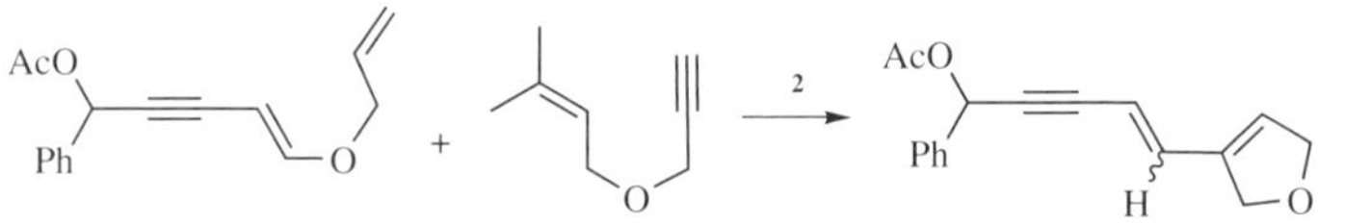

55%

Regioselective cross metathesis occurs due to avoidance of steric hindrance is witnessed.[24] The method is complementary to a previous exploitation of a stabilized chelate to allow the more exposed Ru complex to undergo reaction.[25]

OTBDPS + COOMe —3, CH_2Cl_2, 40°→ OTBDPS ... COOMe 80%

OAc —[Ru]→ [O O [Ru] [Ru]] —R→ OAc R

Based on cross-metathesis 1,4-pentadien-3-yl 3-butenoate is a useful building block for δ-alkenyl-β,γ-unsaturated δ-lactones.[26] Regardless the order of RCM and intermolecular cross metathesis the final products are the same. Cyclic (5-, 6-, 7-membered) β-amino acid esters are prepared from methionine, allylglycine, and serine.[27]

$C_{11}H_{23}$ + O O —2, CH_2Cl_2, 40°→ $C_{11}H_{23}$ O O 75%

MeS H_2N COOH → → SMe Z N H COOMe → Z N H COOMe → Z N H COOMe

Z = COOBn

H_2N COOH → → Z N H COOMe → Z N H COOMe

OH H_2N COOH → → Z N H COOMe → Z N H COOMe

The enyne metathesis is of practical importance because novel 1,3-disubstituted dienes (e.g., 1-ethoxy-3-alkyl-1,3-butadienes) for Diels-Alder reactions are readily produced.[28-30] Reaction at room temperature is promoted by ethane.[28,29] A paper delineates six procedures for diene synthesis based on the cross enyne metathesis and important aspects of scope and limitations are summarized.[31]

Metathetic ring closure. Again, new cyclic structures have been made to test the potential and limits of ring-closing metathesis. Preparation of heterocyclic fluoroalkenes is interesting.[32] Heterocycles that can be open easily to furnish bifunctional alkenes continue to rivet attention. For example, there is no problem in constructing those containing an N–O bond.[33] *N,N′*-Disulfonyl diamines in which the two nitrogen atoms are attached to alkenyl and alkynyl chains undergo RCM and cross-metathesis.[34]

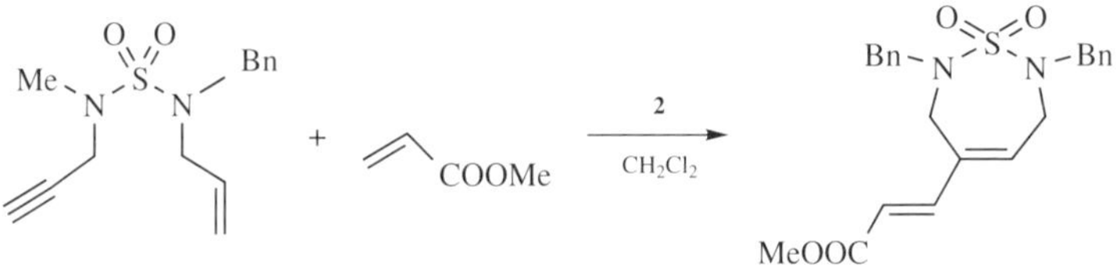

Me N S(O)(O) N Bn + COOMe —2, CH_2Cl_2→ Bn-N S(O)(O) N-Bn, MeOOC

83%

Quadruple RCM reaction can be very efficient.[35] The five-membered rings are formed more rapidly.

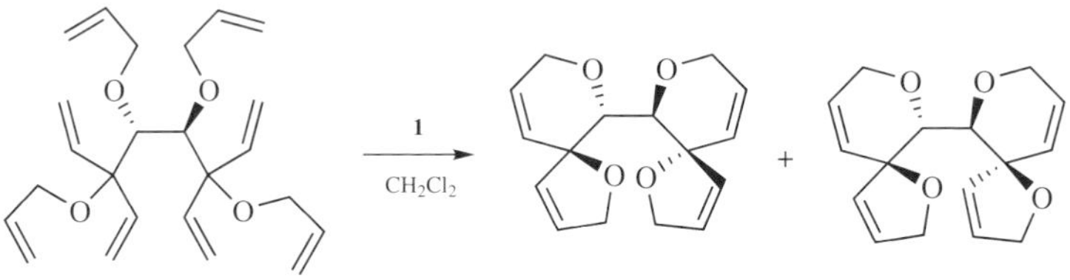

—1, CH_2Cl_2→

65% + 12%

Macrocyclic methylenecycloalkenes are accessible from ring-closing enyne metathesis aided by ethene.[36] The reaction involves two steps, the first one is hydroethylenation of the triple bond which can use both Grubbs I-II catalysts but Grubbs I catalyst is incapable of effecting the RCM step.

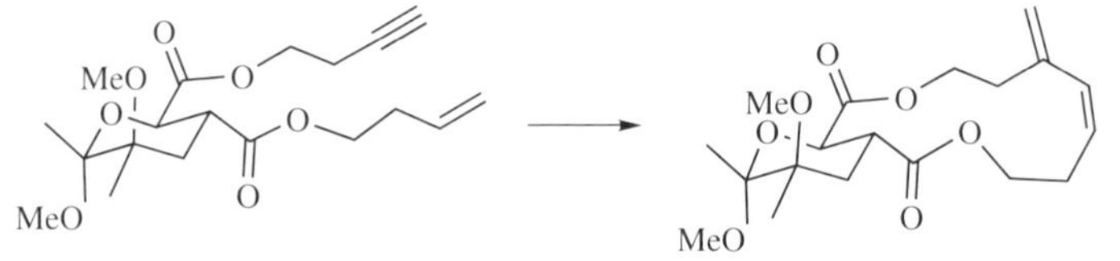

MeO, O, MeO → MeO, O, MeO

61%

The yields of certain unsaturated macrolides synthesized by the method using Grubbs II catalyst are higher at elevated temperatures and lower concentrations of the substrates.[37]

Tandem reactions. Substrates for cross-metathesis can be designed to generate dienes, many of which are suitable for the Diels-Alder reaction. Accordingly, with proper dienophiles in the reaction media more complex structures are synthesized in one operation.[38]

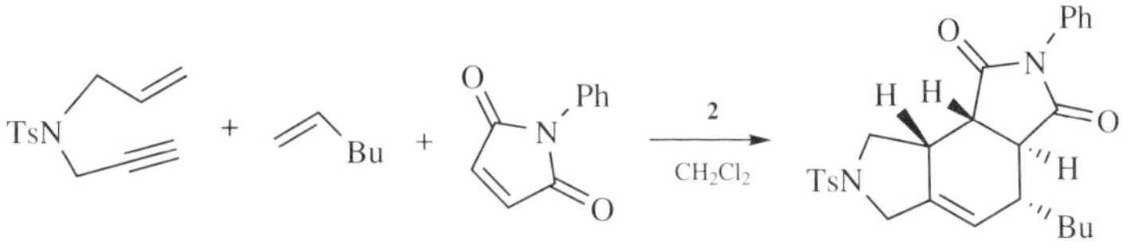

83%

After formation of 6-membered cyclic ethers by RCM reaction the addition of *i*-PrOH and NaOH empowers the catalyst to perform isomerization of the double bond to deliver dihydropyrans.[39]

Cycloalkenes obtained from RCM can be saturated in situ. Thus, addition of NaH activates the Ru-carbene complexes to catalyze hydrogenation.[40]

5-Tosyl-7-heteranorbornenes (X = N-Boc, O) undergo regioselective ring-opening/cross-metathesis with alkenes to give 2-alkenyl-3-tosyl-5-vinyl pyrrolidines and tetrahydrofurans.[41,42]

Formation of the complete skeleton of erythrina alkaloids from a substituted tetrahydroisoquinoline[43,44] in one step and of an advanced intermediate of securinine[45] further attests to the power of the RCM method. In another approach to securinine closure of the butenolide unit by RCM is followed by Pd-catalyzed coupling to create the cyclohexene.[46]

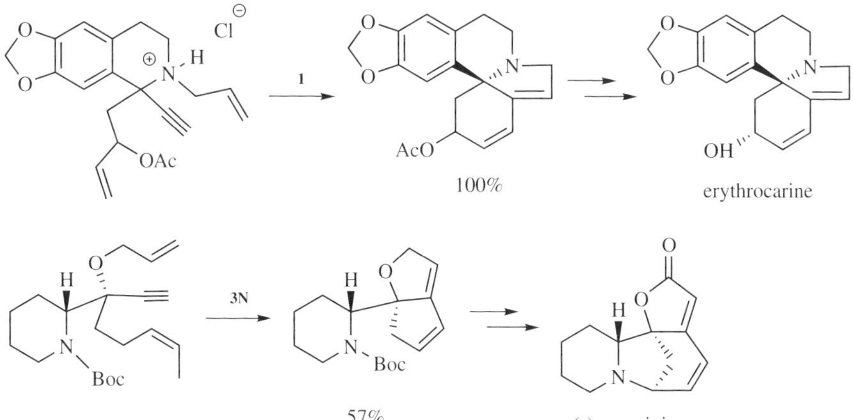

100%

erythrocarine

57%

(-)-securinine

Addition to triple bonds. Enol esters are prepared by careful choice of the catalytic system so that alkyne dimerization is suppressed. Complexes **9**[47] and **10**[48] favor the addition reaction.

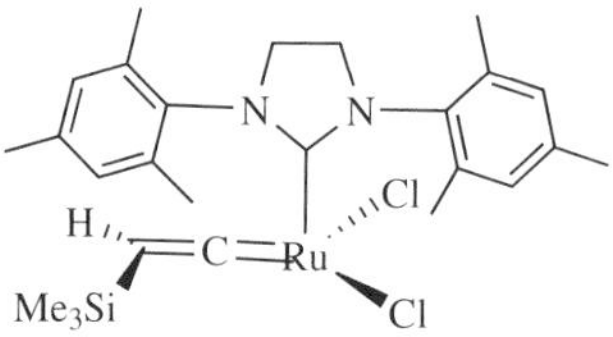

9 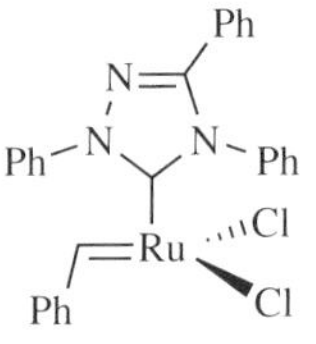 **10**

Grubbs I complex also catalyzes hydrosilylation of alkynes to afford predominantly (*Z*)-1-silyl-1-alkenes.[49]

Intramolecular cycloaddition involving alkylidenecyclopropane and alkyne units leads to bicyclic 4-methylenecyclopentenes.[50]

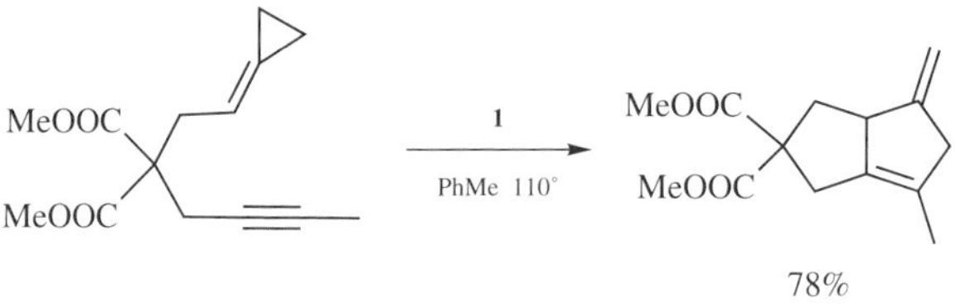

78%

Deallylation. Extension of the method for *N*-allylamine cleavage to *N*-allyllactams needs an additional oxidative cleavage step (using $RuCl_3$-$NaIO_4$) because the enamides are stable.[51]

Redox condensation. Ru-catalyzed hydrogen transfer from benzylic alcohols to vinyltrimethylsilane in the presence of a stabilized phosphonium ylide gives cinnamic esters which are further reduced.[52]

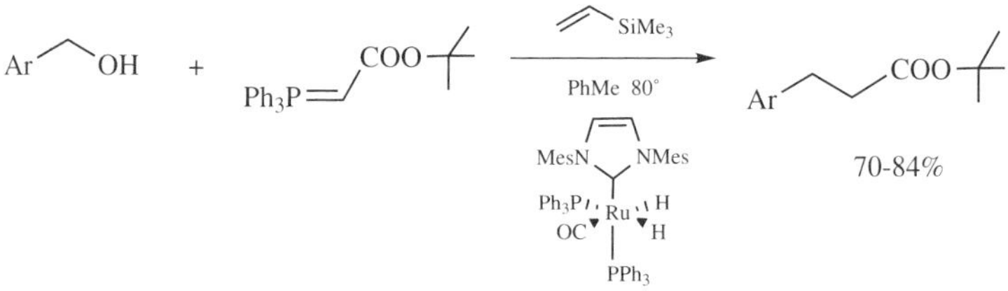

70-84%

Cycloisomerization of 1,6-dienes and *N*-allyl-2-vinylanilines in the presence of $CH_2{=}CHOSiMe_3$ to afford heterocyclic products (e.g., 3-methyleneindolines) is accomplished by Grubbs II catalyst.[53]

[1]Chatterjee, A.K., Choi, T.-L., Sanders, D.P., Grubbs, R.H. *JACS* **125**, 11360 (2003).
[2]Grela, K., Kim, M. *EJOC* 963 (2003).
[3]Yao, Q., Zhang, Y. *JACS* **126**, 74 (2004).
[4]Yao, Q., Motta, A.R. *TL* **45**, 2447 (2004).
[5]Yao, Q., Zhang, Y. *ACIEE* **42**, 3395 (2003).
[6]Gillingham, D., Katoka, O., Garber, S., Hoveyda, A. *JACS* **126**, 12288 (2004).
[7]Dunne, A.M., Mix, S., Blechert, S. *TL* **44**, 2733 (2003).
[8]Romero, P., Piers, W., McDonald, R. *ACIEE* **43**, 6161 (2004).
[9]Taber, D.F., Frankowski, K.J. *JOC* **68**, 6047 (2003).
[10]Thanh, G.V., Loupy, A. *TL* **44**, 9092 (2003).
[11]Balan, D., Adolfsson, H. *TL* **45**, 3089 (2004).
[12]Westhus, M., Gonthier, E., Brohm, D., Breinbauer, R. *TL* **45**, 3141 (2004).
[13]Rivard, M., Blechert, S. *EJOC* 2225 (2003).
[14]Funk, T.W., Efskind, J., Grubbs, R.H. *OL* **7**, 187 (2005).
[15]Pietraszuk, C., Marciniec, B., Fischer, H. *TL* **44**, 7121 (2003).
[16]Michrowska, A., Bieniek, M., Kim, M., Klajn, R., Grela, K. *T* **59**, 4525 (2003).
[17]Demchuk, O.M., Pietrusiewicz, K.M., Michrowska, A., Grela, K. *OL* **5**, 3217 (2003).
[18]Bisaro, F., Gouverneur, V. *TL* **44**, 7133 (2003).
[19]Bisaro, F., Gouverneur, V. *T* **61**, 2395 (2005).
[20]Morrill, C., Grubbs, R.H. *JOC* **68**, 6031 (2003).
[21]Dewi, P., Randl, S., Blechert, S. *TL* **46**, 577 (2005).
[22]Kang, B., Kim, D., Do, Y., Chang, S. *OL* **5**, 3041 (2003).
[23]Hansen, E.C., Lee, D. *OL* **6**, 2035 (2004).
[24]BouzBouz, S., Simmons, R., Cossy, J. *OL* **6**, 3465 (2004).
[25]BouzBouz, S., Cossy, J. *OL* **3**, 1451 (2001).
[26]Virolleaud, M.-A., Bressy, C., Piva, O. *TL* **44**, 8081 (2003).
[27]Gardiner, J., Anderson, K.H., Downard, A., Abell, A.D. *JOC* **69**, 3375 (2004).
[28]Giessert, A.J., Snyder, L., Markham, J., Diver, S.T. *OL* **5**, 1793 (2003).
[29]Lee, H.-Y., Kim, B.G., Snapper, M.L. *OL* **5**, 1855 (2003).
[30]Giessert, A.J., Brazis, N.J., Diver, S.T. *OL* **5**, 3819 (2003).
[31]Diver, S.T., Giessert, A.J. *S* 466 (2004).
[32]Salim, S.S., Bellingham, R.K., Satcharoen, V., Brown, R.C.D. *OL* **5**, 3403 (2003).
[33]Yang, Y.-K., Tae, J. *SL* 1043 (2003).
[34]Salim, S.S., Bellingham, R.K., Brown, R.C.D. *EJOC* 800 (2004).
[35]Wallace, D.J. *TL* **44**, 2145 (2003).
[36]Hansen, E., Lee, D. *JACS* **126**, 15074 (2004).
[37]Yamamoto, K., Biswas, K., Gaul, C., Danishefsky, S.J. *TL* **44**, 3297 (2003).
[38]Lee, H.-Y., Kim, H.Y., Tae, H., Kim, B.G., Lee, J. *OL* **5**, 3439 (2003).
[39]Schmidt, B. *CC* 742 (2004).
[40]Schmidt, B., Pohler, M. *OBC* **1**, 2512 (2003).
[41]Weeresakare, G.M., Liu, Z., Rainler, J.D. *OL* **6**, 1625 (2004).
[42]Liu, Z., Rainler, J.D. *OL* **7**, 131 (2005).
[43]Shimizu, K., Takimoto, M., Mori, M. *OL* **5**, 2323 (2003).
[44]Fukumoto, H., Esumi, T., Ishihara, J., Hatakeyama, S. *TL* **44**, 8047 (2003).
[45]Honda, T., Namiki, H., Kaneda, K., Mizutani, H. *OL* **6**, 87 (2004).
[46]Alibes, R., Ballbe, M., Busque, F., de March, P., Elias, L., Figueredo, M., Font, J. *OL* **6**, 1813 (2004).
[47]Opstal, T., Verpoort, F. *SL* 314 (2003).
[48]Melis, K., De Vos, D., Jacobs, P., Verpoort, F. *JOMC* **671**, 131 (2003).
[49]Arico, C.S., Cox, L.R. *OBC* **2**, 258 (2004).

[50]Lopez, F., Delgado, A., Rodriguez, J.R., Castedo, L., Mascarenas, J.L. *JACS* **126**, 10262 (2004).
[51]Alcaide, B., Almendros, P., Alonso, J.M. *TL* **44**, 8693 (2003).
[52]Edwards, M.G., Jazzar, R.F.R., Paine, B.M., Shermer, D.J., Whittlesey, M.K., Williams, J.M.J., Edney, D.D. *CC* 90 (2004).
[53]Terada, Y., Arisawa, M., Nishida, A. *ACIEE* **43**, 4063 (2004).

Ruthenium(III) chloride. 13, 268; **14**, 271–272; **19**, 289–290; **20**, 324; **21**, 376; **22**, 387

Derivatization. Very small amounts of $RuCl_3 \cdot 3H_2O$ (0.1 mol%) are used in converting RCHO into acetals,[1,2] and acylation of various RXH (X = O, S, NH).[3]

Oxidation. Different combinations of $RuCl_3$ and inexpensive oxidants have been explored for conversion of alkenes to acyloins: Oxone ($K_2S_2O_8$),[4] $KHSO_5$;[5] to 1,2-diols: $NaIO_4(H_2SO_4)$.[6,7] A set of dihydroxylation conditions has been defined in terms of substrate structure, additive and solvent system.[7] That Bronsted acids accelerate hydrolysis of ruthenate esters is advantageous but certain functionalities (e.g., siloxy groups) are not compatible. Highly substituted alkenes are not reactive enough to be dihydroxylated.

Certain enantiopure acyloins are obtained by a selective benzylic oxidation of chiral diols with $RuCl_3$ - $K_2S_2O_8$.[8]

Ar, OH, OH, Cl → ($RuCl_3$ -Oxone / $NaHCO_3$ / EtOAc - MeCN - H_2O) → Ar, O, OH, Cl

88% (99%ee)

The $RuCl_3$-$NaIO_4$ couple causes scission of 3a-hydroxy perhydrobenzofurans to mesocyclic keto lactones.[9]

OH, O, H → ($RuCl_3$ -$NaIO_4$ / H_2O -CCl_4 -MeCN) → O, O, O

81%

Oxidation of tertiary amines to *N*-oxides calls for another combination: $RuCl_3$-bromamine.[10] Oxidative cyanation at an α-position (preferably a methyl group) of tertiary amines occurs under oxygen and in the presence of NaCN.[11]

Cyclization. Tetralin and hetercyclic analogues containing unsaturated side chain are formed via intramolecular alkylation.[12]

$RuCl_3 \cdot xH_2O$ / AgOTF / DCE 60°

X = O, NTf, CH_2

R = H 56%
X = O

$RuCl_3 \cdot xH_2O$ / AgOTF / DCE 60°

99%

Ethylene glycol monoallyl ethers are cyclized to give 2-alkyl-1,3-dioxolanes on heating with $RuCl_3 \cdot 3H_2O$ at 80°, whereas the less acidic $RuClH(CO)(PPh_3)_3$ causes isomerization of the double bond.[13]

Beckmann rearrangement.[14] The Lewis acidity of $RuCl_3$ proves sufficiently strong to convert ArC(=NOH)R to ArNHCOR.

Glycosyl amides.[15] Glycosyl azides react with AcSH to provide RNHAc when catalyzed by $RuCl_3$. Formation of 5-hydroxy-1,2,3,4-thiatriazole intermediates by 1,3-dipolar cycloaddition is implicated.

Hydroxyapatite-supported complexes. Treatment of a hydroxyapatite with aq. $RuCl_3$ gives RuHAP. One such catalyst in combination with $NaIO_4$ is useful for dihydroxylation.[16] A cationic species generated by reaction with AgOTf is active as a Lewis acid for promoting aldol condensation and Diels-Alder reaction.[17]

[1]Qi, J.-Y., Ji, J.-X., Yueng, C.-H., Kwong, H.-L., Chan, A.S.C. *TL* **45**, 7719 (2004).
[2]De, S.K., Gibbs, R.A. *TL* **45**, 8141 (2004).
[3]De, S.K. *TL* **45**, 2919 (2004).
[4]Plietker, B. *JOC* **68**, 7123 (2003); *OL* **6**, 289 (2004).
[5]Plietker, B. *JOC* **69**, 8287 (2004).
[6]Plietker, B., Niggemann, M. *OL* **5**, 3353 (2003).
[7]Plietker, B., Niggemann, M., Pollrich, A. *OBC* **2**, 1116 (2004).
[8]Plietker, B. *OL* **6**, 289 (2004).
[9]Ferraz, H.M.C., Longo, Jr., L.S. *OL* **5**, 1337 (2003).
[10]Sharma, V.B., Jain, S.L., Sain, B. *TL* **45**, 4281 (2004).
[11]Murahashi, S.-I., Komiya, N., Terai, H., Nakae, T. *JACS* **125**, 15312 (2003).
[12]Youn, S.W., Pastine, S.J., Sames, D. *OL* **6**, 581 (2004).
[13]Urbala, M., Kuznik, N., Krompiec, S., Rzepa, J. *SL* 1203 (2004).
[14]De, S.K. *SC* **34**, 3431 (2004).
[15]Fazio, F., Wong, C.-H. *TL* **44**, 9083 (2003).
[16]Ho, C.-M., Yu, W.-Y., Che, C.-M. *ACIEE* **43**, 3303 (2004).
[17]Mori, K., Hara, T., Mizugaki, T., Ebitani, K., Kaneda, K. *JACS* **125**, 11460 (2003).

Ruthenium(III) chloride–triphenylphosphine–tin(II) chloride.

2-Substituted indoles.[1] Regioselective epoxide opening by $ArNH_2$ followed by cyclization leads to indole derivatives. Cyclization involves 1,2-Dianilinoalkane intermediates.

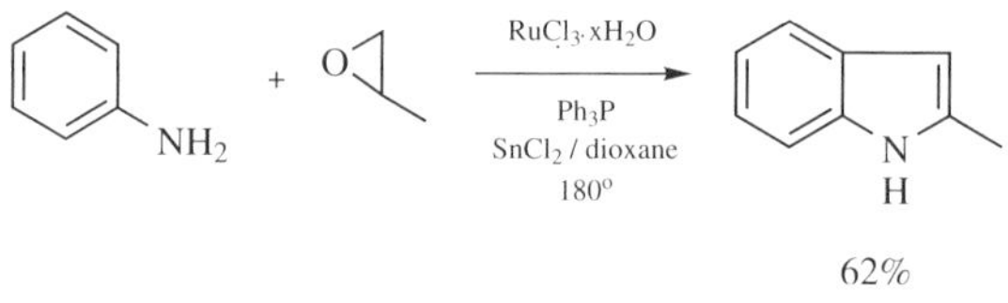

NH_2 + O

$RuCl_3 \cdot xH_2O$

Ph_3P

$SnCl_2$ / dioxane

180°

N H

62%

[1]Cho, C.S., Kim, J.H., Choi, H.-J., Kim, T.-J., Shim, S.C. *TL* **44**, 2975 (2003).

Ruthenium hydroxide.

Amides.[1] The alumina-supported $Ru(OH)_n$ assists efficient hydration of nitriles in water.

[1]Yamaguchi, K., Matsushita, M., Mizuno, N. *ACIEE* **43**, 1576 (2004).

Ruthenium(IV) oxide.

Oxidation. RuO_2/ZSM-5, which is prepared by stirring $RuCl_3 \cdot nH_2O$ and zeolite ZSM-5 in aq. NaOH, oxidizes alcohols to carbonyl compounds in air,[1] while Oxone recycles RuO_2 for oxidative cleavage of alkynes to carboxylic acids.[2]

[1]Qian, G., Zhao, R., Lu, G., Qi, Y., Suo, J. *SC* **34**, 1753 (2004).
[2]Yang, D., Chen, F., Dong, Z.-M., Zhang, D.-W. *JOC* **69**, 2221 (2004).

S

Samarium. 14, 275; **17**, 305–307; **18**, 311; **19**, 291; **20**, 325–326; **21**, 378; **22**, 388

Debromination. Reduction of 1,1-dibromo-1-alkenes (e.g., dibromostyrene) with Sm in MeOH is controllable in terms of product formation. When limited to 1.5 equivalents of Sm, monobromoalkenes are the major products. Complete removal of bromine is achieved with 3 equivalents, and saturated hydrocarbons with 5.5 equivalents.[1]

Benzoins. By way of a carbene rearrangement $Ar_2C{=}O$ adds to DMF to afford benzoins when promoted with Sm-Me_3SiCl.[2] Benzoin formation from ArCOCl in DMF at room temperature can be modified by varying the reaction conditions. At 60° deoxybenzoins are obtained. Substoichiometric Sm (0.35 equiv.) causes production of diacyloxystilbenes.[3] Interestingly, reductive diacylation of acrylic esters and MVK is observed under essentially the same conditions.[4]

ArCOCl + CH2=CHCOOR —(Sm, DMF)→ ArCO–CH2–CH(COOR)–COAr

The reduction trend of $ArSO_2Cl$ is analogous.[5] Either diaryl disulfones, arenethiol sulfonates, or diaryl disulfides are obtained according to quantities of Sm and reaction temperature.

[1]Wang, L., Li, P., Xie, Y., Ding, Y. *SL* 1137 (2003).
[2]Liu, Y., Xu, X., Zhang, Y. *T* **60**, 4867 (2004).
[3]Liu, Y., Wang, X., Zhang, Y. *SC* **34**, 4009 (2004).
[4]Liu, Y., Zhang, Y. *T* **59**, 8429 (2003).
[5]Liu, Y., Zhang, Y. *TL* **44**, 4291 (2003).

Samarium – metal halides. 21, 378; **22**, 389

1,1-Diarylmethyl ketones.[1] Diaryl ketones on reduction by low-valent titanium species generated from Sm and $TiCl_4$ couple with 1-acylbenzotriazoles. 1-Acylbenzotriazoles alone give 1,2-diketones under such conditions.

Reductive cyclization.[2] Dimethyl 2,8-octadienedioate gives methyl perhydroindan-2-one-1-carboxylate on treatment with Sm – SmI_2. The reaction is stereoselective, the (*E,E*)-diester affords predominantly the product with a *trans* ring junction, and the (*E,Z*)-diester, the *cis*-product.

Fiesers' Reagents for Organic Synthesis, Volume 23. Edited by Tse-Lok Ho

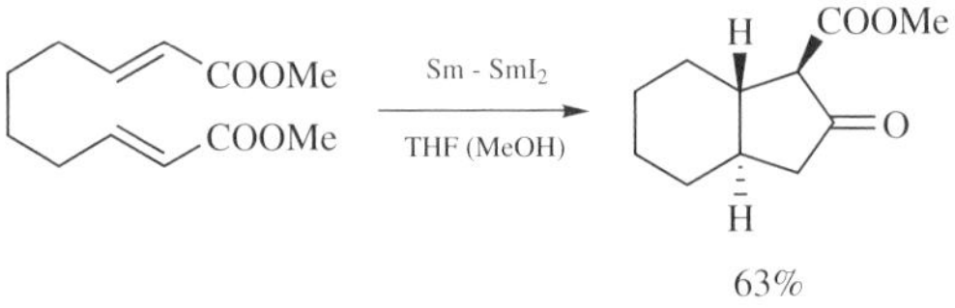

COOMe
COOMe
Sm - SmI_2
THF (MeOH)
H COOMe
O
H
63%

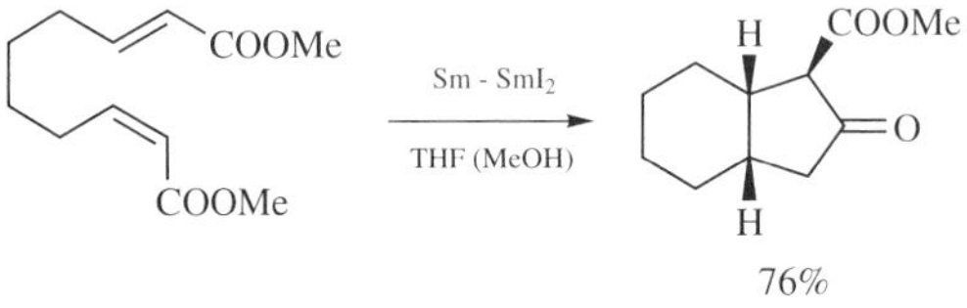

COOMe
COOMe
Sm - SmI_2
THF (MeOH)
H COOMe
O
H
76%

[1]Wang, X., Zhang, Y. *SC* **33**, 2627 (2003).
[2]Shinohara, I., Okue, M., Yamada, Y., Nagaoka, H. *TL* **44**, 4649 (2003).

Samarium(II) bromide. 22, 390

Pinacol coupling. Preparation of $SmBr_2$ from Sm and 1,2-dibromoethane in THF at room temperature is expedient, the reagent effects coupling of carbonyl compounds. It can be used catalytically by adding mischmetall.[1]

[1]Helion, F., Lannou, M.-I., Namy, J.-L. *TL* **44**, 5507 (2003).

Samarium(III) chloride.

Epoxide opening. Iodohydrins and azidohydrins are formed in high yields when epoxides are exposed to $SmCl_3.6H_2O$ and NaX (X = I, N_3).

[1]Bhaumik, K., Mali, U.W., Akamanchi, K.G. *SC* **33**, 1603 (2003).

Samarium(II) hexamethyldisilazide.

Alkynylimines. Insertion of isocyanides into 1-alkynes to give alkynylimines[1] is catalyzed by $[(Me_3Si)_2N]_3Sm$. Pentylamine is a useful additive for the reaction

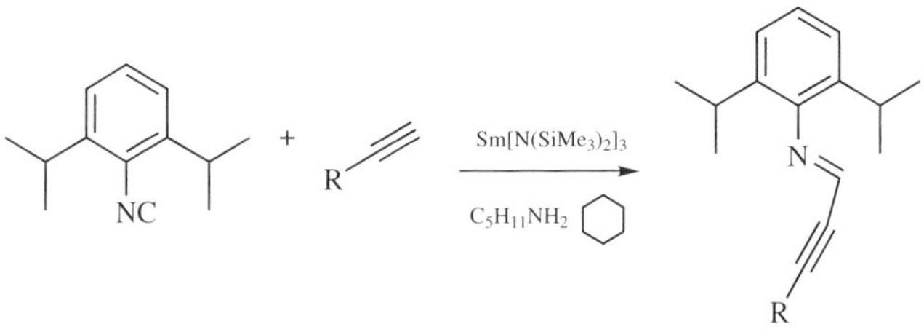

NC
+
R
$Sm[N(SiMe_3)_2]_3$
$C_5H_{11}NH_2$
N
R

[1]Komeyama, K., Sasayama, D., Kawabata, T., Takehira, K., Takaki, K. *CC* 634 (2005).

Samarium(II) iodide. **13**, 270–272; **14**, 276–281; **15**, 282–284; **16**, 294–300; **17**, 307–311; **18**, 312–316; **19**, 292–296; **20**, 327–335; **21**, 379–385; **22**, 390–396

Reagent preparation. The title reagent can be prepared from Sm and iodoform in THF.[1] Reagent in MeCN is obtained by sonication of Sm with 1,2-diiodoethane in that solvent.[2]

Reduction. Ketones, amines and conjugated esters are rapidly reduced to alcohols, amines, and saturated esters respectively, by SmI_2 and Et_3N, with water as proton source.[3] The remarkable rates of reduction is shown by comparison with that in MeOH (<10 sec. vs 24 h). Similar systems reduce stilbene to dibenzyl and conjugated dienes to alkenes (cycloheptatriene to 1,3-cycloheptadiene in <0.2 min. then to cycloheptene in 5 min. in almost quantitative yields)[4] and cleave allyl ethers.[5]

The diastereoselectivity for reduction of aldols to 1,3-diols is highly dependent on solvents. In DME the *syn*-isomers are strongly favored but in MeCN the *anti*-isomers are formed preferentially.[6]

Effective reduction of alkyl halides[7] may be less significant in comparison with cyclization from *o*-iodoaryl allyl (and propargyl) ethers to give dihydrobenzofurans,[8] notwithstanding its synthetic usefulness.

SmI_2

SmI_2 / H_2O, amine / <1min

SmI_2 / H_2O, amine / 5 min

α-Chloro enones apparently form alkenylsamarium reagents that add to carbonyl compounds via allenyloxysamarium intermediates, giving hydroxyl ketones[2] inaccessible by the Baylis-Hillman reaction because of β-substitution. The closely related substrates, α-chloro esters, are readily prepared in a two steps from dichloroacetic esters via 2,2-dichloro-3-hydroxyalkanoic esters, which require only the treatment of with SmI_2.[9]

The SmI_2-mediated aldol reaction between α-halo-α'-sulfinyl ketones and aldehydes is diastereoselective, products of very high optical purity are obtained from ketones containing a chiral sulfinyl group.[10]

PhCHO + ... SmI_2 / THF / $-100°$

96 : 4

61%

SmI_2 is capable of reducing organic sulfides[11] and the acyl radical intermediates produced from thioesters can be trapped in a synthetically useful way (e.g., to obtain potential protease inhibitors).[12]

Reductive cleavage of N-O bond by SmI_2 in THF finds application in generation of 1,3-amino alcohols from isoxazolidines,[13] and conversion of $ArNO_2$ to ArNHCOR in the presence of an ester.[14] A synthesis of 2-aryl-2,3-dihydro-4(1*H*)-quinazolinones from 2-nitrobenzamide and ArCHO involves reduction, Schiff base formation, and cyclization.[15] Hydrazides are cleaved by SmI_2 after *N*-trifluoroacetylation.[16]

Cleavage of 2-acylaziridines generates samarium enolates which are excellent donors for the aldol reaction. An access to β-amino-β'-hydroxy ketones is established.[17]

Treatment of α,β-epoxy esters with SmI_2 in THF followed by D_2O quench furnishes 2,3-dideuterio esters.[18] Aliphatic α,β-epoxy amides are transformed into α-hydroxy amides in MeOH,[19] and (*E*)-α-hydroxy-β,γ-unsaturated amides in THF using 0.2–0.5 equiv. of SmI_2.[20] That the epoxide opening is regiochemically different from the epoxy esters is due fast and strong chelation of Sm to the amide moiety before the C-O bond homolysis.

In the deoxygenation leading to cinnamides the stereochemistry of the products is highly dependent on the degree of substitution at both α- and β-carbon atoms.[21] Cyclopropanation can be carried out in tandem with the deoxygenation step, taking advantage of the capability of $Sm\text{-}CH_2I_2$ to perform such a transformation.[22]

SmI_2; $Sm\text{-}CH_2I_2$ → CONEt₂ cyclopropane

R = Ph 73% (>98% de)

2,2,3,3-Tetradeuteriocarboxylic acids and their derivatives (esters, amides) are expediently prepared from the corresponding 2-alkynoic acids by reduction with SmI_2 in D_2O.[23]

Coupling reactions. Baylis-Hillman adducts undergo reductive dimerization on exposure to SmI_2 in THF at −20°. However, reaction in refluxing THF affords mainly the deoxygenated compounds (more stable isomers) due to more rapid transformation of the radical-samarium enolates to C-bonded diiodosamarium species.[24]

−20°, 90min	82%	17%
Δ, 5min	10%	80%

2,3-Disubstituted indoles can be prepared from *N*,2-diacylanilines by a deoxygenative coupling process.[25] *N*-*t*-Butanesulfinyl aldimines undergo diastereoselective coupling in the presence of HMPA at −78° to give C_2-symmetrical 1,2-diamine derivatives.[26] Treatment of α-cyano imines with SmI_2 is a convenient method for preparing certain α-diketimines.[27]

Diastereoselective cross-coupling occurs between $Cr(CO)_3$-complexes of aromatic nitrones and carbonyl compounds,[28] but the reaction with α,β-unsaturated esters provides γ-hydroxylamino adducts.[29] The nitrone functionality is retained but shifted after the reaction if an alkoxy group is placed in its α-position.[30]

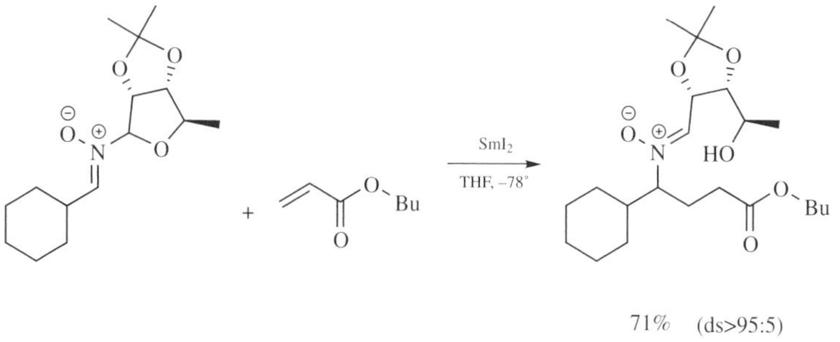

SmI_2
THF, −78°

+

71% (ds>95:5)

Reductive α,β-diacylation of acrylic esters and tertiary amides with ArCOCl is achieved in DMF.[31] Participation of DMF in the reduction of diaryl ketones via cross coupling results in benzoins after a rearrangement step.[32]

Conjugated enolates generated from reductive deacetoxylation of γ-acetoxy α,β-unsaturated esters are trapped by carbonyl compounds at the α-position.[33] Strong solvent effects are found in certain intramolecular reductive alkylation.[34]

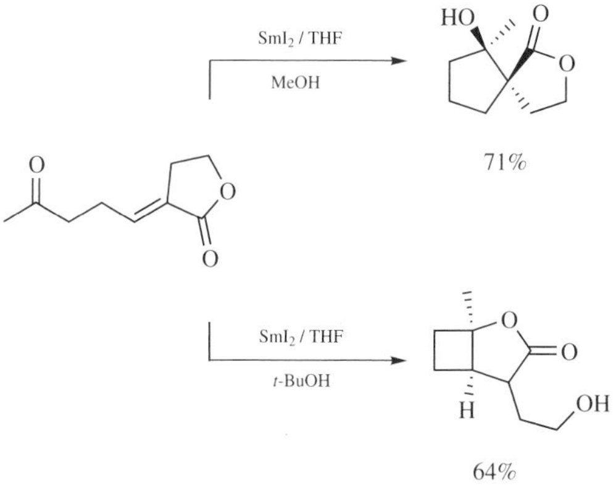

SmI_2 / THF
MeOH

71%

SmI_2 / THF
t-BuOH

64%

A method for construction of spiroannulated oxindoles is based on intramolecular CC bond formation between an isocyanate unit and an enone β-carbon during reduction of the enone.[35]

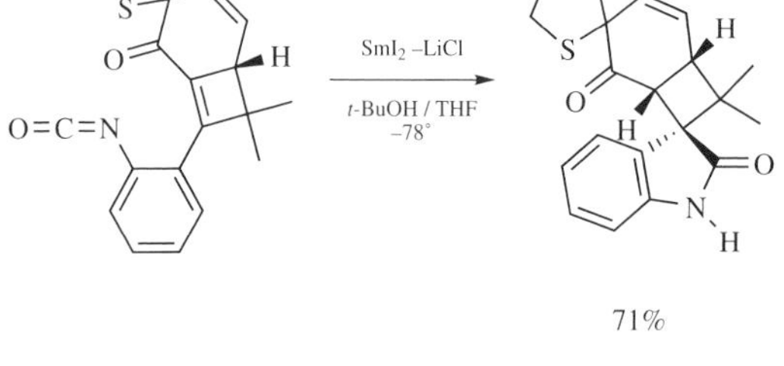

SmI_2–LiCl
t-BuOH / THF
−78°

71%

The coupling of carbonyl compounds with methoxyallene using SmI_2–HMPA–*t*-BuOH in THF leads to 4-hydroxy-1-methoxy-1-alkenes.[36]

Conjugated esters and amides cross-couple with nitrones (free radical reaction) to afford γ-amino acid derivatives.[37]

Condensation. A Strecker-type reaction to prepare α-amino phosphonates is realized by using SmI_2 as catalyst.[38]

Addition reactions. Disulfides and diselenides are converted to $RXSmI_2$ which serve as Michael donors to conjugated alkynones. Carbonyl compounds can be used as agent to deliver α-(chalcogenoalkylidene)-β-hydroxy ketones.[39]

Michael addition between ketene silyl acetals and enones is brought about by SmI_2. Addition of imines then gives the Mannich products.[40]

Synthetically significant annulations are those involving ketyl radical addition to form an eight-membered ring (isoschizandrin)[41] and addition to a methylenecyclopropane unit which can be elaborated further.[42]

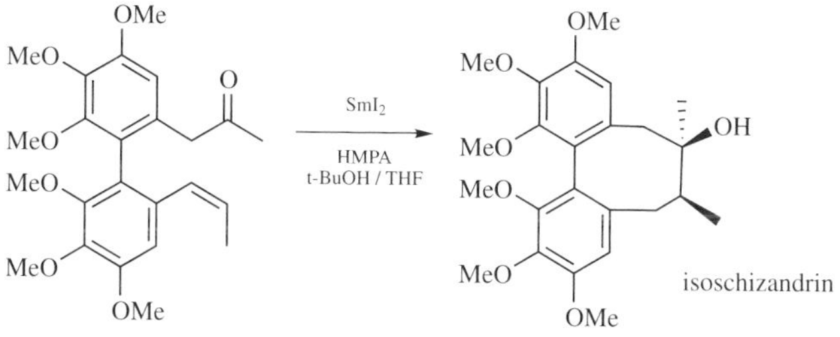

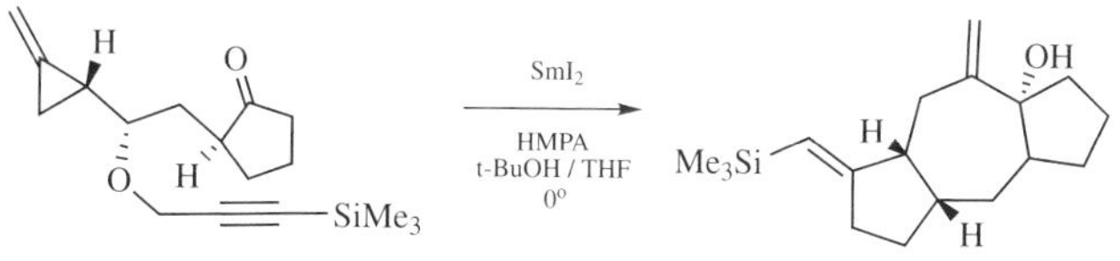

The carbon radical generated from an intramolecular addition to a cyclobutene unit can be reduced and then undergo nucleophilic attack.[43]

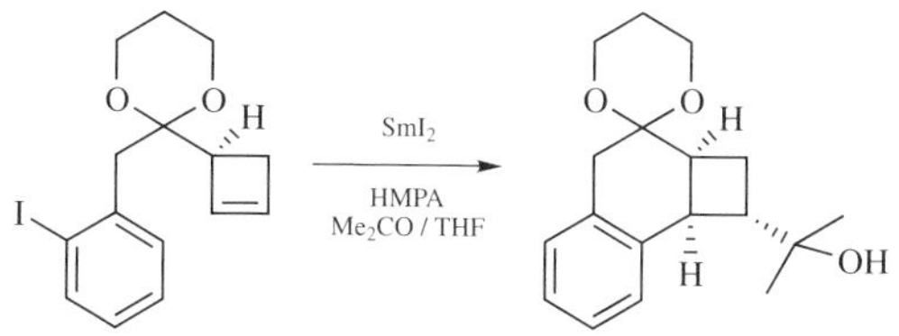

Rearrangement-reduction. Epoxy carbinols capable of simultaneously binding Sm with the two oxygen functionalities are subject to isomerization. The carbonyl products are promptly reduced.[44]

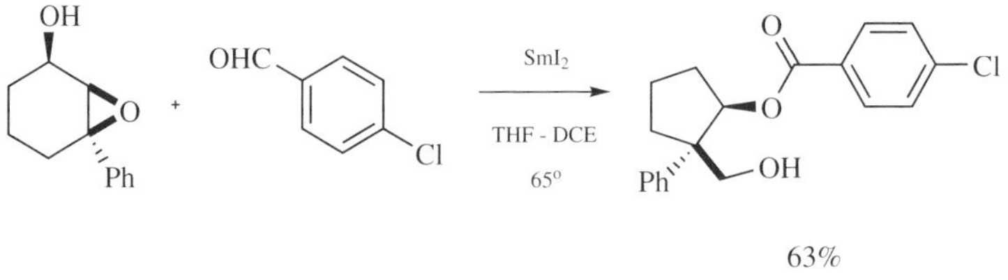

SmI_2, THF - DCE, 65°

63%

[1]Concellon, J.M., Rodriguez-Solla, H., Bardales, E., Huerta, M. *EJOC* 1775 (2003).
[2]Concellon, J.M., Bernad, P.L., Huerta, M., Garcia-Granda, S., Diaz, M.R. *CEJ* **9**, 5343 (2003).
[3]Dahlen, A., Hilmersson, G. *CEJ* **9**, 1123 (2003).
[4]Dahlen, A., Hilmersson, G. *TL* **44**, 2661 (2003).
[5]Dahlen, A., Sundgren, A., Lahmann, M., Oscarson, S., Hilmersson, G. *OL* **5**, 4085 (2003).
[6]Chopade, P.R., Davis, T.A., Prasad, E., Flowers II, R.A. *OL* **6**, 2685 (2003).
[7]Dahlen, A., Hilmersson, G., Knettle, B.W., Flowers, II, R.A. *JOC* **68**, 4870 (2003).
[8]Dahlen, A., Petersson, A., Hilmersson, G. *OBC* **1**, 2423 (2003).
[9]Concellon, J.M., Huerta, M., Llavona, R. *TL* **45**, 4665 (2004).
[10]Obringer, M., Colobert, F., Neugnot, B., Solladie, G. *OL* **5**, 629 (2003).
[11]Yoda, H., Kohata, N., Takabe, K. *SC* **33**, 1087 (2003).
[12]Blakskjaer, P., Hoj, B., Riber, D., Skrydstrup, T. *JACS* **125**, 4030 (2003).
[13]Revuelta, J., Cicchi, S., Brandi, A. *TL* **45**, 8375 (2004).
[14]Wang, X., Guo, H., Xie, G., Zhang, Y. *SC* **34**, 3001 (2004).
[15]Cai, G., Xu, X., Li, Z., Weber, W.P., Lu, P. *JHC* **39**, 1271 (2003).
[16]Ding, H., Friestad, G.K. *OL* **6**, 637 (2004).
[17]Mukaiyama, T., Ogawa, Y., Kuroda, K. *CL* **33**, 1472 (2004).
[18]Concellon, J.M., Bardales, E., Llavona, R. *JOC* **68**, 1585 (2003).
[19]Concellon, J.M., Bardales, E. *OL* **5**, 4783 (2003).
[20]Concellon, J.M., Bernad, P.L., Bardales, E. *CEJ* **10**, 2445 (2004).
[21]Concellon, J.M., Bardales, E. *JOC* **68**, 9492 (2003).
[22]Concellon, J.M., Huerta, M., Bardales, E. *T* **60**, 10059 (2004).
[23]Concellon, J.M., Rodriguez-Solla, H., Concellon, C. *TL* **45**, 2129 (2004).
[24]Li, J., Qian, W., Zhang, Y. *T* **60**, 5793 (2004).
[25]Fan, X., Zhang, Y. *T* **59**, 1917 (2003).
[26]Zhong, W.-W., Izumi, K., Xu, M.-H., Lin, G.-Q. *OL* **6**, 4747 (2004).
[27]Thakur, A.J., Prajapati, D., Sandhu, J.S. *CL* **33**, 102 (2004).
[28]Chavarot, M., Rivard, M., Rose-Munch, F., Rose, E., Py, S. *CC* 2330 (2004).
[29]Masson, G., Cividino, P., Py, S., Vallee, Y. *ACIEE* **42**, 2265 (2003).
[30]Johannesen, S.A., Albu, S., Hazell, R.G., Skrydstrup, T. *CC* 1962 (2004).
[31]Liu, Y., Zhang, Y. *T* **59**, 8429 (2003).
[32]Liu, Y., Xu, X., Zhang, Y. *T* **60**, 4867 (2004).
[33]Otaka, A., Yukimasa, A., Watanabe, J., Sasaki, Y., Oishi, S., Tamamura, H., Fujii, N. *CC* 1834 (2003).
[34]Hutton, T.K., Muir, K.W., Proctor, D.J. *OL* **5**, 4811 (2003).

[35]Ready, J.M., Reisman, S.E., Hirata, M., Weiss, M.M., Tamaki, K., Ovaska, T.V., Wood, J.L. *ACIEE* **43**, 1270 (2004).
[36]Holemann, A., Reissig, H.-U. *OL* **5**, 1463 (2003).
[37]Riber, D., Skrydstrup, T. *OL* **5**, 229 (2003).
[38]Xu, F., Luo, Y., Deng, M., Shen, Q. *EJOC* 4728 (2003).
[39]Zheng, X., Xu, X., Zhang, Y. *SL* 2061 (2003).
[40]Jaber, N., Assie, M., Fiaud, J.-C., Collin, J. *T* **60**, 3075 (2004).
[41]Molander, G.A., George, K.M., Monovich, L.G. *JOC* **68**, 9533 (2003).
[42]Underwood, J.J., Hollingworth, G.J., Horton, P.N., Hursthouse, M.B., Kilburn, J.D. *TL* **45**, 2223 (2004).
[43]Rivkin, A., Nagashima, T., Curran, D.P. *OL* **5**, 419 (2003).
[44]Fan, C.-A., Hu, X.-D., Tu, Y.-Q., Wang, B.-M., Song, Z.-L. *CEJ* **9**, 4301 (2003).

Samarium(III) iodide.

Aldol-Michael reactions. Using SmI_3 as catalyst a one-step 3-component condensation of ArCHO, active methylene compounds and phenacyl bromide furnishes functionalized phenethyl aryl ketones.

[1]Ma, Y., Zhang, Y. *SC* **33**, 711 (2003).

Scandium(III) triflate. **18**, 317–318; **19**, 300–302; **20**, 335–337; **21**, 387–389; **22**, 398

Functional group transformations. Taking advantage of the Lewis acidity of $Sc(OTf)_3$ several common transformations have been carried out: ether exchange of ROH and $CH(OMe)_2$,[1] 1,3-oxathiolane formation,[2] cyanosilylation of carbonyl compounds,[3] methoxycarbonylation of amines,[4] and transamidation.[5] In its presence a simple preparation of *N*-(*t*-butyldimethylsilyl)hydrazones[6] from carbonyl compounds using 1,2-bis(*t*-butyldimethylsilyl)hydrazine is realized and whereby several traditional reactions employing hydrazones (Wolff-Kishner reduction, Barton vinyl iodide synthesis, etc.) are practically improved.

Substitutions. Aryl C-glycosides are readily obtained in a reaction of glycosyl acetates with phenols in the presence of $Sc(OTf)_3$.[7] Friedel-Crafts acylation with RCOOH can be carried out with $Sc(OTf)_3$ [or $Bi(OTf)_3$] and TFAA.[8] Mixed carbonates (and thionocarbonates) of $Ph(CH_2)_nOH$ with 2-hydroxybenzthiazoles[9] are suitable electrophiles for arenes to catalysis by $Sc(OTf)_3$.

Additions. Aromatic substitution is accomplished by the catalyzed reaction of η^2-coordinated anisole (with exposed diene, and hence nonaromatic units) with enones[10] followed by treatment with $CuBr_2$, which acts as decomplexant and oxidant.

The $Sc(OTf)_3$ catalyst for allylation of aldehydes, aldimines, and ring opening of styrene oxides by stannanes is recyclable when using PEG as reaction medium.[11] Asymmetric conjugate addition to *N*-(2-alkenoyl)oxazolidinones is effectively promoted by $Sc(OTf)_3$ or $ZrCl_4$.[12]

Hydroamination. Alkylidenecyclopropanes react with sulfonamides to give ring-opened products or pyrrolidines.[13] [A cationic Sc complex (**1**) is shown to promote intramolecular hydroamination.[14]]

Ar Ar + $TsNH_2$ $\xrightarrow[\text{DCE } 85°]{Sc(OTf)_3}$ Ar Ar NHTs

75%

+ $TsNH_2$ $\xrightarrow[\text{DCE } 85°]{Sc(OTf)_3}$ NTs

46%

$B(C_6F_5)_3$

(**1**)

[1]Karimi, B., Ma'mani, L. *TL* **44**, 6051 (2003).
[2]Karimi, B., Ma'mani, L. *S* 2503 (2003).
[3]Karimi, B., Ma'mani, L. *OL* **6**, 4813 (2004).
[4]Distao, M., Quaranta, E. *T* **60**, 1531 (2004).
[5]Eldred, S.E., Stone, D.A., Gellman, S.H., Stahl, S.S. *JACS* **125**, 3422 (2003).
[6]Furrow, M.E., Myers, A.G. *JACS* **126**, 5436 (2004).
[7]Ben, A., Yamauchi, T., Matsumoto, T., Suzuki, K. *SL* 225 (2004).
[8]Matsushita, Y., Sugamoto, K., Matsui, T. *TL* **45**, 4723 (2004).
[9]Mukaiyama, T., Kamiyama, H., Yamanaka, H. *CL* **32**, 814 (2003).
[10]Smith, P.L., Keane, J.M., Shankman, S.E., Chordia, M.D., Harman, W.D. *JACS* **126**, 15543 (2004).
[11]Choudary, B.M., Jyothi, K., Madhi, S., Kantam, M.L. *SL* 231 (2004).
[12]Williams, D.R., Mullins, R.J., Miller, N.A. *CC* 2220 (2003).
[13]Chen, Y., Shi, M. *JOC* **69**, 426 (2004).
[14]Lauterwasser, F., Hayes, P.G., Bräse, S., Piers, W.E., Schafer, L.L. *OM* **23**, 2234 (2004).

Selenium. 18, 318; **20**, 337; **21**, 389; **22**, 398

Organoselenium compounds. Elemental selenium is reduced by $NaBH_4$ and immediate reaction with RBr provides RSeR, but the addition of HCOOH prior to RBr, RSeH are obtained.[1]

Dibenzyl diselenides are readily prepared from ArCHO on reaction with the Se-CO-H_2O system.[2] After reaction of lithium enolates with Se, exposure to atmospheric pressure CO leads to selenocarboxylates which are convertible to the seleno esters RCOCH(R')COSeMe.[3]

A synthesis of alkenyl selenides involves preparation of lithium organoselenides from RLi and Se, and addition to alkynes (mainly in the anti-Markovnikov fashion).[4]

An access to β-keto selenoesters from ketones proceeds via selenation of enolates followed by carbonylation of β-keto selenolithiums and rearrangement.[5]

Reduction. Selenium catalyzes reduction of enones and conjugated dienones (both termini aromatic) to saturated ketones by CO-H_2O in DMF.[6] The same reducing system converts dinitroarenes to nitroanilines.[7]

N,N′-Dipyridylureas. Unsymmetrical ureas are obtained by reductive condensation of nitropyridines with aminopyridines under CO in the presence of selenium (or selenium dioxide).[8] Nitrenes and isocyanates are the intermediates.

[1]Krief, A., Trabelsi, M., Dumont, W., Derock, M. *SL* 1751 (2004).
[2]Tian, F., Yu, Z., Lu, S. *JOC* **69**, 4520 (2004).
[3]Fujiwara, S., Nishiyama, A., Shin-ike, T., Kambe, N., Sonoda, N. *OL* **6**, 453 (2004).
[4]Zeni, G., Stracke, M.P., Nogueira, C.W., Braga, A.L., Menezes, P.H., Stefani, H.A. *OL* **6**, 1135 (2004).
[5]Fujiwara, S., Nishiyama, A., Shin-ike, T., Kambe, N., Sonoda, N. *OL* **6**, 453 (2004).
[6]Tian, F., Lu, S. *SL* 1953 (2004).
[7]Liu, X., Lu, S. *CL* **32**, 1142 (2003).
[8]Chen, J., Ling, G., Lu, S. *T* **59**, 8251 (2003).

Silica chloride.

Condensations. The compound is a heterogeneous catalyst for esterification[1] and tosylation[2] of alcohols using RCOOH and TsOH, respectively.

Functional group transformations. Conversion of aldehydes to nitriles and ketones into amides (via Beckmann rearrangement) is accomplished on reaction with $NH_2OH \cdot HCl$ with microwave irradiation and catalysis by silica chloride.[3] On wet silica gel carbonyl compounds are regenerated from oximes, hydrazones and semicarbazones.[4]

[1]Srinivas, K.V.N.S., Mahender, I., Das, B. *S* 2479 (2003).
[2]Das, B., Reddy, V.S., Reddy, M.R. *TL* **45**, 6717 (2004).
[3]Srinivas, K.V.N.S., Mahender, I., Das, B. *CL* **32**, 738 (2003).
[4]Shirini, F., Zolfigol, M.A., Khaleghi, M., Mohammadpoor-Baltork, I. *SC* **33**, 1839 (2003).

Silica gel. 15, 282; **18**, 319; **19**, 303–304; **20**, 338–339; **21**, 390–391; **22**, 398–400

Bound reagents. Reactants or ligands can be attached to the surface of silica pellets by exchange reaction with a trialkoxysilane end group via linker.[1]

Reaction medium. Wet silica gel finds use as reaction medium (no other solvent) for the allylation of aldehydes with tetraallylstannane.[2] Condensation of amines with β-diketones and β-keto esters,[3] and of *o*-aminoarenethiols with enones to form 1,5-benzothiazepines[4] are also readily conducted by this simple method. Synthesis of thiol esters[5] from alkyl halides using silica-supported RCOSK has some advantages.

Acid catalysts. Silica supported $NaHSO_4$ has found use in cleaving acetonides,[6] trityl ethers,[7] phenolic MOM ethers,[8] and catalyzing formation of *p*-hydroxybenzyl ethers.[9]

Perchloric acid adsorbed on silica has catalytic activity for acetylation of alcohols, phenols, arenethiols and anilines.[10] It is also useful to bring about Ferrier rearrangement.[11]

Silica impregnated with phosphoric acid is a reusable catalyst for assembly of β-amino esters from silyl ketene acetals, aldehydes and amines.[12]

Oxidation. Silica-supported *t*-BuOOH or Oxone can be used to oxidize glycosyl sulfides to sulfoxides,[13] conditions are compatible with many functional groups. Dehydrogenation of 2-substituted imidazolines[14] and cleavage of semicarbazones and phenylhydrazones[15] are facile on a supported $KMnO_4$. Mediation of epoxidation (using Oxone) by an *N*-acylated α-fluorotropinone anchored on amorphous silica has moderate success in asymmetric induction.[16]

[1]Tomofte, R.S., Woodward, S. *TL* **45**, 39 (2004).
[2]Jin, Y.Z., Yasuda, N., Furano, H., Inanaga, J. *TL* **44**, 8765 (2003).
[3]Gao, Y., Zhang, W., Xu, J. *SC* **34**, 909 (2004).
[4]Kodomari, M., Noguchi, T., Aoyama, T. *SC* **34**, 1783 (2004).
[5]Aoyama, T., Takido, T., Kodomari, M. *SC* **33**, 3817 (2003).
[6]Mahender, G., Ramu, R., Ramesh, C., Das, B. *CL* **32**, 734 (2003).
[7]Das, B. , Mahender, G., Kumar, V.S., Chowdhury, N. *TL* **45**, 6709 (2004).
[8]Ramesh, C., Ravindranath, N., Das, B. *JOC* **68**, 7101 (2003).
[9]Ramu, R., Nath, N.R., Reddy, M.R., Das, B. *SC* **34**, 3135 (2004).
[10]Chakraborti, A.K., Gulhane, R. *CC* 1896 (2003).
[11]Agarwal, A., Rani, S., Vankar, Y.D. *JOC* **69**, 6317 (2004).
[12]Lock, S., Miyoshi, N., Wada, M. *CL* **33**, 1308 (2004).
[13]Chen, M.-Y., Patkar, L.N., Chen, H.-T., Lin, C.-C. *CR* **338**, 1327 (2003).
[14]Mohammadpoor-Baltork, I., Zolfigol, M.A., Abdollahi-Alibeik, M. *TL* **45**, 8687 (2004).
[15]Hajipour, A.R., Adibi, H., Ruoho, A.E. *JOC* **68**, 4553 (2003).
[16]Sartori, G., Armstrong, A., Maggi, R., Mazzacani, A., Sartorio, R., Bigi, F., Dominguez-Fernandez, B. *JOC* **68**, 3232 (2003).

Silicon tetrachloride.

α-Amino nitriles.[1] From aldimines the addition reaction employs KCN and $SiCl_4$ is quite efficient.

Ether cleavage.[2] Aromatic ethers, esp. ArOMe, are cleaved by the combination of LiI, $SiCl_4$, and $BF_3{\cdot}HOAc$. That *o*-(methylthio)anisole gives the phenol that retains the MeS group shows excellent chemoselectivity (HSAB relationship).

[1]El-Ahl, A.-A.S. *SC* **33**, 989 (2003).
[2]Zewge, D., King, A., Weissman, S., Tschaen, D. *TL* **45**, 3729 (2004).

Silver acetate. 21, 392; 22, 400

β-Bromostyrenes.[1] 2,3-Dibromo-3-arylpropanoic acids undergo bromodecarboxylation on brief treatment with AgOAc in HOAc under microwave irradiation.

Pyrroles.[2] Homopropargylic amines and *C*-silyl derivatives undergo cyclization by the influence of AgOAc.

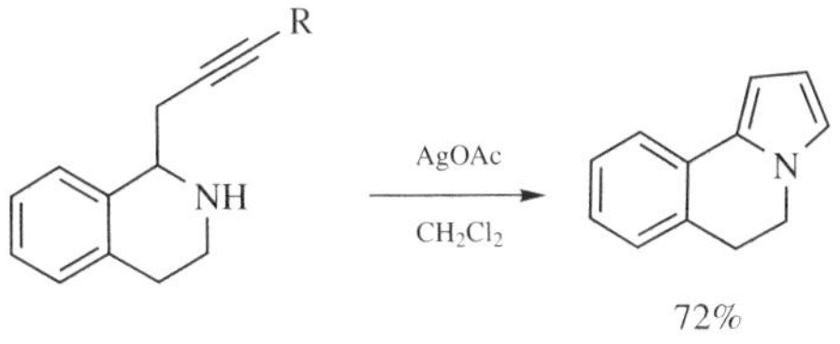

AgOAc, CH_2Cl_2

72%

[1]Kuang, C., Senboku, H., Tokuda, M. *T* **61**, 637 (2005).
[2]Agarwal, S., Knolker, H. *OBC* **2**, 3060 (2004).

Silver benzoate.

Wolff-Cope rearrangment tandem.[1] 2-Alkenylcyclopropyl diazoketones undergo rearrangement on exposure to AgOBz. In the absence of nucleophiles a Cope rearrangement occurs to deliver cycloheptadienones.[1] Photochemically the substrates pursue another reaction pathway to give 4-alkenyl-2-cyclopentenones.

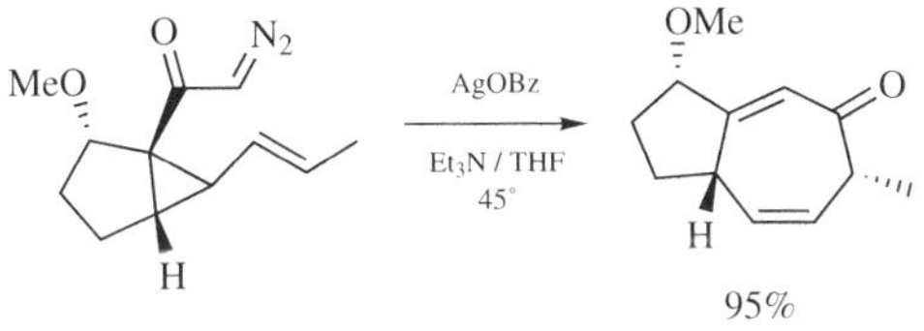

AgOBz, Et_3N / THF, 45°

95%

[1]Sarpong, R., Su, J.T., Stoltz, B.M. *JACS* **125**, 13624 (2003).

Silver bis(trifluoromethane)imide.

[2+2]Cycloaddition.[1] Siloxyalkynes and electron-deficient alkenes are united in the presence of $AgNTf_2$ to afford cyclobutenes.

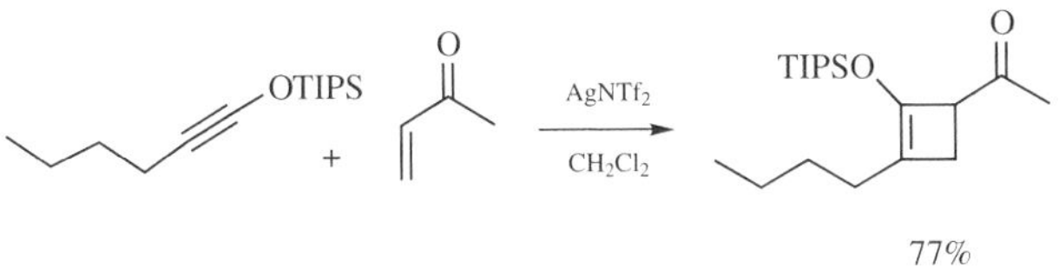

$AgNTf_2$, CH_2Cl_2

77%

[1]Sweis, R.F., Schramm, M.P., Kozmin, S.A. *JACS* **126**, 7442 (2004).

Silver iodide. 19, 305

Propargylic amines.[1] A 3-component coupling involving RCHO, 1-alkynes, and amines is achieved in boiling water in the presence of AgI.

[1]Wei, C., Li, Z., Li, C.-J. *OL* **5**, 4473 (2003)

Silver nitrate. 18, 320; **19**, 305–306; **20**, 340; **21**, 393; **22**, 400–401

Aziridines.[1] The disilver(I) complex derived from $AgNO_3$ and tri-(*t*-butyl)-terpyridine effectively promotes aziridination of alkenes with PhI=NTs.

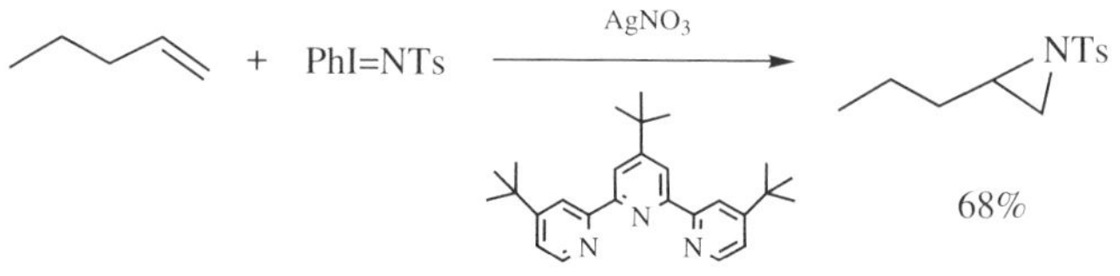

Benzazepinones.[2] A 3-step ring expansion of isoquinolines involves *N*-benzylation, addition of tribromomethide ion and ring expansion induced by $AgNO_3$.

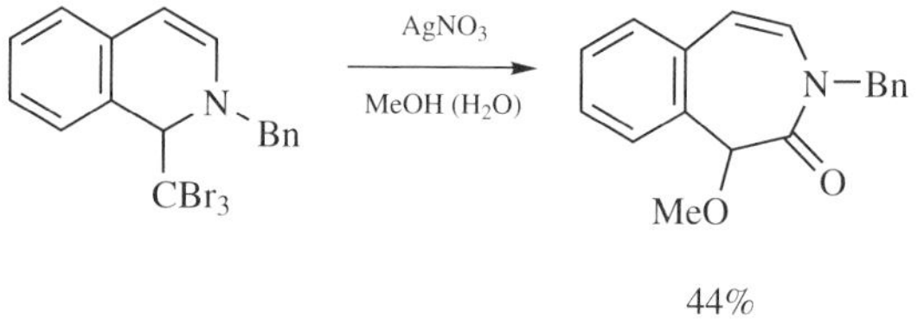

Dethioacetalization.[3] The $AgNO_3$-Br_2 combination can be used to liberate α-ketophosphonates from the corresponding dithianes.

[1]Cui, Y., He, C. *JACS* **125**, 16202 (2003).
[2]Pauvert, M., Collet, S., Guingant, A. *TL* **44**, 4203 (2003).
[3]Afarinkia, K., Faller, A., Twist, A.J. *S* 357 (2003).

Silver(I) oxide. 18, 321; **20**, 341; **21**, 393; **22**, 401

O-Tosylation.[1] Application of the method (TsCl, KI-Ag_2O) for monotosylation of diols to pyranosides is successful.

[1]Wang, H., She, J., Zhang, L.-H., Ye, X.-S. *JOC* **69**, 5774 (2004).

Silver phosphate.

Silylene transfer.[1] 7,7-Di-*t*-butylsilabicyclo[4.1.0]heptane is decomposed by Ag_3PO_4 and the *t*-Bu_2Si unit is transferred to alkynes. If a copper salt and carbonyl compounds are also present oxasilacyclopentenes are formed.

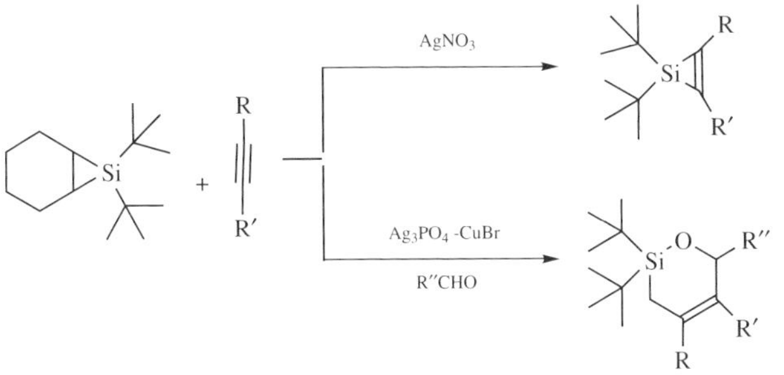

[1]Clark, T.B., Woerpel, K.A. *JACS* **126**, 9522 (2004).

Silver tetrafluoroborate. 18, 322; 21, 394

Furans.[1] Formation of substituted furans from 2-oxyfunctionalized 3-alkynones under the influence of $AgBF_4$ involves isomerization and cyclization.

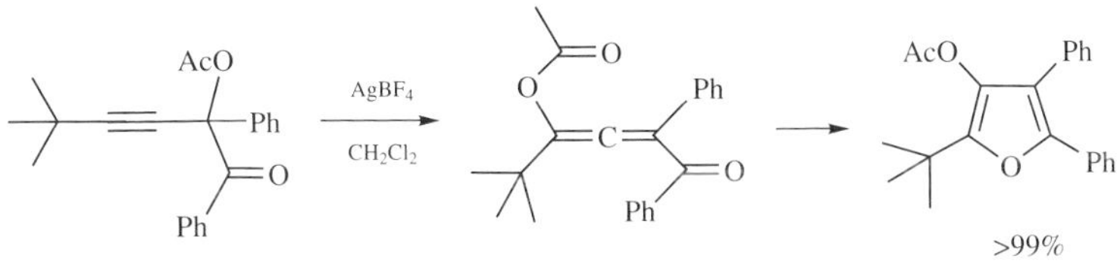

Hydroarylation.[2] The union of unactivated arenes and alkenes catalyzed by an Ag-Pt system is due to activation of the alkenes by the Pt complex, but the important role of silver ion is unclear. Product distributions indicate a Friedel-Crafts alkylation pathway is followed.

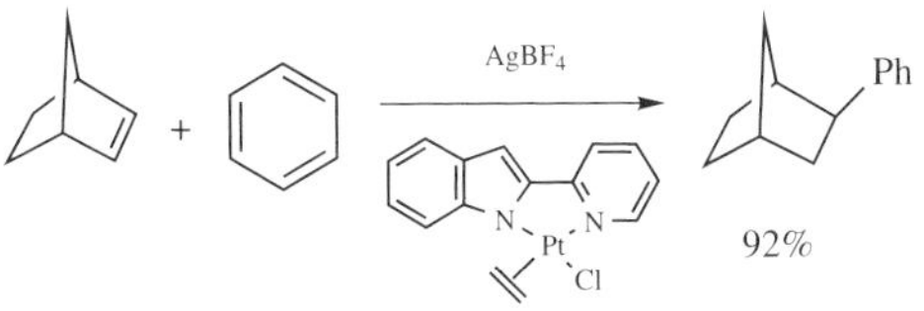

[1]Sromek, A.W., Kel'in, A.V., Gevorgyan, V. *ACIEE* **43**, 2280 (2004).
[2]Karshtedt, D., Bell, A.T., Tilley, T.D. *OM* **23**, 4169 (2004).

Silver trifluoromethanesulfonate. 13, 274–275; 14, 282–283; 16, 302; 17, 314; 18, 322–323; 19, 306; 20, 342; 21, 394; 22, 402

Glycosylation. Silver triflate is an effective promoter for glycosylation involving glycosyl trichloroacetimidates[1] and thiazoline-2-sulfides[2] as donors. Reaction of unactivated glycosyl sulfides requires a combination of AgOTf and NIS.[3]

[1]Wei, G., Gu, G., Du, Y. *JCC* **22**, 385 (2003).
[2]Demchenko, A.V., Pornsuriyasak, P., De Meo, C., Malysheva, N.N. *ACIEE* **43**, 3069 (2004).
[3]Gadikota, R.R., Callam, C.S., Wagner, T., Del Fraino, B., Lowary, T.L. *JACS* **125**, 4155 (2003)

Silver 2,4,6-trinitrobenzenesulfonate.

Ar-*Alkylation.* The B-ring closure in an approach to diterpene alkaloids was designed to employ an intramolecular Friedel-Crafts alkylation with a bridgehead bromide. The best conditions involve reaction with the title reagent in $MeNO_2$.

MeN, Br, OMe, OAc, OH, AcO; reagent: O_2N, NO_2, SO_3Ag, NO_2; CH_3NO_2; product: MeN, OMe, OAc, OH, AcO

53%

[1]Williams, C.M., Mander, L.N. *OL* **5**, 3499 (2003).

Sodium. **13**, 277; **18**, 323–324; **20**, 342–343; **21**, 395; **22**, 403

Reductive cleavage. 3,4,5-Trimethoxystilbenes are reduced by Na in THF with loss of the central methoxy group.[1] Depending on the quencher, the double bond may be retained or saturated.

OMe, MeO, OMe, Ph; Na / THF, 7 h; quench; MeO, OMe, Ph; MeO, OMe, Ph

quencher:			
	H_2O	75%	—
	$BrCH_2CH_2Br$	—	78%

Coupling reactions. Reductive coupling of chlorodialkoxyboron with Na/Hg is a convenient method for the preparation of tetraalkoxydiboranes.[2] Pinacol coupling with Na and catalytic PhBr at room temperature can be performed without solvent.[3]

[1]Azzena, U., Dettori, G., Idini, M.V., Pisano, L., Sechi, G. *T* **59**, 7961 (2003).
[2]Anastasi, N.R., Waltz, K.M., Weerakoon, W.L., Hartwig, J.F. *OM* **22**, 365 (2003).
[3]Zhao, H., Li, D.-J., Deng, L., Liu, L., Guo, Q.-X. *CC* 506 (2003).

Sodium acetoxyborohydride.

Reduction. *N*-(α-Benzenesulfonylalkyl) amides which are readily obtained from a reaction of primary amides with RCHO and $PhSO_2H$ in the presence of $MgSO_4$, are reduced to secondary amines.

[1]Mataloni, M., Petrini, M., Profeta, R. *SL* 1129 (2003).

Sodium *t*-butoxide.

Deacylation.[1] The Boc group of RNHBoc is removed by refluxing with *t*-BuONa in wet THF. Ordinarily, *N*-Boc cleavage is performed in acidic media.

[1]Tom, N.J., Simon, W.M., Frost, H.N., Ewing, M. *TL* **45**, 905 (2004).

Sodium bis(2-methoxyethoxy)aluminum hydride, Red-Al®. **15**, 290; **21**, 396; **22**, 403

β-Hydroxy amides. α,β-Epoxy amides are reduced efficiently with Red-Al in DMF in the presence of 15-crown-5.[1] Conversion of the same substrates to the corresponding α-hydroxy amides needs only change the reducing system to $(dba)_3Pd_2 - CHCl_3$, Bu_3P, HCOOH and Et_3N.

(dba)$_3$Pd$_2$.CHCl$_3$
Bu$_3$P, Et$_3$N
HCOOH, THF
94%
Red-Al

[1]Kakei, H., Nemoto, T., Ohshima, T., Shibasaki, M. *ACIEE* **43**, 317 (2004).

Sodium borohydride. **13**, 278–279; **15**, 290; **16**, 304; **18**, 326–327; **19**, 307–309; **20**, 344–345; **21**, 397–398; **22**, 403–405

Special techniques. γ-Cyclodextrin-bicapped C_{60} mediates asymmetric reduction of ketones by $NaBH_4$.[1] In diglyme at 162° carboxylic acids and esters (including the extremely hindered *t*-amyl 2-chlorobenzoate) are reduced to the alcohols.[2]

Reduction. Reduction of β-keto esters alcoholic solvents can incur transesterification.[3] Reduction of α-keto esters to 1,2-diols by the $NaBH_4$-Me_3SiCl system in THF is rendered enantioselective with a polymer-bound *N*-sulfonyl-α,α-diphenylprolinol.[4]

Among the three different carbonyl groups in oxothiazolidinones with an ester side chain, only the ester group is reduced by $NaBH_4$ because the others form an extended anionic chain and are thereby protected.[5]

COOEt S O O N H Ph → $NaBH_4$ / EtOH → HO S O O N H Ph

64%

α-Substituted acrylic esters obtained from the Baylis-Hillman reaction are reduced to saturated diols.[6] 1,2-*O*-Arylidene pyranoses are formed when the 2-aroylglycosyl bromides are treated with $NaBH_4$-KI.[7] Reduction occurs on the dioxolanyl cation intermediates.

Using $NaBH_4$ to convert RCOOH to RCH_2OH in THF at room temperature it needs to add catalytic amount of 3,4,5-trifluorophenylboronic acid to form acyloxyboron chelates that are susceptible to reduction.[8]

Monoalkylation of *N,N*'-disubstituted ethylenediamines is readily accomplished via reduction of imidazolidine derivatives which are formed with aldehydes.[9] Conversion of various primary amides to alcohols is achieved in two steps: formation of *N,N*-di-Boc derivatives and reduction with $NaBH_4$.[10]

N-Halomagnesio imines generated from Grignard reaction of nitriles are subject to borohydride reduction.[11] Properly protected cyanohydrins of aldoses are converted into aminoglycosides.[12]

OBn BnO BnO OH CN OBn → $NaBH_4$ / EtOH → OBn BnO BnO O NH_2 OBn

70% (β only)

4-Hydroxy-2-alkynoic esters are reduced to the corresponding (*E*)-alkenoic esters by $NaBH_4$ at low temperature,[13] and 4-keto-2-alkynoic ester can be reduced directly.[14]

Tandem reactions. Reduction of α-phosphonylthio and α-phosphonylseleno ketones is followed by elimination. Substituted alkenes are produced.[15] The use of RTeNa which are generated by borohydride reduction of diorgano ditellurides to add to propargylic amines gives (*Z*)-allylic amines.[16]

Iminocyclopropanes generated in situ are easily reduced by $NaBH_4$ – $BF_3.OEt_2$ (borane) to afford cyclopropylamines.[17]

Condensations. A new version of the Perkin reaction[18] consists of heating ArCHO and HOAc with $NaBH_4$ in NMP at 180–190°. $NaB(OAc)_4$ is presumed to be the nucleophile.

[1]Nishimura, T., Nakajima, M., Maeda, Y., Uemura, S., Takekuma, S., Takekuma, H., Yoshida, Z. *BCSJ* **77**, 2047 (2004).
[2]Zhu, H.-J., Pittman, Jr, C.U. *SC* **33**, 1733 (2003).
[3]Padhi, S.K., Chadha, A. *SL* 639 (2003).
[4]Wang, G., Hu, J., Zhao, G. *TA* **15**, 807 (2004).
[5]Markovic, R., Baranac, M., Stojanovic, M. *SL* 1034 (2004).
[6]Patra, A., Batra, S., Bhaduri, A.P. *SL* 1611 (2003).
[7]Suzuki, K., Mizuta, T., Yamamura, M. *JCC* **22**, 143 (2003).
[8]Tale, R.H., Patil, K.M., Dapurkar, S.E. *TL* **44**, 3427 (2003).
[9]Salerno, A., Figueroa, M.A., Perillo, I.A. *SC* **33**, 3193 (2003).
[10]Ragnarsson, U., Grehn, L., Monteiro, L.S., Maia, H.L.S. *SL* 2386 (2003).
[11]van den Nieuwendijk, A.M.C.H., Ghisaidoobe, A.B.T., Overkleeft, H.S., Brussee, J., van der Gen, A. *T* **60**, 10385 (2004).
[12]Dorsey, A.D., Barbarow, J.E., Trauner, D. *OL* **5**, 3237 (2003).
[13]Meta, C.T., Koide, K. *OL* **6**, 1785 (2004).
[14]Naka, T., Koide, K. *TL* **44**, 443 (2003).
[15]Maciagiewicz, I., Dybowski, P., Skrowronska, A. *T* **59**, 6057 (2003).
[16]Zeni, G., Barros, O.S.do R., Moro, A.V., Baraga, A.L., Peppe, C. *CC* 1258 (2003).
[17]Yoshida, Y., Umezu, K., Hamada, Y., Atsumi, N., Tabuchi, F. *SL* 2139 (2003).
[18]Chiriac, C.I., Tanasa, F., Onciu, M. *TL* **44**, 3579 (2003).

Sodium borohydride – iodine.

Reductions. The combination is useful for deoxygenation, e.g., of sulfoxides and of lactams without disturbing a *gem*-dicarboxylic ester unit in the same molecule.[2]

α-Oximino esters are converted into 1,2-amino alcohols in THF, but in this case many other additives can be used instead of iodine.[3]

[1]Karimi, B., Zareyee, D *S* 335 (2003).
[2]Haldar, P., Ray, J.K. *TL* **44**, 8229 (2003).
[3]Periasamy, M., Sivakumar, S., Reddy, M.N. *S* 1965 (2003).

Sodium borohydride – metal salt. **21**, 398–399; **22**, 405–406

Deallylation. Prenyl ethers but not the esters are cleaved by $NaBH_4$-$ZrCl_4$ in CH_2Cl_2.[1]

O O NaBH$_4$ ZrCl$_4$ CH$_2$Cl$_2$ OH 96%

Reduction. α-Amino acids are reduced to amino alcohols with $NaBH_4$-$TiCl_4$.[2] In a convenient way conjugate reduction uses the $NaBH_4$-$NiCl_2$ system (nickel boride).[3] The $NaBH_4$-$InCl_3$ combination selectively saturates the α,β-position of dienones, but only conjugate systems with aromatic rings at both ends have been tested.[4] The same system effects β-elimination of β,γ-epoxy bromides to provide allylic alcohols.[5]

O Br NaBH$_4$ InCl$_3$ MeCN OH 78%

Cyclocarbonylation of alkynols is effected by $NaBH_4$-$K_2[Ni(CN)_4]$ in the presence of stoichiometric KCN which provides the CO unit.[6]

R HO R′ NaBH$_4$ KCN; H$^+$ R O O R′

[1]Babu, K.S., Raju, B.C., Srinivas, P.V., Rao, J.M. *TL* **44**, 2525 (2003).
[2]Zhang, C.-H., Yang, N.-F., Yang, L.-W. *CJOC* **24**, 343 (2004).
[3]Khurana, J.M., Sharma, P. *BCSJ* **77**, 549 (2004).
[4]Ranu, B.C., Samanta, S. *JOC* **68**, 7130 (2003).
[5]Ranu, B.C., Banerjee, S., Das, A. *TL* **45**, 8579 (2004).
[6]Gutierrez, J.L.C., Jimenez-Cruz, F., Espinosa, N.R. *TL* **46**, 803 (2005).

Sodium bromate. 18, 330; **20**, 347; **21**, 400; **22**, 406

Oxidation. Solvent-free oxidation of benzylic alcohols to carbonyl compounds by $NaBrO_3$/ion-exchange resin.[1] The sulfide to sulfoxide conversion is chemoselective in a solvent like hexane in which oxidation of aryl sulfides is inhibited.[2]

[1]Shaabani, A., Lee, D.G. *SC* **33**, 1255 (2003).
[2]Shaabani, A., Bazgir, A., Soleimani, K., Salehi, P. *SC* **33**, 2935 (2003).

Sodium cyanoborohydride. 21, 401; **22**, 407

Reductive amination.[1] Because imines are more reactive toward $NaBH_3CN$ synthesis of aza-bridged 9,10-dihydroanthracenes from the *cis*-dialdehyde and primary amines encounters no problem.

2-Arylethanols.[2] A combination of $NaBH_3CN$ and ZnI_2 is effective to open styrene oxides regioselectively.

Radical reactions.[3] Generation of Bu_3SnH in situ from Bu_3SnCl and $NaBH_3CN$ in *t*-BuOH to conduct radical reactions has several advantages.

[1]Stoy, P., Rush, J., Pearson, W.H. *SC* **34**, 3481 (2004).
[2]Finkielsztein, L.M., Aguirre, J.M., Latano, B., Alesso, E.N., Iglesias, G.Y.M. *SC* **34**, 895 (2004).
[3]Renaud, P., Beaufils, F., Feray, L., Schenk, K. *ACIEE* **42**, 4230 (2003).

Sodium chlorite.

Dethioacetalization.[1] Carbonyl group is generated from the 1,3-dithiane unit on reaction with $NaClO_2$ and NaH_2PO_4 in aq. MeOH.

[1]Ichige, T., Miyake, A., Kanoh, N., Nakata, M. *SL* 1686 (2004).

Sodium hexamethyldisilazide.

Tandem condensations.[1] 4-Silyl-3-butenonitrile epoxide undergoes elimination and Brook rearrangement to generate 4-siloxybutenonitrile anion. The nucleophilic site is moved, as shown by a Michael reaction of such an anion and further reaction.

Si O CN + Br COOEt —NaN(SiMe3)2→ Si O H COOEt CN

66%

Cyclization.[2] α-Elimination of a terminal alkenyl chloride to generate an alkylidene carbene is employed in the construction of the A-ring of an ingenol model.

ClCH H O Me3Si O —NaN(SiMe3)2, Et2O r.t.→ O Me3Si O

[1]Matsumoto, T., Masu, H., Yamaguchi, K., Takeda, K. *OL* **6**, 4367 (2004).

Sodium hydride. **14**, 288; **16**, 307–308; **18**, 333; **19**, 312–313; **20**, 349–350; **21**, 402–404; **22**, 408–409

Elimination. In situ elimination following alkylation of 3-tosyl-3,4-dihydro-α-pyridones using NaH as base is synthetically expedient.[1]

Ring expansion. 4-(*p*-Toluenesulfinyl)-2,5-cyclohexadienones undergo intramolecular Michael addition and the sulfinyl group is expelled when the norcaradienolate intermediates isomerize to afford tropones.[2]

[1]Lin, C.H., Tsai, M.R., Wang, Y.S., Chang, N.C. *JOC* **68**, 5688 (2003).
[2]Carreno, M.C., Sanz-Cuesta, M.J., Rigaborda, M. *CC* 1007 (2005).

Sodium hypochlorite. **15**, 293; **16**, 308; **17**, 316; **18**, 334–335; **19**, 313; **20**, 350; **21**, 404; **22**, 409

α-Chloro ketones. α-Chloro ketones are formed when alkenyl chlorides are treated with NaOCl-HOAc.[1]

[1]VanBrunt, M.P., Ambenge, R.O., Weinreb, S.M. *JOC* **68**, 3323 (2003).

Sodium iodide. **21**, 404; **22**, 409–410

Reduction. Aryl azides are reduced by NaI in HCOOH or HOAc, and the amine products are acylated by the carboxylic acid.[1]

HWE-reaction.[2] The (*Z*)-selectivity of the condensation between RCHO and diarylphosphonoacetic esters is increased by excess Na^+. Thus, addition of NaI to the reaction medium is beneficial.

Halogen exchange. 4-Chloroquinolines and 2-chloropyridine/4-chloropyridine are readily transformed into the iodo analogues by treatment with NaI after protonation (activation).[3]

[1]Kamal, A., Ramana, A.V., Reddy, K.S., Ramana, K.V., Babu, A.H., Prasad, B.R. *TL* **45**, 8187 (2004).
[2]Pihko, P.M., Salo, T.M. *TL* **44**, 4361 (2003).
[3]Wolf, C., Tumambac, G.E., Villalobos, C.N. *SL* 1801 (2003).

Sodium methoxide.

Cleavage of aryl t-butyl carbonates.[1] Removal of Boc group with a base avoids certain side reactions.

Baylis-Hillman reaction.[2] Introduction of a carbinol unit at the α-position of cycloalkenones is accomplished using MeONa/MeOH as the catalytic system.

[1]Nakamura, K., Nakajima, T., Kayahara, H., Nomura, E., Taniguchi, H. *TL* **45**, 495 (2004).
[2]Luo, S., Mi, X., Xu, H., Wang, P.G., Cheng, J.-P. *JOC* **69**, 8413 (2004).

Sodium nitrate.

Hydration. Heating nitriles with $NaNO_3$-impregnated fluorapatite and water leads to primary amides.

[1] Solhy, A., Smahi, A., El Badaoui, E., Elaabar, B., Amoukal, A., Sebti, S. *TL* **44**, 4031 (2003).

Sodium nitrite. 21, 405; **22**, 411

Nitrosation. *C*-Nitrosation at a benzylic position simultaneous to a Nef reaction on the adjacent carbon atom using $NaNO_2$–HOAc affords α-oximino ketones.[1]

Deprotection. *O,S*-Acetals and thioacetals are cleaved by treatment with $NaNO_2$–AcCl to furnish the parent carbonyl compounds.[2]

Hydrodeamination.[3] Diazotization of $ArNH_2$ by the $NaNO_2$–H_2SO_4 system goes further in the presence of EtOAc which acts as a reducing agent.

Oxidation.[4] Urazoles are conveniently dehydrogenated with $NaNO_2$–silica sulfuric acid at room temperature.

[1]Ran, C., Yang, G., Wu, T., Xie, M. *TL* **44**, 8061 (2003).
[2]Khan, A.T., Mondal, E., Sahu, P.R. *SL* 377 (2003).
[3]Bacherikov, V.A., Wang, M.-J., Cheng, S.-Y., Chen, C.-H., Chen, K.-T., Su, T.-L. *BCSJ* **77**, 1027 (2004).
[4]Zolfigol, M.A., Chehardoli, G., Mallakpour, S.E. *SC* **33**, 833 (2003).

Sodium perborate. 14, 290–291; **16**, 310; **18**, 337–338; **19**, 314; **20**, 351; **22**, 411

Oxidation of aldehydes.[1] With V_2O_5 as catalyst, $NaBO_3$ renders a mixture of ArCHO and MeOH into ArCOOMe.[1]

Fluorous phenyliodine(III) diacetate.[2] A convenient preparation of the fluorous reagent is by oxidation of the corresponding iodides with $NaBO_3$.

[1]Gopinath, R., Barkakaty, B., Talukdar, B., Patel, B.K. *JOC* **68**, 2944 (2003).
[2]Rocaboy, C., Gladysz, J.A. *CEJ* **9**, 88 (2003).

Sodium periodate. 15, 294; **18**, 338–339; **19**, 315; **21**, 405; **22**, 411–412

Oxidation. Metal halides such as LiBr can be oxidized in situ by $NaIO_4$ to provide positive halogen. By the technique alkenes are transformed into vicinal dibromides, bromohydrins, etc.[1]

Alkyl halides are converted into carbonyl compounds on heating with $NaIO_4$/DMF.[2]

[1]Dewkar, G.K., Narina, S.V., Sudalai, A. *OL* **5**, 4501 (2003).
[2]Das, S., Panigrahi, A.K., Maikap, G.C. *TL* **44**, 1375 (2003).

Sodium tetrachloroaurate. 22, 412

Indoles. 2-Alkynylanilines undergo cyclization on treatment with $NaAuCl_4$.[1] It is also possible to introduce a halogen atom at C-3 in the same operation.

Conjugate addition to enones by indoles at the β-position is also catalyzed by the gold salt.[2]

Pyridines. Carbonyl compounds condense with propargylic amines leading to substituted pyridines.[3]

NMe =O + H_2N — $NaAuCl_4.2H_2O$, EtOH 100° → NMe N

66%

[1]Arcadi, A., Bianchi, G., Marinelli, F. *S* 610 (2004).
[2]Arcadi, A., Bianchi, G., Chiarini, M., D'Anniballe, C., Marinelli, F. *SL* 944 (2004).
[3]Abbiati, G., Arcadi, A., Bianchi, G., Di Giuseppe, S., Marinelli, F., Rossi, E. *JOC* **68**, 6959 (2004).

Strontium.

Barbier reaction.[1] Metallic Sr and RX form organostrontium reagents that add to aldehydes.

[1]Miyoshi, N., Kamiura, K., Oka, H., Kita, A., Kuwata, R., Ikehara, D., Wada, M. *BCSJ* **77**, 341 (2004).

***O*-(*N*-Succinimidyl)-*N,N,N′N′*-tetramethyluronium tetrafluoroborate.**

Dialkyl 3-oxoglutarates.[1] Malonic monoesters undergo self-condensation on treatment with the title reagent (**1**) in the presence of a tertiary amine.

1

[1]Ryu, Y., Scott, A.I. *TL* **44**, 7499 (2003).

Sulfamic acid.

Activation of oxygen compounds. Cyclic ethers undergo acetolysis[1] with HOAc, Ac_2O in the presence of H_2NSO_3H. THP ether formation,[2] Beckmann rearrangement,[3] von Pechmann reaction,[4] as well as imino-Diels-Alder reaction[5] can be promoted by sulfamic acid. Some of these reactions are easy to work up. For example, in a Beckmann rearrangement that is conducted in MeCN dispersion of the reaction mixture in ether precipitates sulfamic acid which can be filtered and reused.

[1]Wang, B., Gu, Y., Gong, W., Kang, Y., Yang, L., Suo, J. *TL* **45**, 6599 (2004).
[2]Wang, B., Yang, L.-M., Suo, J.-S. *SC* **33**, 3929 (2003).
[3]Wang, B., Gu, Y., Luo, C., Yang, L., Suo, J. *TL* **45**, 3369 (2004).
[4]Singh, P.R., Singh, D.U., Samant, S.D. *SL* 1909 (2004).
[5]Nagarajan, R., Magesh, C.J., Perumal, P.T. *S* 69 (2004).

Sulfur. **15**, 297; **18**, 341–342; **20**, 353–354; **22**, 414

Dithio esters. A convenient synthesis of ArC(=S)SR from $ArCH_2Cl$ involves formation of the sulfones $ArCH_2SO_2Ph$ and reaction with sulfur in the presence of *t*-BuOK, followed by addition of an alkyl halide.[1]

[1]Abrunhosa, I., Gulea, M., Masson, S. *S* 928 (2004).

Sulfuric acid.

Fischer indole synthesis.[1] Aqueous H_2SO_4 is a proper reagent for transforming $ArNHNH_2 \cdot HCl$ and cyclic enol ethers into 3-hydroxyalkylindoles in $AcNMe_2$.

N-Acylation. Sulfonamides are acylated by acid anhydrides in the presence of 96% H_2SO_4.[2]

[1]Campos, K.R., Woo, J.C.S., Lee, S., Tillyer, R.D. *OL* **6**, 79 (2004).
[2]Martin, M.T., Roschangar, F., Eaddy, J.F. *TL* **44**, 5461 (2003).

Sulfuryl chloride. **13**, 284; **14**, 291–292; **16**, 311; **18**, 342; **20**, 354; **22**, 414–415

Disulfides. A simple preparation of disulfides from thiols is by treatment with SO_2Cl_2 at room temperature.[1]

Chlorination. Chlorination of phenols by SO_2Cl_2 in nonpolar solvents and in the presence of *t*-butylaminomethyl-polystyrene as catalyst is highly *o*-selective (for phenol: *o/p* ratio is 13:1).[2]

[1]Leino, R., Lonnqvist, J.-E. *TL* **45**, 8489 (2004).
[2]Gnaim, J.M., Sheldon, R.A. *TL* **45**, 8471 (2004).

T

Tantalum(V) chloride. 21, 407; **22**, 416

Allylation.[1] Allylantalum chloride species are evidently involved in the allylation of imines with allylstannane reagent in the presence of $TaCl_5$. Imines react much faster than aldehydes, therefore the reaction can be carried out with a mixture of amines and aldehydes.

Epoxide opening.[2] Silica-supported $TaCl_5$ promotes reaction of *N*-tosylaziridines with arylamines at room temperature.

[1]Shibata, I., Nose, K., Sakamoto, K., Yasuda, M., Baba, A. *JOC* **69**, 2185 (2004).
[2]Chandrasekhar, S., Prakash, S.J., Shyamsunder, T., Ramachandar, T. *SC* **34**, 3865 (2004).

Tellurium.

Cycloaromatization.[1] Enediynes aromatize under mild conditions that involves Te as mediator. Telleracycloheptatrienes are the most probable intermediates.

SiMe$_3$ / SiMe$_3$ — Te; NaOH, N_2H_4, $NaBH_4$, Aliquat464 → [R, Te, R] → 73%

[1]Landis, C.A., Payne, M.M., Eaton, D.L., Anthony, J.E. *JACS* **126**, 1338 (2004).

Terbium(III) triflate.

Friedel-Crafts acylation.[1] 3-Arylpropanoic acids cyclize on heating with catalytic $Tb(OTf)_3$ to afford indanones in moderate to good yields. However, the reaction temperature is 250°.

[1]Cui, D.-M., Zhang, C., Kawamura, M., Shimada, S. *TL* **45**, 1741 (2004).

Tetrabutylammonium bromide. 20, 356; **21**, 407; **22**, 416

Michael reaction. The reaction between stannyl enol ethers and α,β-unsaturated esters is catalyzed by Bu_4NBr.[1] A combination of Bu_4NBr and $BF_3 \cdot OEt_2$ constitutes an effective catalyst system for the conjugate addition of carbamates to enones.[2]

Fiesers' Reagents for Organic Synthesis, Volume 23. Edited by Tse-Lok Ho

[1]Yasuda, M., Chiba, K., Ohigashi, N., Katoh, Y., Baba, A. *JACS* **125**, 7291 (2003).
[2]Xu, L.-W., Li, L., Xia, C.-G., Zhou, S.-L., Li, J.-W., Hu, X.-X. *SL* 2337 (2003).

Tetrabutylammonium fluoride, TBAF. 13, 286–287; **14**, 293–294; **15**, 298, 304; **17**, 324–326; **18**, 344–345; **19**, 319–321; **20**, 357–359; **21**, 407–409; 22, 417–418

Desilylation. Enolate generation from silyl enol ethers and Bu_4NF in the presence of formalin completes hydroxymethylation.[1]

Desilylation occurs after Si-to-C migration of certain silylalkenes on instigation of Bu_4NF.[2,3]

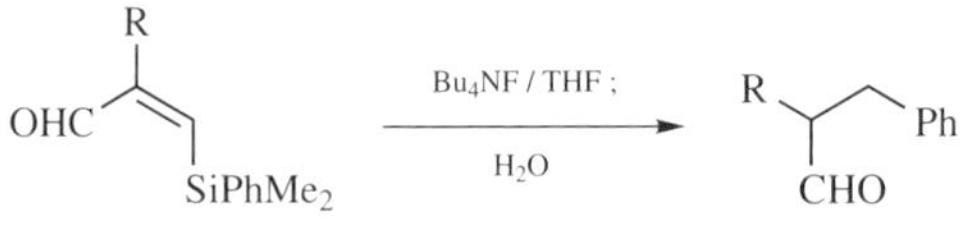

Coupling reactions. Applicability of triallyl(aryl)silanes to coupling with ArCl under Pd catalysis depends on activation. Such is provided by Bu_4NF.[4]

Defunctionalization. Selective cleavage of carbamates by Bu_4NF makes it quite valuable synthetically.[5]

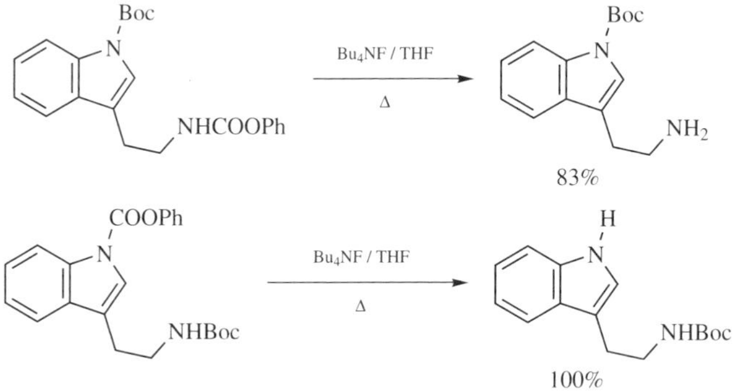

α-Trifluoromethylalkanoic esters. After α-bis(methylthio)methylenation of esters and conversion into α-bromo-α-methanesulfonyldifluoromethyl derivatives by consecutive reactions with BrF_3 and HOF, treatment with Bu_4NF completes addition of another fluorine atom and removal of Br atom and Ms group.[6]

β-Hydroxy-β-allyl carboxylic acids.[7] Treatment with TBAF triggers allyl shift of β-allyldimethylsilyl-α,β-unsaturated carboxylic acid derivatives (an internal Michael reaction). Oxidative removal of the fluorodimethylsilyl group in the products generates the alcohols. The rearrangement step is subject to stereocontrol by a γ-substituent.

Bu_4NF
DMF - H_2O (10:1);
NaOH / H_2O

Cyclization.[8] As a base TBAF induces cyclization of cycloalkanones that bear an alkynoic ester sidechain at the α-position. A different reaction pathway is pursued in the presence of *t*-BuOK.

Bu_4NF
THF 20°
62%
t-BuOK
THF 50°
45%

Fluoropyridines. Replacement of the nitro group in a nitropyridine is achieved by reaction with Bu_4NF in DMF.[9]

[1]Ozasa, N., Wadamoto, M., Ishihara, K., Yamamoto, H. *SL* 2219 (2003).
[2]Aronica, L.A., Raffa, P., Caporusso, A.M., Salvadori, P. *JOC* **68**, 9292 (2003).
[3]Trost, B.M., Ball, Z.T. *JACS* **126**, 13942 (2004).
[4]Sahoo, A.K., Nakao, Y., Hiyama, T. *CL* **33**, 632 (2004).
[5]Jacquemard, U., Beneteau, V., Lefoix, M., Routier, S., Merour, J.-Y., Coudert, G. *T* **60**, 10039 (2004).
[6]Hagooly, A., Rozen, S. *CC* 594 (2004).
[7]Trost, B.M., Ball, Z.T. *JACS* **126**, 13942 (2004).
[8]Klein, A., Miesch, M. *TL* **44**, 4483 (2003).
[9]Kuduk, S.D., DiPardo, R.M., Bock, M.G. *OL* **7**, 577 (2005).

Tetrakis[(1-benzyl-1,2,3-triazol-4-yl)methyl]amine.

1,2,3-Triazoles. The cycloaddition of organoazides with 1-alkynes is catalyzed by Cu(I) species. The title compound stabilizes the Cu(I) and also enhances its catalytic activity.

[1]Chan, T.R., Hilgraf, R., Sharpless, K.B., Fokin, V.V. *OL* **6**, 2853 (2004).

Tetrakis[chloro(pentamethylcyclopentadienyl)ruthenium(I)].

Carroll rearrangement. With bipyridine the title complex forms Cp*Ru(bpy)Cl to promote decomposition of allyl acetoacetates into 5-alken-2-ones.[1]

[1]Burger, E.C., Tunge, J.A. *OL* **6**, 2603 (2004).

Tetrakis(dimethylamino)ethene.

Trifluoromethyl sulfides/selenides. The title reagent reduces CF_3I to give trifluoromethide anion for reaction with PhXXPh (X = S, Se) in DMF.[1] The reaction is very efficient because $[PhX]^-$ that is released can also attack CF_3I to form CF_3XPh.

[1]Pooput, C., Medebielle, M., Dolbier Jr, W.R. *OL* **6**, 301 (2004).

Tetrakis(triphenylphosphine)palladium(0). 13, 289–294; **14**, 295–299; **15**, 300–304; **16**, 317–323; **17**, 327–331; **18**, 347–349; **19**, 324–331; **20**, 362–368; **21**, 411–417; **22**, 419–427

Addition reactions. Splitting ArSSAr and ArSeSeAr with $(Ph_3P)_4Pd$ in the presence of alkynes (no solvent) leads to *cis*-1,2-bis(arylchalcogeno)alkenes.[1]

A synthesis of conjugated dienes uses alkenylboronic acids to add to allenes.[2] Intramolecular addition leads to heterocycles when the substrates are 1-bromoallenes, the intermediary semicyclic π-allylpalladium complexes are susceptible to nucleophilic substitution.[3]

Br
Ts N OH
$(Ph_3P)_4Pd$
NaOMe
OMe
Ts N O
61%

BnO + $(HO)_2B$ Bu
$(Ph_3P)_4Pd$
HOAc - dioxane
BnO Bu
89%

Different patterns are observed for the hydroarylation of alkynes with $ArB(OH)_2$.[4,5] Both electronic and steric effects are responsible for the regioselectivity.

$PhB(OH)_2$ + OH
$(Ph_3P)_4Pd$
HOAc - dioxane
80°
OH
Ph
93%

$PhB(OH)_2$ + R OH
$(Ph_3P)_4Pd$
HOAc - dioxane
80°
Ph OH R
Ph OH R
R = Me 65% R = t-Bu 75%

5,6-Alkadien-1-ynes undergo reductive cyclization (HCOOH to supply hydride) to afford 1,2-dimethylenecyclopentanes by initial formation of vinylpalladium species from reaction at the allene moiety.[6] Acidic hydroxyl compounds (e.g., PhOH, RCOOH) add to conjugated dienes to give allylically oxygenated products.[7]

Substitution reactions. Deblocking of allyl ethers and allyl carbonates by Pd-catalyzed reaction is well known. Solid-supported barbituric acid has been tested as nucleophile.[8]

Allylation using allylic alcohols as electrophiles is assisted by carboxylic acids.[9,10] Also the combination of allylic alcohols with organoboronic acids and elimination of boric acid result in a net displacement of the OH group.[11]

2-Vinylcyclopropane-1,1-diphosphonic esters afford open-chain unsaturated products when react with amines under Pd-catalysis.[12] A facile route to a synthon for strychnos alkaloid synthesis is by intramolecular substitution of alkenyl iodides.[13]

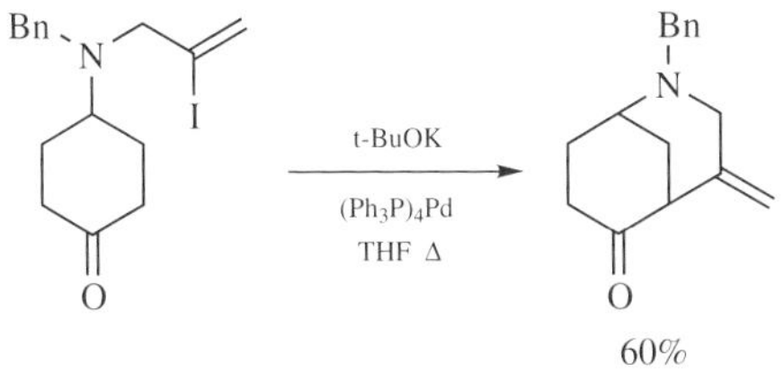

t-BuOK; $(Ph_3P)_4Pd$; THF Δ

60%

Coupling reactions. Homocoupling of 1-iodoalkynes catalyzed by $(Ph_3P)_4Pd$ represents a simple way to access symmetrical 1,3-diynes.[14]

Suzuki coupling of polar substrates in aqueous media has been carried out with $(Ph_3P)_4Pd$ supported on reverse-phase glass beads.[15] For synthesis of 3-arylpyridines, 3-pyridylboroxin serves well as coupling partner.[16] Suzuki coupling of some unusual compounds such as 2-(*N*-[pyrid-1-yl]amino)pyrid-5-yl bromide[17] encounters no problems. Arylation of indoles by the coupling method employs best indolyl bromides and $ArB(OH)_2$ and not the alternative combination of indoleboronic acids and ArBr because the reaction generally gives better yields and is not critically influenced by the *N*-substituent (or free indole).[18]

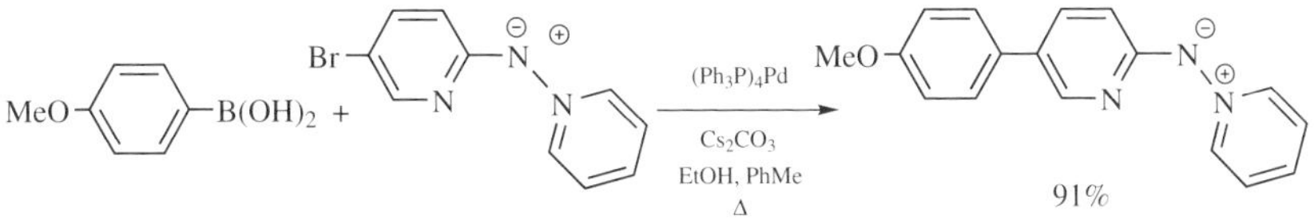

MeO–C$_6$H$_4$–B(OH)$_2$ + Br–pyridyl–N⁻–N⁺-pyridinium; $(Ph_3P)_4Pd$; Cs_2CO_3; EtOH, PhMe; Δ

91%

Halobenzyl halides undergo Suzuki coupling at the side chain first and then the nuclear site.[19] α-Bromo sulfoxides[20] and arenesulfonyl chlorides[21] are found to couple with $ArB(OH)_2$, in the latter reaction SO_2 is eliminated.

cis-1,2-Diarylcyclopropanes are assembled by Suzuki coupling of potassium *trans*-2-arylcyclopropyltrifluoroborates with ArX. The salts are prepared from (*E*)-styrylboronic acid in three steps.[22]

Ph H H BF_3K + Br O; $(Ph_3P)_4Pd$; $K_3PO_4.3H_2O$; H_2O, PhMe; Ph H H O; 88%

Allenyl carbinols and organoboronic acids combine and give off boric acid under the influence of Pd(0) catalyst, generating conjugated dienes.[23]

OH + $B(OH)_2$; $(Ph_3P)_4Pd$; dioxane 80°C; 99%

Negishi coupling of alkenyl halides with alkynylzinc halides as a method for conjugated enyne synthesis gives better yields than the Sonagashira coupling.[24] In this way chain extension of 2,4-pentadienal by alkynyl units is readily achieved using 5-bromo-2,4-pentadienal for coupling.[25] Coupling of lithiated alkynylarenes directly or after Li/Zn exchange affords different substituted allenes.[26]

Ph; base; $(Ph_3P)_4Pd$; PhI; Ph Ph; Ph Ph

base:			
	BuLi	55%	100 : 0
	LDA, $ZnBr_2$	84%	0 : 100

Allenylarenes are conveniently prepared from ArI and $Bu_3SnC{=}C{=}CH_2$ with promotion by $(Ph_3P)_4Pd$ - LiCl.[27] Access to β,γ-unsaturated acids in two steps involves hydrostannylation of 3-butynoic acid and Stille coupling.[28] Introduction of propargyl and allyl groups to the double bond of arylethylidenemalononitriles[29] in one operation is feasible. Similarly, β-styrylgermanes are prepared from an ethynyltriorganogermane via hydrostannylation and Stille coupling.[30]

Ph NC CN + Bu_3Sn + Cl; $(Ph_3P)_4Pd$; PhMe; Ph NC NC; 78%

Heteroaromatic thioethers are useful for Stille coupling with organostannanes in the presence of (salen)CuMe as mediator.[31]

There are demonstrations of successful cross-coupling of organoindium (e.g., Me_4InLi)[32] and organobismuth compounds (e.g., **1**)[33] with ArX.

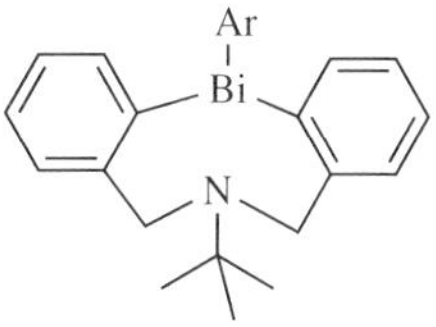

1

For synthesis of functionalized alkenes such as 2-phenylseleno-1,3-dienes, selective coupling of alkenylzirconocenes with 1-phenylseleno-1-bromoalkenes[34] is probably the method of choice. α-Fluorostyrenes are also conveniently prepared from 1-fluoro-1-triorganosilylethenes (with CsF as an essential additive to generate reactive coupling agents).[35]

Well-designed cyclization to accompany coupling reactions is synthetically attractive. An example is the reaction of 1,2,7-trienes with ArI that gives either monocyclic, bicyclic[36] or tricyclic products under somewhat varied conditions.[37]

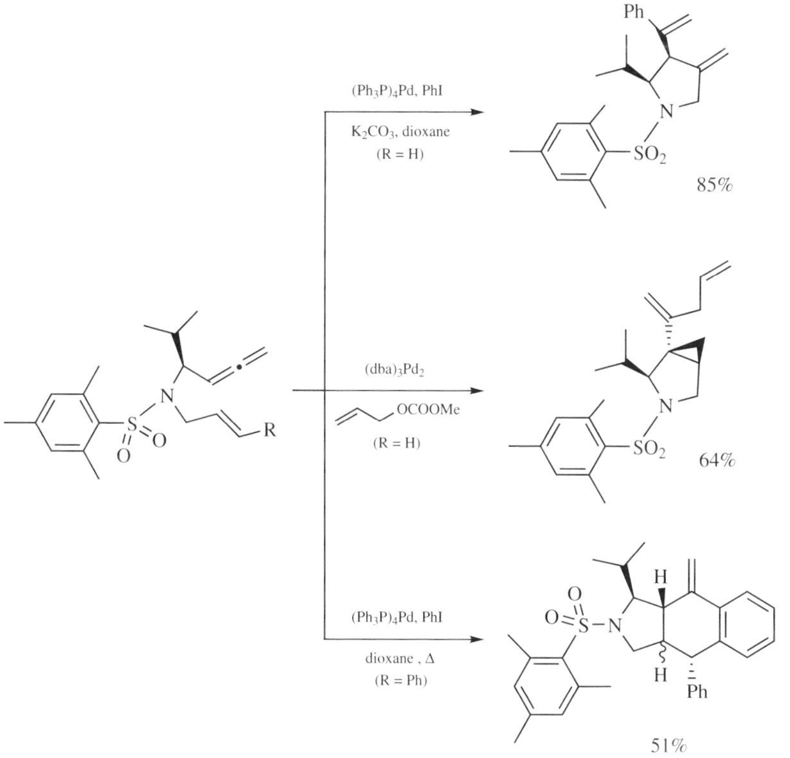

$(Ph_3P)_4Pd$, PhI
K_2CO_3, dioxane
(R = H)
85%

$(dba)_3Pd_2$
OCOOMe
(R = H)
64%

$(Ph_3P)_4Pd$, PhI
dioxane , Δ
(R = Ph)
51%

Suzuki coupling also delivers five-membered rings containing two adjacent unsaturated side chains that are derived from allyl chloride and alkyne/allene units.[38] Closure to give cyclopropanes is also observed in the reaction of allenylmethylmalonic esters.[39]

$PhB(OH)_2$ + [Bu-alkyne / O / Cl substrate] —$(Ph_3P)_4Pd$, KF / PhMe, 25°→ [Bu, Ph tetrahydrofuran product] 83%

$ArB(OH)_2$ + [allene / X / Cl substrate] —$(Ph_3P)_4Pd$, K_3PO_4→ [Ar, H, H, X product]

PhI + [C_6H_{13} allene, MeOOC, COOMe substrate] —$(Ph_3P)_4Pd$, Bu_4NBr, MeCN→ [Ph, H, C_6H_{13}, MeOOC, COOMe cyclopropane] 93% (*trans/cis* 94:6)

2,3-Alkadienyl amines in which the allenyl moiety has a terminal bromine atom undergo cyclization to afford aziridines.[40] If the amino group is tertiary and one of the substituents contains active hydrogen (OH or NH) a larger ring may be created.

[Ts, N, Br, OH substrate] —$(Ph_3P)_4Pd$, NaOMe→ [Ts, N, O, OMe product] 73%

Reductive cyclization occurs when haloarenes bearing an unsaturated side chain in the ortho position are treated with $(Ph_3P)_4Pd$ in the presence of Et_3SiH.[41] Thus, 2-(3-butenyl)phenyl bromide affords 1-methyleneindane.

Tricycle containing a cyclobutenediol moiety is the product from a remarkable Stille cross-coupling.[42]

[HO, HO, $SiMe_3$ substrate] + [vinyl $SnBu_3$] —$(Ph_3P)_4Pd$, PhH Δ→ [HO, OH, $SiMe_3$, H tricycle] 62%

Acylation and allylation. Chalcones are obtained from either combination of ArCOCl and ArCH=CHB$(OH)_2$ or ArB$(OH)_2$ and ArCH=CHCOCl.[43]

N-Cinnamylation of arylamines can use 1-arylpropynes in a catalytic system of $(Ph_3P)_4Pd$ – PhCOOH.[44] It involves reversible hydropalladation to generate allenylarenes which are converted into π-allylpalladium species for further reaction. Since allyl alcohols are activated by $[(PhO)_3P]_4Pd$ in forming π-allyl(hydroxo)-palladium species allylation of alcohols with the system is readily realized.[45]

2-Pyridylpyrroles.[46] 2-Methyleneaziridines and acetylpyridines (acetyl group at any position) are condensed under the influence of $(Ph_3P)_4Pd$. Formation of $PyCOCH_2PdH$ to hydropalladate the exocyclic methylene group of the aziridines is implicated. Reductive elimination of the adducts is followed by isomerization (ring expansion) and dehydration.

$(Ph_3P)_4Pd$, 120°

[1]Ananikov, V.P., Beletskaya, I.P. *OBC* **2**, 284 (2004).
[2]Oh, C.H., Ahn, T.W., Reddy, R. *CC* 2622 (2003).
[3]Ohno, H., Hamaguchi, H., Ohata, M., Kosaka, S., Tanaka, T. *JACS* **126**, 8744 (2004).
[4]Oh, C.H., Jung, H.H., Kim, K.S., Kim, N. *ACIEE* **42**, 805 (2003).
[5]Kim, N., Kim, K.S., Gupta, A.K., Oh, C.H. *CC* 618 (2004).
[6]Oh, C.H., Jung, S.H., Park, D.I., Choi, J.H. *TL* **45**, 2499 (2004).
[7]Utsunomiya, M., Kawatsura, M., Hartwig, J.F. *ACIEE* **42**, 5865 (2003).
[8]Tsukamoto, H., Suzuki, T., Kondo, Y. *SL* 1101 (2003).
[9]Manabe, K., Kobayashi, S. *OL* **5**, 3241 (2003).
[10]Patil, N.T., Yamamoto, Y. *TL* **45**, 3101 (2004).
[11]Tsukamoto, H., Sato, M., Kondo, Y. *CC* 1200 (2004).
[12]Moreau, P., Maffei, M. *TL* **45**, 743 (2004).
[13]Sole, D., Diaba, F., Bonjoch, J. *JOC* **68**, 5746 (2003).
[14]Damle, S.V., Seomoon, D., Lee, P.H. *JOC* **68**, 7085 (2003).
[15]Daku, K.M.L., Newton, R.F., Pearce, S.P., Vile, J., Williams, J.M.J. *TL* **44**, 5095 (2003).
[16]Cioffi, C.L., Spencer, W.T., Richards, J.T., Herr, R.J. *JOC* **69**, 2210 (2004).
[17]Reyes, M.J., Izquierdo, M.L., Alvarez-Builla, J. *TL* **45**, 8713 (2004).
[18]Prieto, M., Zurita, E., Rosa, E., Munoz, L., Llyod-Williams, P., Giralt, E. *JOC* **69**, 6812 (2004).
[19]Langle, S., Abarbri, M., Duchene, A. *TL* **44**, 9255 (2003).
[20]Rodriguez, N., Cuenca, A., de Arellano, C.R., Medio-Simon, M., Asensio, G. *OL* **5**, 1705 (2003).
[21]Dubbaka, S.R., Vogel, P. *OL* **6**, 95 (2004).
[22]Fang, G.-H., Yan, Z.-J., Deng, M.-Z. *OL* **6**, 357 (2004).
[23]Yoshida, M., Gotou, T., Ihara, M. *CC* 1124 (2004).
[24]Negishi, E., Qian, M., Zeng, F., Anastasia, L., Babinski, D. *OL* **5**, 1597 (2003).
[25]Vicart, N., Saboukoulou, G.-S., Ramondenc, Y., Ple, G. *SC* **33**, 1509 (2003).
[26]Ma, S., He, Q. *ACIEE* **43**, 988 (2004).

[27]Huang, C.-W., Shanmugasundaram, M., Chang, H.-M., Cheng, C.-H. *T* **59**, 3635 (2003).
[28]Thibonnet, J., Abarbri, M., Parrain, J.-L., Duchene, A. *T* **59**, 4433 (2003).
[29]Jeganmohan, M., Shanmugasundaram, M., Cheng, C.-H. *JOC* **69**, 4053 (2004).
[30]David-Quillot, F., Marsacq, D., Balland, A., Thibonnet, J., Abarbri, M., Duchene, A. *S* 448 (2003).
[31]Egi, M., Liebeskind, L.S. *OL* **5**, 801 (2003).
[32]Lee, P.H., Lee, S. W., Seomoon, D. *OL* **5**, 4963 (2003).
[33]Shimada, S., Yamazaki, O., Tanaka, T., Rao, M.L.N., Suzuki, Y., Tanaka, M. *ACIEE* **42**, 1845 (2003).
[34]Cai, M.-Z., Huang, J.-D., Peng, C.-Y. *JOMC* **681**, 98 (2003).
[35]Hanamoto, T., Kobayashi, T. *JOC* **68**, 6354 (2003).
[36]Ohno, H., Takeoka, Y., Kadoh, Y., Miyamura, K., Tanaka, T. *JOC* **69**, 4541 (2004).
[37]Ohno, H., Miyamura, K., Takeoka, Y., Tanaka, T. *ACIEE* **42**, 2647 (2003).
[38]Zhu, G., Zhang, Z. *OL* **5**, 3645 (2003); **6**, 4041 (2004).
[39]Ma, S., Jiao, N., Yang, Q., Zheng, Z. *JOC* **69**, 6463 (2004).
[40]Ohno, H., Hamaguchi, H., Ohata, M., Tanaka, T. *ACIEE* **42**, 1749 (2003).
[41]Oh, C.H., Park, S.J. *TL* **44**, 3785 (2003).
[42]Salem, B., Klotz, P., Suffert, J. *OL* **5**, 845 (2003).
[43]Eddarir, S., Cotelle, N., Bakkour, Y., Rolando, C. *TL* **44**, 5359 (2003).
[44]Patil, N.T., Wu, H., Kadota, I., Yamamoto, Y. *JOC* **69**, 8745 (2004).
[45]Kayaki, Y., Koda, T., Ikariya, T. *JOC* **69**, 2595 (2004).
[46]Siriwardana, A.I., Kathriarachchi, K.K.A.D.S., Nakamura, I., Gridnev, I.D., Yamamoto, Y. *JACS* **126**, 13898 (2004).

Tetrakis(triphenylphosphine)palladium(0) – carbon monoxide. 22, 427

Carbonylation. Carbonylative coupling between ArI and triorganoindium reagents under CO is promoted by $(Ph_3P)_4Pd$.[1] *N,N*-Disubstituted ureas are electrochemically produced[2] while under the influence of the Pd(0) catalyst.

Enones and Me_3SiCl form π-(siloxyallyl)palladium complexes which react with CO to give acylpalladium species for coupling with RZnX that results in 1,4-diketones.[3] For preparation of 3-phenylseleno-2-alkenamides by the Pd-catalyzed reaction their components are acquired from PhSeSePh, 1-alkynes and $PhSNR_2$.[4]

+ $PhS\text{-}NMe_2$
+ PhSe-SePh
$(Ph_3P)_4Pd$
CO / PhH
PhSe O
NMe_2
55%

Carbonylative Stille coupling of sterically hindered reactants is improved by adding a Cu(I) salt.[5]

OTf
+
Bu_3Sn
$SiMe_3$
$(Ph_3P)_4Pd$
LiCl, CuBr
CO, THF
65°C
O
$SiMe_3$
90%

[1]Pena, M.A., Sestelo, J.P., Sarandeses, L.A. *S* 780 (2003).
[2]Chiarotto, I., Feroci, M. *JOC* **68**, 7137 (2003).
[3]Yuguchi, M., Tokuda, M., Orito, K. *JOC* **69**, 908 (2004).
[4]Knapton, D.J., Meyer, T.Y. *OL* **6**, 687 (2004).
[5]Mazzola, R.D., Giese, S., Benson, C.L., West, F.G. *JOC* **69**, 220 (2004).

Tetrakis(triphenylphosphine)palladium(0) – copper(I) salt. 18, 349–350; **20**, 369; **21**, 417; **22**, 427–428

Coupling reactions. There is solvent and temperature dependence on the regioslectivity in Stille coupling of 3,5-dibromo-2-pyrone.[1]

+ $PhSnBu_3$ —($(Ph_3P)_4Pd$, CuI)→ +

PhMe, 100°	94%	–
DMF, 50°	–	75%

Copper(I) salts and fluoride ion have synergistic effect on Stille coupling.[2] For the synthesis of 1,1-diarylethenes by Stille coupling it is advisable to use β-trimethylsilyl-α-tributylstannylstyrenes as reaction partners to avoid *cine*-substitution.[3]

Ketones are synthesized from thiol esters and *B*-alkyl-9-BBN in the Pd-catalyzed reaction and Cu(I) 2-thiophenecarboxylate as mediator.[4] Since hydrostannylation of 2-alkynoic esters introduces a tin residue at C-2 and subsequent addition of organohalogen compounds completes synthesis of enones (also more complex structures such as the cross-conjugated dienones).[5]

O, Br + Bu_3Sn COOMe —($(Ph_3P)_4Pd$, CuCl)→ O, COOMe 61%

N-Methoxycarbonyl-3-allylindoles are assembled from 2-(alkynyl)aryl isocyanates and allyl methyl carbonate.[6] The catalyst system derives activity from Lewis acidity and transition metallic properties.

A very satisfactory application of the Stille coupling of alkenyl halides with Bu_3SnCN is in a total synthesis of borrelidin.[7]

[1]Kim, W.-S., Kim, H.-J., Cho, C.-G. *JACS* **125**, 14288 (2003).
[2]Mee, S.P.H., Lee, V., Baldwin, J.E. *ACIEE* **43**, 1132 (2004).
[3]Belema, M., Nguyen, V.N., Zusi, F.C. *TL* **45**, 1693 (2004).

[4]Yu, Y., Liebeskind, L.S. *JOC* **69**, 3554 (2004).
[5]Kerr, D.J., Metje, C., Flynn, B.L. *CC* 1380 (2003).
[6]Kamijo, S., Yamamoto, Y. *JOC* **68**, 4764 (2003).
[7]Duffey, M.O., LeTiran, A., Morken, J.P. *JACS* **125**, 1458 (2003).

Tetrakis(triphenylphosphine)platinum(0). 20, 369–370; **21**, 417–418; **22**, 429–430

Addition. Splitting of selenoesters by $(Ph_3P)_4Pt$ provides components for carboselenation of alkynes, a reaction it also catalyzes.[1] Thiocarbonylation of 1-alkynes with RSH and CO produces α-alkylacrylic thiolesters.[2]

C_6H_{13}–C≡CH + PhSe–C(=O)–Ph → [$(Ph_3P)_4Pt$, PhMe Δ] → C_6H_{13}(PhSe)C=CH–Ph 89%

C_6H_{13}–C≡CH + PhSH + CO → [$(Ph_3P)_4Pt$, MeCN 120°] → C_6H_{113}C(=CH_2)–C(=O)SPh 83%

[1]Hirai, T., Kuniyasu, H., Kato, T., Kurata, Y., Kambe, N. *OL* **5**, 3871 (2003).
[2]Kawakami, J., Mihara, M., Kamiya, I., Takeba, M., Ogawa, A., Sonoda, N. *T* **59**, 35219 (2003).

***N,N,N′,N′*-Tetramethylfluoroformamidinium hexafluorophosphate.**

Acylation.[1] For acylation of alcohols, thiols, and dithiocarbamates with carboxylic acids, $FC[(NMe_2)_2]PF_6$ is an excellent mediator.

[1]Pittelkow, M., Kamounah, F.S., Boas, U., Pesersen, B., Christensen, J.B. *S* 2485 (2004).

***N,N,N′,N′*-Tetramethylguanidine. 22**, 430

Deprotection.[1] Aryl silyl ethers and acetates can be cleaved using the title reagent.

[1]Oyama, K., Kondo, T. *OL* **5**, 209 (2003).

Thiourea.

Halohydrins. Thiourea catalyzes opening of epoxides by halogens.

[1]Sharghi, H., Eskandari, M.M. *T* **59**, 8509 (2003).

Tin. 13, 298; **17**, 333–334; **18**, 352; **20**, 372–373; **21**, 421; **22**,

Allylation. Barbier-type reaction of allyl bromides with aldehydes in water is mediated by Sn (or Zn) and $NaBF_4$[1] or nanoparticles of Sn.[2]

[1]Zha, Z., Xie, Z., Zhou, C., Chang, M., Wang, Z. *NJC* **27**, 1297 (2003).
[2]Zha, Z., Qiao, S., Jiang, J., Wang, Y., Miao, Q., Wang, Z. *T* **61**, 2521 (2005).

Tin(II) chloride. **13**, 298–299; **15**, 309–310; **16**, 329; **18**, 353–354; **19**, 337–338; **20**, 373; **21**, 422–423; **22**, 430–431

Allylations. With $SnCl_2$ to promote allylation in water, variation of conditions includes application of ultrasound,[1] combination with other metal halides ($TiCl_3$,[2] $PdCl_2$[3]).

Acetonides. $SnCl_2$ catalyzes the insertion of acetone into an epoxide ring.[4]

Quinolines. Reduction followed by condensation converts a mixture of 2-formylnitroarenes and ketones into 2,3-disubstituted quinolines. The overall transformation is accomplished by $SnCl_2$ and $ZnCl_2$.[5]

[1]Wang, J., Yuan, G., Dong, C.-Q. *CL* **33**, 286 (2004).[4]
[2]Tan, X.-H., Shen, B., Deng, W., Zhao, H., Liu, L., Guo, Q.-X. *OL* **5**, 1833 (2003).
[3]Tan, X.-H., Hou, Y.-Q., Shen, B., Liu, L., Guo, Q.-X. *TL* **45**, 5525 (2004).
[4]Vyvyan, J.R., Meyer, J.A., Meyer, K.D. *JOC* **68**, 9144 (2003).
[5]McNaughton, B.R., Miller, B.L. *OL* **5**, 4257 (2003).

Tin(IV) chloride. **13**, 300–301; **14**, 304–306; **15**, 311–313; **17**, 335–340; **18**, 354–356; **19**, 338–339; **20**, 373–375; **21**, 422–423; **22**, 431–433

N-Alkylation. Chlorosulfonium salts derived from reaction of sulfides with NCS undergo ionization on treatment with $SnCl_4$. Attack by a well-positioned azide results in N—C bond formation and subsequent decomposition gives imino sulfides.[1]

N_3 N COOMe SPh —NCS, $SnCl_4$, 0°→ [H COOMe N N N_2^+ H SPh] → H COOMe N N H SPh 49%

Allylation.[2] Allyl bromides and mesylates are converted into allyltrichlorostannanes for the Barbier-type reaction with carbonyl compounds using a combination of $SnCl_4$ and Bu_4NI. The system does not reduce aldehydes like $TiCl_4$ - Bu_4NI.

[1]Magnus, P., Matthews, K.S., Lynch, V. *OL* **5**, 2181 (2003).
[2]Masuyama, Y., Suga, T., Watabe, A., Kurusu, Y. *TL* **44**, 2845 (2003).

Tin(II) triflate. **13**, 301–302; **14**, 306–307; **15**, 313–314; **17**, 341–344; **18**, 357–358; **19**, 340; **20**, 376; **21**, 424; **22**, 433–434

Deprotection.[1] Selective removal of the *N*-Boc group is accomplished using $Sn(OTf)_2$.[1] *O*-Benzyl carbamates are stable.

[1]Bose, D.S., Kumar, K.K., Reddy, A.V.N. *SC* **33**, 445 (2003).

Titanium(IV) bromide. 22, 434

4-Bromotetrahydropyrans. Homoallyl vinyl ethers undergo condensation with carbonyl compounds to form *cis*-2-(β-hydroxyalkyl)-4-bromotetrahydropyrans in good yields.[1,2]

$TiBr_4$ CH_2Cl_2 –78°

79%

[1]Patterson, B., Marumoto, S., Rychnovsky, S.D. *OL* **5**, 3163 (2003).
[2]Patterson, B., Rychnovsky, S.D. *SL* 543 (2004).

Titanium(IV) chloride. 13, 304–309; **14**, 309–311; **15**, 317–320; **16**, 332–337; **17**, 344–347; **18**, 359–361; **19**, 341–344; **20**, 377–379; **21**, 425–427; **22**, 434–439

Addition reactions. To prepare imines/enamines from alkynes the hydroamination is catalyzed by $TiCl_4$.[1] 1,2,3-Trisubstituted indoles are acquired from arylhydrazines and alkynes via hydrohydrazination. The adducts tautomerize to give arylhydrazones, which undergo indolization.[2]

Hydroamination of norbornene with arylamines (including fluoroaniline, cyanoaniline) proceeds by heating the reaction components with $TiCl_4$ in toluene.[3]

Intramolecular Prins reaction involving an alkenylcyclopropane moiety leads to homoallylic chlorides due to cleavage of the three-membered ring.[4]

$TiCl_4$ CH_2Cl_2 –78° → 0°

74%

Condensations. Converting α-amino acid methyl ester HCl salts to arylimines without racemization the promotion by $TiCl_4$ – Et_3N is effective.[5] A condensation route to certain functionalized α-amino acid derivatives involves catalysis by $TiCl_4$.[6]

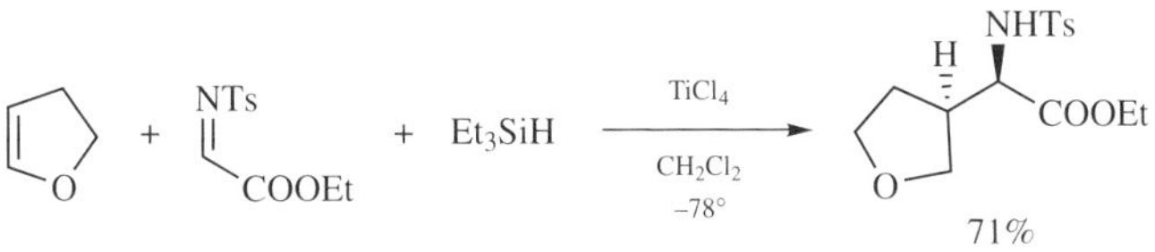

NTs, COOEt + Et_3SiH → ($TiCl_4$, CH_2Cl_2, –78°) NHTs, H, COOEt, O

71%

By the Mukaiyama-aldol reaction catalyzed by $TiCl_4$ β-hydroxy-α,α-difluoroacylsilanes are readily synthesized.[7] Double aldol reaction of 1,3-bis(trimethylsiloxy)-1,3-dienes with nonenolizable 1,3-diketones leads to cross-conjugated cyclohexadienones or phenols.[8] A new catalyst system for the Baylis-Hillman reaction contains $TiCl_4$ and bridged compound **1**.[9]

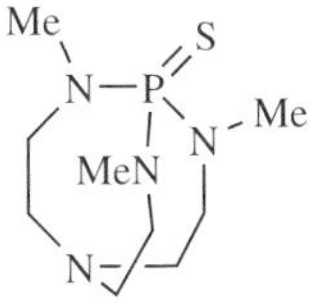

Me, S, N—P, N - Me, MeN, N

1

Condensation of 2-trimethylsilylcyclopropyl ketones with aldehydes affords aldols that bear an allyl group at the α-carbon,[10] due to fragmentative generation of titanium enolates. In the presence of alkynes instead of aldehydes the reaction products are cyclopentenes.[11]

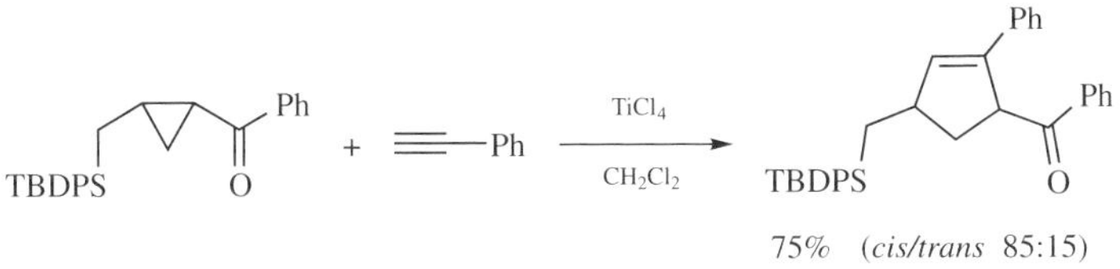

TBDPS, O, Ph + ≡—Ph → ($TiCl_4$, CH_2Cl_2) TBDPS, Ph, Ph, O

75% (*cis/trans* 85:15)

Baylis-Hillman reaction is speeded up by both a detachable Michael donor (e.g., Me_2S) and a Lewis acid ($TiCl_4$). Synthesis of 4-chloro-3-hydroxyalkyl-3-buten-2-ones takes full advantage of the combination.[12]

Titanium enolates are formed when α-bromo thiol esters are treated with a mixture of $TiCl_4$ and Ph_3P. These enolates condense with unsubstituted thiol esters to furnish β-keto thiol esters.[13] To suppress self-condensation bulky thiol esters (e.g., 2,4,6-triisopropylphenylthio esters) of the α-bromo acids are employed as donors.

The Ugi reaction involving aromatic aldehydes is generally too slow and acceleration by $TiCl_4$/MeOH is a useful observation.[14]

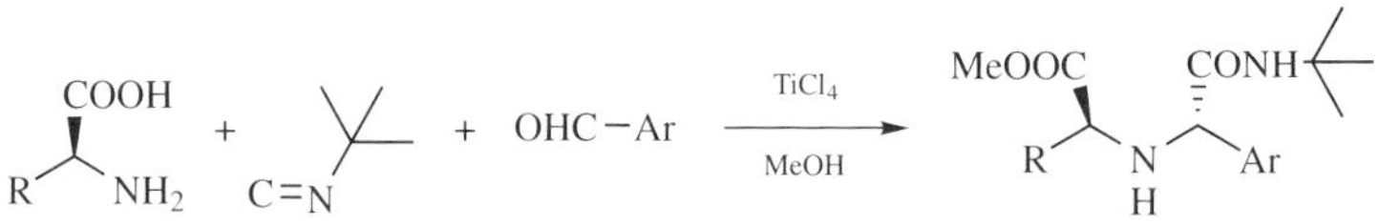

Friedel-Crafts reactions. Promoted by $TiCl_4$ phenols are o-acylated with RCOCl in the neat.[15] A key step in the synthesis of olivacine is Friedel-Crafts acylation of 2,4,6-trimethoxypyridine with *N*-benzylindole-2,3-dicarboxylic anhydride.[16]

The apparent *p*-alkylation of $ArNR_2$ by arylacetic esters actually is due to complexation of the amino group by $TiCl_4$ such that the aromatic ring is susceptible to attack by the titanium enolates of the esters.[17]

Ring closure. 1,3-Dithietan-2-ones are formed unexpectedly from α-pivaloxyalkyl xanthates [RCH(OPiv)SC(=S)OEt] when treated with $TiCl_4$.[18]

Intramolecular oxidative diaryl coupling leading to substituted dihydrodibenzo[*a,c*]cycloheptenes is efficient when the 1,3-diarylpropanes are exposed to a mixture of $TiCl_4$ and $MoCl_5$.[19]

Allylation. Different stereoselectivities are manifested in the allylation with tin reagents[20] as catalyzed by $TiCl_4$ or $BF_3 \cdot OEt_2$.

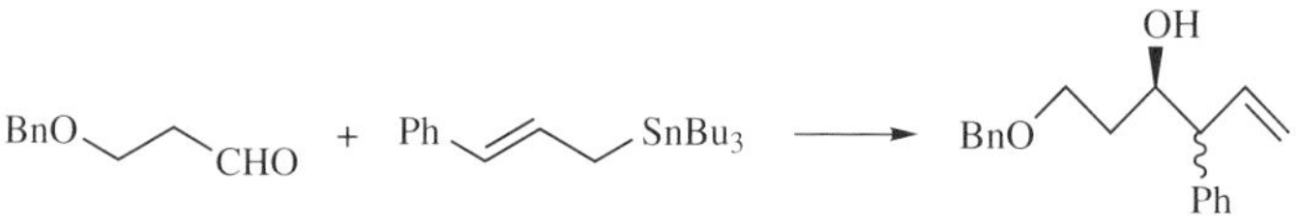

$BF_3 \cdot OEt_2$	55%	syn/anti 96 : 4
$TiCl_4$	100%	*syn/anti* 4 : 96

[1]Ackermann, L. *OM* **22**, 4367 (2003).
[2]Ackermann, L., Born, R. *TL* **45**, 9541 (2004).
[3]Ackermann, L., Kaspar, L.T., Gschrei, C.J. *OL* **6**, 2515 (2004).
[4]Yu, C.-M., Yoon, S.-K., Hong, Y.-T., Kim, J. *CC* 1840 (2004).
[5]Leggio, A., Le Pera, A., Liguori, A., Napoli, A., Romeo, C., Siciliano, C., Sindona, G. *SC* **33**, 4331 (2003).
[6]Ghosh, A.K., Xu, C.-X., Kulkarni, S.S., Wink, D. *OL* **7**, 7 (2005).
[7]Chung, W.J., Ngo, S.C., Higashiya, S., Welch, J.T. *TL* **45**, 5403 (2004).
[8]Langer, P., Bose, G. *ACIEE* **42**, 4033(2003).
[9]You, J., Xu, J., Verkade, J.G. *ACIEE* **42**, 5054 (2003).
[10]Yadav, V.K., Balamurugan, R. *OL* **5**, 4281 (2003).
[11]Yadav, V.K., Sriramurthy, V. *ACIEE* **43**, 2669 (2004).
[12]Kinoshita, S., Kinoshita, H., Iwamura, T., Watanabe, S., Kataoka, T. *CEJ* **9**, 1496 (2003).
[13]Hashimoto, Y., Konishi, H., Kikuchi, S. *SL* 1264 (2004).
[14]Godet, T., Bonvin, Y., Vincent, G., Merle, D., Thozet, A., Ciufolini, M.A. *OL* **6**, 3281 (2004).
[15]Bensari, A., Zaveri, N.T. *S* 267 (2003).
[16]Miki, Y., Tsuzaki, Y., Hibino, H., Aoki, Y. *SL* 2206 (2004).
[17]Periasamy, M., KishoreBabu, N., Jayakumar, K.N. *TL* **44**, 8939 (2003).

[18]Quiclet-Sire, B., Sanchez-Jimenez, G., Zard, S.Z. *CC* 1408 (2003).
[19]Kramer, B., Waldvogel, S.R. *ACIEE* **43**, 2446 (2004).
[20]Nishigaichi, Y., Takuwa, A. *TL* **44**, 1405 (2003).

Titanium(IV) chloride – indium. 22, 440

Reduction of sulfoxides.[1] Various types of sulfoxides (diaryl, dialkyl, alkyl/aryl) are reduced by the reagent system in THF at room temperature.

[1]Yoo, B.W., Choi, K.H., Kim, D.Y., Choi, K.I., Kim, J.H. *SC* **33**, 53 (2003).

Titanium(IV) chloride - zinc. 20, 381; **21**, 430; **22**, 440

Quinazolin-4-(3H)-ones.[1] Reduction of 2-nitrobenzamides by the $TiCl_4$-Zn system in the presence of $HC(OEt)_3$ gives the heterocycles directly.

[1]Shi, D., Rong, L., Wang, J., Zhuang, Q., Wang, X., Hu, H. *SC* **34**, 1759 (2004).

Titanium(IV) iodide. 21, 430–431; **22**, 441–442

α-Imino ketones. Reductive coupling of acyl chlorides and alkanonitriles occurs on treatment with TiI_4.[1] The reaction is best quenched with Ac_2O.

[1]Shimizu, M., Manabe, N., Goto, H. *CL* **32**, 1088 (2003).

Titanium tetraisopropoxide. 13, 311–313; **14**, 311–312; **15**, 322; **16**, 339; **17**, 347–348; **18**, 363–364; **19**, 346–347; **20**, 381–382; **21**, 431; **22**, 442–443

Epoxidation. 2-*endo*-Furyl-2-*exo*-hydroperoxobornane has been investigated for its ability to epoxidize allylic alcohols by catalysis with $(i\text{-PrO})_4Ti$. Results are not encouraging (ee 24-46%).[1]

Amides. Chlorotitanium amide/imide [i.e., $Cl_2TiN(SiMe_3)_2$, $ClTi{=}NSiMe_3$] are generated when nitrogen is absorbed into low-valent titanium complexes derived from $(i\text{-PrO})_4Ti$, Li, and Me_3SiCl. They react with acylpalladium species to form amides.[2]

CO
$Pd(OAc)_2$ / $t\text{-}Bu_3P$
K_2CO_3, HMPA
$ClTi{=}NSiMe_3$
$Cl_2TiN(SiMe_3)_2$
$(i\text{-PrO})_4Ti$ + Li + Me_3SiCl
59%

Amines. For imine formation $(i\text{-PrO})_4Ti$ is easier to handle then $TiCl_4$. For example, reaction of carbonyl compounds with RNH_2 or NH_3 can be carried out in an alcoholic solvent.[3,4] Generally the imines are reduced immediately with $NaBH_4$ for preparation of amines.

1,4-Cyclohexadien-3-yltitanium triisopropoxide.[5] Lithiation of 1,4-cyclohexadiene with *s*-BuLi-TMEDA followed by treatment with $(i\text{-PrO})_4Ti$ gives the bisallylic titanium compound which reacts with aldehydes stereoselectively.

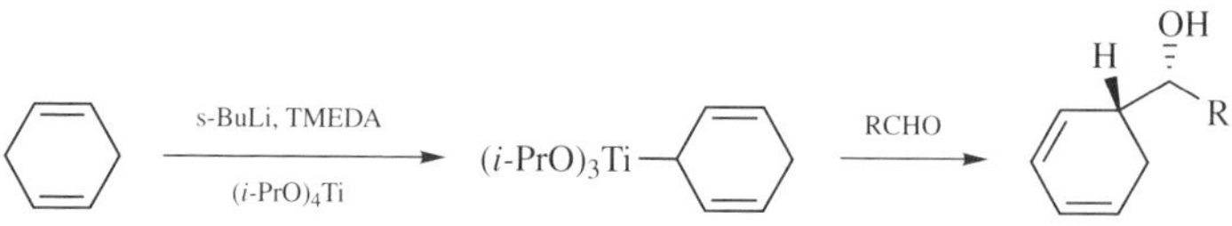

[1]Lattanzi, A., Iannece, P., Vicinanza, A., Scettri, A. *CC* 1440 (2003).
[2]Ueda, K., Mori, M. *TL* **45**, 2907 (2004).
[3]Kumpaty, H.J., Bhattacharyya, S., Rehr, E.W., Gonzalez, A. *S* 2206 (2003).
[4]Miriyala, B., Bhattacharyya, S., Williamson, J.S. *T* **60**, 1463 (2004).
[5]Schleth, F., Studer, A. *ACIEE* **43**, 313 (2004).

Titanocene bis(triethyl phosphite). 20, 383; **21**, 432–433; **22**, 443–444

Alkenation. Based on carbonyl alkenation mediated by titanocene bis(triethyl phosphite), enol ethers are synthesized from bis(phenylthio)methyltributylstannane by conversion into *O,S,S*-orthoformates and the reaction with carbonyl compounds.[1] A more direct method involves condensation of carbonyl compounds with chloromethyl ethers by the same reagent.[2]

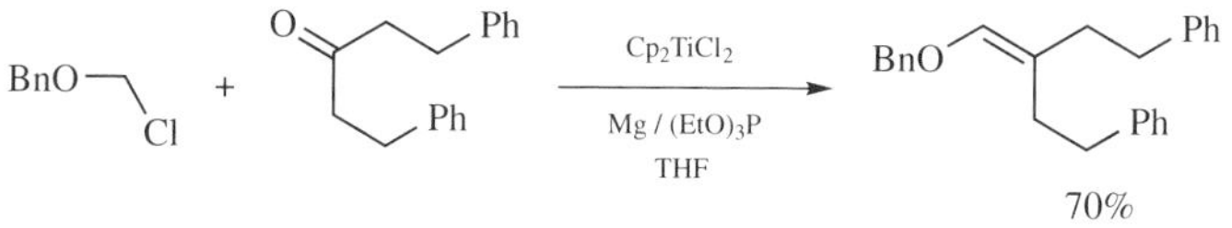

N-***Arylpyrrolidines.*** Reductive coupling of *N*-[3,3-bis(phenylthio)propyl]anilides occurs to give 2-substituted *N*-arylpyrrolidines in one step using the titanocene reagent.[3]

[1]Takeda, T., Sato, K., Tsubouchi, A. *S* 1457 (2004).
[2]Takeda, T., Shono, T., Ito, K., Sasaki, H., Tsubouchi, A. *TL* **44**, 7897 (2003).
[3]Takeda, T., Saito, J., Tsubouchi, A. *TL* **44**, 5571 (2003).

Titanocene dichloride – magnesium. 22, 445

2-Substituted benzofurans.[1] Synthesis of these compounds from 2-(2-trimethylsiloxyaryl)-1,3-dithianes involves conversion to titanocene ylides with Cp_2TiCl_2–Mg and then reaction with esters. Indole derivatives are similarly prepared.

[1]Macleod, C., McKiernan, G.J., Guthrie, E.J., Farrugia, L.J., Hamprecht, D.W., Macritchie, J., Hartley, R.C. *JOC* **68**, 387 (2003).

Titanocene dichloride - manganese. 20, 384; **21**, 434; **22**, 445

Reformatsky reaction.[1] Good yields of β-hydroxy esters are obtained using Cp_2TiCl_2–Mn as condensing agent instead of Zn.

Cyclization. Iodohydrin allyl ethers undergo intramolecular radical addition to give tetrahydrofuran derivatives[2] on treatment with Cp_2TiCl_2 – Mn and Me_3SiCl. The same reagent system catalyzes cyclization of epoxypolyprenes such that carbon skeletons corresponding to natural terpenes are assembled in reasonable yields.

Carbon radicals are also generated from cleavage of epoxides, and an ensuing addition to a properly distanced double bond eventually leads to formation of a tetrahydrofuran ring.[4]

EtOOC, EtOOC, O — Cp_2TiCl_2 - Zn / collidine-HCl, THF → EtOOC, EtOOC, H, O, H, H, H — 63%

[1]Parrish, J.D., Shelton, D.R., Little, R.D. *OL* **5**, 3615 (2003).
[2]Zhou, L., Hirao, T. *JOC* **68**, 1633 (2003).
[3]Justicia, J., Rosales, A., Bunuel, E., Oller-Lopez, J.L., Valdivia, M., Haidour, A., Oltra, J.E., Barrero, A.F., Cardenas, D.J., Cuerva, J.M. *CEJ* **10**, 1778 (2004).
[4]Gansäuer, A., Rinker, B., Pierobon, M., Grimme, S., Gerenkamp, M., Mück-Lichtenfeld, C. *ACIEE* **42**, 3687 (2003).

Titanocene dichloride - zinc. 20, 384–385; **21**, 435; **22**, 445–446

Allylation. Allyltitanium reagents derived from allyl bromide and the title reagent react with aldehydes readily, and chemoselectively in the presence of ketones.[1]

Cyclization. Intramolecular radical addition from α-bromo-β-propargyloxy esters affords 4-methylenetetrahydrofuran-3-carboxylic esters.[2]

δ,ε-Epoxy-α,β-unsaturated esters and amides suffer C–O bond cleavage but the carbon radicals add to the double bond to form cyclopropane derivatives. The homologous compounds give cyclobutanes.[3]

O, $CONMe_2$ — Cp_2TiCl_2 - Zn / collidine-HCl, THF → OH, $CONMe_2$ — 88%

[1]Jana, S., Guin, C., Roy, S.C. *TL* **45**, 6575 (2004).
[2]Jana, S., Guin, C., Roy, S.C. *TL* **46**, 1155 (2005).
[3]Gansäuer, A., Lauterbach, T., Geich-Gimbel, D. *CEJ* **10**, 4983 (2004).

p-Toluenesulfonyl azide – dimethyl acetonylphosphonate.

Alkyne synthesis.[1] A precursor of dimethyl diazomethylphosphonate for the transformation of aldehydes to homologous alkynes is generated in situ from the reagent pair in a suspension of K_2CO_3 in MeCN. Addition of aldehydes and MeOH initiates deacylation and the dehydrative homologation.

[1]Roth, G.J., Liepold, B., Müller, S.G., Bestmann, H.J. *S* 59 (2004).

p-Toluenesulfonylethyne.

Thiol protection.[1] The reagent forms adducts [(Z)-TsCH=CHSR] with thiols in the presence of Et_3N at 0° while alcohols and phenols do not react. Deprotection is achieved using pyrrolidine or alkanethiolate salts.

[1]Arjona, O., Medel, R., Rojas, J., Costa, A.M., Vilarrasa, J. *TL* **44**, 6369 (2003).

Trialkylaluminum. **15**, 341–342; **17**, 372–375; **18**, 365–367; **19**, 369–370; **20**, 401–402; **21**, 452–453; **22**, 463–464

Reaction with carbamates. With Me_3Al carbamates are converted to acetamides.[1]

Cyclization. Acyclic cross-conjugated ketones afford cyclohexenones in the presence of both Lewis acid (Me_3Al) and base.[2]

SO_2Ph O — Me_3Al, OH, Et_3N, PhMe → SO_2Ph O

91%

Rearrangement and fragmentation. Ring contraction of cyclic β,γ-epoxy alcohols mediated by Et_3Al is stereoselective and the formyl group is reduced within the aluminum chelates after the rearrangement.[3] On the other hand, γ,δ-epoxy alcohols undergo fragmentation to generate allylic alcohols from the epoxy moiety.[4]

HO O — R_3Al, THF → HO

89%

1-Trimethylsilyl-2-alkanones. Peterson reaction of aldehydes RCHO with $(Me_3SiCH_2)_3Al \cdot 3LiBr$ at room temperature gives $RCOCH_2SiMe_3$.[5]

N-Alkylation of hydrazones.[6] Me_3Al activates aldehydes to condense with hydrazones and donates a methyl group to the ensuing iminium ions.

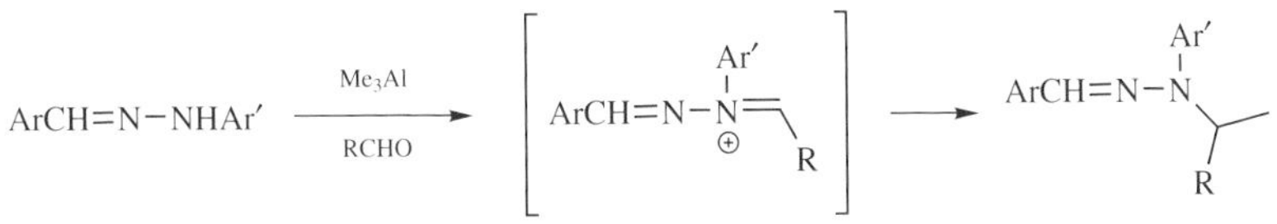

[1]El Kaim, L., Grimaud, L., Lee, A., Perroux, Y., Tirla, C. *OL* **6**, 381 (2004).
[2]Magomedov, N.A., Ruggiero, P.L., Tang, Y. *OL* **6**, 3373 (2004).
[3]Li, X., Wu, B., Zhao, X.Z., Jia, Y.X., Tu, Y.Q., Li, D.R. *SL* 623 (2003).
[4]Wang, F., Wang, S.H., Tu, Y.Q., Ren, S.K. *TA* **14**, 2189 (2003).
[5]Abedi, V., Battiste, M.A. *JOMC* **689**, 1856 (2004).
[6]El Kaim, L., Grimaud, L., Perroux, Y., Tirla, C. *JOC* **68**, 8733 (2003).

Triallylborane.

1-Boraadamantane.[1] Compound **1** is the product from a cycloaddition process.[1] The synthesis was inspired by Russian reports on the preparation of 2-spiro-1-boraadamantane derivatives.

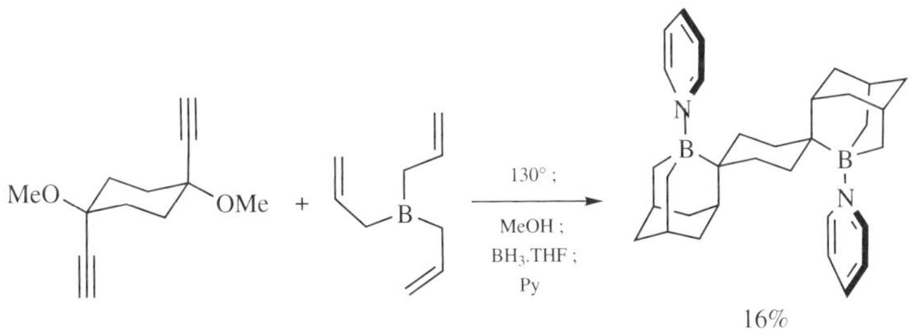

[1]Wagner, C.E., Shea, K.J *OL* **6**, 313 (2004).

Triarylbismuth dichloride - DBU. 22, 449

Oxidation.[1] For oxidation of alcohols to carbonyl compounds by the triarylbismuth dichloride-DBU reagent pair the reaction rates depend on the substitution of the aryl groups. Presence of an electron-withdrawing group at the *p*-position and/or an *o*-methyl group increases the oxidizing power.

Arylation.[2] In arylation of conjugated carbonyl compounds at the α-position Ar_3BiCl_2 are useful donors. The reaction is catalyzed by R_3P.

[1]Matano, Y., Hisanaga, T., Yamada, H., Kusukabe, S., Nomura, H., Imahori, H. *JOC* **69**, 8676 (2004).
[2]Koech, P.K., Krische, M.J. *JACS* **126**, 5350 (2004).

Trialkylphosphines. 20, 387–388; **21**, 350–351, 438; **22**, 355–356, 449–450

β-Azido ketones.[1] Addition of azide ion to conjugated ketones in water is catalyzed by Bu_3P, although pyridine and *N*-methylimidazole are equally effective and more convenient Lewis bases.

Heterocycles. 2-Alkenylfurans are formed from the reaction of 2-alken-4-ynones and aldehydes in the presence of Bu_3P.[2] 2-Methyl-2,3-alkadienoic esters and imines are caused to combine in a stepwise cyclization process leading to tetrahydropyridines.[3]

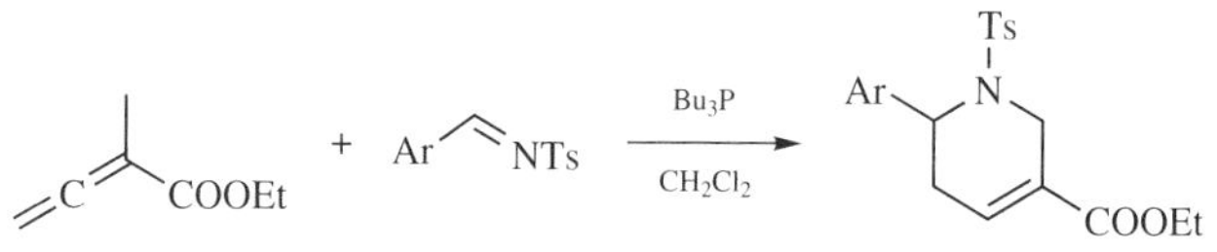

COOEt + Ar–CH=NTs —(Bu_3P, CH_2Cl_2)→ Ts, Ar, N, COOEt

R=Ph 98%

Conjugated dienes. Alkenylphosphonic esters serve as 1,2-dianion synthons that reaction with aldehydes in the presence of Bu_3P furnishes conjugated dienes.[4] Initiated by the addition of Bu_3P to generate α-phosphonyl carbanions to condense with the aldehydes the process ends in ylide formation and Wittig reaction.

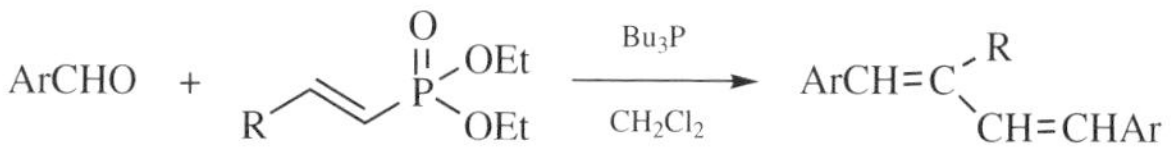

ArCHO + R–CH=CH–P(O)(OEt)(OEt) —(Bu_3P, CH_2Cl_2)→ ArCH=C(R)CH=CHAr

R = CN, COOBu

Baylis-Hillman reaction. One of three different cyclization products is obtained exclusively or predominantly from a bicyclic compound containing both an unsaturated ester and an enone unit using Me_3P as promoter, by varying the solvent.[5]

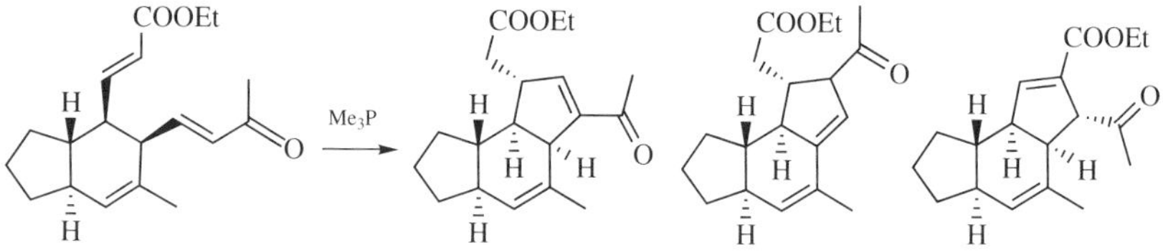

				yield
CF_3CH_2OH	—	—	(100)	78%
HMPA	—	(90)	(10)	65%
THF-H_2O(3:1)	(100)	—	—	74%

Addition reactions. Carbamates add to conjugated ketones in the presence of Bu_3P and Me_3SiCl.[6] Hydroalkoxylation of the activated alkenes with the aid of Me_3P is reported.[7] Me_3P performs conjugate addition to generate enolates that are used to deprotonate alcohols.

α-Hydrazinostyrenes. Deoxygenative hydrazination of acetophenones occurs on mixing with Bu_3P and an azodicarboxylic ester.[8] It is interesting that the Mitsunobu reagent complex reacts with the ketones and the adducts decompose by elimination of Bu_3PO in a nonconventional manner.

Reduction. For reduction of sulfoxides, sulfonyl chlorides, *N*-oxides, and organo-azides it is advantageous to us tris(2-carboxyethyl)phosphine because water solubility of any excess reagent and/or the phosphine oxide byproduct makes purification simple.[9]

[1]Xu, L.-W., Xia, C.-G., Li, J.-W., Zhou, S.-L. *SL* 2246 (2003).
[2]Kuroda, H., Hanaki, E., Izawa, H., Kano, M., Itahashi, H. *T* **60**, 1913 (2004).
[3]Zhu, X.-F., Lan, J., Kwon, O. *JACS* **125**, 4716 (2003).
[4]Liu, Y.-X., Xu, C.-F., Cao, R.-Z., Liu, L.-Z. *SC* **33**, 3561 (2003).
[5]Methot, J.L., Roush, W.R. *OL* **5**, 4223 (2003).
[6]Xu, L.-W., Xia, C.-G. *TL* **45**, 4507 (2004).
[7]Stewart, I.C., Bergman, R.G., Toste, F.D. *JACS* **125**, 8696 (2003).
[8]Liu, Y., Xu, C., Liu, L. *S* 1335 (2003).
[9]Faucher, A.-M., Grand-Maitre, C. *SC* **33**, 3503 (2003).

Tributylgermanium hydride.

Radical generation. The replacement of Bu_3SnH with Bu_3GeH for radical reactions is advantageous.[1] The reagent is less toxic, stable, and the reactions are easier to workup.

[1]Bowman, W.R., Krintel, S.L., Schilling, M.B. *OBC* **2**, 585 (2004).

Tributyltin hydride– 2,2′-azobis(isobutyronitrile). 19, 353–357; **20**, 391–394; **21**, 439–444; **22**, 450–453

Group exchange. Radicals derived from alkenyl bromides and ArBr on exposure to Bu_3SnH-AIBN are readily trapped in situ by $(MeO)_3P$, therefore the reaction is useful for synthesizing unsaturated phosphonates.[1]

Addition reactions. α-(*N*-Allyl)aminoacetaldehyde oxime ethers undergo hydrostannylation with formation of a five-membered ring. The reaction gives 3-alkoxyamino-4-tributylstannylmethylpyrrolidines.[2]

Tether tactic to direct synthesis of 1,3-amino alcohols from *N*-allylamines is quite efficient.[3] Thus, *N*-bromomethylsilylation installs the radical precursor with which a heterocycle is easily formed and eventually oxidatively cleaved.

Ring formation. Furans are synthesized by intramolecular addition of stannyloxy radical to conjugated esters.[4]

Bu3SnH - AIBN
PhH, 80°
COOEt
COOEt
63%

Elaboration of 4- to 8-membered lactams[5] from *N*-alkynylimines involves addition of Bu_3Sn radical to the alkyne linkage and capture of alkenyl radicals by CO with C-N bond formation to follow. The acyl radical intermediates exhibit azaphilicity.

+ CO
Bu3SnH - AIBN
PhH, 90°
Bu3SnCH
75%

Radical cyclization toward the indole nucleus can take two courses, depending on the degree of saturation.[6]

Bu3SnH - AIBN
58%

Bu3SnH - AIBN
55%

Deoxygenation. Alcohols are deoxygenated after derivatized into mixed phosphites in which one of the alkyl groups is 2-(*o*-iodophenyl)ethyl group. The key step involves heating such derivatives with Bu_3SnH-AIBN in benzene.[7]

[1]Jiao, X.-Y., Bentrude, W.G. *JOC* **68**, 3303 (2003).
[2]Miyabe, H., Tanaka, H., Naito, T. *CPB* **52**, 74 (2004).

[3]Blaszykowski, C., Dhimane, A.-L., Fensterbank, L., Malacria, M. *OL* **5**, 1341 (2003).
[4]Tae, J., Kim, K.-O. *TL* **44**, 2125 (2003).
[5]Ryu, I., Miyazato, H., Kuriyama, H., Matsu, K., Tojino, M., Fukuyama, T., Minakata, S., Komatsu, M. *JACS* **125**, 5632 (2003).
[6]Flanagan, S.R., Harrowven, D.C., Bradley, M. *TL* **44**, 1795 (2003).
[7]Zhang, L., Koreeda, M. *JACS* **126**, 13190 (2004).

2,2,2-Trichloroethyl chlorosulfate.

Aryl sulfates.[1] The title reagent is useful for derivatizing phenols. Such mixed sulfates as $ArOSO_2OCH_2CCl_3$ are decomposed to ammonium salts by treatment with $HCOONH_4$ with Zn or catalytic Pd/C.

[1]Liu, Y., Lien, I.-F.F., Ruttgaizer, S., Dove, P., Taylor, S.D *OL* **6**, 209 (2004).

(3,3,3-Trichloropropyl)triphenylphosphonium chloride.

Wittig reaction.[1] Wittig reagent generated from the title compound by treatment with NaHMDS (1.4 equiv.) can be used to synthesize (*Z*)-1,1,1-trichloro-3-alkenes from aldehydes. Dehydrochlorination of the products with DBU in CH_2Cl_2 at room temperature affords (*Z*)-1,1-dichloro-1,3-alkadienes. The presence of excess NaHMDS (4 equiv.) in the Wittig reaction causes a twofold dehydrochlorination of the primary products, (*Z*)-1-chloro-3-alken-1-ynes are obtained.

[1]Karatholuvhu, M.S., Fuchs, P.L. *JACS* **126**, 14314 (2004).

Triethylborane. 20, 395; **21**, 445–446; **22**, 454–456

Addition reactions. Carbon radical addition to glyoxylic oxime ethers is useful for synthesis of many different α-amino acid derivatives,[1] and the free radicals are easily generated from iodides with the aid of Et_3B. Excellent results are obtained from enantioselective addition of carbon radicals to *N*-acylhydrazones.[2]

Dehalogenative cyclization of iodoalkenes is induced by Et_3B while using $HGaCl_2$ or $HInCl_2$ as mediator.[3] Acrylic esters are hydroxytrifluoromethylated on mixing with CF_3I, KF and Et_3B in aqueous THF.[4] Owing to the popularity of coupling reactions, acquisition of stereodefined alkenylmetallic reagents becomes important. Hydrostannylation,[5] hydrogallation and hydroindation[6] of 1-alkynes, and also hydrostannylation of fluorine-containing internal alkynes[7] mediated by Et_3B proceed well.

Et_3B contributes an ethyl group to enones during its mediation of an aldol reaction. Its admixture with MVK and PhCHO in THF produces 4-hydroxy-3-propyl-2-alkanones.[8] Different kinds of cyclic structures are made by the addition followed by intramolecular trapping.[9,10]

NOBn; Et_3B, PhMe, Δ → NHOBn (41%) + NHOBn (14%)

$PhSO_2N(Cl)$–Cl + Ph; Et_3B, PhH, Δ → $PhSO_2N$ (Cl, Cl, Ph) 96%

Ethyne group transfer from a temporary oxasilyl tether to a carbon center also involves intramolecular addition (atom transfer).[11]

I, O–Si, $SiMe_3$; Et_3B → O–Si, I, $SiMe_3$; Bu_4NF → OH

Promotion of radical addition to glyoxylic oxime ethers (radicals from secondary alkyl iodides) by Et_3B usually gives good yields of α-amino acid derivatives.[12]

Lewis acid-promoted radical addition to conjugated systems is a synthetic process worthy of exploitation. For example, isopropyl radical generated from *i*-PrI in the presence of Bu_3SnH, Et_3B and O_2 is well behaved.[13]

1,4-Diketones, γ-ketocarboxylic esters and amides and 3-perfluoroalkyl ketones are assembled in a free radical addition under oxidative conditions from active iodides and 2-silylalkenes.[14]

$MeSiPh_2$, R + I, R′, O; Et_3B, O_2, NH_4Cl, H_2O → R, O, R′, O

Hydroxyalkylation at C-2 of tetrahydrofuran ring with aldehydes involves treatment with Et_3B and then *t*-BuOOH.[15] High *syn*-selectivity is observed with ArCHO.

Chain lengthening. α-Sulfonyl carbanions react with Et_3B to give α-ethylated diethylboranes which isomerizes on thermolysis. Further processing with oxidative workup results in chain lengthening by two carbon units.[16] [Other trialkylboranes react to lengthen carbon chain accordingly.]

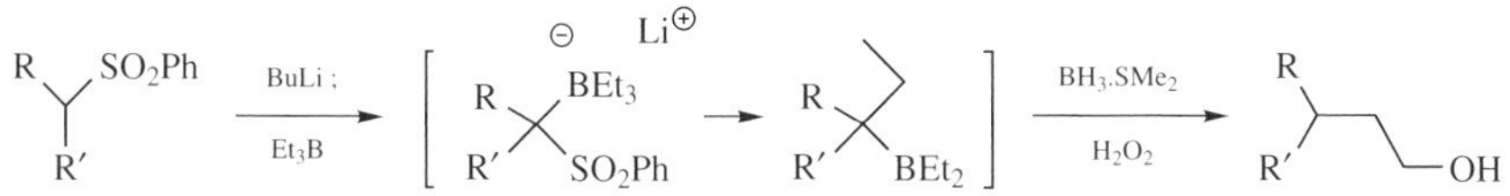

Ethyl group transfer also occurs when Et_3B reacts with unstabilized iodonium ylides.[17]

[1]McNabb, S.B., Ueda, M., Naito, T. *OL* **6**, 1911 (2004).
[2]Friestad, G.K., Shen, Y., Ruggles, E.L. *ACIEE* **42**, 5061 (2003).
[3]Takami, K., Mikami, S., Yorimitsu, H., Shinokubo, H., Oshima, K. *T* **59**, 6627 (2003).
[4]Yajima, T., Nagano, H., Saito, C. *TL* **44**, 7027 (2003).
[5]Miura, K., Wang, D., Matsumoto, Y., Fujisawa, N., Hosomi, A. *JOC* **68**, 8730 (2003).
[6]Takami, K., Mikami, S., Yorimitsu, H., Shinokubo, H., Oshima, K. *JOC* **68**, 6627 (2003).
[7]Kanda, M., Konno, T., Chae, J., Ishihara, T., Yamanaka, H. *JFC* **120**, 185 (2003).
[8]Chandrasekhar, S., Narsihmulu, C., Reddy, N.R., Reddy, M.S. *TL* **44**, 2583 (2003).
[9]Miyabe, H., Ueda, M., Fujii, K., Nishimura, A., Naito, T. *JOC* **68**, 5618 (2003).
[10]Tsuritani, T., Shinokubo, H., Oshima, K. *JOC* **68**, 3246 (2003).
[11]Sukeda, M., Ichikawa, S., Matsuda, A., Shuto, S. *JOC* **68**, 3465 (2003).
[12]McNabb, S.B., Ueda, M., Naito, T. *OL* **6**, 1911 (2004).
[13]Srikanth, G.S.C., Castle, S.L. *OL* **6**, 449 (2004).
[14]Kondo, J., Shinokubo, H., Oshima, K. *ACIEE* **42**, 825 (2003).
[15]Yoshimitsu, T., Arano, Y., Nagaoka, H. *JOC* **68**, 625 (2003).
[16]Billaud, C., Goddard, J.-P., Le Gall, T., Mioskowski, C. *TL* **44**, 4451 (2003).
[17]Ochiai, M., Tuchimoto, Y., Higashiura, N. *OL* **6**, 1505 (2004).

Trifluoroacetic acid, TFA. **14**, 322–323; **15**, 338–339; **18**, 375–376; **20**, 395–396; **21**, 446–447; **22**, 457

Epimerization.[1] On exposure to CF_3COOH and $PhSO_3H$ aromatic *C*-nucleosides undergo epimerization to favor the β-epimers (54–78%).

Addition to imines.[2] β-Silyl ketene dithioacetals react with imines on acid catalysis.

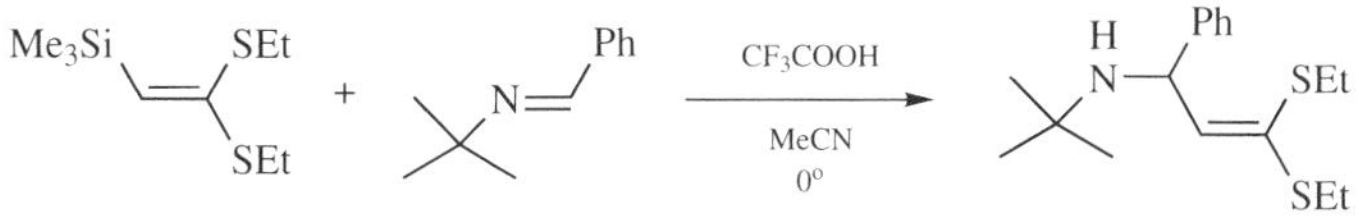

4-Hydroxytetrahydropyrans. Homoallylic alcohols and aldehydes condense to give 4-hydroxytetrahydropyrans.[3] 2-Silylmethylcyclopropyl carbinols are also transformed into related compounds.[4]

HO SiR3 Ph R = Si(tBr)Ph2 — CF_3COOH, CHO → HO Ph O SiR3 72%

[3 + 2]Cycloaddition. Azomethane ylides are generated from *N*-trimethylsilylmethyl-*N*-methoxymethylamines with CF_3COOH. The ylides and isatins combine to give oxindoles spiroannulated to an oxazolidine ring.[5]

(N-Ph isatin) + Me_3Si–N(Bn)–OMe — CF_3COOH, CH_2Cl_2 → spiro-oxazolidine oxindole (NBn, N-Ph) 93%

[1]Liang, Y.L., Stivers, J.T. *TL* **44**, 4051 (2003).
[2]Yahiro, S., Shibata, K., Saito, T., Okauchi, T., Minami, T. *JOC* **68**, 4947 (2003).
[3]Barry, C.St.J., Crosby, S.R., Harding, J.R., Hughes, R.A., King, C.D., Parker, G.D., Willis, C.L. *OL* **5**, 2429 (2003).
[4]Yadav, Y.K., Kumar, N.V. *JACS* **126**, 8652 (2004).
[5]Nair, V., Mathai, S., Augustine, A., Viji, S., Radhakrishnan, K.V. *S* 2617 (2004).

Trifluoroacetic anhydride, TFAA. 18, 376–377; **19**, 361; **20**, 396–397; **21**, 447–448; **22**, 458

N-Tosylimines. Condensation of ArCHO with $TsNH_2$ is promoted by TFAA.[1]

[1]Lee, K.Y., Lee, C.G., Kim, J.N. *TL* **44**, 1231 (2003).

Trifluoroamine oxide.

Fluorination.[1] The title compound, which can be prepared in 10–15% yield by electric discharge of an equimolar mixture of oxygen and NF_3 at $-196°$, is a selective reagent for fluorination of β-diketones and β-keto esters.

[1]Gupta, O.D., Shreeve, J.M. *TL* **44**, 2799 (2003).

Trifluoromethanesulfonic acid (triflic acid). 14, 323–324; **15**, 339; **18**, 377; **19**, 362–363; **20**, 398–399; **21**, 448–449; **22**, 458–459

Friedel-Crafts reactions. Triflic acid promotes acylation of arenes by cinnamic acids. 3-Arylindanones are formed when the cinnamic acids are electron-deficient, but the cyclization step is stopped when they are electron-rich.[1] 1,3-Diarylpropynones cyclize to afford 3-arylindenones by short exposure to TfOH.[2]

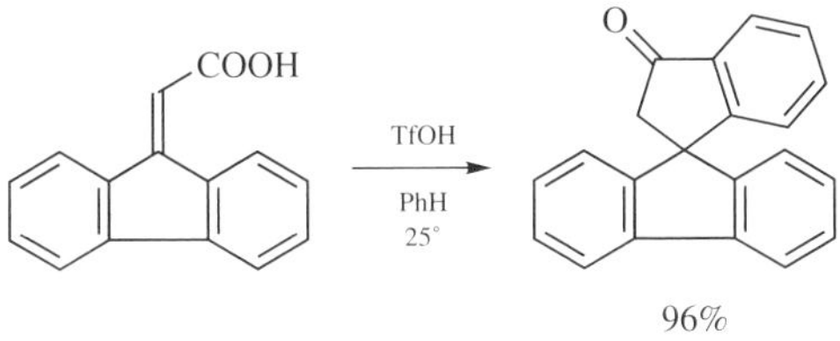

COOH

TfOH

PhH
25°

96%

An expedient preparation of trifluoromandelic esters employs TfOH to bring about the alkylation of arenes with ethyl trifluoropyruvate.[3]

Unsaturated amines are protonated at two sites. Such reactive species are Friedel-Crafts alkylating agents and the formation of 2-(α-phenylethyl)pyridine in excellent yield from 2-vinylpyridine indicates that species bearing two closely spaced positive charges are readily generated in the strong acid.[4] Cyclization of 2-nitro-3-arylpropanoic esters to 4*H*-1,2-benzoxazines[5] is observed on brief heating with TfOH in $CHCl_3$.

Glycosylation. Promotion of glycosylation using glycosyl fluorides was reported with new information concerning ionic liquid as reaction medium.[6] A new glycosyl donor suitable for TfOH-catalyzed reactions is the 6-nitro-2-benzothiazoate.[7]

Prenylation. A prenyl group is delivered to aldehydes from 6,6-dimethyl-7-octene-1,4-diol in a stereoselective process so that intermolecular chirality transfer is realized.[8] (An oxonium-Cope rearrangement is plausibly involved.)

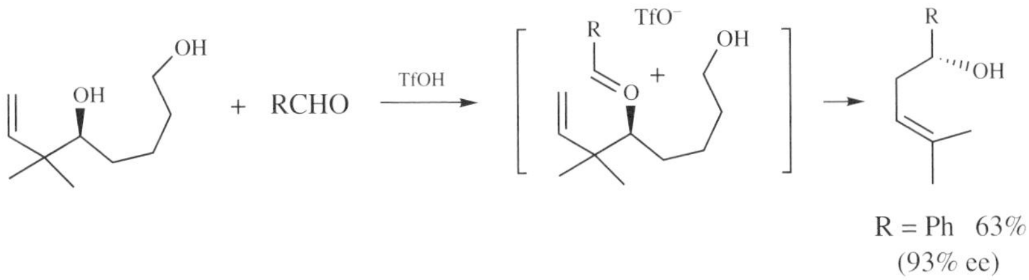

OH OH + RCHO TfOH [R TfO⁻ OH O +] R OH

R = Ph 63%
(93% ee)

Addition to imines. Protonation of imines initiates attack from nucleophiles such as silyl enol ethers[9] and ethyl diazoacetate,[10] in the latter case aziridines are produced.

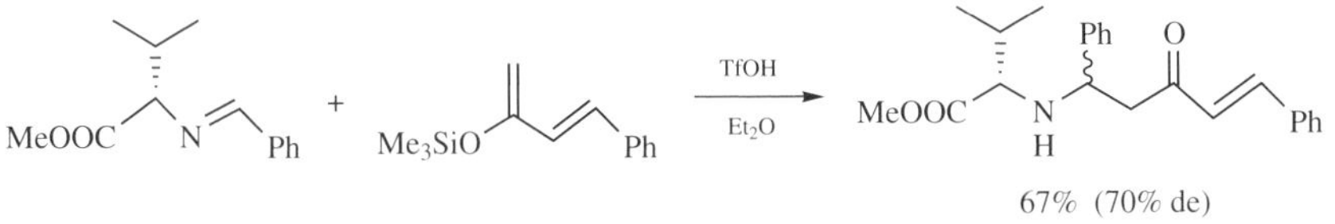

[1]Rendy, R., Zhang, Y., McElrea, A., Gomez, A., Klumpp, D.A. *JOC* **69**, 2340 (2004).
[2]Vasilyev, A.V., Walspurger, S., Haouas, M., Sommer, J., Pale, P., Rudenko, A.P. *OBC* **2**, 3483 (2004).
[3]Prakash, G.K.S., Yan, P., Török, B., Olah, G.A. *SL* 527 (2003).
[4]Zhang, Y., McElrea, A., Sanchez, Jr, G.V., Do, D., Gomez, A., Aguirre, S.L., Rendy, R., Klumpp, D.A. *JOC* **68**, 5119 (2003).
[5]Nakamura, S., Uchiyama, M., Ohwada, T. *JACS* **125**, 5282 (2003).
[6]Sasaki, K., Matsumura, S., Toshima, K. *TL* **45**, 7043 (2004).
[7]Hashihayata, T., Mandai, H., Mukaiyama, T. *BCSJ* **77**, 169 (2004).
[8]Cheng, H.-S., Loh, T.-P. *JACS* **125**, 4990 (2003).
[9]Ishimaru, K., Kojima, T. *TL* **44**, 5441 (2003).
[10]Williams, A.L., Johnston, J.N. *JACS* **126**, 1612 (2004).

Trifluoromethanesulfonic anhydride (triflic anhydride). **13**, 324–325; **14**, 324–326; **15**, 339–340; **16**, 357–358; **18**, 377–378; **19**, 363–365; **20**, 399; **21**, 449; **22**, 459–461

Oxy group activation. 4-Hydroxy-1-alkenyl esters cyclize to cyclopropanecarbaldehydes stereoselectively on treatment with Tf_2O – 2,6-lutidine.[1]

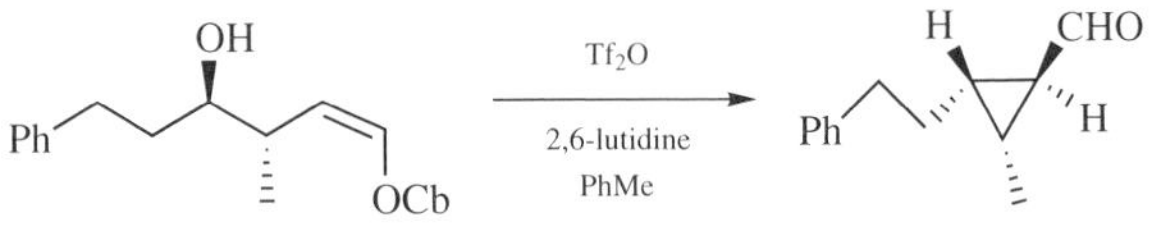

Iminotriflates are readily formed from amides using Tf_2O – pyridine and their reaction with aqueous $(NH_4)_2S$ leads to thioamides.[2] 5-Hydroxypyrrolidinones undergo ring opening on exposure to Tf_2O – pyridine and subsequent recyclization gives furan derivatives.[3]

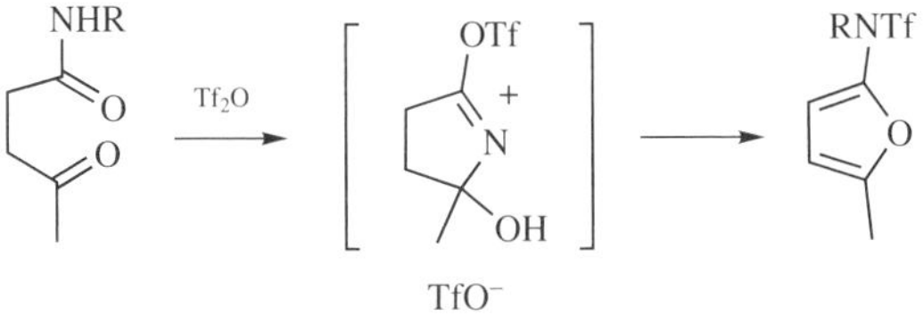

Carbonyl activation of thioesters by Tf_2O in the presence of nitriles gives rise to 4-thiopyrimidines.[4] A method of aromatic thiolation involves electrophilic substitution with activated bis(2-methoxycarbonylethyl) sulfoxide followed by elimination (double retro-Michael reaction).[5]

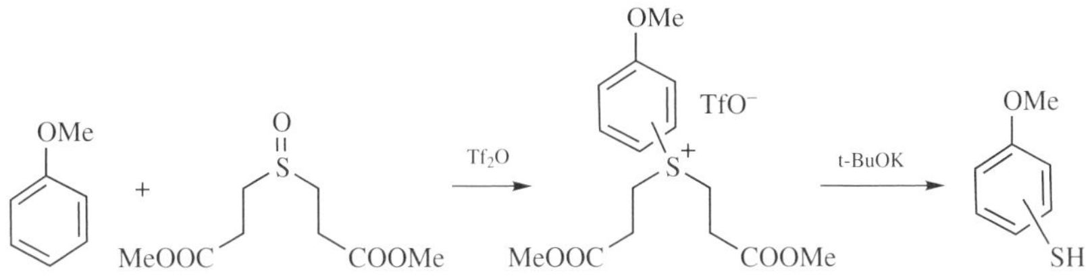

Facile generation of Pummerer rearrangement intermediates for intramolecular cyclization has synthetic values and a refined synthesis of chiral tetrahydroisoquinolines is based on the method.[6] A formal intramolecular S_N2' substitution of 2-indolyl sulfoxide is instigated by its activation with Tf_2O.[7] Sulfonamides are synthesized from organosulfonate salts on brief reaction with Tf_2O and Ph_3PO followed by the addition of amines.[8]

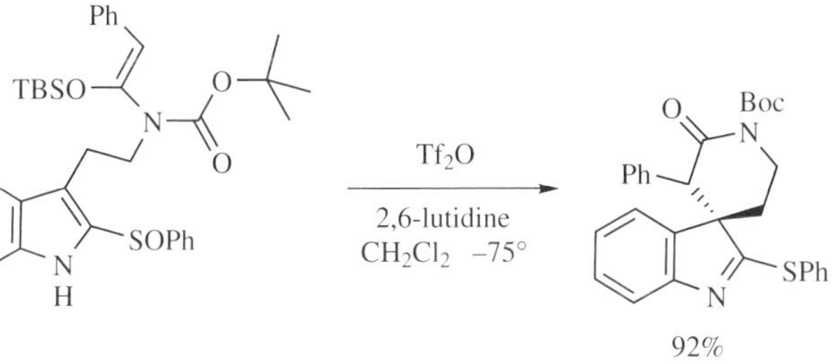

S-Benzyloxydiphenylsulfoxonium ions are strong benzylating agents (e.g., for sodium malonate) which are readily made from benzyl alcohols by activation with the Ph_2SO - Tf_2O duet at low temperature.[9] *N*-Benzenesulfinylpiperidine also becomes a highly electrophilic agent in the presence of Tf_2O – *i*-PrN_2Et, and certain alkenols cyclize on initiation by benzenesulfenylation.[10]

Through oxime triflate formation and Beckmann rearrangement amidines and enamines are synthesized on quenching with amines and stabilized carbon nucleophiles, respectively.[11]

Dethioacetalization. Dithioacetals are hydrolyzed nonoxidatively by the combined action of Tf_2O and *N*-benzenesulfinylpiperidine with aqueous workup.[12]

[1]Risatti, C.A., Taylor, R.E. *ACIEE* **43**, 6671 (2004).
[2]Charette, A.B., Grenon, M. *JOC* **68**, 5792 (2003).
[3]Padwa, A., Rashatasakhon, P., Rose, M. *JOC* **68**, 5139 (2003).
[4]Herrera, A., Martinez-Alvarez, R., Ramiro, P. *T* **59**, 7331 (2003).

[5]Becht, J.-M., Wagner, A., Mioskowski, C. *JOC* **68**, 5758 (2003).
[6]Saitoh, T., Shikiya, K., Horiguchi, Y., Sano, T. *CPB* **51**, 667 (2003).
[7]Feldman, K.S., Vidulova, D.B. *OL* **6**, 1869 (2004).
[8]Caddick, S., Wilden, J.D., Judd, D.B. *JACS* **126**, 1024 (2004).
[9]Takuwa, T., Onishi, J.Y., Matsuo, J., Mukaiyama, T. *CL* **33**, 8 (2004).
[10]Crich, D., Surve, B., Sannigrahi, M. *H* **62**, 827 (2004).
[11]Takuwa, T., Minowa, T., Onishi, J.Y., Mukaiyama, T. *CL* **33**, 322 (2004).
[12]Crich, D., Picione, J. *SL* 1257 (2003).

Trifluoromethanesulfonic anhydride – diphenyl sulfoxide.

Glycosylation. Thioglycosides are activated by Tf_2O - Ph_2SO such that their donor properties are revealed.[1]

Alkene activation. Functionalization of styrenes, e.g., to α-imidostyrenes,[2] via formation of diphenylstyrylsulfonium salts is synthetically useful.

Tf_2O- Ph_2SO
CH_2Cl_2 –78°
87%

With α-alkylstyrenes the activation results in formation of allylic sulfonium species that are liable to attack by nucleophiles. Accordingly, allylic amines are readily obtained.[3]

[1]Codee, J.D.C., Litjens, R.E.J.N., den Heeten, R., Overkleeft, H.S., van Boom, J.H., van der Marel, G.A. *OL* **5**, 1519 (2003).
[2]Yamanaka, H., Mukaiyama, T. *CL* **32**, 1192 (2003).
[3]Yamanaka, H., Matsuo, J., Kawana, A., Mukaiyama, T. *CL* **32**, 626 (2003).

Trifluoromethanesulfonyl azide.

Diazo transfer.[1] Activated methylene compounds such as cyanoacetic esters and benzenesulfonylacetic esters are readily converted into the diazo derivatives by TfN_3 at 0°. The reagent is available from mixing aqueous $(Bu_4N)HSO_4$ and NaN_3 with Tf_2O and extraction into hexane for immediate use.

[1]Wurz, R.P., Lin, W., Charette, A.B. *TL* **44**, 8845 (2003).

S-Trifluoromethyldibenzothiophenium tetrafluoroborate.

Trifluoromethylation.[1] The commercially available reagent is useful for trifluoromethylation of β-keto esters

[1]Ma, J.-S., Cahard, D. *JOC* **68**, 8726 (2003).

2-Trifluoromethyl-1,3-dithianylium triflate.

Pentafluoroethylation.[1] The title compound is readily prepared by adding triflic acid to the condensation product of 1,3-propanedithiol and TTFA. After reaction with nucleophiles subsequent difluorodesulfurization affords compounds with a terminal C_2F_5 group.

[1]Sevenard, D.V., Kirsch, P., Lork, E., Röschenthaler, G.-V. *TL* **44**, 5995 (2003).

Trifluoromethyl(trimethyl)dioxirane.

Oxidation. The title reagent is highly effective in oxidizing 2-alkyn-1,4-diols to the diketones without affecting the triple bond.

[1]D'Accolti, L., Fiorentino, M., Fusco, C., Crupi, P., Curci, R. *TL* **45**, 8575 (2004).

(Trifluoromethyl)trimethylsilane. **15**, 341; **18**, 378–379; **19**, 366–367; **20**, 400; **21**, 450–451; **22**, 461–462

Trifluoromethylation. As a convenient source of a trifluoromethyl group Me_3SiCF_3 is easy to handle. Transfer to the carbonyl function is catalyzed by by LiOAc,[1] and Me_3NO.[2]

C-Trifluoromethylation of cyclic imides[3] is very facile with Me_3SiCF_3 - BuNF. When ATPH is present enones are forced to accept the CF_3 group at the β-position,[4] but yields of this reaction are relatively low.

Halide substitution is promoted by CuI - KF in ionic liquids,[5] even *p*-diiodobenzene is converted into 4-iodo-α,α,α-trifluorotoluene in 38% yield

[1]Mukaiyama, T., Kawano, Y., Fujisawa, H. *CL* **34**, 88 (2005).
[2]Prakash, G.K.S., Mandal, M., Panja, C., Mathew, T., Olah, G.A. *JFC* **123**, 61 (2003).
[3]Hoffmann-Röder, A., Seiler, P., Diederich, F. *OBC* **2**, 2267 (2004).
[4]Sevenard, D.V., Sosnovskikh, V.Ya., Kolomeitsev, A.A., Konigsmann, M.H., Röschenthaler, G.-V. *TL* **44**, 7623 (2003).
[5]Kim, J., Shreeve, J.M. *OBC* **2**, 2728 (2004).

1,3,5-Trihydroxyisocyanuric acid.

Aerobic oxidation.[1] The title compound is an excellent radical-producing catalyst for the Co/Mn mediated oxidation of toluene derivatives. Besides the known preparation the reagent can be made by hydrogenolysis of its tribenzyl ether which is the condensation product of *O*-benzylhydroxylamine and diphenyl carbonate.

[1]Hirai, N., Kagayama, T., Tatsukawa, Y., Sakaguchi, S., Ishii, Y. *TL* **45**, 8277 (2004).

Trimethylsilyl azide. **13**, 24–25; **14**, 25; **15**, 342–343; **16**, 17; **18**, 379–380; **19**, 371–372; **20**, 403; **21**, 453; **22**, 464–465

2-Deoxyglycosyl azides. The Me_3SiN_3 - Me_3SiONO_2 system provides $[H/N_3]$ addends to react with glycals, furnishing deoxyglycosyl azides.[1]

2-Azidoalkanols.[2] Epoxide opening by Me_3SiN_3 (or Me_3SiCN) is assisted by Bu_4NF via nucleophlic attack by fluoride ion on the silicon atom.

Addition reactions. Conjugated imines undergo dual addition with Me_3SiN_3 and tetraallylstannane in the presence of $SnCl_4 \cdot 5H_2O$. In situ hydrolysis of the imino group apparently precedes attack by the tin reagent. The products are 6-azido-4-hydroxy-1-alkenes.[3]

Ph N Ph R + Sn 4 Me_3SiN_3 - $SnCl_4$ HOAc, CH_2Cl_2 N_3 OH R

Synthesis of *N*-unsubstituted 1,2,3-triazoles is conveniently performed by cycloaddition of Me_3SiN_3 with alkynes, catalyzed by CuI.[4]

cis-1,2-Diazidation of electron-rich alkenes is best performed with $PhI(OAc)_2$ - Me_3SiN_3.[5] Iodine azide gives mainly the *trans*-diazides.

O N N $PhI(OAc)_2$ Me_3SiN_3,Et_4NI O N N N_3 N_3

41%

[1]Reddy, B.G., Madhusudanan, K.P., Vankar, Y.D. *JOC* **69**, 2630 (2004).
[2]Konno, K., Toshiro, E., Hinoda, N. *S* 2161 (2003).
[3]Shimizu, M., Nishi, T. *SL* 889 (2004).
[4]Jin, T., Kamijo, S., Yamamoto, Y. *EJOC* 3789 (2004).
[5]Chung, R., Yu, E., Incarvito, C.D., Austin, D.J. *OL* **6**, 3881 (2004).

Trimethylsilyl chloride. **15**, 89; **16**, 85–86; **18**, 381; **19**, 374–375; **20**, 404–405; **21**, 453–454; **22**, 465–466

O-Protection. *N*-acylation of nucleotides preceded by *O*-silylation of free hydroxyl groups in the sugar moieties is expedient.[1]

Functionalized ketones. Addition of carbamates to enones to afford β-amino ketones[2] is promoted by Me_3SiCl and Bu_3P. A convenient method for α-oximation of ketones with amyl nitrite uses Me_3SiCl.[3]

Heterocycles. The 3-component condensation of activated carbonyl compounds, aldehydes, and urea to furnish 3,4-dihydropyrimidin-2(1*H*)-ones[4] proceeds readily in the presence of Me_3SiCl.

3-Azetidinone acetals are synthesized[5] on electroreduction of imino derivatives of α-amino esters using Me_3SiCl as protecting agent.

Me_3SiN_3 - Bu_4NClO_4 / e / THF ; BzCl, Et_3N

62% (88% ee)

Nitroarenes.[6] Arylboronic acids are converted into nitroarenes (functional group replacement) by reaction with a nitrate salt (NH_4NO_3, $AgNO_3$) and Me_3SiCl.

Disulfides.[7] Oxidative dimerization of RSH by DMSO is catalyzed by Me_3SiCl.

[1]Zhu, X.-F., Williams, H.J., Scott, A.I. *SC* **33**, 1233 (2003).
[2]Xu, L.-W., Xia, C.-G. *TL* **45**, 4507 (2004).
[3]Mohammed, A.H.A., Nagendrappa, G. *TL* **44**, 2753 (2003).
[4]Zhu, Y., Pan, Y., Huang, S. *SC* **34**, 3167 (2004).
[5]Kise, N., Ozaki, H., Moriyama, N., Kitagishi, Y., Ueda, N. *JACS* **125**, 11591 (2003).
[6]Prakash, G.K.S., Panja, C., Mathew, T., Surampudi, V., Petasis, N.A., Olah, G.A. *OL* **6**, 2205 (2004).
[7]Karimi, B., Hazarkhani, H., Zareyee, D. *S* 2513 (2002).

Trimethylsilyl cyanide. 13, 87–88; **14**, 107; **15**, 102–104; **17**, 89; **18**, 381–382; **19**, 375; **20**, 405; **21**, 455–456; **22**, 466

Strecker-type reactions. In the preparation of *O*-trimethylsilyl cyanohydrins from carbonyl compounds and Me_3SiCN many effective catalysts have been found: K_2CO_3,[1] $LiClO_4$,[2] both under solvent-free conditions, CsF,[3] $[Ph_3PMe]I$,[4] NMO[5] are used in CH_2Cl_2. Amine oxides used in conjunction with a chiral (salen)-Ti complex[6] are beneficial while a chiral prolinol ligand also containing the *N*-oxide unit[7] possesses moderate ability of asymmetric induction. Cocatalysts for the addition are Me_3SiBr and 1,2-bis(cyanomercurio)-3,4,5,6-tetrafluorobenzene,[8] and (salen)-Al complex with an amine oxide.[9]

By a Strecker-type reaction the α-(*O*-trimethylsiloxyamino) nitriles are obtained.[10] Chiral sulfilimines possess an internal stereocenter to direct diastereoselective addition. The nucleophilicity of Me_3SiCN is enhanced by CsF.[11]

Addition reactions. Ketene silyl acetals and Me_3SiCN are two kinds of nucleophiles to add to conjugated imines in good order. A catalyst for the double addition is $AlCl_3$.[12] (Note the different chemoselectivity of the carbon nucleophiles in a similar reaction in an entry under Me_3SiN_3.) *N,N'*-Diarylthioureas constitute another class of catalysts in the cyanation of nitrones by virtue of their activation in forming hydrogen bonds with the substrates.[13]

Aryl cyanides.[14] Pd-catalyzed substitution of aryl halides can use Me_3SiCN as cyanide source.

[1]He, B., Li, Y., Feng, X., Zhang, G. *SL* 1776 (2004).
[2]Azizi, N., Saidi, M.R. *JOMC* **688**, 283 (2003).

[3]Kim, S.S., Rajagopal, G., Song, D.H. *JOMC* **689**, 1734 (2004).
[4]Cordoba, R., Plumet, J. *TL* **44**, 6157 (2003).
[5]Kim, S.S., Kim, D.W., Rajagopal, G. *S* 213 (2004).
[6]He, B., Chen, F., Li, Y., Feng, X., Zhang, G. *EJOC* 4657 (2004).
[7]Shen, Y., Feng, X., Li, Y., Zhang, G., Jiang, Y. *EJOC* 129 (2004).
[8]King, J.B., Gabbai, F.P. *OM* **22**, 1275 (2003).
[9]Chen, F.-X., Liu, X., Qin, B., Zhou, H., Feng, X., Zhang, G. *S* 2266 (2004).
[10]Heydari, A., Mehrdad, M., Tavakol, H. *S* 1962 (2003).
[11]Li, B.-F., Yuan, K., Zhang, M.-J., Wu, H., Dai, L.-X., Wang, Q.R., Hou, X.-L. *JOC* **68**, 6264 (2003).
[12]Shimizu, M., Kamiya, M., Hachiya, I. *CL* **32**, 606 (2003).
[13]Okino, T., Hoashi, Y., Takemoto, Y. *TL* **44**, 2817 (2003).
[14]Sundermeier, M., Mutyala, S., Zapf, A., Spannenberg, A., Beller, M. *JOMC* **684**, 50 (2003).

Trimethylsilyldialkylamines. **18**, 382; **19**, 376; **20**, 407; **21**, 457; **22**, 466–467

Elimination. Cyclic acetals and oxazolidines undergo ring opening with Et_2NSiMe_3. The products are enol ethers and enamines/imines, respectively.[1]

Trifluoromethylating agent. Adduct obtained from heating 1,1,1-trifluoroacetophenone with Me_2NSiMe_3 at 110° is a stable reagent for trifluoromethylation.[2] It releases the nucleophilic $[CF_3^-]$ by fluoride ion.

Silylations.[3] The reagent Et_2NSiMe_3 is for attachment of a trimethylsilyl group to terminal alkynes which is promoted by $ZnCl_2$.

[1]Iwata, A., Tang, H., Kunai, A., Oshita, J., Yamamoto, Y., Matui, C. *JOC* **67**, 5170 (2002).
[2]Motherwell, W.B., Storey, L.J. *SL* 646 (2002).
[3]Andreev, A.A., Konshin, V.V., Komarov, N.V., Rubin, M., Brouwer, C., Gevorgyan, V. *OL* **6**, 421 (2004).

Trimethylsilyldiazomethane. **20**, 405–406; **22**, 467–468

Furans.[1] On treatment with Me_3SiCHN_2 acyl isocyanates form 4-trimethylsiloxyoxazoles which can be used to react with propynoic esters or acetylenedicarboxylic esters to afford furan derivatives.

R–C(=O)–N=C=O + Me_3SiCHN_2 —MeCN→ [2-R-4-(OSiMe₃)oxazole] —X–C≡C–X, Δ→ 2-R-3,4-X₂-furan

Cyclopropanation.[2] Using Me_3SiCHN_2 improves diastereoselectivity in cyclopropanation via copper-catalyzed decomposition of diazoalkanes, owing to steric bulk of the trimethylsilyl group.

[1]Hari, Y., Iguchi, T., Aoyama, T. *S* 1359 (2004).
[2]France, M.B., Milojevich, A.K., Stitt, T.A., Kim, A.J. *TL* **44**, 9287 (2003).

Trimethylsilyl iodide. 16, 188–189; **18**, 383; **19**, 376–377; **20**, 407; **21**, 458; **22**, 470

Michaelis-Arbuzov rearrangement. Warming R_2POR' with Me_3SiI (or Me_3SiBr) affords $R_2R'P{=}O$.[1]

Glycosyl iodides. Glycosyl acetates are converted into the iodides which are excellent glycosyl donors.[2-4] Other acetoxy groups in the sugars are much less reactive and therefore can be kept intact.

3-Iodoketene silyl acetals. Addition of Me_3SiI to alkyl propynoates gives the functionalized allenes which undergo hydroxyalkylation with aldehydes.[5]

COOMe; Me_3SiI, CH_2Cl_2 → Me_3SiO, OMe, C, I; PhCHO, MgI_2 → OH, O, Ph, OMe, I; 85%

Prins reaction. Unsaturated *O*-acetyl hemiacetals ionize by the aid of Me_3SiI (or Me_3SiBr) to form tetrahydropyrans.[6]

AcO, Cl, O; Me_3SiI, lutidine, CH_2Cl_2 → Cl, I, O

[1]Renard, P.-Y., Vayron, P., Mioskowski, C. *OL* **5**, 1661 (2003).
[2]Lam, S.N., Gervay-Hague, J. *OL* **5**, 4219 (2003).
[3]Miquel, N., Vignando, S., Russo, G., Lay, L. *SL* 341 (2004).
[4]Kobashi, Y., Mukaiyama, T. *CL* **33**, 874 (2004).
[5]Deng, G.-H., Hu, H., Wei, H.-X., Pare, P.W. *HCA* **86**, 3510 (2003).
[6]Jasti, R., Vitale, J., Rychnovsky, S.D. *JACS* **126**, 9904 (2004).

4-(Trimethylsilylmethyl)benzyl trichloromethylacetimidate.

Alcohol protection.[1] By a $Sc(OTf)_3$-catalyzed substitution various kinds of alcohols are etherified. The protecting group can be removed with DDQ (in preference to *p*-methoxybenzyl group). On the other hand, PMB ethers are selectively cleaved by $ZrCl_4$ in MeCN. [Both ethers are susceptible to hydrogenolysis.]

[1]Reddy, C.R., Chittiboyina, A.B., Kache, R., Jung, J.-C., Watkins, E.B., Avery, M.A. *T* **61**, 1289 (2005).

2-Trimethylsilylthiazole.

β-Lactam cleavage.[1] 4-Formyl-β-lactams undergo an uncommon cleavage of the N(1)-C(4) bond by reaction with the title reagent.

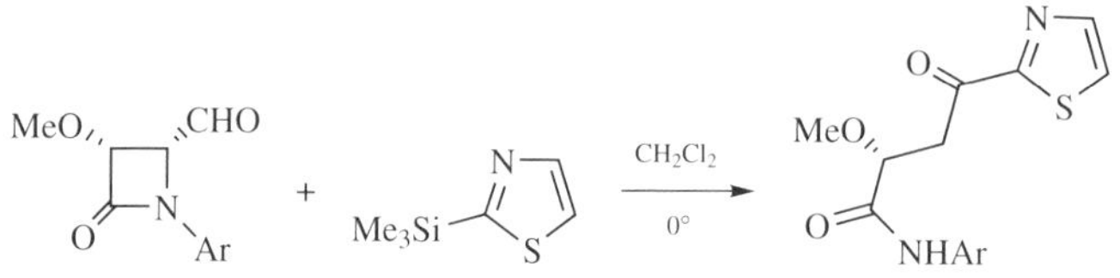

[1]Alcaide, B., Almendros, P., Redondo, M.C. *OL* **6**, 1765 (2004).

Trimethylsilyl trifluoromethanesulfonate. **13**, 329–331; **14**, 333–335; **15**, 346–350; **16**, 363–364; **17**, 379–386; **18**, 383–384; **19**, 379–381; **20**, 408–410; **21**, 460–462; **22**, 470–473

Functional group manipulation. Acetalization of carbonyl compounds by diols[1] is catalyzed by Me_3SiOTf in the presence of $ROSiMe_3$. A number of 6-substituted *O*-isopropylguaicols are converted to guaicol acetates via group exchange when treated with stoichiometric Me_3SiOTf – Ac_2O in MeCN.[2]

N-Vinylation of chiral oxazolidinones is conveniently accomplished in two steps: α-Ethoxyethylation by acetaldehyde diethyl acetal followed by treatment with Me_3SiOTf – Et_3N.[3]

Glycosylation. Several glycosyl donors are activated by Me_3SiOTf. These are benzyl phthalates,[4] phosphites[5] and phosphates.[6]

Pyrroles. Nitriles insert into push-pull cyclopropanes to form substituted pyrroles, after dehydrative aromatization, when catalyzed by Me_3SiOTf.[7]

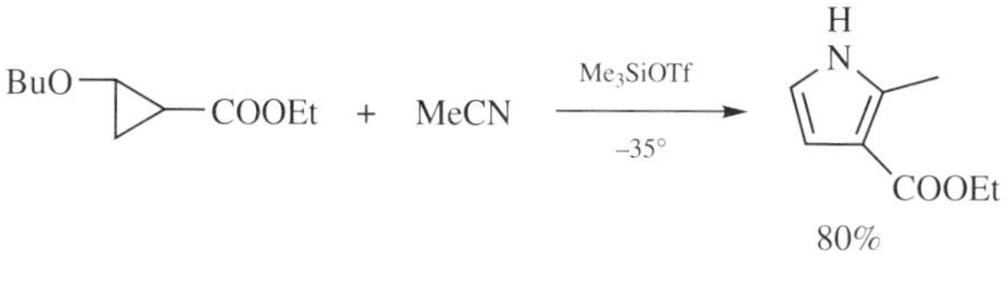

Tropinones. The conjugated ketene silyl acetal derived from an acetoacetic ester smoothly react with *N*-protected 2,5-dialkoxypyrrolidines to form tropinones.[8]

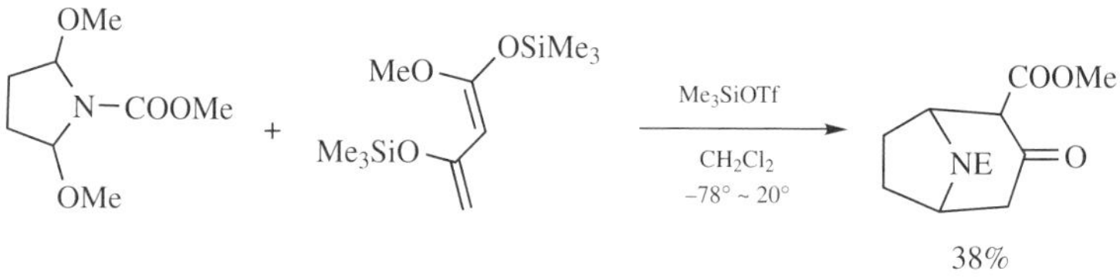

Epoxide opening. A powerful reagent system for alkylative scission of epoxides is R_3Al - Me_3SiOTf.[9] The products are silyl ethers.

Allylation. When allenylsilanes condense with carbonyl compounds under the influence of Me_3SiOTf, 2-hydroxyalkyl-1,3-dienes result. A synthesis of 3,4-dimethylenetetrahydropyrans based on this reaction (intermolecular[10] and intramolecular[11]) is significant.

SiMe3 HO H O — Me3SiOTf, Et2O, −78° ~ rt → O H

Rearrangements. The Michaelis-Arbuzov rearrangement of alkyl phosphinites proceeds at room temperature on exposure to catalytic amount of Me_3SiOTf,[12] whereas elevated temperatures are required using Me_3SiX ($X = Br, I$) as promoter.

Schmidt reaction of enones with alkyl azides gives enaminones.[13] Thus, 2-cyclohexenones afford 2-alkylaminomethylenecyclopentanones.

O + BnN_3 — Me3SiOTf, CH2Cl2 → O NHBn

68%(*Z/E* 3:2)

Isomerization of β,γ-epoxyalkyl silyl ethers to β-siloxy aldehydes is a key step in an alternative route to protected aldols. Another siloxy group whose oxygen atom can participate in tetrahydrofuran formation must be bulky in order to avoid such event.[14]

[1]Kurihara, M., Hakamata, W. *JOC* **68**, 3413 (2003).
[2]Williams, C.M., Mander, L.N. *TL* **45**, 667 (2004).
[3]Gaulon, C., Dhal, R., Dujardin, G. *S* 2269 (2003).
[4]Kim, K.S., Lee, Y.J., Kim, H.Y., Kang, S.S., Kwon, S.Y. *OBC* **2**, 2408 (2004).
[5]Tsuda, T., Sato, S., Nakamura, S., Hashimoto, S. *H* **59**, 509 (2003).
[6]Tsuda, T., Nakamura, S., Hashimoto, S. *T* **60**, 10711 (2004).
[7]Yu, M., Pagenkopf, B.L. *OL* **5**, 5099 (2003).
[8]Albrecht, U., Armbrust, H., Langer, P. *SL* 143 (2004).
[9]Shanmugam, P., Miyashita, M. *OL* **5**, 3265 (2003).
[10]Cho, Y.S., Karupaiyan, K., Kang, H.J., Pae, A.N., Cha, J.H., Koh, H.Y., Chang, M.H. *CC* 2346 (2003).
[11]Kang, H.J., Kim, S.H., Pae, A.N., Koh, H.Y., Chang, M.H., Choi, K., Han, S.-Y., Cho, Y.S. *SL* 2545 (2004).
[12]Renard, P.-Y., Vayron, P., Leclerc, E., Valleix, A., Mioskowski, C. *ACIEE* **42**, 2389 (2003).

[13]Reddy, D.S., Judd, W.R., Aube, J. *OL* **5**, 3899 (2003).
[14]Jung, M.E., Hoffmann, B., Rausch, B., Contreras, J.-M. *OL* **5**, 3159 (2003).

Trimethyltin hydroxide.

Ester hydrolysis. With Me_3SnOH for the hydrolysis of esters conditions are very mild. Many functional groups including azido function and silyl ethers are preserved.

[1]Nicolaou, K.C., Estrada, A., Zak, M., Lee, S., Safina, B. *ACIEE* **44**, 1378 (2005).

4,4,6-Trimethyl-2-vinyl-1,3,2-dioxaborinane.

Heck reaction. The title reagent is much more stable and easier to purify than the corresponding pinacolato analogue because the latter tends to form azeotropes with solvents. The effectiveness of both in the Heck reaction is similar.

[1]Lightfoot, A.P., Maw, G., Thirsk, C., Twiddle, S.J.R., Whiting, A. *TL* **44**, 7645 (2003).

Triphenylbismuth carbonate.

Triazolin-3,5-diones.[1] When triazolinediones are required to participate in the Diels-Alder reaction, they can be generated from urazoles by oxidation with $(Ph_3Bi)CO_3$.

[1]Menard, C., Doris, E., Mioskowski, C. *TL* **44**, 6591 (2003).

Triphenylphosphine. 18, 385–386; **19**, 382–383; **20**, 411–412; **21**, 462–463; **22**, 475

Aza-Baylis-Hillman reaction.[1] With Ph_3P as catalyst, δ-formyl α,β-enones are converted into 5-hydroxycyclopentenyl ketones. Such products are of obvious synthetic utility.

O
Ph CHO
Ph_3P
MeCN
O OH
Ph
99%

Acrylic esters. Alcoholysis of maleic anhydride in refluxing toluene in the presence of Ph_3P gives rise to acrylic esters.[2] This reaction shows good sense in terms of the HSAB principle.

Coupling reactions. Activated methylene compounds are alkylated with 2-alkynoic esters in the anti-Michael sense in the presence of Ph_3P – HOAc – NaOAc.[3] [A superior catalyst is Bu_3P.[4]]

COOEt + O OH
Ph_3P
HOAc, NaOAc
PhMe, Δ
EtOOC O OH

A synthesis of functionalized cyclopentene is based on Ph_3P-catalyzed annulation with activated allylic halides.[5]

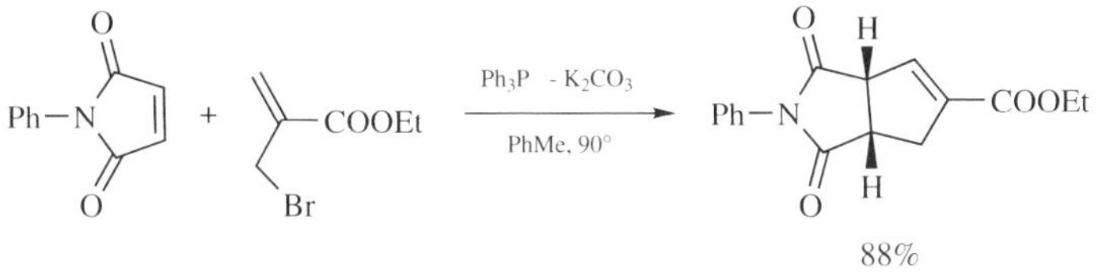

Ph_3P - K_2CO_3, PhMe, 90°

88%

Reduction. A Lewis acid such as $AlBr_3$ in combination with Ph_3P promotes reduction of α-dicarbonyl compounds to α-ketols.[6] α-Keto thiol esters are also reduced.

Heterocycles. Epoxides take up CO_2 on heating with Ph_3P – PhOH – NaI to generate 1,3-dioxolan-2-ones.[7] γ-Acyloxyalkynoic esters undergo cyclization to afford furans on heating with Ph_3P.[8]

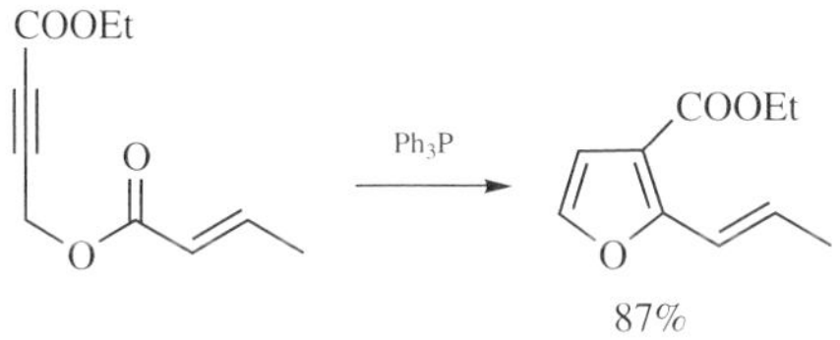

Ph_3P

87%

Analogues. Byproducts such as $Ph_3P{=}O$ often cause difficulties in purification of reaction mixtures, the problem is particularly serious when the desired products are delicate. 2-Trimethylsilylethyl *p*-diphenylphosphinobenzoate is a surrogate of Ph_3P that its use facilitates product isolation after the Mitsunobu reaction.[9]

[1]Yeo, J.E., Yang, X., Kim, H.J., Koo, S. *CC* 236 (2004).
[2]Adair, G.R.A., Edwards, M.G., Williams, J.M.J. *TL* **44**, 5523 (2003).
[3]Hanedanian, M., Loreau, O., Taran, F., Mioskowski, C. *TL* **45**, 7035 (2004).
[4]Hanedanian, M., Loreau, O., Sawicki, M., Taran, F. *T* **61**, 2287 (2005).
[5]Du, Y., Lu, X., Zhang, C. *ACIEE* **42**, 1035 (2003).
[6]Kikuchi, S., Hashimoto, Y. *SL* 1267 (2004).
[7]Huang, J.-W., Shi, M. *JOC* **68**, 6705 (2003).
[8]Jung, C.-K., Wang, J.-C., Krische, M.J. *JACS* **126**, 4118 (2004).
[9]Yoakim, C., Guse, I., O'Meara, J.A., Thavonekham, B. *SL* 473 (2003).

Triphenylphosphine – diethyl azodicarboxylate. **13**, 332; **14**, 336–337; **17**, 389–390; **18**, 387; **19**, 384–385; **20**, 413; **21**, 463–464; **22**, 475–476

Alkyl thiocyanates.[1] By the agency of Ph_3P - DEAD alcohols are rapidly converted to RSCN by NH_4SCN.[1] Carboxylic acids yield acyl isocyanates under the same conditions.

Carbamate esters.[2] Alcohols and amines are joined together on heating (90° –100°) with the Misunobu reagent in DMSO while CO_2 is bubbled through the solution. The key step is a reaction between $[RNHCOO]^-$ and $[R'OPPh]^+$.

[1]Iranpoor, N., Firouzabadi, H., Akhalghinia, B., Azadi, R. *S* 92 (2004).
[2]Chaturvedi, D., Kumar, A., Ray, S. *TL* **44**, 7637 (2003).

Triphenylphosphine – 2,3-dichloro-5,6-dicyano-1,4-benzoquinone

Transformation of alcohols. The reagent system activates alcohols for substitution and their conversion into alkyl nitrites,[1] cyanides,[2] and azides[3] by reaction with Bu_4NX ($X = NO_2$, CN, N_3) is quite simple. Diethyl α-halophosphonates and α-azidophosphonates are similarily prepared from the corresponding α-hydroxyphosphonates.[4]

β-Halohydrins.[5] Epoxides are opened under the same conditions using Bu_4NX as halide source.With double the quantity of Bu_4NBr dibromides are obtained.

[1]Akhlaghinia, B., Pourali, A.R. *S* 1747 (2004).
[2]Iranpoor, N., Firouzabadi, H., Akhlaghinia, B., Nowrouzi, N. *JOC* **69**, 2562 (2004).
[3]Iranpoor, N., Firouzabadi, H., Akhlaghinia, B., Nowrouzi, N. *TL* **45**, 3291 (2004).
[4]Firouzabadi, H., Iranpoor, N., Sobhani, S. *T* **60**, 203 (2004).
[5]Iranpoor, N., Firouzabadi, H., Aghapour, G., Nahid, A. *BCSJ* **77**, 1885 (2004).

Triphenylphosphine – diisopropyl azodicarboxylate. **15**, 352–353; **17**, 390; **18**, 387–388; **19**, 385; **20**, 413–414; **21**, 464; **22**, 476–477

Amine synthesis.[1] Mitsunobu reaction of *t*-butyl *N*-diethylphosphonylcarbamate with alcohols gives *N*-alkyl derivatives which on treatment with TsOH in ethanol results in tosylates of primary amines.

Esterification.[2] Only aliphatic alcohols are activated by the reagent couple therefore carboxylic acids and alcohols containing free phenolic OH can be used to form esters.

[1]Klepacz, A., Zwierzak, A. *SC* **31**, 1683 (2001).
[2]Appendino, G., Minassi, A., Daddario, N., Bianchi, F., Tron, G.C. *OL* **4**, 3839 (2002).

Triphenylphosphine – halogen/perhalocarbons. **22**, 477–478

Dehydration. Tertiary alcohols undergo dehydration with Ph_3P-I_2 at room temperature to form the most stable alkenes.[1] Cyclodehydration occurs to afford 2-substituted 4-chloroimidazoles when *N*-cyanomethylcarboxamides are exposed to Ph_3P–CCl_4.[2] Conversion of allylic carbamates to 1,3-transposed isocyanates[3] involves dehydration to cyanate esters and [3.3]sigmatropic rearrangement.

Halides. Alkenyl chlorides are prepared from ketones via enol phosphates by treating the latter compounds with Ph_3P-Cl_2.[4] For conversion of α-cyanohydrins into α-bromo nitriles the combination of Ph_3P with 2,4,4,6-tetrabromo-2,5-cyclohexadienone shows good activity.[5]

Methods for activation of sugar derivatives for glycosylation are numerous. Glycosyl bromides are formed using Ph_3P-CBr_4 in DMF and they afford α-glycosides in high stereoselectivity.[6] The same reagent also activates glycosyl anisoylamides.[7]

Ring expansion of 2-(α-hydroxyalkyl)indolines to 3-chlorotetrahydroquinolines[8] on reaction with Ph_3P-CCl_4 is stereoselective.

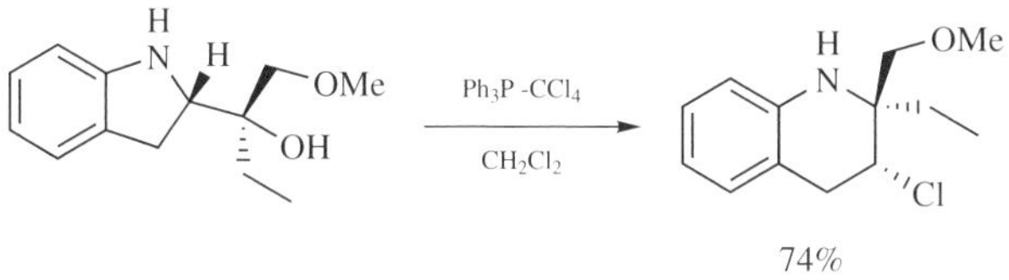

Highly regioselective Wittig reaction of glycosyl acetates is observed when they are subjected to microwave heating with Ph_3P - Cl_2 – KCl in toluene. The enol ether group is stable during deacetylation of other protecting groups by NaOMe but removable under mild acidic conditions.[9]

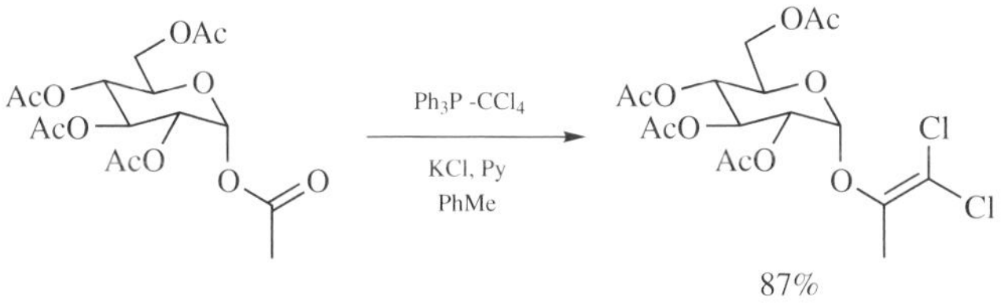

Transformation of carboxylic acids into acyloxyphosphonium salts or acyl halides by various combinations of Ph_3P and halogens/pseudohalogens for further reaction with alcohols and amines is well exploited. A system consisting of Cl_3CCN is equally effective[10] and that of NBS is said to be useful for monoacylating diamines.[11] A newer application is in preparation of α-diazoketones.[12]

Nitrogen-containing π-deficient heteroaromatic lactams such as 2(1*H*)-quinolinone, quinazolinone and quinoxalinone are converted into chloro compounds by a reagent derived from Ph_3P and isocyanuric chloride.[13]

Condensation. A synthesis of ureas involves heating amines with Ph_3P-CCl_4 in CH_2Cl_2 followed by addition of Et_3N and bubbling in CO_2.[14]

[1]Alvarez-Manzaneda, E.J., Chahboun, R., Torres, E.C., Alvarez, E., Alvarez-Manzaneda, R., Haidour, A., Ramos, J. *TL* **45**, 4453 (2004).
[2]Zhong, Y.-L., Lee, J., Reamer, R.A., Askin, D. *OL* **6**, 929 (2004).
[3]Ichikawa, Y., Ito, T., Nishiyama, T., Isobe, M. *SL* 1034 (2003).
[4]Kamei, K., Maeda, N., Tatsuoka, T. *TL* **46**, 229 (2005).
[5]Matveeva, E.D., Podrugina, T.A., Tishkovskaya, E.V., Zefirov, N.S. *MC* 260 (2003).
[6]Nishida, Y., Shingu, Y., Dohl, H., Kobayashi, K. *OL* **5**, 2377 (2003).
[7]Pleuss, N., Kunz, H. *ACIEE* **42**, 3174 (2003).
[8]Ori, M., Toda, N., Takami, K., Tago, K., Kogen, H. *ACIEE* **42**, 2540 (2003).

[9]de Figueiredo, R.M., Bailliez, V., Dubreuil, D., Olesker, A., Cleophax, J. *S* 2831 (2003).
[10]Jang, D.O., Cho, D.H., Kim, J.-G. *SC* **33**, 2885 (2003).
[11]Bandgar, B.P., Bettigeri, S.V. *SC* **34**, 2917 (2004).
[12]Cuevas-Yanez, E., Garcia, M.A., de la Mora, M.A., Muchowski, J.M., Cruz-Almanza, R. *TL* **44**, 4815 (2003).
[13]Sugimoto, O., Tanji, K. *H* **65**, 181 (2005).
[14]Porwanski, S., Menuel, S., Marsura, X., Marsura, A. *TL* **45**, 5027 (2004).

Triphenylsilyl vanadate. 22, 478–9

Isomerization-condensation. Using $VO(OSiPh)_3$ to promote condensation of allenyl carbinols with imines leads to β′-amino-α,β-unsaturated ketones.[1] Products possessing an *anti*- α′,β′-configuration are favored.

OH + NCOOMe, Ph — $(Ph_3SiO)_3VO$, CH_2Cl_2 → O, NHCOOMe, Ph

79%

[1]Trost, B.M., Jonasson, C. *ACIEE* **42**, 2063 (2003).

Triphosgene. 18, 388; **19**, 386; **20**, 415–416; **21**, 464–465; **22**, 479

Dealkylation. Tertiary benzyl amines are selectively cleaved on reaction with triphosgene.[1] The products (R_2NCOCl) can be used for other purposes. Other tertiary almines can be transformed into dialkylcarbamoyl azides on treatment with triphosgene and immediately quenched with NaN_3.[2]

Desulfurization. Alkyl thionocarbamates undergo desulfurization and transesterification on reaction with triphosgene, followed by adding alcohols or phenols.[3]

[1]Banwell, M.G., Coster, M.J., Harvey, M.J., Moraes, J. *JOC* **68**, 613 (2003).
[2]Gumaste, V.K., Deshmukh, A.R.A.S. *TL* **45**, 6571 (2004).
[3]Joshi, U.M., Patkar, L.N., Rajappa, S. *SC* **34**, 33 (2004).

Triruthenium dodecacarbonyl. 18, 308; **19**, 386–387; **20**, 416–417; **21**, 465–467; **22**, 479–481

Addition to multiple CC bonds. 1-Alkynes add to acrylic esters to provide 4-alkynoic esters[1] under the influence of $Ru_3(CO)_{12}$ and in the presence of bis(triphenylphosphine)iminium chloride, an additive providing dissociated chloride ion. Homologous amides are formed from 2-formamidopyridine through hydrocarbamoylation.[2] Picolyl formate is homologated by alkynes to afford picolyl 2-alkenoates.[3]

Allylation. In situ reduction of $ArNO_2$ by CO and attachment of an [ArNH] residue to an allylic position of a nonactivated alkene are catalyzed by the Ru complex.[4] The nitroarenes may contain electron-withdrawing or electron-donating groups.

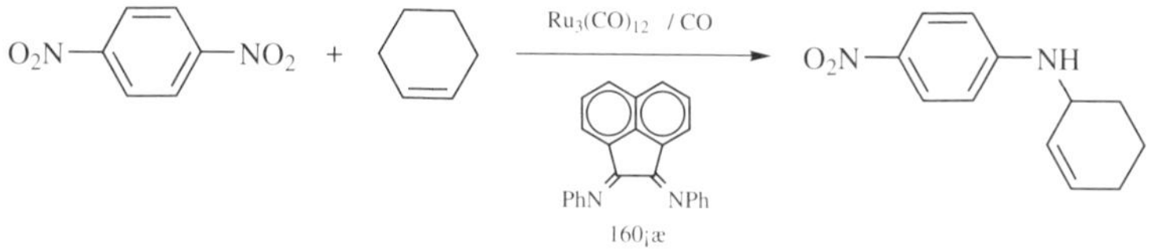

Reductive allylation of aldehydes by a distant allylic ester creates 2-vinylcycloalkanols.[5]

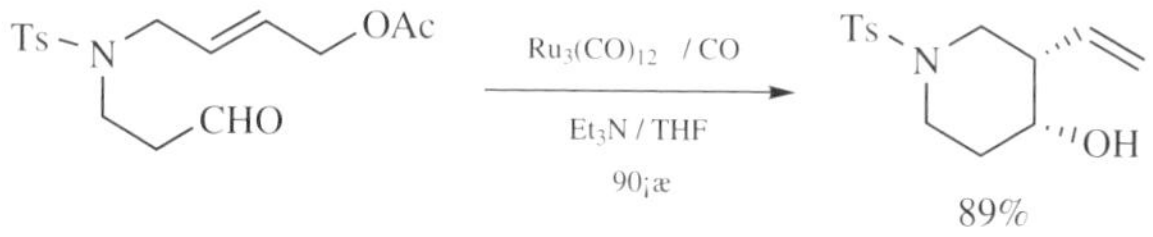

o-Functionalization. Comparing to 2-aryloxazoles, 2-arylthiazoles and 1-methyl-3-arylpyrazoles, 1-arylpyrazoles are particularly reactive toward propionation with CO and ethene at the *o*-position of the aryl groups due to strong chelation effect.[6] Electron-donating group(s) in the aryl moiety favors the reaction and dimethylacetamide is the solvent of choice. Similarly, silylation of arylazoles, arylimines, and arylpyridines with R_3SiH is also completely regioselective because Ru insertion into the arene C—H occurs after coordination with a nitrogen atom.[7]

Coupling reactions. α-Picolyl esters are acyl donors that form ketones with organoboron compounds. The acyl-oxygen bond cleavage is followed by hydrodecarbonylation in the presence of a hydride source ($HCOONH_4$).[8] In conjunction with various Pd salts and complexes $Ru_3(CO)_{12}$ transforms α-picolyl formate and ArX into picolyl aroates.[9]

Cyclocarbonylation. Allenyl alcohols and amines incorporate CO with formation of unsaturated lactones and lactams[10] on heating with $Ru_3(CO)_{12}$ and Et_3N under CO. Conjugated imines combine with CO and 1-alkenes to give α,α-disubstituted β,γ-unsaturated γ-lactams.[11]

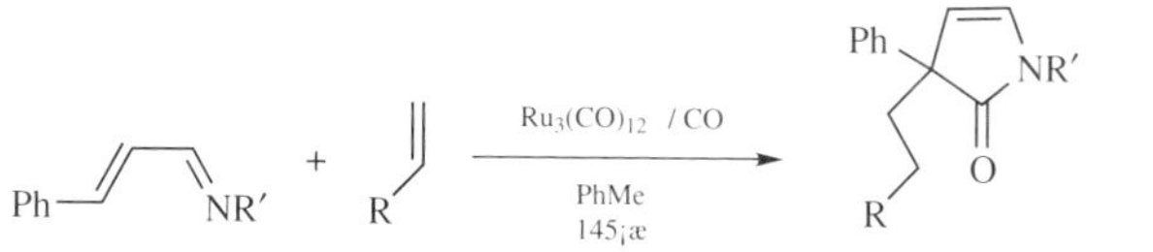

[1]Nishimura, T., Washitake, Y., Nishiguchi, Y., Maeda, Y., Uemura, S. *CC* 1312 (2004).
[2]Ko, S., Han, H., Chang, S. *OL* **5**, 2687 (2003).
[3]Na, Y., Ko, S., Hwang, L.K., Chang, S. *TL* **44**, 4475 (2003).
[4]Ragaini, F., Cenini, S., Turra, F., Caselli, A. *T* **60**, 4989 (2004).
[5]Yu, C.-M., Lee, S., Hong, Y.-T., Yoon, S.-K. *TL* **45**, 6557 (2004).
[6]Asaumi, T., Matsuo, T., Fukuyama, T., Ie, Y., Kakiuchi, F., Chatani, N. *JOC* **69**, 4433(2004).

[7]Kakiuchi, F., Matsumoto, M., Tsuchiya, K., Igi, K., Hayamizu, T., Chatani, N., Murai, S. *JOMC* **686**, 134 (2003).
[8]Tatamidani, H., Yokota, K., Kakiuchi, F., Chatani, N. *JOC* **69**, 5615 (2004).
[9]Ko, S., Lee, C., Choi, M.-G., Na, Y., Chang, S. *JOC* **68**, 1607 (2003).
[10]Yoneda, E., Zhang, S.-W., Zhou, D.-Y., Onitsuka, K., Takahashi, S. *JOC* **68**, 8571 (2004).
[11]Dönnecke, D., Imhof, W. *T* **59**, 8499 (2003).

Tris(acetonitrile)cyclopentadienylruthenium(I) hexafluorophosphate. 21, 467–468; **22**, 481–482

Addition to alkynes. Propargylic alcohols undergo regioselective Ru-catalyzed hydrosilylation.[1] On oxidative desilylation β-hydroxy ketones ensue, therefore propargylic alcohols are latent aldols.

Allylic substitutions. Cinnamyl phenol ethers are formed in a Ru-catalyzed reaction, but regioselectivity and enantioselectivity (in the presence of a chiral bisoxazoline ligand) are moderate.[2] Allylic ethers are cleaved efficiently using the Ru complex in conjunction with quinaldic acid.[3]

Cyclization. Diynes (1,6- and 1,7-) give cycloalkenyl ketones on heating with the Ru complex in aqueous acetone.[4] For unsymmetrical substrates the carbonyl group of the products is placed in a less hindered site.

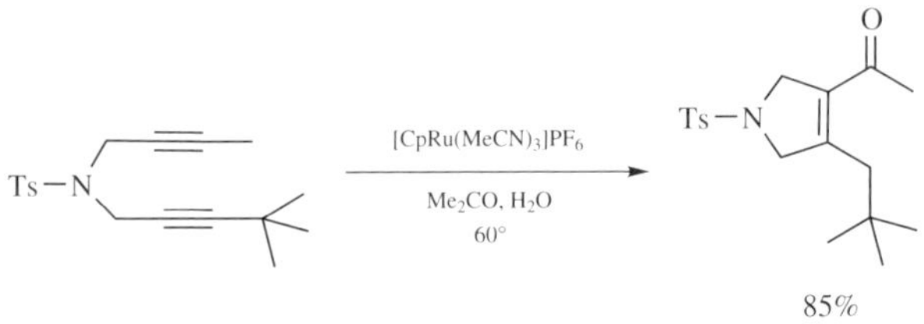

Ts−N ... $[CpRu(MeCN)_3]PF_6$ / Me_2CO, H_2O, 60° → Ts−N ... O

85%

Cycloisomerization of enynes in which the double bond is part of an allyl silyl ether shows stereochemical correlation with the Ru complexes (Cp vs. Cp* ligand).[5]

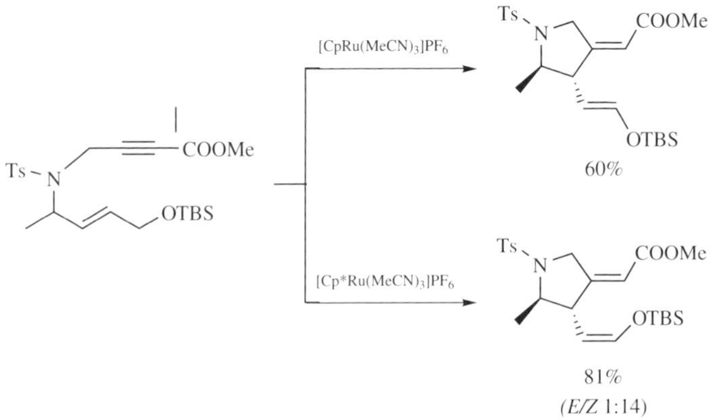

Ts−N ... COOMe ... OTBS

$[CpRu(MeCN)_3]PF_6$ → Ts N COOMe OTBS 60%

$[Cp^*Ru(MeCN)_3]PF_6$ → Ts N COOMe OTBS 81% (*E/Z* 1:14)

Cycloaddition. High diastereoselectivity is observed for the Ru-catalyzed intramolecular [5+2]cycloaddition to form a 5:7-fused ring system from substrates containing both an alkenylcyclopropane and an alkyne unit.[6]

Dimerization.[7] Two allenyl carbinol molecules are joined from head to center with acetoxylation of one molecule when treated with the Ru complex and $Cu(OAc)_2$. Also present in the reaction milieu are HOAc and i-Pr_2NEt.

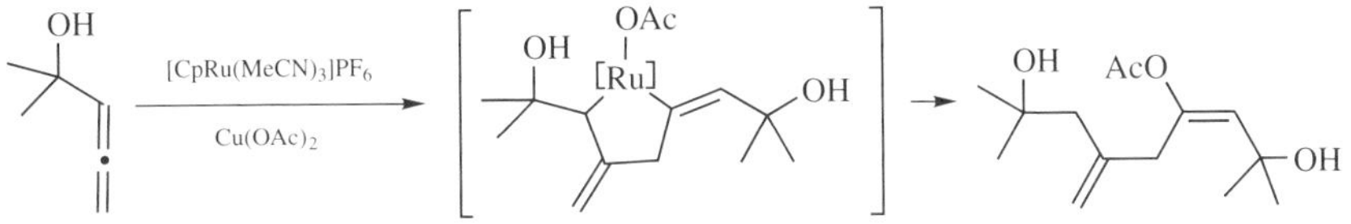

$[CpRu(MeCN)_3]PF_6$ / $Cu(OAc)_2$

Hydrosilylation. The Cp* analogues of the title reagent catalyzes hydrosilylation of alkynes. With protodesilylation by AgF in aq. THF-MeOH to succeed the addition reaction a method for stereoselective preparation of (E)-alkenes is established.[8]

[1]Trost, B.M., Ball, Z.T., Joge, T. *ACIEE* **42**, 3415 (2003).
[2]Mbaye, M.D., Renaud, J.-L., Demerseman, B., Bruneau, C. *CC* 1870 (2004).
[3]Tanaka, S., Saburi, H., Ishibashi, Y., Kitamura, M. *OL* **6**, 1873 (2004).
[4]Trost, B.M., Rudd, M.T. *JACS* **125**, 11516 (2003).
[5]Trost, B.M., Surivet, J.P., Toste, F.D. *JACS* **126**, 15592 (2004).
[6]Trost, B.M., Shen, H.C., Schulz, T., Koradin, C., Schirok, H. *OL* **5**, 4149 (2003).
[7]Yoshida, M., Gotou, T., Ihara, M. *TL* **44**, 7151 (2003).
[8]Lacombe, F., Radkowski, Seidel, G., Furstner, A. *T* **60**, 7315 (2004).

Tris(dibenzylideneacetone)dipalladium. **14**, 339; **15**, 353–355; **16**, 372; **17**, 394; **18**, 389–393; **19**, 388–390; **20**, 417–420; **21**, 469–473; **22**, 483–485

N-Arylation. Solvent-free protocol[1], assistance by microwaves,[2] and use of aryl nonaflates[3] for N-arylation are reported. Annulated N-substituted indoles such as tetrahydrocarbazoles are synthesized by sequential N-arylation and alkenation.[4] By modification of catalyst in situ it is possible to control reaction selectivity in the synthesis of N-aryl-2-benzylindolines from 2-allylanilines.[5] In other words, the initial N-arylation is favored by a bulky electron-rich monodentate ligand (2-di-t-butylphosphinobiphenyl) and in the second stage addition of the chelating ligand bis(2-diphenylphosphinophenyl) ether promotes alkene insertion in preference to reductive elimination.

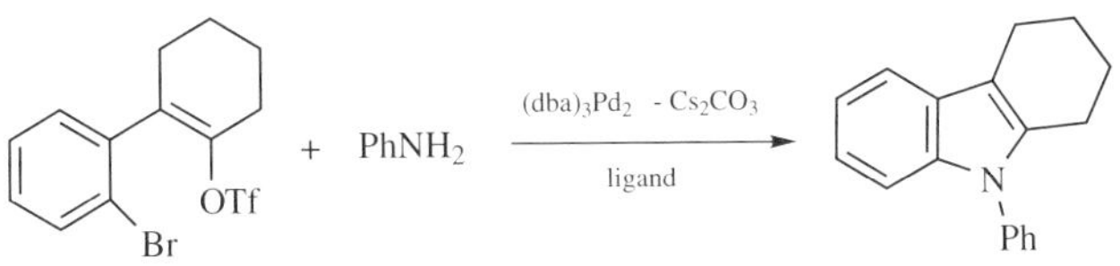

+ $PhNH_2$ — $(dba)_3Pd_2$ - Cs_2CO_3 / ligand →

Enamine/imine synthesis via coupling of amines with alkenyl halides is quite effective.[6,7] Nitrogen-containing components for the coupling now include chiral lactams,[8] oxazolidinones,[9] ureas,[10] carbamates[11] and sulfonamides.[12]

Coupling reactions. Suzuki coupling with ArF,[13] using as ligands (ferrocenylphenyl)diphenylphosphine,[14] 2-(dicyclohecylphosphino)benzamides,[15] and t-Bu_2PH with an *N*-acyl-*N*-heterocyclic carbene precursor[16] are some of the new developments. Of particular significance is the discovery of the phosphatrioxaadamantane ligands (**1**) which are inexpensive to make (from 2,4-pentanedione and PH_3, followed by *P*-arylation), air-stable, recoverable and reusable, and whose performance compares well with t-Bu_3P.[17] As expected, **1** finds use in *N*-arylation.[18] Another universal catalyst system for Suzuki coupling containing the (2',6'-dimethoxy-2-bipenyl)dicyclohecylphosphine ligand has been identified.[19]

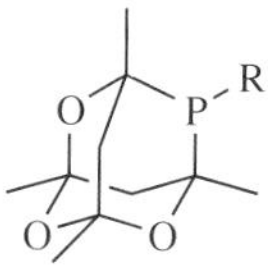

1

For Stille coupling with ArCl, caged phosphintriamide ligands are useful.[20,21] Desulfonylative coupling of arylsulfonyl chlorides with organostannanes requires also a copper(I) salt.[22]

The Pd-complex together with tricyclopentylphosphine and *N*-methylimidazole constitutes a catalyst system for Negishi coupling of unactivated halides and tosylates, and it is applicable to different kinds of organozinc reagents.[23] A report deals with microwave assistance in Negishi coupling and Kumada coupling of ArCl.[24]

The coupling between ArX and 4-pentenol forms 2-benzyltetrahydrofurans.[25] An unusual intramolecular alkene insertion into the alkoxypalladium intermediates is involved.

Preparation of ArCN by the Pd-catalyzed reaction of ArX with KCN is promoted by small amounts of an organotin compound (e.g., Bu_3SnCl).[26] Alkynylgermanium azatris-(ethoxide)s undergo Pd-catalyzed coupling with ArX.[27]

Allylation. Allylating agents for aldehydes are generated from allyl esters and bis(pinacolato)diboron in the presence of $(dba)_3Pd_2$.[28] Triorganoindium reagents find use in S_N2 reaction of cyclic allylic esters.[29]

Ketone synthesis. Organostannanes react with acyl chlorides[30] or thiol esters[31] by Pd catalysis and Cu promotion.

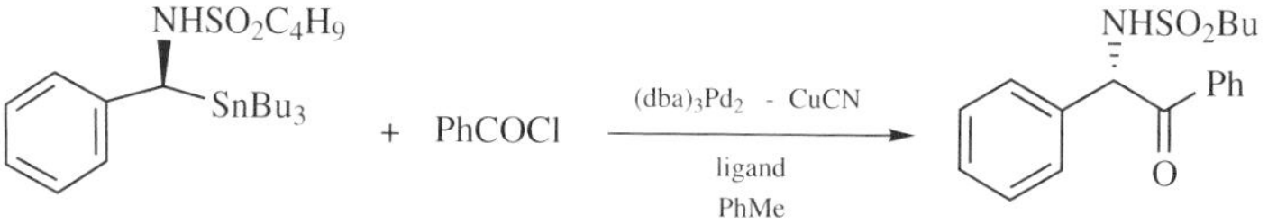

$NHSO_2C_4H_9$ / $SnBu_3$ + PhCOCl —($(dba)_3Pd_2$ - CuCN; ligand; PhMe)→ $NHSO_2Bu$ / Ph / O

[1]Artamkina, G.A., Ermolina, M.V., Beletskaya, I.P. *MC* 158 (2003).
[2]Wang, T., Magnin, D.R., Hamann, L.G. *OL* **5**, 897 (2003).
[3]Anderson, K.W., Mendez-Perez, M., Priego, J., Buchwald, S.L. *JOC* **68**, 9563 (2003).
[4]Willis, M., Brace, G., Holmes, I. *ACIEE* **44**, 403 (2005).
[5]Lira, R., Wolfe, J.P. *JACS* **126**, 13906 (2004).
[6]Barluenga, J., Fernandez, M.A., Aznar, F., Valdes, C. *CEJ* **10**, 494 (2004).
[7]Barluenga, J., Fernandez, M.A., Aznar, F., Valdes, C. *CC* 1400 (2004).
[8]Browning, R.G., Badarinarayana, V., Mahmud, H., Lovely, C.J. *T* **60**, 359 (2004).
[9]Ghosh, A., Sieser, J.E., Riou, M., Cai, W., Rivera-Ruiz, L. *OL* **5**, 2207 (2003).
[10]Sergeev, A.G., Artamkina, G.A., Beletskaya, I.P. *TL* **44**, 4719 (2003).
[11]Wallace, D.J., Klauber, D.J., Chen, C., Volante, R.P. *OL* **5**, 4749 (2003).
[12]Burton, G., Cao, P., Li, G., Rivero, R. *OL* **5**, 4373 (2003).
[13]Widdowson, D.A., Wilhelm, R. *CC* 578 (2003).
[14]Kwong, F.Y., Chan, K.S., Yeung, C.H., Chan, A.S.C. *CC* 2336 (2004).
[15]Dai, W.-M., Li, Y., Zhang, Y., Lai, K.W., Wu, J. *TL* **45**, 1999 (2004).
[16]Palencia, H., Garcia-Jimenez, F., Takacs, J.M. *TL* **45**, 3849 (2004).
[17]Gerristma, D., Brenstrum, T., McNulty, J., Capretta, A. *TL* **45**, 8319 (2004).
[18]Adjabeng, G., Brenstrum, T., Wilson, J., Frampton, C., Robertson, A., Hillhouse, J., McNulty, J., Capretta, A. *OL* **5**, 953 (2003).
[19]Walker, S.D., Barder, T.E., Martinelli, J.R., Buchwald, S.L. *ACIEE* **43**, 1871 (2004).
[20]Su, W., Urgaonkar, S., Verkade, J.G. *OL* **6**, 1421 (2004).
[21]Su, W., Urgaonkar, S., McLaughlin, P.A., Verkade, J.G. *JACS* **126**, 16433 (2004).
[22]Dubbaka, S.R., Vogel, P. *JACS* **125**, 15292 (2003).
[23]Zhou, J., Fu, G.C. *JACS* **125**, 12527 (2003).
[24]Walla, P., Kappe, C.O. *CC* 564 (2004).
[25]Wolfe, J.P., Rossi, M.A. *JACS* **126**, 1620 (2004).
[26]Yang, C., Williams, J.M. *OL* **6**, 2837 (2004).
[27]Faller, J.W., Kultyshev, R.G., Parr, J. *TL* **44**, 451 (2003).
[28]Sebelius, S., Wallner, O.A., Szabo, K.J. *OL* **5**, 3065 (2003).
[29]Baker, L., Minehan, T. *JOC* **69**, 3957 (2004).
[30]Kells, W K., Chong, J.M. *JACS* **126**, 15666 (2004).
[31]Wittenberg, R., Srogl, J., Egi, M., Liebeskind, L.S. *OL* **5**, 3033 (2003).

Tris(dibenzylideneacetone)dipalladium - chloroform. 19, 390–392; **20**, 420–422; **21**, 474–477; **22**, 485–489

Addition reactions. The catalytic activity of the title complex in promoting additions is evidenced from its use in the splitting silylstannanes and delivering the components to propargylic alcohols and their ethers (but not acetates),[1] and ring closure of alkynyl triflamides (via intramolecular addition to π-allylpalladium complexes which are formed by hydropalladation of the alkyne moiety, isomerization and complexation).[2]

The conversion of 2-methyleneaziridines to α-amidoacetones[3] proceeds from hydropalladation of the double bond with RCOOPdH, reductive elimination to O$\rightarrow$N transacylation with ring opening. This reaction pattern differs from that involves Bronsted acids.

N-Acyliminium salts that are formed in situ are prone to add a vinyl group from vinyltributylstannane through the palladized intermediates.[4] The result is simultaneous *C*-vinylation and *N*-acylation.

Ring formation. Cycloisomerization of 7-alkyn-2-enols catalyzed by the Pd-catalyst is analogous to the ene reaction. Aldehydes are obtained.[5]

$(dba)_3Pd_2 \cdot CHCl_3$, HCOOH, DCE

1-Alkynes, allyl carbonates, and Me_3SiN_3 combine to give 4-substituted 2-allyl-1,2,3-triazoles in the Pd-catalyzed, Cu-promoted reaction.[6] π-Allyl(azido)palladium and alkynylcopper species intervened in the [3 + 2]cycloaddition.

π-Allylpalladium complexes are involved in the transformation of 4-hydroxy-2-alkenyl carbonates into 4-vinyl-1,3-dioxolan-2-ones.[7] In the case of the alkynyl analogues the reaction in the presence of phenols leads to O-C bond coupling.[8] In both series the carbon dioxide unit is retained in the products, yet homopropargylic alcohols behave differently.[9]

+ ArOH, $(dba)_3Pd_2 \cdot CHCl_3$, DPPE, dioxane 50°

+ ArOH, $(dba)_3Pd_2 \cdot CHCl_3$, DPPE, dioxane 50°

Reductive diyne cyclization is the key step for a synthesis of (+)-streptazolin.[10] This Pd-catalyzed reaction is performed in the presence of Et_3SiH and HCOOH.

$(dba)_3Pd_2 \cdot CHCl_3$, HCOOH, Et_3SiH, PhMe 80°

(+)-streptazolin

Under a CO atmosphere 1,6-enynes undergo carbonylative cyclization to provide 2-methylenecyclopentaneacetic acid and heterocyclic analogues.[11]

Coupling reactions. Molecular strain of 1-(3-acyloxypropynyl)cyclobutanols becomes a major factor in influencing product formation.[12] Coupling with the phenols takes place after ring expansion. 1,3-Butadienylated cyclobutanols undergo Heck reaction (with assistance of Ag_2CO_3) to afford 2-(3-aryl-1*Z*-propenyl)cyclopentanones.[13]

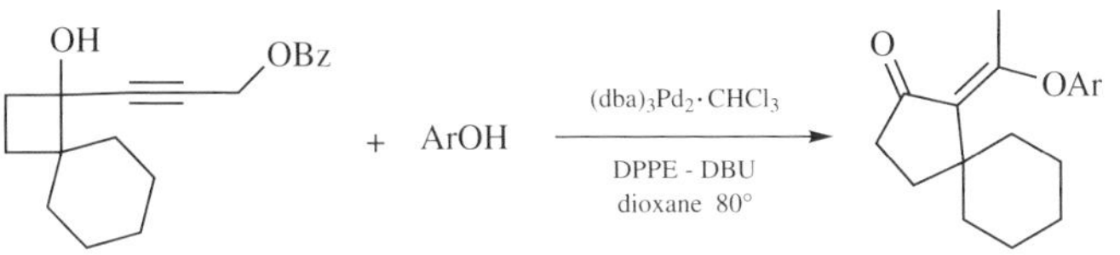

In situ activation of allylic alcohols by Ts_2O or Ms_2O (but not Tf_2O) for coupling with butyl acrylate solves the problem of handling unstable allyl sulfonates.[14]

Heck reaction of cyclic enol ethers often is complicated by double bond isomerization in the products (suppression requires excess Ag_2CO_3), and removal of unreacted ArI. Employment of aryldiazonium salts obviates the problem.[15] Cross-coupling involving ArI and 2-indolyldimethylsilanols to give 2-arylindoles using the Pd complex (and CuI as additive) has been reported.[16]

Allylation. Cyclic β-keto esters and β-diketones are allylated with propyn-1-ylarenes and 2-butynoic esters using the catalyst system $(dba)_3Pd_2 \cdot CHCl_3$ /PhCOOH.[17]

Substitution reactions. *N*-Heterocyclic carbene ligands are highly effective in the Pd-catalyzed allylic substitution.[18]

[1]Nielson, T.E., LeQuement, S., Tanner, D. *S* 1381 (2004).
[2]Lutete, L.M., Kadota, I., Yamamoto, Y. *JACS* **126**, 1622 (2004).
[3]Oh, B.H., Nakamura, I., Yamamoto, Y. *JOC* **69**, 2856 (2004).
[4]Davis, J.L., Dhawan, R., Arndtsen, B.A. *ACIEE* **43**, 590 (2004).
[5]Kressierer, C.J., Müller, T.J.J. *SL* 655 (2004).
[6]Kamijo, S., Jin, T., Huo, Z., Yamamoto, Y. *JACS* **125**, 7786 (2003).
[7]Yoshida, M., Ohsawa, Y., Ihara, M. *JOC* **69**, 1590 (2004).
[8]Yoshida, M., Fujita, M., Ishii, T., Ihara, M. *JACS* **125**, 4874 (2003).
[9]Yoshida, M., Morishita, Y., Fujita, M., Ihara, M. *TL* **45**, 1861 (2004).
[10]Trost, B.M., Chung, C.K., Pinkerton, A.B. *ACIEE* **43**, 4327 (2004).
[11]Aggarwal, V.K., Butters, M., Davies, P.W. *CC* 1046 (2003).
[12]Yoshida, M., Komatsuzaki, Y., Nemoto, H., Ihara, M. *OBC* **2**, 3099 (2004).
[13]Yoshida, M., Sugimoto, K., Ihara, M. *OL* **6**, 1979 (2004).
[14]Tsukada, N., Sato, T., Inoue, Y. *CC* 2404 (2003).
[15]Schmidt, B. *CC* 1656 (2003).
[16]Denmark, S.E., Baird, J.D. *OL* **6**, 3649 (2004).
[17]Patil, N.T., Yamamoto, Y. *JOC* **69**, 6478 (2004).
[18]Sato, Y., Yoshino, T., Mori, M. *OL* **5**, 31 (2003).

Tris(dimethylamino)phosphine.

Michael reaction.[1] Rapid 1,4-addition of pronucleophiles such as cyano-acetic esters to Michael acceptors occurs at room temperature in the presence of the title reagent.

[1]Grossman, R.B., Comesse, S., Rasne, R.M., Hattori, K., Delong, M.N. *JOC* **68**, 871 (2003).

Tris(pentafluorophenyl)borane. 20, 422; **21**, 478; **22**, 489–490

Substitutions. Using $(C_6F_5)_3B$ as catalyst glycal acetates are converted into 2,3-anhydroglycosyl amides and sulfonamides in an aza-version of Ferrier rearrangement.[1] Propargylic chloroacetates and allylsilanes enter reaction to afford 1,5-enynes.[2]

Nonactivated aziridines are opened by amines to afford *trans*-1,2-diamines in the presence of $(C_6F_5)_3B$.[3]

[1]Chandrasekhar, S., Reddy, C.R., Chandrasekhar, G. *TL* **45**, 6481 (2004).
[2]Schwier, T., Rubin, M., Gevorgyan, V. *OL* **6**, 1999 (2004).
[3]Watson, I.D.G., Yudin, A.K. *JOC* **68**, 5160 (2003).

Tris(pentafluorophenyl)silyl triflate.

Enol triflates.[1] The title reagent, prepared from treatment of phenyl or allyltris (pentafluorophenyl)silane with TfOH, is a potent silylating agent for carbonyl compounds and esters/lactones.

[1]Levin, V.V., Dilman, A.D., Belyakov, P.A., Korlyukov, A.A., Struchkova, M.I., Tartakovsky, V.A. *EJOC* 5141 (2004).

Tris(trimethylsilyl)silane. 19, 393; **20**, 423; **22**, 490–491

Cyclization. Unusually high diastereoselectivity in radical cyclization is observed using the reagent system $(Me_3Si)_3SiH$ - AIBN,[1] which is superior to Bu_3SnH-AIBN. Even more interesting is the desulfurative cyclization to give a lignan compound.[2]

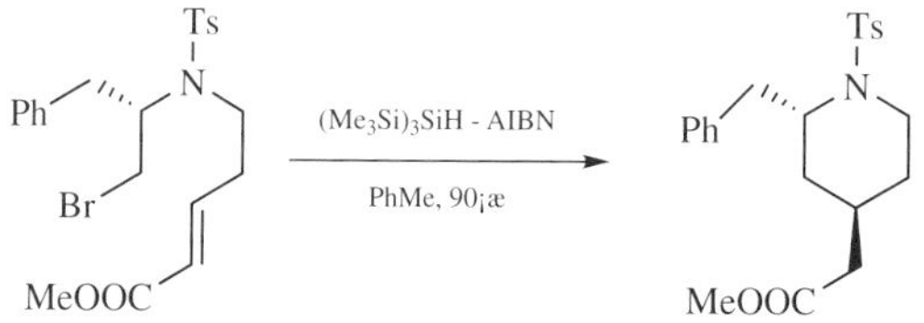

Ts N Ph Br MeOOC → $(Me_3Si)_3SiH$ - AIBN, PhMe, 90¡æ → Ts N Ph MeOOC

76% (*trans/cis* 92:8)

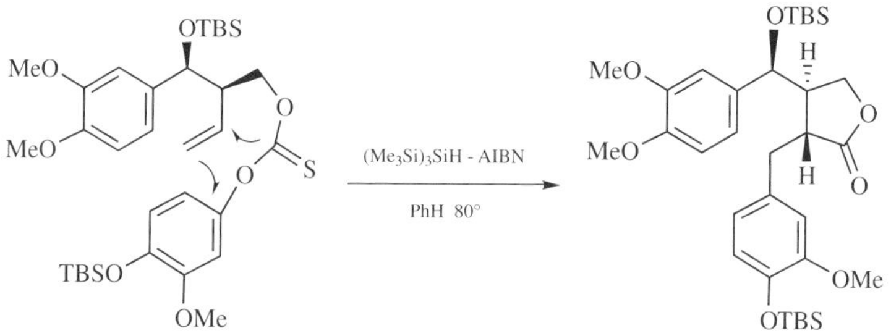

OTBS MeO MeO O S O TBSO OMe → $(Me_3Si)_3SiH$ - AIBN, PhH 80° → OTBS H MeO MeO O H O OMe OTBS

44%

Ring-fused quinolines are very efficiently assembled from carbamates, thiocarbamates, thioamides and thioureas by tandem radical additions.[3]

$(Me_3Si)_3SiH$ - AIBN, hν PhH

88%

***N*-Alkoxyamines.**[4] Preparation of these compounds from alkyl halides by a radical process uses *t*-butyl hyponitrite as thermal radical initiator (decomposition to generate *t*-butoxy radical) to abstract the hydrogen from $(Me_3Si)_3SiH$. In turn the silyl radical reacts with the halide and finally combination of the ensuing carbon radical with the nitroxide reagent completes the reaction.

[1]Gandon, L.A., Russell, A.G., Snaith, J.S. *OBC* **2**, 2270 (2004).
[2]Fischer, J., Reynolds, A.J., Sharp, L.A., Sherburn, M.S. *OL* **6**, 1345 (2004).
[3]Du, W., Curran, D.P. *OL* **5**, 1765 (2003).
[4]Braslau, R., Tsimelzon, A., Gewandter, J. *OL* **6**, 2233 (2004).

Tungsten carbene and carbyne complexes. 20, 424–425; **21**, 480–481; **22**, 492

Cyclopentannulation.[1] Diastereoselectivity of the process is greatly influenced by solvent.

THF	81%	---
Et_2O	---	84%

Alkyne metathesis.[2] A synthesis of tribenzocyclines from *o*-di(1-propynyl)arenes is accomplished by warming with $(t\text{-BuO})_3W{\equiv}C{-}CMe_3$.

[1]Barluenga, J., Alonso, J., Fananas, F.J. *JACS* **125**, 2610 (2003).
[2]Miljanic, O.S., Vollhardt, K.P.C., Whitener, G. *SL* 29 (2003).

Tungsten carbonyl heterocomplexes.

Distannylation.[1] Using 4-nitrobenzonitrile-$W(CO)_5$ complex as catalyst alkynes undergo vicinal distannylation with R_3SnH.

[1]Braune, S., Kazmaier, U. *ACIEE* **42**, 306 (2003).

Tungsten(VI) chloride. 19, 395; **20**, 425; **21**, 480-481; **22**, 492

Deoxychlorination.[1] Cyclic ketones are converted into alkenyl chlorides and/or *gem*-dichlorides by WCl_6 in CH_2Cl_2. For example, 2,2-dichloroadamantane is obtained from adamantanone in 54%. But camphor affords only 2-chloro-*p*-cymene (80%) and norcamphor gives 1,2(*exo*)-dichloronorbornane (41%).

[1]Jung, M.E., Wasserman, J.I. *TL* **44**, 7273 (2003).

Tungsten hexacarbonyl. 22, 492–493

Cyclization. The $W(CO)_6$-catalyzed photochemical cyclization of silyl enol ethers involving an allenyl chain pursues an endo mode.[1]

OTBS; $W(CO)_6$, hν, THF,H_2O; O; H; 94%

OTBS; $W(CO)_6$, hν, THF,H_2O; O; H; 89%

4-Alkynols undergo cyclization to give dihydropyran derivatives.[2] Cyclic enol ethers are formed preferentially by *endo*-cyclization.[3]

OTBDPS; OH; O; O; $W(CO)_6$, hν, Et_3N / THF; RO; H; O; O; O; + RO; H; O; O; O; (95 : 5)

[1]Miura, T., Kiyota, K., Kusama, H., Lee, K., Kim, H., Kim, S., Lee, P.H., Iwasawa, N. *OL* **5**, 1725 (2003).
[2]Wipf, P., Graham, T.H. *JOC* **68**, 8798 (2003).
[3]Alcazar, E., Pletcher, J., McDonald, F. *OL* **6**, 3877 (2004).

Tungsten pentacarbonyl tetrahydrofuran. 22, 493

2-Arylquinolines.[1] Vinylidene complex formation from (thf)W$(CO)_5$ and ethynylimines instigates electrocyclization to afford 2-arylquinolines after oxidative demetallation.

N
Ph
(thf)W$(CO)_5$
THF
N
Ph
W$(CO)_5$
NMO
N
Ph
70%

[1]Sangu, K., Fuchibe, K., Akiyama, T. *OL* **6**, 353 (2004).

U

Ultrasound.

Reduction.[1] A report on ultrasound-promoted C=O group reduction of enones by $NaBH_4$ in THF is somewhat unusual.

Aldol and Michael reactions. NaOH and KOH are used in aldol[2] and Michael reactions,[3] with ultrasound irradiation. Note that improvement of certain reactions requires high-intensity ultrasound.

[1]Zeynizadeh, B., Yahyaei, S. *ZN(B)* **59**, 699 (2004).
[2]Cravotto, G., Demetri, A., Nano, G.M., Palmisano, G., Penoni, A., Tagliapietra, S. *EJOC* 4438 (2003).
[3]Li, J.-T., Cui, Y., Chen, G.-F., Cheng, Z.-L., Li, T.-S. *SC* **33**, 353 (2003).

Urea.

Amidation.[1] Direct preparation of carboxamides from acids and urea with microwave heating is promoted by imidazole.

[1]Khalafi-Nezhad, A., Mokhtari, B., Rad, M.N.S. *TL* **44**, 7325 (2003).

Urea-hydrogen peroxide, UHP. 22, 494

Halogenation. After brief heating of ArI with UHP at about 85°, subsequent addition of conc. HCl completes the transformation into (dichloroiodo)arenes.[1] Iodination of arenes, including anilines, haloarenes, and aroic acids, by iodine and UHP, is usually accomplished under mild conditions.[2]

[1]Zielinska, A., Skulski, L. *TL* **45**, 1087 (2004).
[2]Lulinski, P., Kryska, A., Sosnowski, M., Skulski, L. *S* 441 (2004).

Urea nitrate.

Nitrocarbazoles.[1] Very rapid nitration of carbazoles at room temperature in HOAc is reported.

[1]Nagarajan, R., Muralidharan, D., Perumal, P.T. *SC* **34**, 1259 (2004).

Fiesers' Reagents for Organic Synthesis, Volume 23. Edited by Tse-Lok Ho

V

Vanadium pentoxide.

Oxidation.[1] Vanadium pentoxide catalyzes the conversion of ArCHO into esters by sodium perborate or percarbonate in alcoholic solvents.

[1]Gopinath, R., Barkakaty, B., Talukdar, B., Patel, B.K. *JOC* **68**, 2944 (2003).

Fiesers' Reagents for Organic Synthesis, Volume 23. Edited by Tse-Lok Ho

Water.

Condensation reactions. To replace organic solvents with water in many reactions has been quite successful. Some are more obvious than others because the substrates are highly polar or the reactions are thermodynamically driven. Thus, derivatization of oxazolidinone-fused sugars into urea glycosides by reaction with amines,[1] preparation of imidazolidines from aldehydes and ethylenediamines,[2] aldol reaction of salicylaldehydes with vinyl acetate,[3] and the 1,3-dipolar cycloaddition between organoazides and electron-deficient alkynes[4] are easily conceived. Efficient and clean aldol reaction is performed with Na_2CO_3 in water.[5] More unusual are the Michael additions,[6,7] and particularly Wittig reaction that shows (*E*)-preference.[8]

$C_{11}H_{23}CH=PPh_3$ + OHC–C_6H_4–OMe $\xrightarrow[\text{18-crown-6, }\Delta]{K_2CO_3}$ $C_{11}H_{23}CH=CH$–C_6H_4–OMe

	Z:*E*
in PhMe	81 : 19
in H_2O	27 : 73

Water greatly enhances the rate of the Baylis-Hillman reaction.[9] Aza-Diels-Alder reaction of Danishefsky diene and imines can be conducted in water with NaOTf as catalyst.[10]

Substitution reactions. Displacement of allylic esters by various pronucleophiles (*O*-, *S*-, *N*-, *C*-based) in H_2O–DMF (1:1) does not require a Pd catalyst if a hydrophilic ligand is present.[11]

Superheated reactions. With microwave irradiation, ethynylarenes are hydrated in water to deliver acetophenones[12] and monoalkylated malonic esters and β-keto esters lose the ester residue.[13] Hydrolysis of secondary alkyl halides in alkaline solution at 250° and 5 MPa largely follows an S_N2 pathway.[14] [Other reactions may not require such vigorous conditions, e.g., RCH_2I to RCH_2NO_2 conversion by $AgNO_2$ proceeds at room temperature.[15]]

In supercritical water various types of organosilanes undergo C-Si bond cleavage.[16] Heck reaction of PhI with styrene to form stilbene in scH_2O does not require any heavy metal catalyst.[17]

[1]Ichikawa, Y., Matsukawa, Y., Isobe, M. *SL* 1019 (2004).
[2]Jurcik, V., Wilhelm, R. *T* **60**, 3205 (2004).
[3]Kim, J.H., Lee, S., Kwon, M.-G., Park, S.S., Choi, S.-K., Kwon, B.-M. *SC* **34**, 1223 (2004).
[4]Li, Z., Seo, T.S., Ju, J. *TL* **45**, 3143 (2004).
[5]Zhang, Z., Dong, Y.-W., Wang, G.-W. *CL* **32**, 966 (2003).
[6]Naidu, B.N., Sorensen, M.E., Connolly, T.P., Ueda, Y. *JOC* **68**, 10098 (2003).
[7]Miranda, S., Lopez-Alvarado, P., Giorgi, G., Rodriguez, J., Avendano, C., Menendez, J.C. *SL* 2159 (2003).
[8]Pandolfi, E.M., Lopez, G.V., Dias, E., Seoane, G.A. *SC* **33**, 2187 (2003).
[9]Cai, J., Zhou, Zhao, G., Tang, C. *OL* **4**, 4723 (2002).
[10]Loncaric, C., Manabe, K., Kobayashi, S. *CC* 574 (2003).
[11]Chevrin, C., Le Bras, J., Henin, F., Muzart, J. *TL* **44**, 8099 (2003).
[12]Vasudevan, A., Verzal, M.K. *SL* 631 (2004).
[13]Curran, D.P., Zhang, Q. *ASC* **3**, 329 (2003).
[14]Yamasaki, Y., Hirayama, T., Oshima, K., Matsubara, S. *CL* **33**, 864 (2004).
[15]Ballini, R., Barboni, L., Carlo, G. *JOC* **69**, 6907 (2004).
[16]Itami, K., Terakawa, K., Yoshida, J., Kajimoto, O. *JACS* **125**, 6058 (2003).
[17]Zhang, R., Sato, O., Zhao, F., Sato, M., Ikushima, Y. *CEJ* **10**, 1501 (2004).

Y

Ytterbium. 14, 348; **15**, 366; **16**, 384; **18**, 401; **19**, 400; **20**, 431; **21**, 487; **22**, 497

Coupling reactions. Reductive coupling of diaryl ketones with aldimines is effected at room temperature by Yb.[1]

Allylation.[2] Yb-mediated displacement of the cyano group of many RCOCN provides 1-alken-4-ones. Crotyl and prenyl bromides react to give more highly substituted products.

[1]Su, W., Yang, B. *SC* **33**, 2613 (2003).
[2]Gohain, M., Gogoi, B.J., Prajapati, D., Sandhu, J.S. *NJC* **27**, 1038 (2003).

Ytterbium(III) triflate. 18, 402–403; **19**, 401–402; **20**, 431–433; **21**, 487–489; **22**, 498–500

Hydroxyl derivatization. $Yb(OTf)_3$ promotes formation of many derivatives from alcohols: Tosylates,[1] substituted diphenylmethyl ethers,[2] acetates of secondary alcohols from 1,2-alkanediols.[3]

Glycosylation catalyzed by $Yb(OTf)_3$ does not affect the acid-labile dimethoxytrityl group that coexists.[4] Addition of DME improves the α/β ratio.[5]

Substitutions. $Yb(OTf)_3$ enables formation of glycosyl cyanides from pyranosyl or furanosyl 1,2-*O*-sulfinates on reaction with NaCN in HMPA.[6] It also catalyzes transpositional sulfinylation of alkenes with arenesulfinamides in the presence of Me_3SiCl.[7]

NH_2 + → ($Yb(OTf)_3$; Me_3SiCl; CH_2Cl_2)

79%

Transformation of α-fluoroalkyl sulfides (obtained from sulfides and DAST) into *O,S*-acetals by reaction with alcohols is readily achieved in the presence of $Yb(OTf)_3$ and a hindered pyridine base at low temperature.[8] This is an important method relating to synthesis of natural polyethers.

Heterocycles. Pictet-Spengler reaction,[9] benzimidazole and quinoxaline-2,3-dione synthesis from 1,2-diaminoarenes with alehydes[10] and dialkyl oxalates,[11] respectively, are

Fiesers' Reagents for Organic Synthesis, Volume 23. Edited by Tse-Lok Ho

catalyzed by $Yb(OTf)_3$. 4-Methylenepiperidine is expeditiously prepared in a silyl version of the imino-ene reaction.[12]

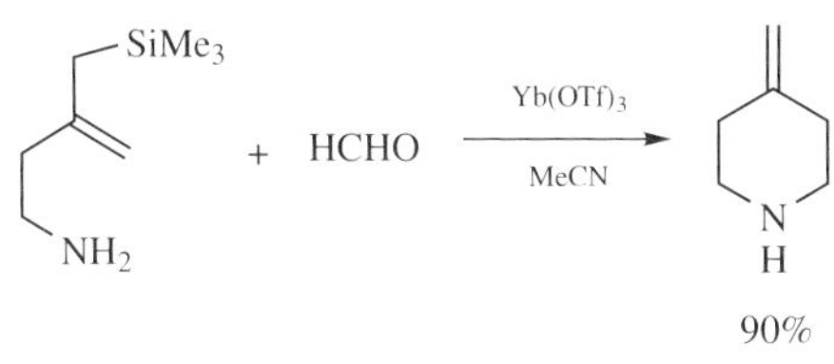

Insertion into cyclopropane ring by an activated C=O unit[13] or nitrone[14] leads to 5- and 6-membered heterocycles.

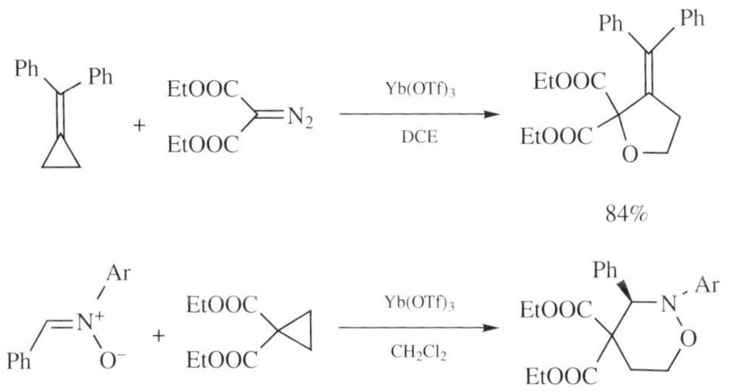

1-Arenylalkenes. The aldol reaction between a symmetrical ketone and ArCHO followed by elimination of carboxylic acid affords 1-arylalkene product.[15] This reaction tandem is promoted by Lewis acids, but $Yb(OTf)_3$ usually gives better results and no solvent is needed.

Note that under different conditions (presence of Hünig base) the products are 1,3-diol monoaroates. The reaction course follows an aldol-Tishchenko reaction tandem.[16]

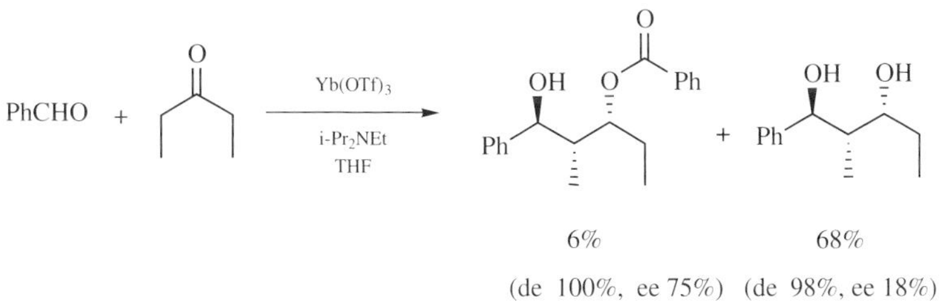

Condensations. CC bond-forming processes involving hydrogen transfer obeys the principle of atom economy. Several such reactions are mediated by $Yb(OTf)_3$, e.g., formation of β-aryl enones from RCHO and alkynylarenes.[17] Imino-ene reaction requires Me_3SiOTf,[18] and cyclization of β-dicarbonyl compounds (to form 6-, 7-, and 8-membered ketones) that contain a double bond, $(MeCN)_2PdCl_2$,[19] in addition to $Yb(OTf)_3$, respectively.

$$ArCHO + \equiv\!-Ph \xrightarrow{Yb(OTf)_3} ArC(O)CH{=}CHPh$$

Acylzirconocene chlorides add to imines to furnish α-amino ketones when catalyzed by $Yb(OTf)_3$ – Me_3SiOTf.[20]

[1]Comagic, S., Schirrmacher, R. *S* 885 (2004).
[2]Sharma, G.V.M., Prasad, T.R., Srinivas, B. *SC* **34**, 941 (2004).
[3]Ikejiri, M., Miyashita, K., Tsunemi, T., Imanishi, T. *TL* **45**, 1243 (2004).
[4]Adinolfi, M., Iadonisi, A., Schiattarella, M. *TL* **44**, 6479 (2003).
[5]Adinolfi, M., Iadonisi, A., Ravida, A., Schiattarella, M. *TL* **45**, 4485 (2004).
[6]Benksim, A., Beaupere, D., Wadouachi, A.. *OL* **6**, 3913 (2004).
[7]Alajarin, M., Pastor, A., Cabrera, J. *SL* 995 (2004).
[8]Inoue, M., Yamashita, S., Hirama, M. *TL* **45**, 2053 (2004).
[9]Srinivasan, N., Ganesan, A. *CC* 916 (2003).
[10]Curini, M., Epifano, F., Montanari, F., Rosati, O., Taccone, S. *SL* 1832 (2004).
[11]Wang, L., Liu, J., Tian, H., Qian, C. *SC* **34**, 1349 (2004).
[12]Furman, B., Dziedzie, M. *TL* **44**, 8249 (2003).
[13]Shi, M., Xu, B. *TL* **44**, 3839 (2003).
[14]Young, I.S., Kerr, M.A. *ACIEE* **42**, 3023 (2003).
[15]Curini, M., Epifano, F., Maltese, F., Marcotullio, M.C. *EJOC* 1631 (2003).
[16]Mlynarski, J., Mitura, M. *TL* **45**, 7549 (2004).
[17]Curini, M., Epifano, F., Maltese, F., Rosati, O. *SL* 552 (2003).
[18]Yamanaka, M., Nishida, A., Nakagawa, M. *JOC* **68**, 3112 (2003).
[19]Yang, D., Li, J.-H., Gao, Q., Yan, Y.-L. *OL* **5**, 2869 (2003).
[20]Kakuuchi, A., Taguchi, T., Hanzawa, Y. *EJOC* 116 (2003).

Z

Zeolites. 15, 367; **18**, 405–406; **19**, 403–404; **20**, 434; **21**, 491; **22**, 501

Group removals. Catalytic dehalogenation of halophenols with Na_2SO_3 in MeOH is a novel application of Hβ zeolite.[1] Certain esters are cleaved on heating with zeolite as catalyst, e.g., allyl esters.[2] Symmetrical diacetates furnish monoesters by methanolysis in the presence of HY zeolite, while diols are acetylated by HOAc in $CHCl_3$.[3]

Friedel-Crafts reactions. Acetylation of ArOR over Hβ zeolite is quite convenient, although substrates cannot contain OH, NH_2, CHO groups.[4] Nitration of moderately deactivated arenes is accomplished with liquid NO_2 and molecular O_2, and regioselectivity is affected by zeolite.[5]

Pictet-Spengler reaction is promoted by zeolite catalysts to give 1-substituted tetrahydroisoquinolines.[6] To perform the carbonyl-ene reaction with formaldehyde, encapsulation of the latter substance in zeolite NaY provides a long-lived and highly activated reagent.[7]

Allylation. Reaction of aldehydes with allylsilanes in $MeNO_2$ is catalyzed by zeolite RE-Y.[8]

N-Methylation. Primary aliphatic amines are methylated by dimethyl carbonate at 130–150° in the presence of alkali metal ion exchanged Y-faujasites, with selectivity in the 92–99% range.[9]

Racemization. Acid zeolites (e.g., Hβ-type) are useful as alcohol racemization catalysts, thus they have found applications in dynamic kinetic resolution.[10]

[1]Adimurthy, S., Ramachandraiah, G., Bedekar, A.V. *TL* **44**, 6391 (2003).
[2]Pandey, R.K., Kadam, V.S., Upadhyay, R.K., Dongare, M.K., Kumar, P. *SC* **33**, 3017 (2003).
[3]Srinivas, K.V.N.S., Reddy, E.B., Das, B. *SL* 2419 (2003).
[4]Smith, K., El-Hiti, G.A., Jayne, A.J., Butters, M. *OBC* **1**, 1560 (2003).
[5]Peng, X., Fukui, N., Mizuta, M., Suzuki, H. *OBC* **1**, 2326 (2003).
[6]Hegedus, A., Hell, Z. *TL* **45**, 8553 (2004).
[7]Okachi, T., Onaka, M. *JACS* **126**, 2306 (2004).
[8]Sasidharan, M., Tatsumi, T. *CL* **32**, 624 (2003).
[9]Selva, M., Tundo, P. *TL* **44**, 8139 (2003).
[10]Wuyts, S., De Temmerman, K., De Vos, D.E., Jacobs, P.A. *CEJ* **11**, 386 (2005).

Zinc. 13, 346–347; **14**, 349–350; **16**, 386–387; **17**, 406–407; **18**, 406–408; **19**, 404–405; **20**, 435–436; **21**, 491–492; **22**, 501–502

Reduction. 1,1-Diiodo-1-alkenes are reduced predominantly to give (*Z*)-1-iodo-1-alkenes[1] with Zn-Cu with HOAc in a mixture of THF and MeOH at 0°. Interestingly, methyl ketones are formed[2] on treatment of 1,1-dibromo-1-alkenes with Zn in near critical H_2O at 275°.

Fiesers' Reagents for Organic Synthesis, Volume 23. Edited by Tse-Lok Ho

Deoxygenation of heteroaromatic *N*-oxides[3] by the Zn–$HCOONH_4$ system is quite efficient, whereas Zn with dilute acid reduces nitrocyclopropanes,[4] and α-oximino-β-dicarbonyl compounds are deoximated.[5]

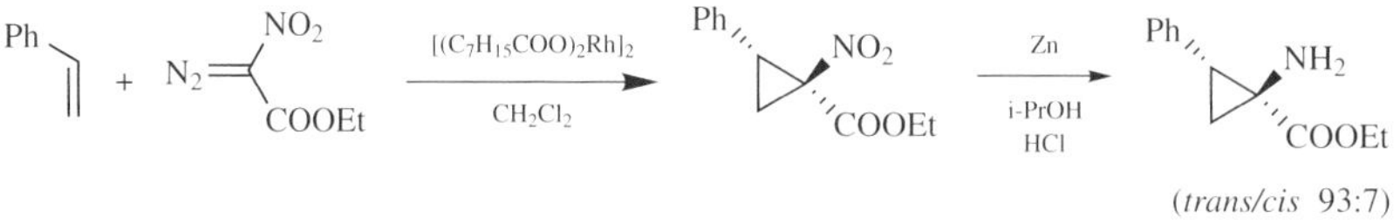

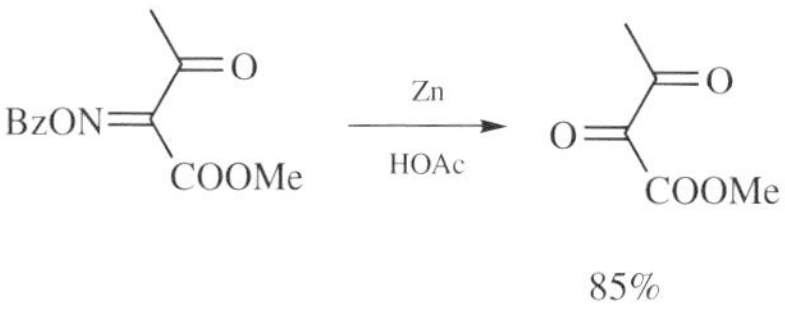

(trans/cis 93:7)

85%

For reductive acetylation of ArCHO, the Zn – imidazole – Ac_2O combination is more conventional, other esters can be prepared with Zn – $Yb(OTf)_3$ – $(RCO)_2O$.[6]

β-Keto amides are formed from Passerini adducts (α-keto aldehydes + RNC + HOAc) by photoinduced deacetoxylation with Zn.[7]

Reductive cleavage of benzylamines uses Zn – $HCOONH_4$ in either MeOH or ethylene glycol. Microwaves are also applied with the latter solvent.[8] As expected, *N*-2,2,2-trichloroethoxycarbonyl amines are readily deblocked in the presence of some other protecting groups (e.g., *N*-Boc of another amino group in a diamine).[9]

Zn-Cu couple and HOAc selectively reduces 1,1-diiodo-1-alkenes to leave (*Z*)-1-iodoalkenes.[10] The products are valuable precursors of conjugated dienes containing a double bond with (*Z*)-configuration.

Quinolizidin-1-ones and the vinylogous ketones undergo reductive rearrangement on heating with Zn in HOAc. The reaction is a key step for a synthesis of cephalotaxine.[11]

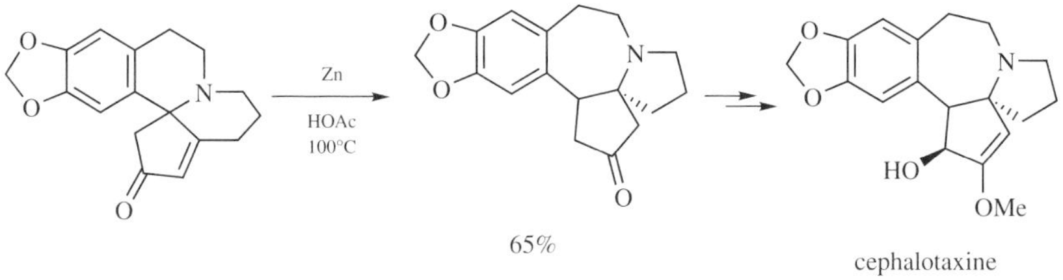

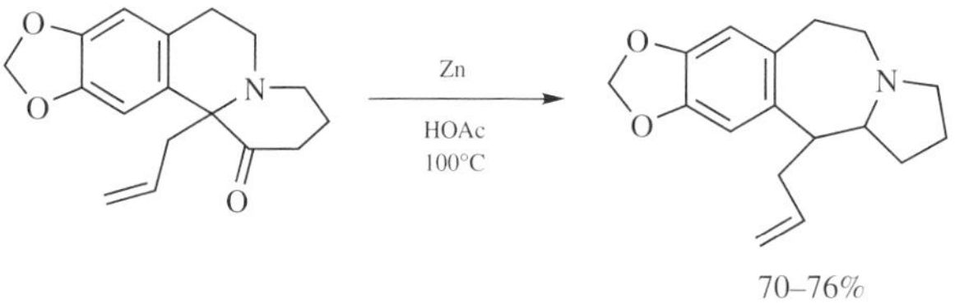

65%

cephalotaxine

70–76%

Addition reactions. Allylation of carbonyl compounds in aqueous media by 3-bromopropenyl acetate is γ-selective, therefore it affords unsaturated vicinal diol derivatives (favoring *syn* isomers).[12] For propargylation via sonochemical Barbier-type reaction the Zn metal is activated with 1,2-diiodoethane.[13]

Barbier-type allylation of trifluoromethyl aldimines proceeds well with Zn-Me_3SiCl in DMF at room temperature.[14] Zinc also mediates radical addition to polyfluorophenyl aldimines in aqueous media.[15] The stepwise process involving preformed organozinc reagents prepared in the presence of Me_3SiCl and 1,3-dimethyl-2-imidazolidinone add successfully to ordinary aldimines.[16] Amine salts, HCHO and allyl bromide are converted into homoallylic amines by Zn/CuI in aq. HOAc.[17]

Zn; NH_4Cl, CH_2Cl_2

85%

Radical addition to glyoxylic aldimines in aqueous media for elaboration of α-amino acids is moderately (but not uniformly) successful.[18]

Cyclopropanes are formed when 5-iodo-2-alkenoic esters are heated with Zn, via an intramolecular radical addition.[19] Tandem additions uniting three reaction partners to form β-keto esters[20] are quite useful. The acyl unit comes from a nitrile.

COOBn + I; Zn / MeCN ; HCl, H_2O; COOBn, O

82%

Friedel-Crafts reactions. Solvent-free acylation of arenes in the presence of zinc powder is assisted with microwave irradiation.[21] Also, Fries rearrangement is accomplished in DMF with zinc.[22]

Reductive couplings. α-Halo ketones form symmetrical 1,4-diketones[23] on treatment with Zn – I_2 in refluxing THF. Pinacol coupling of ArCHO with Zn in acidic aqueous solvent also employs ultrasound.[24] Cross-imino pinacol coupling by Zn-Cu takes advantage of synergistic effects of $BF_3 \cdot OEt_2$ and Me_3SiCl.[25]

Zn(Cu); $BF_3.OEt_2$, Me_3SiCl, MeCN

69% (*syn/anti* 87 : 13)

Synthesis of biaryls from ArX (X = Cl, Br, I) using Zn – $HCOONH_4$ and basic (NaOH) conditions in refluxing MeOH apparently is much milder than the classic Ullmann coupling.[26]

Substitutions. Replacement of the cyano group of an α-amino nitrile by an allyl or alkoxycarbonylmethyl group in the Zn-mediated reaction constitutes a more convenient chain extension process.[27]

Allylic substitution of the acetoxy derivative of a Baylis-Hillman adduct by an alkyl group is realized with the Zn-mediated reaction (+ RX) in saturated NH_4Cl.[28]

[1]Kadota, I., Ueno, H., Ohno, A., Yamamoto, Y. *TL* **44**, 8645 (2003).
[2]Wang, L., Li, P., Yan, J., Wu, Z. *TL* **44**, 4685 (2003).
[3]Balicki, R., Cybulski, M., Maciejewski, G. *SC* **33**, 4137 (2003).
[4]Wurz, R.P., Charette, A.B. *JOC* **69**, 1262 (2004).
[5]Ryu. I., Kuriyama, H., Miyazato, H., Minakata, S., Komatsu, M., Yoon, J.-Y., Kim, S. *BCSJ* **77**, 1407 (2004).
[6]Hirao, T., Santhitikul, S., Takeuchi, H., Ogawa, A., Sakurai, H. *T* **59**, 10147 (2003).
[7]Neo, A.N., Delgado, J., Polo, C., Marcaccini, S., Marcos, C.F. *TL* **46**, 23 (2005).
[8]Srinivasa, G.R., Babu, S.N.N., Lakshmi, C., Gowda, D.C. *SC* **34**, 1831 (2004).
[9]Zhu, X., Schmidt, R.R. *S* 1262 (2003).
[10]Kadota, I., Ueno, H., Ohno, A., Yamamoto, Y. *TL* **44**, 8645 (2003).
[11]Li, W.-D.Z., Wang, Y.-Q. *OL* **5**, 2931 (2003).
[12]Lombardo, M., Morganti, S., d'Ambrosio, F., Trombini, C. *TL* **44**, 2823 (2003).
[13]Lee, A.S.-Y., Chu, S.F., Chang, Y.-T., Wang, S.-H. *TL* **45**, 1551 (2004).
[14]Legros, J., Meyer, F., Coliboeuf, M., Crousse, B., Bonnet-Delpon, D., Begue, J.-P. *JOC* **68**, 6444 (2003).
[15]Liu, X., Zhu, S., Wang, S. *S* 683 (2004).
[16]Iwai, T., Ito, T., Mizuno, T., Ishino, Y. *TL* **45**, 1083 (2004).
[17]Estevam, I.H.S., Bieber, L.W. *TL* **44**, 667 (2003).
[18]Ueda, M., Miyabe, H., Sugino, H., Naito, T. *OBC* **3**, 1124 (2005).
[19]Sakuma, D., Togo, H. *SL* 2501 (2004).
[20]Yamamoto, Y., Nakano, S., Maekawa, H., Nishiguchi, I. *OL* **6**, 799 (2004).
[21]Paul, S., Nanda, P., Gupta, R., Loupy, A. *S* 2877 (2003).
[22]Paul, S., Gupta, M. *S* 1789 (2004).
[23]Ceylan, M., Gurdere, M.B., Badak, Y., Kazaz, C., Secen, H. *S* 1750 (2004).
[24]Yang, J.-H., Li, J.-T., Zhao, J.-L., Li, T.-S. *SC* **34**, 993 (2004).
[25]Shimizu, M., Suzuki, I., Makino, H. *SL* 1635 (2003).
[26]Abiraj, C., Srinivasa, G.R., Gowda, D.C. *TL* **45**, 2081 (2004).
[27]Bernardi, L., Bonini, B.F., Capito, E., Dessole, G., Fochi, M., Comes-Franchini, M., Ricci, A. *SL* 1778 (2003).
[28]Das, B., Banerjee, J., Mahender, G., Majhi, A. *OL* **6**, 3349 (2004).

Zinc – metal salts. 22, 502–503

Reductive alkylation. Disulfides are reduced by Zn-$AlCl_3$ and the zinc thiolates can be used to prepare unsymmetrical sulfides by reaction with epoxides[1] and ROTs.[2]

Allylation. Aqueous protocol for allylation of carbonyl compounds using Zn and $SnCl_2$ or $CdSO_4$ is inexpensive.[3] With Zn and BiI_3 to generate substituted allylbismuth species for allylation, 1,5-stereoinduction is feasible. In allylation using allyl acctates $CoBr_2$ acts as a catalyst.[5]

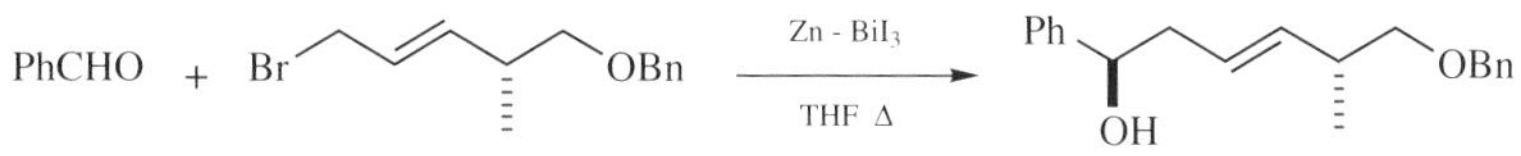

PhCHO + Br ... OBn —(Zn - BiI_3, THF Δ)→ Ph ... OH ... OBn

83% (anti/syn 96 : 4)

Reformatsky reaction. β-Lactams are produced directly in a modified Reformatsky reaction in which Cp_2TiCl_2 is added as an activator for Zn,[6] although its necessity may not be very significant (use of iodine with high-intensity ultrasound also works.[7])

[1]Movassagh, B., Sobhani, S., Kheirdoush, F., Fadaei, Z. *SC* **33**, 3103 (2003).
[2]Movassagh, B., Mossadegh, A. *SC* **34**, 2337 (2004).
[3]Zhou, C., Zhou, Y., Jiang, J., Xie, Z., Wang, Z., Zhang, J. *TL* **45**, 5537 (2004).
[4]Donnelly, S., Thomas, E.J., Fielding, M. *TL* **45**, 6779 (2004).
[5]Gomes, P., Gosmini, C., Perichon, J. *S* 1909 (2003).
[6]Chen, L., Zhao, G., Ding, Y. *TL* **44**, 2611 (2003).
[7]Ross, N.A., MacGregor, R.R., Bartsch, R.A. *T* **60**, 2035 (2004).

Zinc chloride. **13**, 349–350; **15**, 368–371; **16**, 391–392; **18**, 410–411; **19**, 409–410; **20**, 439; **21**, 493; **22**, 505–506

Substitutions. The following reactions are promoted by $ZnCl_2$: thiolysis[1] and aminolysis of epoxides,[2] in water and MeCN, respectively, and preparation of 2,3-anhydroglucopyranosides by the Ferrier rearrangement using various alcohols (including tertiary alcohols).[3]

Addition to imines. Alkynylation of imines with 1-alkynes provides propargylic amines.[4] Attack of pentadienylstannanes is subject to regiocontrol by both Lewis acid catalyst and *N*-substituents.[5]

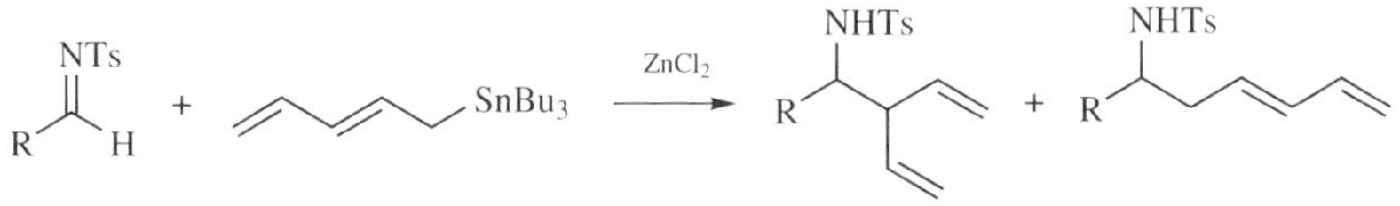

R(H)C=NTs + ... $SnBu_3$ —($ZnCl_2$)→ R–CH(NHTs)– ... + R–CH(NHTs)– ...

(>99 : 1)

N-Tosylimines generated in situ from TsN=S=O and RCHO are trapped by furan in the presence of $ZnCl_2$, thereby an efficient synthesis of 2-(α-tosylaminoalkyl)furans is realized.[6]

The allylic CF_2 group of perfluoroalkenes is reactive toward amines in the presence of $ZnCl_2$ - Et_3N. According to the reactivity a synthesis 2-pefluoroalkylquinolines can be developed.[7]

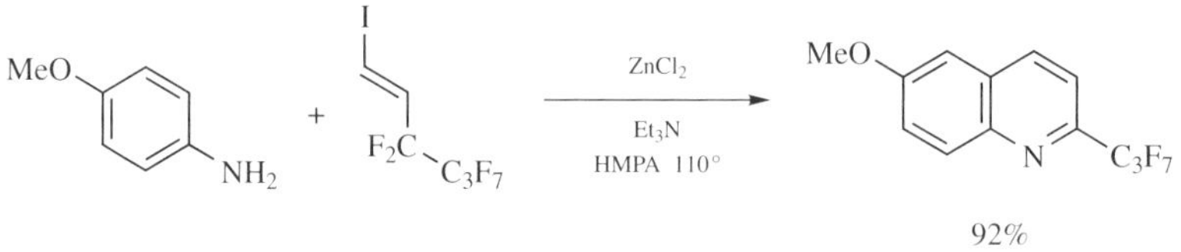

[1]Fringuelli, F., Pizzo, F., Tortoioli, S., Vaccaro, L. *JOC* **68**, 8248 (2003).
[2]Pachon, L.D., Gamez, P., van Brussel, J.J.M., Reedijk, J. *TL* **44**, 6025 (2003).
[3]Bettadaiah, B.K., Srinivas, P. *TL* **44**, 7257 (2003).
[4]Jiang, B., Si, Y.-G. *TL* **44**, 6767 (2003).
[5]Nishigaichi, Y., Ishihara, M., Fushitani, S., Uenoaga, K., Takuwa, A. *CL* **33**, 108 (2004).
[6]Padwa, A., Zanka, A., Cassidy, M.P., Harris, J.M. *T* **59**, 4939 (2003).
[7]Zhao, F., Yang, X., Liu, J. *T* **60**, 9945 (2004).

Zinc fluoride.

Allylation.[1] Allyltrimethoxysilane allylates ArCONHN=CHCOOEt in aqueous THF with the aid of ZnF_2. Moderate asymmetric induction is observed in the presence of a chiral diamine ligand.

[1]Hamada, T., Manabe, K., Kobayashi, S. *ACIEE* **42**, 3927 (2003).

Zinc iodide. 21, 493–494; **22**, 506

Cyclization.[1] Addition of alcohols to diethyl *N*-propargylcarbamoylmethylenemalonates also involves CC bond formation with the alkyne unit. The reaction occurs because chelation of the diester with ZnI_2 promotes a Michael addition.

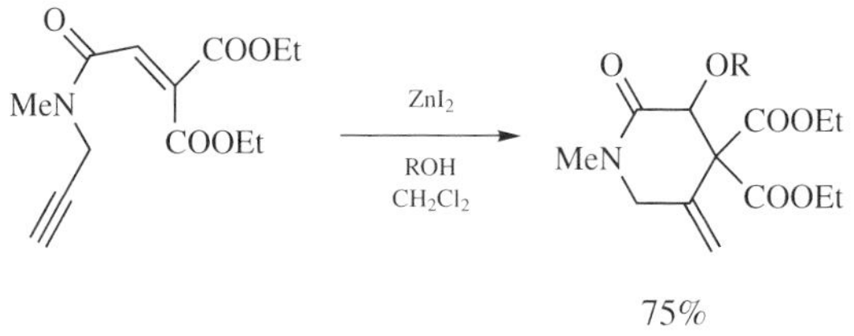

[1]Yamazaki, S., Inaoka, S., Yamada, K. *TL* **44**, 1429 (2003).

Zinc nitrate.

Aldol reaction.[1] A combination of $Zn(NO_3)_2$ and TMEDA is effective to promote aldol reaction of ketones with aldehydes.

[1]Calter, M.A., Orr, R.K. *TL* **44**, 5699 (2003).

Zinc oxide.

Friedel-Crafts acylation.[1] ZnO powder is an economical and efficient catalyst in solvent-free conditions for arene acylation with RCOCl. When CH_2Cl_2 is used as solvent, even prolonged reaction reaction delivers much poorer yield of the product. A demonstration showed reuse of ZnO (3 times) with little loss of activity.

Cyclic ureas.[2] Microwave treatment of urea, diamines and ZnO in DMF leads to formation of cyclic ureas. Amino alcohols react analogously to produce cyclic carbamates.

[1]Sarvari, M.H., Sharghi, H. *JOC* **69**, 6953 (2004).
[2]Kim, Y.J., Varma, R.S. *TL* **45**, 7205 (2004).

Zinc perchlorate.

Acylations. Condensation of carboxylic acids with alcohols by catalysis of $Zn(ClO_4)_2 \cdot 6H_2O$ is carried out without solvent, with water removal by $MgSO_4$.[1] Convenient Boc-protection of arylamines also employs the same catalyst.[2]

β-Enamino esters. The preparation is by a catalyzed condensation of primary and secondary amines with β-keto esters proceeds at room temperature.

[1]Bartoli, G., Boeglin, J., Bosco, M., Locatelli, M., Massaccesi, M., Melchiorre, P., Sambri, L. *ASC* **347**, 33 (2005).
[2]Bartoli, G., Bosco, M., Locatelli, M., Marcantoni, E., Massaccesi, M., Melchiorre, P., Sambri, L. *SL* 1794 (2004).
[3]Bartoli, G., Bosco, M., Locatelli, M., Marcantoni, E., Melchiorre, P., Sambri, L. *SL* 239 (2004).

Zinc triflate - tertiary amine. 21, 494; 22, 506–507

Additions. Propargylic amines are produced from 1-alkynes and imines. Alkynylzinc reagents are formed with $Zn(OTf)_2$ [and presence of Et_3N and *N,N,N′,N′*-tetramethylpropylenediamine] while the imines are activated with acylating agents.[1] Michael addition to conjugate systems in chiral templates produces 4-alkynoic acid derivatives.[2]

A direct synthesis of 3-methylenetetrahydrofuran-4,4-dicarboxylic esters[3] involves Michael addition of propargyl alcohol to alkylidenemalonates and an intramolecular addition to the triple bond that results in cyclization.

EtOOC, Ph, COOEt + OH → $Zn(OTf)_2$, Et_3N → EtOOC, EtOOC, Ph, O

93%

Addition of glycine derivatives to enamines to generate α,β-diamino esters does not require external base.[4]

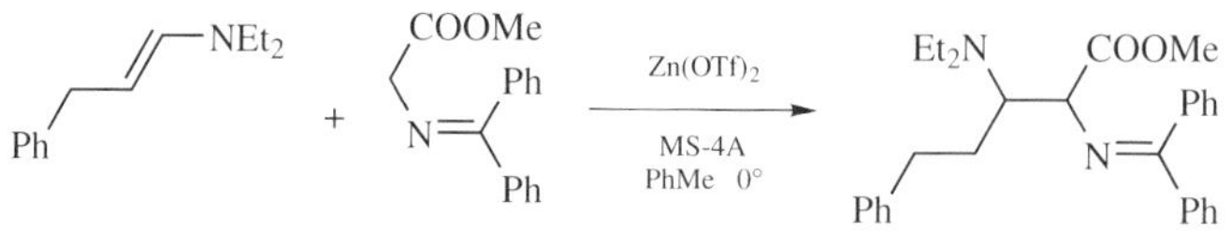

[1]Fischer, C., Carreira, E.M. *OL* **6**, 1497 (2004).
[2]Knopfel, T.F., Boyall, D., Carreira, E.M. *OL* **6**, 2281 (2004).
[3]Nakamura, M., Liang, C., Nakamura, E. *OL* **6**, 2015 (2004).
[4]Kobayashi, J., Yamashiota, Y., Kobayashi, S. *CL* **34**, 268 (2005).

Zirconia, sulfated.

Glycosylation.[1] The solid acid catalyzes reaction between some glycosyl fluorides with alcohols. Solvent effect on the α/β ratio of the products is noted.

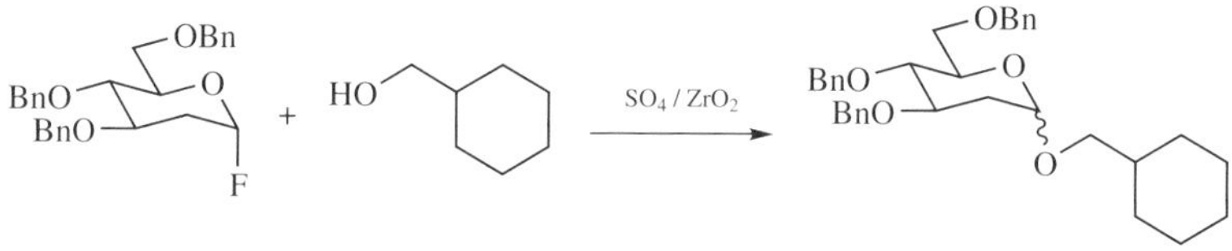

MeCN, 1h α/β 88 : 12
Ms-5A, Et_2O, 15h α/β 19 : 81

Diaryl ketones. Arenes are acylated over sulfated zirconia.[2,3]

[1]Toshima, K., Nagai, H., Kasumi, K., Kawahara, K., Matsumura, S. *T* **60**, 5331 (2004).
[2]Nakamura, H., Arata, K. *BCSJ* **77**, 1893 (2004).
[3]Jin, T.-S., Yang, M.-N., Feng, G.-L., Li, T.-S. *SC* **34**, 479 (2004).

Zirconium (IV) chloride. 22, 508

Functional group manipulations. TBS ethers[1] and *p*-methoxybenzyl ethers[2] are severed on treatment with $ZrCl_4$ in MeCN at room temperature. Tosylation of alcohols with TsOH is catalyzed by $ZrCl_4$ in refluxing CH_2Cl_2.[3] Reports on more routine operations that need a Lewis acid such as acetylation,[4] *t*-butoxycarbonylation[5] and acyl formation[6] are just for confirming the Lewis acidity of $ZrCl_4$. Catalyzed cyclodehydration of 1,2-diacylhydrazines to form 2,5-diaryl-1,3,4-oxadiazoles[7] is also unspectacular.

Halides. Conversion of ROH to RI needs a combination of $ZrCl_4$ and NaI.[8] Chlorination of ketones via silyl enol ethers that employs $ZrCl_4$ and a dichloromalonic ester is rendered asymmetric with chiral alkoxy residues. The role of $ZrCl_4$ is to coordinate the malonate to establish a stable conformation of the chlorinating agent.[9]

Substitutions. Epoxide opening with amines[10,11] and synthesis of pseudoglycals by Ferrier rearrangement[12] have been reported.

Tetrahydroquinolines.[13] Annulation from *N*-arylimines and vinyl ethers is catalyzed by Lewis acid. Use of $ZrCl_4$ in that capacity is realized.

[1]Sharma, G.V.M., Srinivas, B., Krishna, P.R. *TL* **44**, 4689 (2003).
[2]Sharma, G.V.M., Reddy, C.G., Krishna, P.R. *JOC* **68**, 4574 (2003).
[3]Das, B., Reddy, V.S. *CL* **33**, 1428 (2004).
[4]Chakraborti, A.K., Gulhane, R. *SL* 627 (2004).
[5]Sharma, G.V.M., Reddy, J.J., Lakshmi, P.S., Krishna, P.R. *TL* **45**, 6963 (2004).
[6]Smitha, G., Reddy, C.S. *T* **59**, 9571 (2003).
[7]Sharma, G.V.M., Begum, A., Rakesh, Krishna, P.R. *SC* **34**, 2387 (2004).
[8]Firouzabadi, H., Iranpoor, N., Jafarpour, M. *TL* **45**, 7451 (2004).
[9]Zhang, Y., Shibatomi, K., Yamamoto, H. *JACS* **126**, 15038 (2004).
[10]Chakraborti, A.K., Kondaskar, A. *TL* **44**, 8315 (2003).
[11]Swamy, N.R., Goud, T.V., Reddy, S.M., Krishnaiah, P., Venkateswarlu, Y. *SC* **34**, 727 (2004).
[12]Smitha, G., Reddy, C.S. *S* 834 (2004).
[13]Das, B., Reddy, M.R., Reddy, V.S., Ramu, R. *CL* **33**, 1526 (2004).

Zirconium (IV) triflate.

Hydrative rearrangement. Methylenecyclopropanes are transformed into cyclobutanones[1] on treatment with $Zr(OTf)_4$ and an azodicarboxylic ester.

Ph Ph
$Zr(OTf)_4$
PrOOCN=NCOOPr
$MeCN\text{-}H_2O$
Ph Ph
O
96%

Heteroarylation. Lactams are oxidized at the ω-position by oxygen in the presence of $Zr(OTf)_4$ and the resulting species acylate heteroaromatic compounds such as indole.[2]

[1]Shao, L.-X., Shi, M. *EJOC* 426 (2004).
[2]Tsuchimoto, T., Ozawa, Y., Negoro, R., Shirakawa, E., Kawakami, Y. *ACIEE* **43**, 4321 (2004).

Zirconocene. **20**, 441–442; **21**, 496–497; **22**, 509–510

Allylation.[1] Benzyl ethers are transformed into 3-butenylarenes by "Cp_2Zr" if the aromatic ring is *o*-substituted with an alkenyl or alkynyl group.

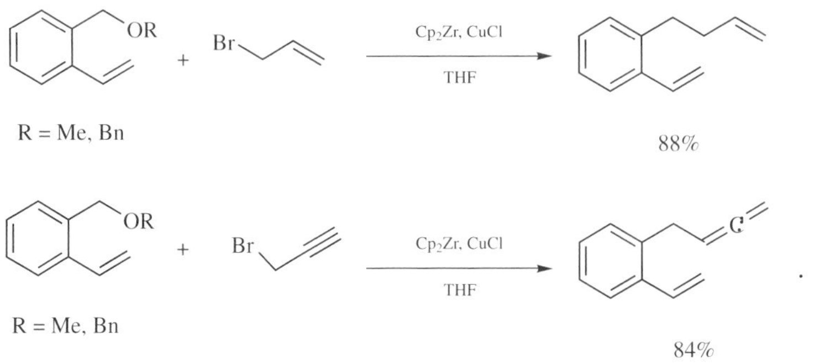

R = Me, Bn

Cp_2Zr, CuCl / THF

88%

R = Me, Bn

Cp_2Zr, CuCl / THF

84%

Hydrosilylation.[2] A wonderful regiocontrol to produce either Markovnikov or anti-Markovnikov adducts is attained by merely changing the relative ratios of two precatalysts (RLi/[Zr]).

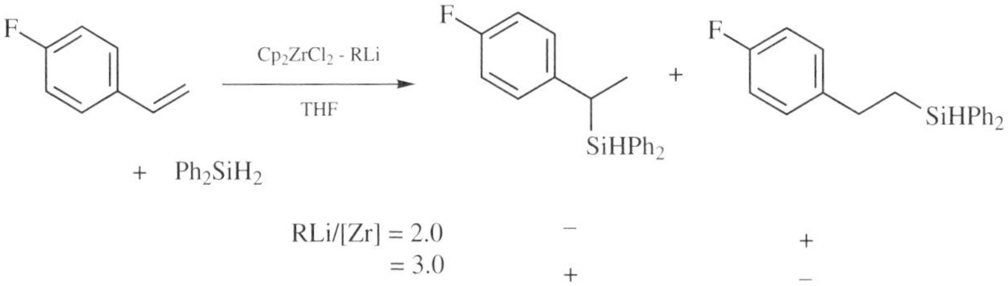

+ Ph_2SiH_2 — Cp_2ZrCl_2 - RLi / THF → $SiHPh_2$ + $SiHPh_2$

RLi/[Zr] = 2.0 – +
= 3.0 + –

Reactions of alkenyl ethers. Homoallylic ethers form cyclopropane derivatives,[3,4] whereas insertion of a C(sp^2)-O bond of an alkenyl ether by [$ZrCp_2$] through double bond migration (formation of zirconacyclopropane intermediates) generates reagents that can be functionalized (in many cases with a Cu catalyst).[5]

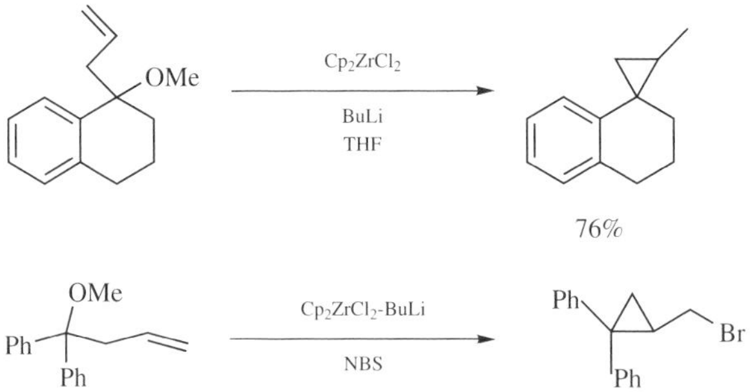

OMe — Cp_2ZrCl_2 / BuLi, THF → 76%

OMe, Ph, Ph — Cp_2ZrCl_2-BuLi / NBS → Ph, Ph, Br

Zr/B exchange. Oxidation of C-Zr bond is inefficient, therefore conversion of such intermediates into alcohols is best performed via reaction with Cy_2BCl and use the established protocol for oxidation.[6]

[1]Ikeuchi, Y., Taguchi, T., Hanzawa, Y. *JOC* **70**, 756 (2005).

[2]Ura, Y., Gao, G., Bao, F., Ogasawara, M., Takahashi, T. *OM* **23**, 4804 (2004).
[3]Gandon, V., Laroche, C., Szymoniak, J. *TL* **44**, 4827 (2003).
[4]Vasse, J.-L., Szymoniak, J. *TL* **45**, 6449 (2004).
[5]Chinkov, N., Majumdar, S., Marek, I. *JACS* **125**, 13258 (2003).
[6]Gorman, J.S.T., Iacono, S.T., Pagenkopf, B.L. *OL* **6**, 67 (2004).

Zirconocene, Zr-alkylated. **15**, 81; **18**, 414; **19**, 412–414; **20**, 442–443; **21**, 496–497; **22**, 510–511

Indenes.[1] When alkynes react with Cp_2ZrEt_2 and then aryl ketones, substituted indenes are formed.

Dehydration.[2] Conversion of primary amides to nitriles by Cp_2ZrMe_2 is via *N*-acylimidozirconocene complex formation which is dependent on chloride anion effect from an additive (e.g., LiCl).

[1]Xi, Z., Guo, R., Mito, S., Yan, H., Kanno, K., Nakajima, K., Takahashi, T. *JOC* **68**, 1252 (2003).
[2]Ruck, R.T., Bergman, R.G. *ACIEE* **43**, 5375 (2004).

Zirconocene dichloride.

Propargylic alcohols. Special attention must be paid in the preparation of 4-hydroxy-2-alkynoic esters, the mild conditions involving Cp_2ZrCl_2 catalysis in the reaction of Ag(I)-based nucleophiles are quite valuable.[1]

Glycosylation.[2] Glycosyl fluorides containing some sensitive groups (azido, 2,2,2-trifluoroethoxysulfonyl) survive the glycosylation protocol mediated by Cp_2ZrCl_2 and $AgClO_4$.

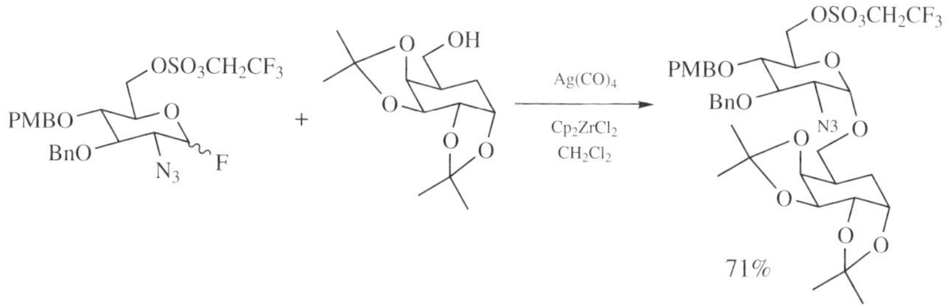

[1]Shahi, S.P., Koide, K. *ACIEE* **43**, 2525 (2004).
[2]Karst, N.A., Islam, T.F., Avci, F.Y., Lindhardt, R.J. *TL* **45**, 6433 (2004).

Zirconocene hydrochloride. **14**, 81; **15**, 80–81; **18**, 416–417; **19**, 415–416; **20**, 445–446; **21**, 497; **22**, 511–512

Deuteration. A facile synthesis of RCDO from RCONR'$_2$ uses $Cp_2Zr(D)Cl$ as reducing agent (9 examples, 70–92%).[1]

Functionalized alkenes. Starting from hydrozirconation of an allenylsilane a subsequent reaction with aldehydes directly leads to substituted 1,3-dienes.[2]

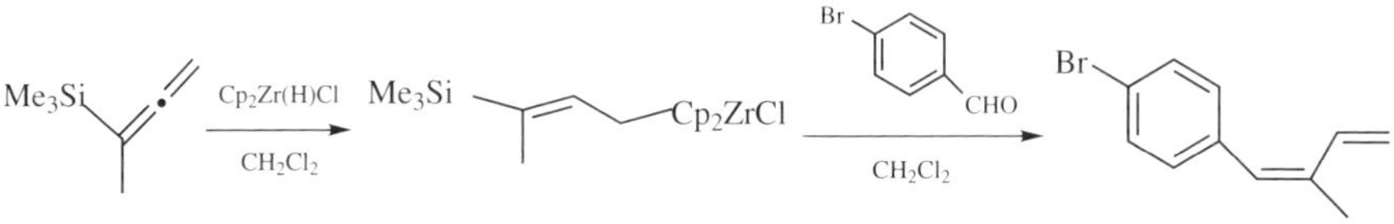

Alkenylzirconocene species act as conjugate addends in the presence of CuI and Me_2S, and the process is useful for synthesis γ,δ-unsaturated carbonyl compounds.[3] On converting to organozinc reagents the alkenylzirconocenes add to α-keto esters and α-imino esters to provide α-hydroxy esters and α-amino esters, respectively.[4]

[1]Spletstoser, J.T., White, J.M., Georg, G.I. *TL* **45**, 2787 (2004).
[2]Pi, J.-H., Huang, X. *TL* **45**, 2215 (2004).
[3]El-Batta, A., Hage, T.R., Plotkin, S., Bergdahl, M. *OL* **6**, 107 (2004).
[4]Wipf, P., Stephenson, C.R.J. *OL* **5**, 2449 (2003).

Zirconocene imides.

C-Alkenylation of imines. $Cp_2Zr{=}NR$ cycloadd to alkynes and insertion of the adducts to the imino C—H bond results in the formation of conjugated imines.

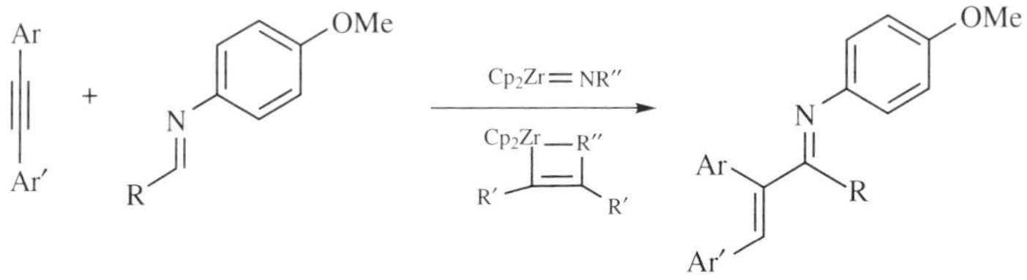

[1]Ruck, R.T., Zuckermann, R.L., Krska, S.W., Bergman, R.G. *ACIEE* **43**, 5372 (2004).

Zirconocene methochloride.

N-Allylation.[1] *N*-Lithiated amines on successive reactions with $Cp_2Zr(Me)Cl$ and vinyl ethers lead to allylamines. Precursors of 2-substituted piperidines are readily accessible from reaction with dihydrofuran.

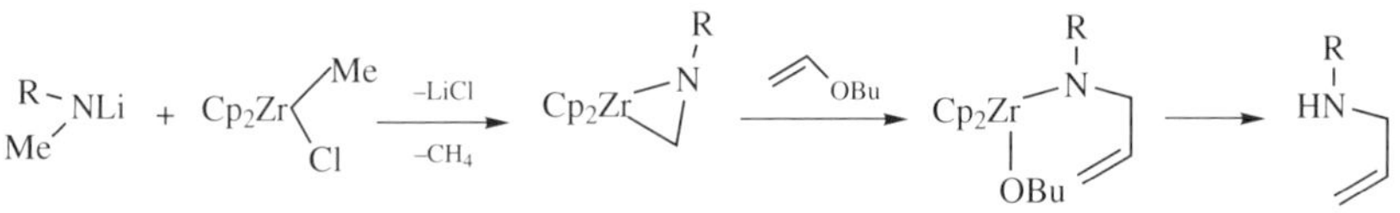

Cyclobutenes and 1,3-dienes.[2] The reaction of methyl(alkenyl)zirconocenes with vinyl bromide affords either cyclobutenes or conjugated dienes, depending on temperature.

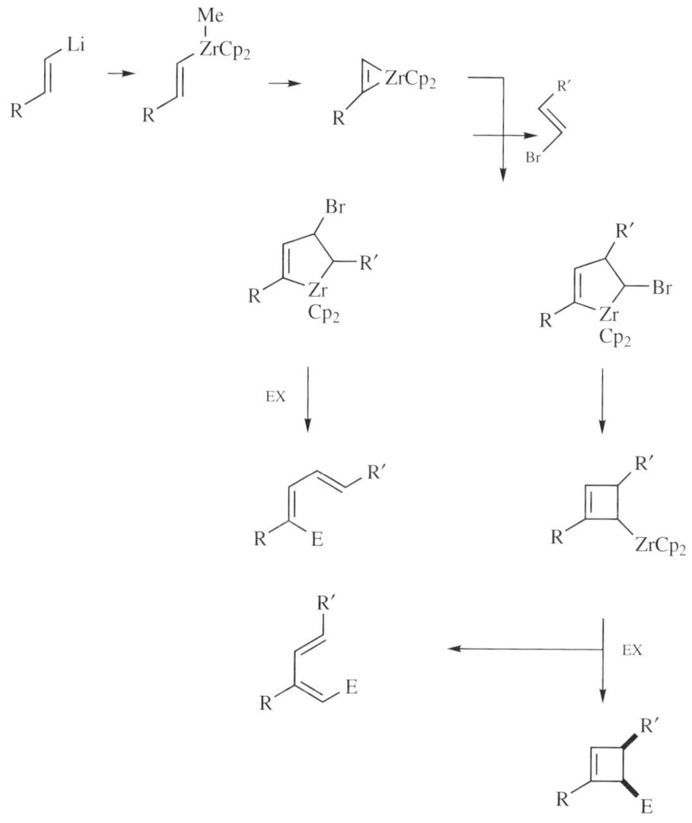

Me ZrCp2 Li R R ZrCp2 R R' Br Br R' R Zr Cp2 R' R Br Zr Cp2 EX R' R E R' R ZrCp2 R' R E EX R' R E

[1]Barluenga, J., Rodriguez, F., Alvarez-Rodrigo, L., Zapico, J.M., Fananas, F.J. *CEJ* **10**, 109 (2004).
[2]Barluenga, J., Rodriguez, F., Alvarez-Rodrigo, L., Fananas, F.J. *CEJ* **10**, 101 (2004).

Zirconocene trimethylphosphine complex.

Alkenephosphonates.[1] The complex prepared from Cp_2ZrBu_2 and Me_3P in THF at room temperature add to 1-alkynes. Quenching the intermediates with chlorophosphate esters gives (*Z*)-2-chlorozirconocenyl-1-alkenephosphonates. The zirconocenyl group can be replaced with various halogen atoms, acyl groups and allyl groups besides hydrogen.

[1]Lai, C., Xi, C., Chen, C., Ma, M., Hong, X. *CC* 2736 (2003).

AUTHOR INDEX

Fiesers' Reagents for Organic Synthesis, Volume 23. Edited by Tse-Lok Ho

SUBJECT INDEX

Fiesers' Reagents for Organic Synthesis, Volume 23. Edited by Tse-Lok Ho

YORK COLLEGE OF PENNSYLVANIA 17403
0 2003 0182512 7

DISCARDED
PENNSYLVANIA
LIBRARY